Communications in Computer and Information Science 1638

More information about this series at https://link.springer.com/bookseries/7899

Haijun Zhang · Yuehui Chen · Xianghua Chu ·
Zhao Zhang · Tianyong Hao · Zhou Wu ·
Yimin Yang (Eds.)

Neural Computing for Advanced Applications

Third International Conference, NCAA 2022
Jinan, China, July 8–10, 2022
Proceedings, Part II

 Springer

Editors
Haijun Zhang (iD)
Harbin Institute of Technology
Shenzhen, China

Xianghua Chu (iD)
Shenzhen University
Shenzhen, China

Tianyong Hao (iD)
South China Normal University
Guangzhou, China

Yimin Yang (iD)
Western University
London, ON, Canada

Yuehui Chen (iD)
University of Jinan
Jinan, China

Zhao Zhang (iD)
Hefei University of Technology
Hefei, China

Zhou Wu (iD)
Chongqing University
Chongqing, China

ISSN 1865-0929   ISSN 1865-0937 (electronic)
Communications in Computer and Information Science
ISBN 978-981-19-6134-2   ISBN 978-981-19-6135-9 (eBook)
https://doi.org/10.1007/978-981-19-6135-9

This Springer imprint is published by the registered company Springer Nature Singapore Pte Ltd.
The registered company address is: 152 Beach Road, #21-01/04 Gateway East, Singapore 189721, Singapore

Preface

Neural computing and artificial intelligence (AI) have become hot topics in recent years. To promote the multi-disciplinary development and application of neural computing, a series of NCAA conferences was initiated on the theme of "make the academic more practical", providing an open platform of academic discussions, industrial showcases, and basic training tutorials. This volume contains the papers accepted for this year's International Conference on Neural Computing for Advanced Applications (NCAA 2022). NCAA 2022 was organized by the University of Jinan, Shandong Jianzhu University, South China Normal University, Harbin Institute of Technology, Chongqing University, and Hefei University of Technology, and it was supported by Springer. Due to the effects of COVID-19, the mainstream part of NCAA 2022 was turned into a hybrid event, with both online and offline participants, in which people could freely connect to live broadcasts of keynote speeches and presentations.

NCAA 2022 received 205 submissions, of which 77 high-quality papers were selected for publication in this volume after double-blind peer review, leading to an acceptance rate of just under 38%. These papers have been categorized into 10 technical tracks: neural network theory and cognitive sciences; machine learning, data mining, data security and privacy protection, and data-driven applications; computational intelligence, nature-inspired optimizers, and their engineering applications; cloud/edge/fog computing, the Internet of Things/Vehicles (IoT/IoV), and their system optimization; control systems, network synchronization, system integration, and industrial artificial intelligence; fuzzy logic, neuro-fuzzy systems, decision making, and their applications in management sciences; computer vision, image processing, and their industrial applications; natural language processing, machine translation, knowledge graphs, and their applications; neural computing-based fault diagnosis, fault forecasting, prognostic management, and system modeling; and spreading dynamics, forecasting, and other intelligent techniques against coronavirus disease (COVID-19).

The authors of each paper in this volume have reported their novel results of computing theory or application. The volume cannot cover all aspects of neural computing and advanced applications, but may still inspire insightful thoughts for the readers. We hope that more secrets of AI will be unveiled, and that academics will drive more practical developments and solutions.

June 2022

Haijun Zhang
Yuehui Chen
Xianghua Chu
Zhao Zhang
Tianyong Hao
Zhou Wu
Yimin Yang

Organization

Honorary Chairs

John MacIntyre   University of Sunderland, UK
Tommy W. S. Chow   City University of Hong Kong, Hong Kong

General Co-chairs

Haijun Zhang   Harbin Institute of Technology, China
Yuehui Chen   University of Jinan, China
Zhao Zhang   Hefei University of Technology, China

Program Co-chairs

Tianyong Hao   South China Normal University, China
Zhou Wu   Chongqing University, China
Yimin Yang   University of Western Ontario, Canada

Organizing Committee Co-chairs

Yongfeng Zhang   University of Jinan, China
Yunchu Zhang   Shandong Jianzhu University, China
Choujun Zhan   Nanfang College Guangzhou, China
Mingbo Zhao   Donghua University, China

Local Arrangement Co-chairs

Menghua Zhang   University of Jinan, China
Weijie Huang   University of Jinan, China
Ming Wang   Shandong Jianzhu University, China
Chengdong Li   Shandong Jianzhu University, China
Xiangping Zhai   Nanjing University of Aeronautics and Astronautics, China

Registration Co-chairs

Yaqing Hou   Dalian University of Technology, China
Jing Zhu   Macau University of Science and Technology, China

Shuqiang Wang	Chinese Academy of Sciences, China
Weiwei Wu	Southeast University, China
Zhili Zhou	Nanjing University of Information Science and Technology, China

Publication Co-chairs

Kai Liu	Chongqing University, China
Yu Wang	Xi'an Jiaotong University, China
Yi Zhang	Fuzhou University, China
Bo Wang	Huazhong University of Science and Technology, China
Xianghua Chu	Shenzhen University, China

Publicity Co-chairs

Liang Feng	Chongqing University, China
Penglin Dai	Southwest Jiaotong University, China
Dong Yang	University of California, Merced, USA
Shi Cheng	Shaanxi Normal University, China
Reza Maleklan	Malmö University, Sweden

Sponsorship Co-chairs

Wangpeng He	Xidian University, China
Bingyi Liu	Wuhan University of Technology, China
Cuili Yang	Beijing University of Technology, China
Jicong Fan	Chinese University of Hong Kong, Shenzhen, China

NCAA Steering Committee Liaison

Jingjing Cao	Wuhan University of Technology, China

Web Chair

Xinrui Yu	Harbin Institute of Technology, China

Program Committee

Dong Yang	City University of Hong Kong, Hong Kong
Sheng Li	University of Georgia, USA
Jie Qin	Swiss Federal Institute of Technology (ETH), Switzerland

Xiaojie Jin	Bytedance AI Lab, USA
Zhao Kang	University of Electronic Science and Technology, China
Xiangyuan Lan	Hong Kong Baptist University, Hong Kong
Peng Zhou	Anhui University, China
Chang Tang	China University of Geosciences, China
Dan Guo	Hefei University of Technology, China
Li Zhang	Soochow University, China
Xiaohang Jin	Zhejiang University of Technology, China
Wei Huang	Zhejiang University of Technology, China
Chao Chen	Chongqing University, China
Jing Zhu	Macau University of Science and Technology, China
Weizhi Meng	Technical University of Denmark, Denmark
Wei Wang	Dalian Ocean University, China
Jian Tang	Beijing University of Technology, China
Heng Yue	Northeastern University, China
Yimin Yang	University of Western Ontario, Canada
Jianghong Ma	City University of Hong Kong, Hong Kong
Jicong Fan	Chinese University of Hong Kong, Shenzhen, China
Xin Zhang	Tianjing Normal University, China
Xiaolei Lu	City University of Hong Kong, Hong Kong
Penglin Dai	Southwest Jiaotong University, China
Liang Feng	Chongqing University, China
Xiao Zhang	South Central University for Nationalities, China
Bingyi Liu	Wuhan University of Technology, China
Cheng Zhan	Southwest University, China
Qiaolin Pu	Chongqing University of Posts and Telecommunications, China
Hao Li	Hong Kong Baptist University, Hong Kong
Junhua Wang	Nanjing University of Aeronautics and Astronautics, China
Yu Wang	Xi'an Jiaotong University, China
Binqiang Chen	Xiamen University, China
Wangpeng He	Xidian University, China
Jing Yuan	University of Shanghai for Science and Technology, China
Huiming Jiang	University of Shanghai for Science and Technology, China
Yizhen Peng	Chongqing University, China
Jiayi Ma	Wuhan University, China
Yuan Gao	Tencent AI Lab, China

Xuesong Tang	Donghua University, China
Weijian Kong	Donghua University, China
Zhili Zhou	Nanjing University of Information Science and Technology, China
Yang Lou	City University of Hong Kong, Hong Kong
Chao Zhang	Shanxi University, China
Yanhui Zhai	Shanxi University, China
Wenxi Liu	Fuzhou University, China
Kan Yang	University of Memphis, USA
Fei Guo	Tianjin University, China
Wenjuan Cui	Chinese Academy of Sciences, China
Wenjun Shen	Shantou University, China
Mengying Zhao	Shandong University, China
Shuqiang Wang	Chinese Academy of Sciences, China
Yanyan Shen	Chinese Academy of Sciences, China
Haitao Wang	China National Institute of Standardization, China
Yuheng Jia	City University of Hong Kong, Hong Kong
Chengrun Yang	Cornell University, USA
Lijun Ding	Cornell University, USA
Zenghui Wang	University of South Africa, South Africa
Xianming Ye	University of Pretoria, South Africa
Reza Maleklan	Malmö University, Sweden
Xiaozhi Gao	University of Eastern Finland, Finland
Jerry Lin	Western Norway University of Applied Sciences, Norway
Xin Huang	Hong Kong Baptist University, Hong Kong
Xiaowen Chu	Hong Kong Baptist University, Hong Kong
Hongtian Chen	University of Alberta, Canada
Gautam Srivastava	Brandon University, Canada
Bay Vo	Ho Chi Minh City University of Technology, Vietnam
Xiuli Zhu	University of Alberta, Canada
Rage Uday Kiran	University of Toyko, Japan
Matin Pirouz Nia	California State University, Fresno, USA
Vicente Garcia Diaz	University of Oviedo, Spain
Youcef Djenouri	Norwegian University of Science and Technology, Norway
Jonathan Wu	University of Windsor, Canada
Yihua Hu	University of York, UK
Saptarshi Sengupta	Murray State University, USA
Wenxiu Xie	City University of Hong Kong, Hong Kong
Christine Ji	University of Sydney, Australia

Jun Yan	Yiducloud, China
Jian Hu	Yiducloud, China
Alessandro Bile	Sapienza University of Rome, Italy
Jingjing Cao	Wuhan University of Technology, China
Shi Cheng	Shaanxi Normal University, China
Xianghua Chu	Shenzhen University, China
Valentina Colla	Scuola Superiore Sant'Anna, Italy
Mohammad Hosein Fazaeli	Amirkabir University of Technology, Iran
Vikas Gupta	LNM Institute of Information Technology, India
Tianyong Hao	South China Normal University, China
Hongdou He	Yanshan University, China
Wangpeng He	Xidian University, China
Yaqing Hou	Dalian University of Technology, China
Essam Halim Houssein	Minia University, Egypt
Wenkai Hu	China University of Geosciences, China
Lei Huang	Ocean University of China, China
Weijie Huang	University of Jinan, China
Zhong Ji	Tianjin University, China
Qiang Jia	Jiangsu University, China
Yang Kai	Yunnan Minzu University, China
Andreas Kanavos	Ionian University, Greece
Zhao Kang	Southern Illinois University Carbondale, USA
Zouaidia Khouloud	Badji Mokhtar Annaba University, Algeria
Chunshan Li	Harbin Institute of Technology, China
Dongyu Li	Beihang University, China
Kai Liu	Chongqing University, China
Xiaofan Liu	City University of Hong Kong, Hong Kong
Javier Parra Arnau	Karlsruhe Institute of Technology, Germany
Santwana Sagnika	Kalinga Institute of Industrial Technology, India
Atriya Sen	Rensselaer Polytechnic Institute, USA
Ning Sun	Nankai University, China
Shaoxin Sun	Chongqing University, China
Ankit Thakkar	Nirma University of Science and Technology, India
Ye Wang	Chongqing University of Posts and Telecommunications, China
Yong Wang	Sun Yat-sen University, China
Zhanshan Wang	Northeastern University, China
Quanwang Wu	Chongqing University, China
Xiangjun Wu	Henan University, China
Xingtang Wu	Beihang University, China
Zhou Wu	Chonqing University, China

Contents – Part II

Contents – Part I

Dynamic Community Detection via Adversarial Temporal Graph Representation Learning

Changwei Gong, Changhong Jing, Yanyan Shen, and Shuqiang Wang[✉]

Shenzhen Institutes of Advanced Technology, Chinese Academy of Sciences,
Shenzhen 518000, China
{cw.gong,ch.jing,yy.shen,sq.wang}@siat.ac.cn

Abstract. Dynamic community detection has been prospered as a powerful tool for quantifying changes in dynamic brain network connectivity patterns by identifying strongly connected sets of nodes. However, as the network science problems and network data to be processed become gradually more sophisticated, it awaits a better method to efficiently learn low dimensional representation from dynamic network data and reveal its latent function that changes over time in the brain network. In this work, an adversarial temporal graph representation learning (ATGRL) framework is proposed to detect dynamic communities from a small sample of brain network data. It adopts a novel temporal graph attention network as an encoder to capture more efficient spatio-temporal features by attention mechanism in both spatial and temporal dimensions. In addition, the framework employs adversarial training to guide the learning of temporal graph representation and optimize the measurable modularity loss to maximize the modularity of community. Experiments on the real-world brain networks datasets are demonstrated to show the effectiveness of this new method.

Keywords: Dynamic community detection · Graph neural networks · Adversarial learning

1 Introduction

Neuroscience is emerging into a generation marked by a large amount of complex neural data obtained from large-scale neural systems [1]. The majority of these extensive data are in the form of data from networks that cover the relationships or interconnections of elements within different types of large-scale neurobiological systems. Significantly, these data often span multiple scales (neurons, circuits, systems, whole brain) or involve different data types in neurobiology (e.g., structural networks expressing anatomical connectivity of nerves, functional networks representing connectivity of distributed brain regions associated with neural activity). The brain network consists of anatomical structures segmenting different brain regions and connecting them by functional networks showing

H. Zhang et al. (Eds.): NCAA 2022, CCIS 1638, pp. 1–13, 2022.
https://doi.org/10.1007/978-981-19-6135-9_1

their complex neuronal communication and signaling patterns. Attributed to advancements in current imaging techniques and advanced methods of medical image processing [2,3], this sophisticated pattern of neural signals may be studied using functional imaging, in which neuronal activity is associated with a variety of behaviors and cognitive functions as well as brain diseases [4–7,11]. At the same time, network science is the study of complex network representation through theories and techniques of computer science and mathematics. With the convergence of two significant scientific developments in recent years, new techniques and analytical methods in the network science field are emerging for evaluating real-world biological networks [8].

Subgraphs, network modules, and communities have been extensively studied in the context of network structures, and in particular, community detection [10,33] methods have been widely used in network neuroscience [9]. Network structure identification or community detection(see schematics in bottom right of Fig. 1) is the partition of nodes in a network into groups in which nodes in the communities are tightly connected, and nodes in different communities are sparsely connected. The organizational principles and operational functions of complex network systems can be revealed and understood through mining the network structure. Furthermore, the creation of comprehensive network maps of neural circuits and systems has resulted from the development of new techniques for mapping the structure and functional connectivity of the brain. A wide range of graph-theoretic tools can be used to examine and analyze the structure of these brain networks. Therefore, methods for detecting modules or network communities in brain networks are of specialized application, and they reveal tightly connected primary building elements or substructures, which frequently relate to particular functional components.

Related Work. Data-driven models have gotten much attention for a long time, and when combined with machine learning techniques, it has led to great success in building pattern recognition models within the field of medical image computing [16,19]. The models have the potential to achieve high accuracy at a low computational cost. Deep learning is currently widely perceived as one of the most significant developments in machine learning in general. The method of deep learning is a new approach for dealing with high-dimensional biological data and learning low-dimensional representations of medical image [14]. The approach based on the generative adversarial methodse [35] and the graph neural network are good instances. Generative adversarial network(GAN) [17], which can bee seen as variational-inference [34] based generative model, is commonly employed in medical image representational analysis [15,20,23]. The utilization of GAN for community detection is inspired by the fact that GANs are often supervised in training, and the newly generated data (in principle) has the same distributioun as real data, allowing for robust, complex data analysis [21,22,24,35]. Convolutional neural network (CNN) [18] approach reduces the dimensions of medical imaging data by pooling and convolution, enabling it to successfully recognize pattern in biomedical task [12,13]. Graph convolutional network (GCN) is developed to extract community features since it derives the CNN capabilities and directly processes on network structured data. However, existing methods to process dynamic network data to obtain temporal graph representations for community detection remain challenging, especially for small sample network datasets.

To end these issues, we developed a novel adversarial temporal graph representation learning (ATGRL) to complete the clustering of brain nodes in dynamic brain networks, detect different communities containing similar brain regions in dynamic brain networks and their evolution, and improve the robustness of the model while handling with small sample network data by employing generative adversarial approaches. The proposed temporal graph attention encoder is efficient to graph representation learning, and more helpful graph embeddings are obtained to complete the clustering to detect more accurate dynamic communities. The detected communities with sound classification effects can be used as biological markers.

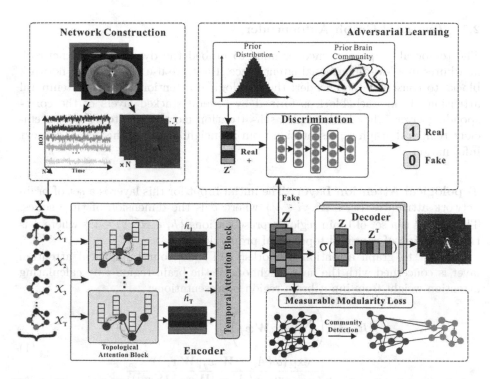

Fig. 1. Proposed adversarial temporal graph representation learning framework for detecting brain communities.

2 Method

Our ATGRL includes two core parts: 1) a temporal graph auto-encoder consists of a temporal graph attention encoder and a decoder, and 2) an adversarial regularizer including a discriminator. The architecture are illustrated in Fig. 1. In the autoencoder, the encoder (E) adopts temporal graph attention networks to transform the time series of brain regions ($\{X^t\}_{t=1}^{T}$) and brain functional connections (A) into the embeddings ($\{Z^t\}_{t=1}^{T}$). Moreover, in the adversarial regularizer, a min-max adversarial game is led between the encoder, which regards as the generator (G), and the discriminator (D) to learn better embeddings. In order to detect communities, the measurable soft modularity loss is employed which optimizes the community assignment matrix P. Therefore, the encoder is trained with triple objectives: a classic reconstruction loss of autoencoder, an adversarial training loss from the discriminator and the measurable modularity loss for detecting community.

2.1 Temporal Graph Autoencoder

The temporal graph autoencoder aims to embed the dynamic brain network attributes in a low-dimensional latent space. First, we use two different network blocks to construct the encoder: the topological attention block and temporal attention block. Each block is formed by several stacked layers of the corresponding layer. They both employ self-attention mechanisms to obtain an efficient temporal graph representation from its neighboring and historical context information.

Topological Attention Layer. The initial input for this layer is a set of brain network attributes $\{x_i \in \mathbb{R}^d, \forall i \in \mathcal{V}\}$ where d is the dimension of time series. The output is a set of brain region representations $\{h_i \in \mathbb{R}^f, \forall i \in \mathcal{V}\}$ where f is the dimension of captured topological properties.

Similar to graph attention networks(GAT) [25], our topological attention layer is concerned with the near neighbors of the brain region i by calculating attention weight from input brain region representations:

$$
h_i = \sigma \left(\sum_{u \in \mathcal{N}_i} \alpha_{ij} \boldsymbol{W} \boldsymbol{x}_j \right),
$$

$$
\alpha_{ij} = \frac{\exp \left(\sigma \left(A_{ij} \cdot [\boldsymbol{W}\boldsymbol{x}_j \| \boldsymbol{W}\boldsymbol{x}_i] \right) \right)}{\sum_{w \in \mathcal{N}_i} \exp \left(\sigma \left(A_{wi} \cdot [\boldsymbol{W}\boldsymbol{x}_w \| \boldsymbol{W}\boldsymbol{x}_i] \right) \right)}
\tag{1}
$$

Here $\mathcal{N}_i = \{j \in \mathcal{V} : (i, j) \in \mathcal{E}\}$ is the set of near neighbor of region i which are linked by functional connection A; $W \in \mathbb{R}^{d \times f}$ is a weight transformation matrix for each region representations; $\sigma(\cdot)$ is sigmoid activation function and $\|$ is the concatenation operation. The learnt coefficients α_{ij}, which is computed by performing softmax on each neighbors, indicates the significance of brain region i to region j. Note that topological attention layer applies on brain region representation at a single timestamp, and multiple topological attention layer can calculate the entire time sequence in parallel.

Temporal Attention Layer. Dynamically capturing constant changing patterns of brain networks is essential for dynamic community detection. When extracting the local timestamp features, it is critical to consider the influence of the global temporal context. The key question is how to capture the temporal alterations in brain networks structure throughout variety of time steps. Temporal attention layer is designed for tackling this issue with the help of the scaled dot-product attention [26]. Its queries, keys, and values are being used to represent the attributes of input brain regions.

We define $H_s = \left\{ h_s^1, h_s^2, ..., h_s^T \right\}$, a representation sequence of a brain region s at continuous timestamps as input, where T is the number of time steps. And the output of the layer is $Z_s = \left\{ z_s^1, z_s^2, ..., z_s^T \right\}$, a new brain network representation sequence for region s at different timestamp.

Using h_s^t as the query, temporal attention layer evaluate its historical representations, inquiring the temporal context of the neighborhood around region s. Hence, temporal self-attention allows the discovery of relationships between time-varying representations of a brain region across several time steps. Formally, the temporal attention layer is computed as:

$$Z_s = \beta_s \left(H_s W_s \right),$$

$$\beta_s^{ij} = \frac{\exp\left(e_s^{ij}\right)}{\sum_{k=1}^{T} \exp\left(e_s^{ik}\right)}, \quad e_s^{ij} = \left(\frac{\left(\left(H_s W_q\right)\left(H_s W_k\right)^T\right)_{ij}}{\sqrt{F'}} \right) \quad (2)$$

where $\beta_s \in \mathbb{R}^{T \times T}$ is the attention coefficient matrix computed by the query-key dot product attention operation; $W_q \in \mathbb{R}^{d \times f}, W_k \in \mathbb{R}^{d \times f}$ and $W_v \in \mathbb{R}^{d \times f}$ are linear projections matrices which transform representations into a particular space.

The two attention blocks are calculated in sequence to obtain the final temporal representation, $i, e.$, the output embeddings Z. And It is utilized to reconstruct the brain network topology in the decoder:

$$\hat{A} = \sigma(ZZ^T) \quad (3)$$

$\hat{A}$ is the reconstructed brain functional connection and $\sigma(\cdot)$ is still sigmoid function.

The classic reconstruction loss is defined by the form of cross entropy:

$$\mathcal{L}_{RE} = \sum \mathbb{E} \left[A_{ij} \log \hat{A}_{ij} + (1 - A_{ij}) \log \left(1 - \hat{A}_{ij} \right) \right] \quad (4)$$

2.2 Adversarial Learning

In this adversarial model, the main objective is enforcing brain network embeddings Z to match the prior distribution. Other naive regularizers push the learned embeddings to conform to the Gaussian distribution rather than capture semantic diversity. As a result, conventional techniques to network embedding cannot effectively profit from adversarial learning. Therefore, we derive the previous distribution of communities by counting different kinds of modules in the functional brain network that have been confirmed by neuroscience. The adversarial model serves as a discriminator by using a three-layer fully connected network to identify whether a latent code drawn from the prior distribution $p_{z'}$ (positive samples) or embeddings z from the temporal graph encoder E (negative samples). The regularizer will eventually enhance the embedding during the minimax competition between the encoder and the discriminator in the training phase.

The loss of the encoder(generator) $\mathcal{L}_G$ and discriminator $\mathcal{L}_D$ in the adversarial model, defined as follows:

$$\mathcal{L}_G = -\mathbb{E}_{x \sim p_{\text{data}}} \log \mathcal{D}_\phi (E_\psi (X, A)) \tag{5}$$

$$\begin{aligned}\mathcal{L}_D = &-\mathbb{E}_{z' \sim p_{z'}} \log \mathcal{D}_\phi(z') \\ &-\mathbb{E}_{x \sim p_{\text{data}}} \log (1 - \mathcal{D}_\phi (E_\psi (X, A)))\end{aligned} \tag{6}$$

in this expression, z' is a latent code sampled from the prior distribution $p_{z'}$ of empirically confirmed brain communities; $\mathcal{D}_\phi(\cdot)$ and $E_\psi(\cdot)$ is the abovementioned discriminator and encoder.

Formally, the objective of this adversarial learning model can be indicated as a minmax criterion:

$$\begin{aligned}\mathcal{L}_{\mathcal{AL}} = &\min_G \max_D \Theta (D, G) \\ = &\min_G \max_D \left(\mathbb{E}_{z' \sim p_{z'}} \log \mathcal{D}(z'; \phi) \right. \\ &+ \mathbb{E}_{x \sim p_{\text{data}}} \left. \log (1 - \mathcal{D} (E (X, A); \phi)))\right.\end{aligned} \tag{7}$$

2.3 Measurable Modularity Loss

Modularity maximization is a technique for community discovery that is commonly used in the detection of brain modules. A partition is regarded high quality (and so has a higher Q score [27]) conceptually if the communities it forms are more dense internally than would be predicted by chance. Thus, the partition that gets the maximum value of Q is considered to be a good estimation of the community structure of a brain network. This intuition may be expressed as follows:

$$Q = \frac{1}{2m} \sum_{ij} [A_{ij} - c_{ij}] \delta (\omega_i, \omega_j) \tag{8}$$

here a_{ij} indicates the number of functional connection between region i and j; $c_{ij} = \frac{k_i k_j}{2m}$ denotes the estimated number of connections based on a null model where $k_i = \sum_j A_{ij}$ is a degree of the region i and $2m = \sum_{ij} A_{ij}$ is overall amount of connections in the brain networks; $\delta(\omega_i, \omega_j) = 1$ if $\omega_i = \omega_j$, which means reigon i and reigon j are in the same community and 0 otherwise.

Inspired by [28], to develop a differentiable objective for optimizing the community assignment matrix $P = softmax(Z) \in \mathbb{R}^{N \times C}$ which represents a matrix of probabilities of brain region attribution to communities, the measurable modularity loss employed by our framework is defined as:

$$\mathcal{L}_{\text{MM}} = \underbrace{-\frac{1}{\sum |A_{ij}|} \operatorname{tr}\left(\mathbf{P}^{\top}\mathbf{B}\mathbf{P}\right)}_{\text{measurable modularity}} + \lambda \underbrace{\left(\sum_i^C \left(\sum_j^N P_{ij} - \frac{1}{C}\right)^2\right)}_{\text{regularization}} \tag{9}$$

where the modularity matrix $B = A - \frac{dd^T}{2m}$; C is the amount of communities and N is the number of regions in the brain networks. The regularization ensures that the model can identify communities of the predicted size.

Thus, the total loss for the encoder optimization in the train process to obtain better embeddings is sum of the above three loss terms, expressed as follows:

$$\mathcal{L}_{total} = \mathcal{L}_{RE} + \mathcal{L}_G + \mathcal{L}_{MM} \tag{10}$$

3 Experiments and Results

In this part, we assess the performance of ATGRL in terms of both dynamic community detection and graph representation learning.

3.1 Dataset Preparation and Implementation Details

Dataset Preparation. We obtained the dynamic brain network dataset required for the experiment by preprocessing long-term functional MRI images of experimental rats. The first preprocessing was carried out in MATLAB utilizing the Statistical Parametric Mapping 8 (SPM8) tool. To adjust for head motion, functional signals were aligned and unwrapped, and the mean motion-corrected image was coregistered with the high-resolution anatomical T2 image. Following that, the functional data were smoothed using a $3mm$ full-width at half-maximum (FWHM) isotropic Gaussian kernel. On the basis of the Wister rat brain atlas, 150 functional network areas were outlined. We used magnitude-squared coherence to assess the spectral relationship between regional time series, resulting in a 150×150 functional connection matrix for each time step, whose members showed the intensity of functional connectivity between all pairs of areas.

Implementation Details. ATGRL was implemented using pytorch backend. The training of the network was accelerated by one Nvidia GeForce RTX 2080 Ti. The training epoch was set at 500, while the learning rate was set to 0.001

during training. To minimize overfitting, Adam [29] was utilized as an optimizer with a weight decay of 0.01. We trained the encoder with 2 topological attention layers and 2 temporal attention layers. We repeat all trials ten times and average the findings. For all datasets and approaches, we set the regularization value to 0.5 and the number of communities at 15.

3.2   Dynamic Community Detection Performance

Baseline. Our approach was compared against the following two kinds of baselines:

GAE. [30] is recently the most common autoencoder-based unsupervised framework for graph data, in which the encoder is composed of two-layer graph convolutional networks to leverage topological information.

ARGA. [31] is an adversarially regularized autoencoder method that employs graph autoencoder to learn the representations, regularizes the latent codes, and forces the latent codes to match a prior distribution; differing from ours, it used simple Gaussian distribution as the prior distribution.

Metrics. For graph-level metrics, we report average community conductance$C$ and modularity$Q$. For ground-truth label correlation analysis, we report normalized mutual information (NMI) between the community assignments and labels and pairwise F-1 score between all node pairs and their associated community pairs.

Ablation Study. As indicated in Table 1, we conducted ablation research on community detection to evaluate the effectiveness of our proposed encoder and adversarial learning, and three significant outcomes were achieved: 1) In the comparison of graph-level metrics, the k-means based method showed impressive performance on community conductance and the modularity loss based method

Table 1. Community detection performance on rat brain networks dataset by graph conductance C, modularity Q, NMI, and pairwise F1 measure.

Method		Metrics(%)			
		C	Q	NMI	F1
K-means based	GAE+K-means	72.4	13.6	50.7	30.6
	ARGA+K-means	**75.0**	22.3	49.6	46.3
Modularity loss based	GAE+$\mathcal{L}_{MM}$	34.4	64.0	51.5	61.5
	Ours(GCN-encoder)	21.5	59.7	45.7	58.2
	Ours	36.0	**68.3**	**68.9**	**63.3**

did better than it on community modularity. This is due to the fact that the two algorithms are fundamentally different in terms of optimization; modularity loss originates with the goal of maximizing modularity. 2) The approach with adversarial regularizer is generally performed well; it represents that adversarial learning does play its role as an auxiliary to graph representation learning. 3) Our algorithm that replaced the proposed encoder with a two-layer graph convolution encoder performs worse; it shows in some way that our proposed encoder may learn better embeddings to make it perform well.

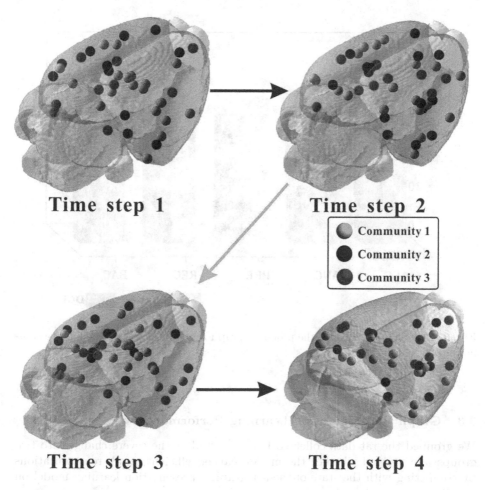

Fig. 2. Visualization of dynamic community detection performance within four time steps.

Visualization of Dynamic Community Detection. We illustrated our result of dynamic community detection by Fig. 2. It shows the changes in the positional distribution of the three major brain communities detected by our approach with increasing time steps. We can see that there is no significant change in the distribution of brain communities at time steps 1 to 2, but there is a more remarkable change at time steps 2 to 3. It is because the rats in the original dataset did change their brain network properties and topology due to experimental factors. Therefore, the outcomes of the experiment are in line with the neuroscientific truth.

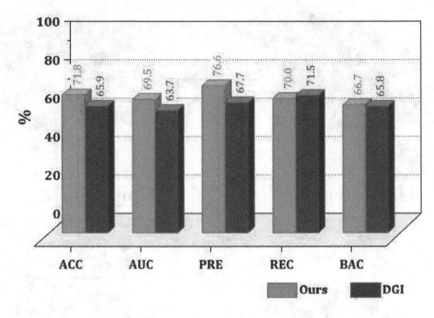

Fig. 3. Classification result of the proposed graph representaion learning and the competing method on baseline.

3.3   Graph Representation Learning Performance

We grouped the rat data collected before and after the severe change into two groups and verified whether the model learned efficient graph representations by competing with the state-of-the-art graph representation learning model on classification performance.

Competing Methods. DGI [32] highlights the importance of cluster and representation learning in combination. We learn unsupervised graph representation with DGI and two algorithms both run SVM on the final representations as the classifier.

Metrics. Evaluation of diagnostic performance is based on quantitative measures in five key areas: To summarize: 1) accuracy (ACC); 2) area under receiver operating characteristic curve (AUC); 3) Precision (PRE); 4) Recall (REC); and 5) balanced accuracy (BAC). Our suggested technique is being evaluated using leave-oneout cross-validation (LOOCV), since we only have a small quantity of data. One of the N individuals is omitted from the testing process, and the $N-1$ subjects that remain are used for training purposes only. It is the greedy search that sets the hyperparameters in each technique to the optimum values.

Prediction Results. As demonstrated in Fig. 3, our approach achieved generally better results on classification performance. In a respect, it verifies that the representations obtained by our method are more strong in the unsupervised learning process.

4 Conclusion

In this research, we propose a novel framework called Adversarial Temporal Graph Representation Learning (ATGRL) for introducing community detection into a deep graph representation learning process directed by an adversarial regularizer. In addition to using temporal graph attention encoder to merge input spatial topology features and temporal contextual representation to represent latent variables, adversarial training with a neuroscientific prior is used to deconstruct the embedding space in the ATGRL framework. Our method outperformed two unsupervised deep embedding and community identification approaches in dynamic brain network datasets. And we obtained better results than the comparison method when using the obtained graph representations for classification, indicating that there is an advantage in graph representation learning that may yield better graph embeddings in the latent space. Detailed model discussions were conducted to investigate the proposed ATGRL and the superiority of the encoder and the adversarial regularizer.

Acknowledgments. This work was supported by the National Natural Science Foundations of China under Grants 62172403 and 81901834, the Distinguished Young Scholars Fund of Guangdong under Grant 2021B1515020019, Guangdong Basic and Applied Basic Research Fundation under Grant 2020A1515010654, the Excellent Young Scholars of Shenzhen under Grant RCYX20200714114641211 and Shenzhen Key Basic Research Projects under Grant JCYJ20200109115641762.

References

1. Bassett, D.S., Sporns, O.: Network neuroscience. Nat. Neurosci. **20**(3), 353–364 (2017)
2. Hu, B., Lei, B., Shen, Y., Liu, Y., Wang, S.: A point cloud generative model via tree-structured graph convolutions for 3D brain shape reconstruction. In: Ma, H., et al. (eds.) PRCV 2021. LNCS, vol. 13020, pp. 263–274. Springer, Cham (2021). https://doi.org/10.1007/978-3-030-88007-1_22

3. Hu, S., et al.: Bidirectional mapping generative adversarial networks for brain MR to PET synthesis. IEEE Trans. Med. Imaging **41**(1), 145–157 (2021)
4. Mo, L., et al.: A variational approach to nonlinear two-point boundary value problems. Nonlin. Anal. Theo. Methods Appl. **71**(12), 834–838 (2009)
5. Pan, J., Lei, B., Shen, Y., Liu, Y., Feng, Z., Wang, S.: Characterization multimodal connectivity of brain network by hypergraph GAN for Alzheimer's disease analysis. In: Ma, H., et al. (eds.) PRCV 2021. LNCS, vol. 13021, pp. 467–478. Springer, Cham (2021). https://doi.org/10.1007/978-3-030-88010-1_39
6. Junren, P., et al.: DecGAN: decoupling generative adversarial network detecting abnormal neural circuits for Alzheimer's disease. arXiv preprint arXiv:2110.05712 (2021)
7. Yu, W., et al.: Morphological feature visualization of Alzheimer's disease via multidirectional perception GAN. IEEE Trans. Neural Netw. Learn. Syst. (2021)
8. Rubinov, M., Sporns, O.: Weight-conserving characterization of complex functional brain networks. Neuroimage **56**(4), 2068–2079 (2011)
9. Michelle, G., Newman, M.E.J.: Community structure in social and biological networks. Proc. Nat. Acad. Sci. **99**(12), 7821–7826 (2002)
10. Fortunato, S.: Community detection in graphs. Phys. Rep. **486**(3–5), 75–174 (2010)
11. Zeng, D., et al.: A GA-based feature selection and parameter optimization for support tucker machine. Procedia Comput. Sci. **111**, 17–23 (2017)
12. Wu, K., Shen, Y., Wang, S.: 3D convolutional neural network for regional precipitation nowcasting. Image Signal Process **7**(4), 200–212 (2018)
13. Wang, S., et al.: Skeletal maturity recognition using a fully automated system with convolutional neural networks. IEEE Access **6**, 29979–29993 (2018)
14. Wang, S., et al.: An ensemble-based densely-connected deep learning system for assessment of skeletal maturity. IEEE Trans. Syst. Man Cybern. Syst. **52**(1), 426–437 (2020)
15. Hu, S., Yuan, J., Wang, S.: Cross-modality synthesis from MRI to PET using adversarial U-net with different normalization. In: 2019 International Conference on Medical Imaging Physics and Engineering (ICMIPE). IEEE (2019)
16. Wang, S., et al.: Classification of diffusion tensor metrics for the diagnosis of a myelopathic cord using machine learning. Int. J. Neural Syst. **28**(02), 1750036 (2018)
17. Ian, G., et al.: Generative adversarial nets. In: Advances in Neural Information Processing Systems, vol. 27 (2014)
18. Bouvrie, J.: Notes on convolutional neural networks (2006)
19. Wang, S., et al.: Prediction of myelopathic level in cervical Spondylotic myelopathy using diffusion tensor imaging. J. Mag. Reson. Imaging **41**(6), 1682–1688 (2015)
20. Yu, W., et al.: Tensorizing GAN with high-order pooling for Alzheimer's disease assessment. IEEE Trans. Neural Netw. Learn. Syst. **33**(9), 4945–4959 (2021)
21. Hu, S., Shen, Y., Wang, S., Lei, B.: Brain MR to PET synthesis via bidirectional generative adversarial network. In: Martel, A.L., et al. (eds.) MICCAI 2020. LNCS, vol. 12262, pp. 698–707. Springer, Cham (2020). https://doi.org/10.1007/978-3-030-59713-9_67
22. Hu, S., et al.: Medical image reconstruction using generative adversarial network for Alzheimer disease assessment with class-imbalance problem. In: 2020 IEEE 6th International Conference on Computer and Communications (ICCC). IEEE (2020)
23. Wang, S., et al.: Diabetic retinopathy diagnosis using multichannel generative adversarial network with semisupervision. IEEE Trans. Autom. Sci. Eng. **18**(2), 574–585 (2020)

24. Lei, B., et al.: Skin lesion segmentation via generative adversarial networks with dual discriminators. Med. Image Anal. **64**, 101716 (2020)
25. Petar, V., et al.: Graph attention networks. arXiv preprint arXiv:1710.10903 (2017)
26. Ashish, V., et al.: Attention is all you need. In: Advances in Neural Information Processing Systems, vol. 30 (2017)
27. Aaron, C., Newman, M.E.J., Moore, C.: Finding community structure in very large networks. Phys. Rev. E **70**(6), 066111 (2004)
28. Ivan, L., Ivanov, S.: Unsupervised community detection with modularity-based attention model. arXiv preprint arXiv:1905.10350 (2019)
29. Kingma, D.P., Ba, J.: Adam: a method for stochastic optimization. arXiv preprint arXiv:1412.6980 (2014)
30. Kipf, T.N., Welling, M.: Variational graph auto-encoders. arXiv preprint arXiv:1611.07308 (2016)
31. Pan, S., et al.: Adversarially regularized graph autoencoder for graph embedding. arXiv preprint arXiv:1802.04407 (2018)
32. Petar, V., et al.: Deep graph infomax. ICLR (Poster) 2(3), 4 (2019)
33. Shen, Y., et al.: Subcarrier-pairing-based resource optimization for OFDM wireless powered relay transmissions with time switching scheme. IEEE Trans. Signal Process. **65**(2), 1130–1145 (2016)
34. Wang, S., et al.: A variational approach to nonlinear two-point boundary value problems. Comput. Math. Appl. **58**(11), 2452–2455 (2009)
35. You, S., et al.: Fine perceptive GANs for brain MR image super-resolution in wavelet domain. IEEE Trans. Neural Netw. Learn. Syst. (2022). https://doi.org/10.1109/TNNLS.2022.3153088

Research on Non-intrusive Household Load Identification Method Applying LightGBM

Zhiwei Kong[1]($\boxtimes$), Rao Fu[1] (iD), Jiachuan Shi[1] (iD), and Wenbin Ci[2]

[1] Shandong Key Laboratory of Intelligent Buildings Technology, School of Information and Electrical Engineering, Shandong Jianzhu University, Jinan 250101, China
2020085125@stu.sdjzu.edu.cn, furao20@sdjzu.edu.cn
[2] State Grid Shandong Electric Power Company, Jinan 250101, China

Abstract. With the continuous advancement of intelligence and big data technology, efficient and accurate non-intrusive load identification is of great significance for load forecasting and user electricity safety. The traditional decision tree algorithm cannot accurately identify loads with similar characteristics. This paper proposed a load identification model applying Light Gradient Boosting Machine (LightGBM) algorithm to the different characteristics of various household loads in the switching and operation process, which improved the performance based on one-sided sampling and mutually exclusive feature bundling. The experimental results showed that the algorithm has a faster operation rate in calculation and identification than the traditional decision tree algorithm, and the identification accuracy rate reaches 97.7%.

Keywords: Non-intrusive · Load identification · LightGBM · One-sided sampling · Mutually exclusive feature bundling

1 Introduction

With the rapid development of smart grids, power grid companies are striving to become more intensive and technological. To establish a smart power consumption system, one of the key progress is the load identification, which is the first step to solve the intelligent load identification and monitoring issues of power users [1]. For electricity users, load identification can grasp the power consumption in real time, so that the electricity may be saved, and hidden danger can be eliminated [2].

At present, the identification methods of household load mainly include intrusive and non-intrusive methods. Intrusive method means that, sensors are installed on electrical devices to monitor them in real time. The measurement data is accurate, but the investment cost be high and the user may share low acceptance. Non-intrusive method means that, monitoring devices are installed at the entrance of the power supply to obtain data such as voltage and current, with low input cost, convenient installation and maintenance, and broad development prospects [3].

Detailed load information extracted through non-intrusive load monitoring can bring benefits to residential users and grid policy making. Load level electricity consumption information can improve residents' electricity consumption habits. And with the improvement of computing power and the continuous development of public data sets, more and more excellent performance algorithms emerge in an endless stream, which is a great source of research in this field. Therefore, our main contributions are summarized as follows:

LightGBM was proposed as a new supervised NILM approach based on machine learning algorithm.

The load characteristics and unknown load types were investigated procedurally.

XGboost and LightGBM algorithms were compared.

This paper was structured as follows: In Sect. 2, the research status at home and abroad were summaried. In Sect. 3, how the model was demonstrated to be trained and tested. In Sect. 4, the experimental results were shown and finally we drew conclusions.

2 Overview

The non-intrusive load identification mainly focuses on the selection of load characteristics and the optimization of the load identification model [4]. Selecting the load characteristics that match the load identification algorithm is the key point. The load characteristics currently used are mainly divided into steady-state characteristics and transient characteristics. The goal of load identification is to identify the type of load. For example, when load switching and other actions are detected, the specific load should be identified on which the action occurs.

Professor Hart from the Massachusetts Institute of Technology first proposed the non-intrusive load monitoring (NILM) technology [5]. Then the American Electric Power Research Institute launched a non-intrusive load identification system project, using transient power changes as features to carry out load monitoring. Limited types of loads could be identified by a single state feature [6].

According to the principle of linear additivity of steady-state current, Yang et al. [7] proposed an optimal objective function method based on differential evolution algorithm, which could decompose multiple types of loads online. With the prevalence of machine algorithms, load characteristics and intelligent identification algorithms were used to better explore electricity consumption patterns. Wang [8] applied the Mean-shift clustering algorithm to cluster data, and completed NILM through non-parametric and multivariate judgments, which had a high degree of recognition, yet low-power electrical equipment was easily interfered by noise and was ignored.

Su et al. [9] adopted Fourier transform and wavelet transform to extract real and reactive power as features for load identification, which further improved the identification accuracy of nonlinear loads. However, this method did not consider the situation of simultaneous switching of multiple loads. Liu [10] proposed an improved nearest neighbor algorithm and Support Vector Machine (SVM) hybrid three-layer classification recognition algorithm for the multi-load mixed operation under high noise interference which was difficult to identify, with relatively slow running rate.

Chen et al. [11] proposed a negative identification model based on Factorial Hidden Markov Models (FHMM), and then used the extended Vitiber algorithm for load identification, which was less sensitive to load fluctuations. Qi et al. [12] used principal component analysis to reduce the dimensionality of LS, as well as Principal Component

Analysis (PCA) combined with Fisher's supervised identification algorithm to solved overlapping load characteristics. Fu et al. [13] proposed a dual-objective optimization NILM algorithm to compensate for the possible errors of a single-objective function, and also proposed a multi-layer decision tree recognition algorithm based on transient features. Li [14] proposed a load decomposition method based on multi-factor decision-making in the decomposition and verification process for the problem of low sampling frequency load decomposition accuracy, and achieved good results.

With the development of the information age and the wide application of artificail intelligence technology, traditional data statistics methods could not meet the complex, heterogeneous and dynamic data processing requirements for the problem of user behavior analysis of meter data. The difficulty is that the target electricity consumption data changes are complex and easily affected by other factors. The above-mentioned classic machine learning methods have made good achievements in the data analysis of small samples. But in large-scale data sets, they faced problems such as huge amount of calculation and poor fitting effect. Therefore, how to improve the computing power of the model is a problem that needs to be solved.

Gradient Boosting Decision Tree (GBDT) is an enduring model in machine learning. It consists of multiple decision trees and is a classic iterative decision tree [15]. eXtreme Gradient Boosting (XGBoost) is an efficient algorithm for implementing GBDT, which requires to traverse all the data during training and iteration. Once the amount of data increases, long running time and low efficiency may be inevitable [16]. In order to solve this issue, an improved LightGBM algorithm was used to realize the identification of unknown loads, which significantly improved the training speed of the model.

3 LightGBM Model Principle

LightGBM is an improved algorithm for implementing the GBDT model. The GOSS (Gradient-based One-side sampling) calculate information gain. Histogram algorithm is used to find the optimal segmentation point, hence improving the speed of traversing high-dimensional data [17]. The leaf growth strategy trains variables iteratively to ensure that the loss function decreases continuously. It had the following advantages: faster training speed, higher classification accuracy, less memory footprint, and distributed support which could process massive data.

3.1 Gradient-Based One-Sided Sampling

The input of the model was the power consumption data of each load in time series, including voltage, current, active power and reactive power. According to the information gain theory in decision trees, samples with larger gradients indicated insufficient training components and provided greater gain. Due to the large amount of household data and the unbalanced distribution of class samples, the GOSS did not train the samples directly, but sorted the features of the training set data according to the gradient. By preset proportion, the GOSS saved all large gradient samples and performed random sampling on small gradient samples. On account of the calculation results, a balance coefficient was introduced when calculating the information gain. In this way, the model could focus on the "undertrained" sample data, so as to complete the full training of massive data.

For features such as current and voltage of the same device, the GOSS algorithm learned the correlation between these data. For different devices, it learned the differences between these data and got the relationship between the features. The larger the absolute value, the stronger the correlation. The calculation criterion of GOSS was to arrange the samples in descending order according to the absolute value of the gradient. Assume that the large gradient samples accounted for a% and the small gradient samples accounted for b%. a% of the samples were selected from the entire sample set to form the large gradient subset A, and then randomly select b% from the remaining samples to form a small gradient subset B. The gradient sum of subset B was multiplied by the balance coefficient $(1 - a)/b$ to eliminate the influence of sampling imbalance on the calculation result. The information gain expression of the jth feature at point d on the subset $A \cup B$ was:

$$\tilde{V}_j(d) = \frac{1}{n}((\sum \{x_l \in A_l\}g_i + \frac{1-a}{b} \sum \{x_l \in B_l\}g_i)^2)/n_l^j(d)$$
$$+ (\sum \{x_r \in A_r\}g_i + \frac{1-a}{b} \sum \{x_r \in B_r\}g_i)^2)/n_r^j(d) \qquad (1)$$

where

$$A_l = \{x_l \in A : x_{ij} \le d\}$$
$$A_r = \{x_r \in A : x_{ij} > d\}$$
$$B_l = \{x_l \in B : x_{ij} \le d\} \qquad (2)$$
$$B_r = \{x_r \in B : x_{ij} > d\}$$

As shown in Eq. 1, x_l was the actual value, x_r was the predicted value, g_i was the structural function of the tree, and n was the number of leaf nodes. After calculating the information gains of all candidate segmentation points, the node $\tilde{d}_j^* = \arg \max_d \tilde{V}_j(d)$ was selected with the largest gain as the optimal segmentation point to complete the segmentation of the data set by the current node.

3.2 Mutually Exclusive Feature Bundle

The spatial distribution of high-dimensional data was relatively sparse, There was a phenomenon of mutual exclusion of features as well, because they did not take non-zero values at the same time, yet take zero values at the same time. For example, the microwave ran on the low-fire setting for a long time, and the characteristics displayed were very similar. It was necessary to bundle many mutually exclusive features into fewer dense feature constraints in order to reduce the number of unnecessary features, avoid unnecessary calculation of zero feature values, improve the calculation speed, and reduce memory consumption.

After the high-dimensional data was bundled with mutually exclusive features, the mutually exclusive features were combined into one feature to construct a feature histogram, which greatly reduced the computational complexity.

3.3 Histogram Algorithm Improvements

In addition, the training data had the characteristics of large span and many outliers. For example, when the laptop was in a long-term working state, the current data jumped repeatedly. The histogram calculation referred to binning the feature values. First, the amount of bins needed to be determined for each feature, and an integer should be assigned to each bin. Then, the discretized numerical features were divided into several intervals, and the number of intervals was equal to the number of boxes. The sample data belonging to the box was updated to the value of the box. And finally represented by a histogram, as shown in Fig. 1, the essence was histogram statistics, and large-scale data was placed in the histogram.

This approach had many advantages, such as convenient data storage, fast calculation rate, strong robust performance, and more stable models. Obviously, this method also had certain limitations. After the feature was discretized, the segmentation points were no longer accurate, so it was supposed to have a particular impact on the results. However, experiments showed that the impact was relatively limited on the final load identification results with the coarse segmentation point, which was also the effect of regularization, and could prevent overfitting.

3.4 Leaf Growth Strategy

Most machine learning algorithms based on decision trees adopt a level-wise splitting strategy. Output "satisfied" and "unsatisfied" was able to separate left and right subtrees according to a certain judgment condition, and each node continues to repeat growing downwards. LightGBM used a leaf-wise splitting strategy to find a single node with the largest information gain among all nodes in each layer, then split the node and cycled it in turn.

This method could reduce more losses than splitting by layers. However, when the amount of data was small, the growth strategy of splitting left may cause overfitting. To reasonably solve this problem, LightGBM reduced the occurrence of overfitting by limiting the maximum depth of the tree.

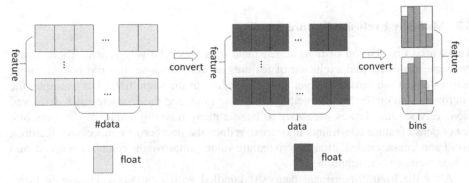

Fig. 1. The bucketing process of the histogram algorithm.

4 Modeling of Non-intrusive Load Identification

4.1 Data Presentation and Analysis

The data of this experiment came from an ordinary household electricity data in the Data Mining Challenge, which collected the time-series electricity consumption information of various electrical equipment. This paper selected microwave, laptop, water dispensers and kettle as the research objects, and built a feature library to identify unknown devices.

In the non-intrusive load monitoring system, the load characteristics are the key to distinguish the types of load equipment, The start and stop of different loads can then be identified in the non-intrusive load identification system. Although different load equipment has different operating characteristics, some loads have multiple characteristics, while some of the characteristics are very similar, which will unavoidably affect the accuracy of load identification [18]. Therefore, selecting a suitable load signature be the first critical step in the identification process.

The current and voltage characteristics of the four electrical equipment were shown in Fig. 2. It could be seen that due to the differences in the internal components of each equipment, the current and voltage wave forms of each equipment were significantly different. Among them, microwave and laptop had obvious periodic fluctuations. The current of the water dispenser would stabilize after a certain period of fluctuations with the switching of the operation, and the current of the kettle would settle to a steady state immediately after the switching operation.

In addition, it could be seen from the coordinate of the current that, the steady-state current of each device had a large difference in amplitude. For example, the steady-state current of the kettle and the water dispenser was 8A and 2A respectively. It was clearly distinguishable and could therefore be used to identify devices and states.

Although the voltage waveform had obvious differences, in fact, its fluctuation range was small. All fluctuate was in the 220 V neighboring rang, and the distinguishability of the equipment was less than obvious. Therefore, this paper did not analyze the voltage separately.

Real power and reactive power were essential data to describe the power consumption of the devices. In Fig. 3, the real power and reactive power and power factor curves of each device showed that when the device was in a steady state, its power characteristics were smooth, and only fluctuate when the state changes, and then the power turned different. It was distinguishable for this. Although the power of microwave ovens and notebook computers was not stable, their periodic frequency bands were also an important feature that could distinguish them from other electrical appliances.

According to the definition of physics, instantaneous power was defined as the product of voltage and current,

$$P_{ij}(t) = U_{ij}(t) \cdot I_{ij}(t) \tag{3}$$

The instantaneous power curve of each device was drawn as shown in Fig. 4, The instantaneous power of each state of each device had obvious differences.

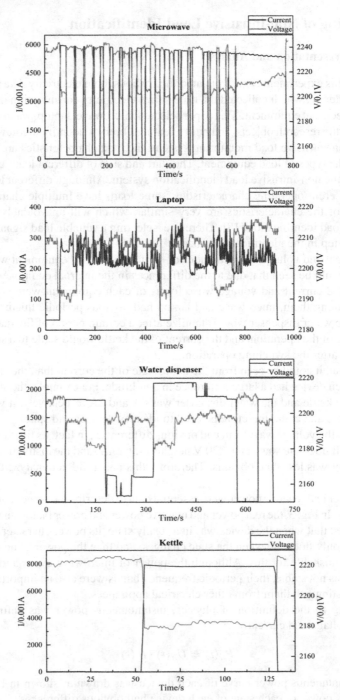

Fig. 2. Current-voltage curves of each device.

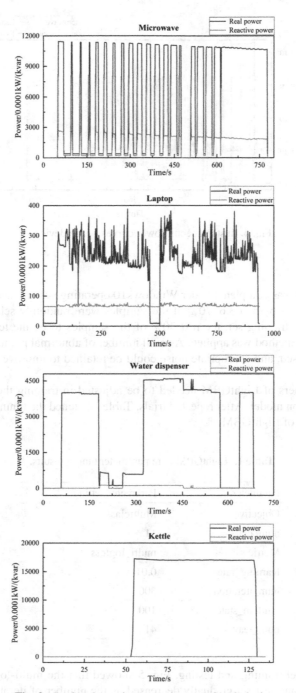

Fig. 3. Real and reactive power curves of each device.

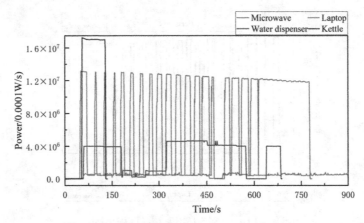

Fig. 4. Instantaneous power curve of each device.

4.2 Modeling

This experiment was completed under Windows10 operating system, and the programming software was python 3.6. 70% of the samples were randomly selected from the data to form the training set, the rest 30% of the samples to be the test set, and the cross-validation method was applied. A small number of abnormal points in the sample were not processed, and appropriate noise could be retained to improve the stability of the model.

The parameters of LightGBM needed to be adjusted in training the non-intrusive load identification model. After repeated trials, Table 1 listened the tuning results of the core parameters of LightGBM.

Table 1. LightGBM core parameter tuning results.

Parameter	Parameter settings
Objective	Multiclass
num_class	11
Metric	multi_logloss
learning_rate	0.01
num_iterations	300
random_state	100
num_leaves	41

During model training and testing, Fig. 5 showed that the multi-logloss index on the training set and test set gradually decreased as the number of iterations increases. The smaller the multi-logloss value was, the better the overall performance of the model would be. When the number of iterations reached at 300 times, the multi-logloss value of the curve had stabilized.

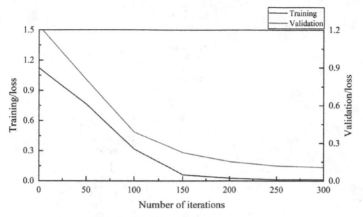

Fig. 5. Multi-logloss change curve.

4.3 The Implementation Process of Household Load Identification

After the required features were determined, extracting and cleaning the relevant feature data should be started. Because the number of samples was unbalanced, the unbalanced number of samples would greatly reduced the effect of the LightGBM classification model. After determining the number of samples, label each device, as shown in Table 2.

Table 2. LightGBM network output value corresponding to load identification.

Output value	Load identification results
1	Microwave
2	Laptop
3	Water dispenser
4	Kettle

Then, the whole data was scrambled and the training set and testing set were extracted according to the ratio of 7:3. The training set was used for supervised learning and training learning, The testing set was used to verify the classification effect of the classifier. The model training process was shown in Fig. 6.

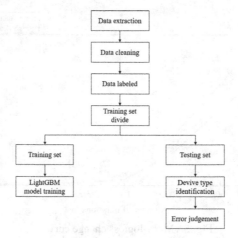

Fig. 6. Model training process.

4.4 Analysis of Experimental Results

To examine the performance of the unknown load recognition on the model test set, Table 3 showed the confusion matrix that used LightGBM algorithm to identify the classification results. The recognition accuracy of unknown device A and unknown device B reaches 98.1% and 97.7%.

Table 3. Confusion matrix for load identification.

Real class	Prediction class	
	Microwave	Kettle
Microwave	266	0
Laptop	0	3
Water dispenser	0	1
Kettle	0	241
Accuracy	98.1%	97.7%

Based on the same data set, this paper chose the XGBoost algorithm and the established model to compare the accuracy and training time. As shown in the Table 4, the LightGBM model had a higher index score, which could fit the data better, and consume less time. The superior load identification ability the model was proved.

Table 4. Comparison of different algorithm.

Algorithm	Accuracy		Time
	Microwave	Kettle	
XGBoost	92.4%	95.0%	69.6 s
LightGBM	98.1%	97.7%	51.7 s

5 Conclusion

This paper aimed household power data to be the research object, selected four types of household loads, analyzed the characteristics of current, voltage, real and reactive power, as well as instantaneous power to identify the differences among all the loads. The four types feature data of known devices and two types of unknown devices were input into the network model. LightGBM re-sorted the correlations between multi-dimensional features and target variables, found the best splitting point on the sorted histogram, The unknown load classification was able to be obtained. LightGBM had the advantages of fast training speed, good classification effect, low memory usage and high stability. The household power load data collected in the actual scene had phenomena such as large amount of data and complex feature dimensions. LightGBM could overcome these problems smoothly, and had important research value for non-intrusive load identification. There is room for further improvement, and LightGBM can be integrated with other algorithms to grant it more advantages.

References

1. Guo, H., Lu, J., Yang, P., Liu, Z.: Review on key techniques of non-intrusive load monitoring. Electric Power Autom. Equip. **41**(01), 135–146 (2021)
2. Zhang, W., Wu, Y., Du, L., Han, W., Chen, Y., Zhang, S.: A survey on the non-intrusive load monitoring. J. Taizhou Univ. **43**(06), 48–54 (2021)
3. Kuang, M., Li, Y., Wei, L., Li, C.: Non-intrusive load identification method based on fusion decision. Data Commun. **04**, 11–16 (2021)
4. Zhou, D., Zhang, H., Zhou, H., Hu, W.: Non-intrusive load event detection method based on state feature clustering. Trans. China Electrotech. Soc. **35**(21), 4565–4575 (2020)
5. Hart, G.: Nonintrusive appliance load. monitoring. Proc. IEEE **80**(12), 1870–1891 (1992)
6. Drenker, S., Kader, A.: Nonintrusive monitoring of electric loads. IEEE Comput. Appl. Power **12**(4), 47–51 (1999)
7. Yang, H., Chang, H., Lin, C.: Design a neural network for features selection in non-intrusive monitoring of industrial electrical loads. In: 2007 11th International Conference on Computer Supported Cooperative Work in Design, pp. 1022–1027 (2007)
8. Wang, Z.: Research on key technologies of low-voltage side power management based on load characteristic identification. PhD thesis, Wuhan University (2011)
9. Su, Y., Lian, K., Chang, H.: Features selection of non-intrusive load monitoring system using STFT and wavelet transform. In: 2011 IEEE 8th International Conference on e-Business Engineering, pp. 293–298 (2011)
10. Liu, R.: Research on house load identification combining with improved nearest neighbor method and support vector machine. Master's thesis, Chongqing University (2014)

11. Chen, S., Gao, F., Liu, T., Zhai, Q., Guan, X.: Load disaggregation method based on factorial hidden markov model and its sensitivity analysis. Autom. Electric Power Syst. **40**(21), 128–136 (2016)
12. Qi, B., Cheng, Y., Wu, X.: Non-intrusive household appliance load identification method based on fisher supervised discriminant. Power Syst. Technol. **40**(08), 2484–2491 (2016)
13. Fu, H.: Comprehensive study on non-intrusive load decomposition algorithm. Master's thesis, Beijing Jiaotong University (2018)
14. Li, T.: Research on non-intrusive power load decomposition method based on multi-load characteristics. Master's thesis, Kunming University of Science and Technology (2018)
15. Yang, Y.: Non-intrusive load identification technology based on decision tree. Sci. Technol. Innov. **13**, 54–55 (2018)
16. Wu, C., Ren, J.: Power system transient stability assessment method based on XGBoost. Electric Power Autom. Equip. **41**(02), 138–143+152 (2021)
17. Li, Z., Yao, X., Liu, Z., Zhang, J.: Feature selection algorithm based on LightGBM. J. Northeastern Univ. (Natl. Sci.) **42**(12), 1688–1695 (2021)
18. Wand, S., Guo, L., Chen, H., Deng, X.: Non-intrusive load identification algorithm based on feature fusion and deep learning. Autom. Electric Power Syst. **44**(09), 103–110 (2020)

Master Multiple Real-Time Strategy Games with a Unified Learning Model Using Multi-agent Reinforcement Learning

Bo Ling[1], Xiang Liu[1], Jin Jiang[1](✉), Weiwei Wu[1], Wanyuan Wang[1], Yan Lyu[1], and Xueyong Xu[2](✉)

[1] Southeast University, Nanjing, China
boling_william@163.com, {xiangliu,weiweiwu,wywang,lvyanly}@seu.edu.cn,
analysis.jinger@gmail.com
[2] North Information Control Research Academy Group Co., Ltd., Nanjing, China
xxyyeah@163.com

Abstract. General artificial intelligence requires an intelligent agent to understand or learn any intellectual tasks like a human being. Diverse and complex real-time strategy (RTS) game for artificial intelligence research is a promising stepping stone to achieve the goal. In the last decade, the strongest agents have either simplified the key elements of the game, or used expert rules with human knowledge, or focused on a specific environment. In this paper, we propose a unified learning model that can master various environments in RTS game without human knowledge. We use a multi-agent reinforcement learning algorithm that uses data from agents in a diverse league played on multiple maps to train the deep neural network model. We evaluate our model in microRTS, a simple real-time strategy game. The results show that the agent is competitive against the strong benchmarks in different environments.

Keywords: Real-time strategy game · Multi-agent reinforcement learning · League learning · Unified model

1 Introduction

Game AI has a history of active research for decades with games often serving as challenge problems, benchmarks and milestones for progress. Muzero, a model-based reinforcement learning agent, masters Go, chess shogi and Atari without rules. Pluribus, a superhuman AI masters six-player poker [14]. OpenAI Five defeated a team of professional Dota 2 players and 99.4% of online players [5]. AlphaStar was rated at Grandmaster level for all three StarCraft races and above 99.8% of officially ranked human players [4]. Tencent developed a superhuman

---

B. Ling and X. Liu—These authors contribute equally.

H. Zhang et al. (Eds.): NCAA 2022, CCIS 1638, pp. 27–39, 2022.
https://doi.org/10.1007/978-981-19-6135-9_3

AI agents that can defeat top esports players on a popular MOBA game, Honor of Kings [18].

To the goal of general artificial intelligence, an intelligence agent is expected to understand or learn any intellectual tasks like a human being. Although these AIs are greatly successful, they have a major limitation, i.e., only suitable for a specific environment. Trying to explore an agent that can master various environments is meaningful and valuable.

In this paper, we study one of the most challenging domains, real-time strategy games. It has real-time, simultaneous, huge action space, huge state space, and multi-agent characteristics. Specifically, we study an algorithm to master multiple real-time strategy games. Multiple games means that the game environment is different in the size of the map, the initial position of the unit, the number of resources, and so on. In different environments, completely different strategies are usually required. Script-based or search-based algorithms usually can be applied to multiple maps but rely on expert experience and domain knowledge. In this paper, we explore a unified learning model applicable to multiple environments in real-time strategy game. We summarize the contributions of this paper as follows.

- To the best of our knowledge, we are the first to explore a unified learning model to play multiple real-time strategy games.
- Under the actor-learner pattern, we design a general neural network model, including the encoding of multi-model inputs and the decoupling of action outputs. The model is suitable for the input of various maps and units, and the output of various action types.
- To improve robustness, we explore a multi-agent reinforcement learning algorithm to enable effective explorations for multi-agent competitive environments.
- We choose three different maps for experiments in the domain of microRTS, a simple RTS game developed for research purposes. Experimental results demonstrate that various environments can apply our proposed model. In addition, the trained AI agent outperforms benchmarks in all maps.

The rest of the paper is organized as follows. Section 2 reviews the related work. In Sect. 3, we introduce the model and formulate the Reinforcement Learning problems. In Sect. 4, we propose a unified learning model to play multiple real-time strategy games. We then conduct experiments in Sect. 5 to validate the performance of the proposed method. Finally, we conclude the paper and discuss the future work in Sect. 6.

2   Related Work

The research methods of RTS games mainly include search-based, rule-based and learning-based methods. In this section, we mainly review these three methods separately.

2.1 Rule-Based Methods

Rule-based research compiles the rules summarized by human players in practice into programs, and the game program selects the corresponding strategy execution according to the game situation during the game.

Silva et al. [16] proposed Strategy Creation via Voting (SCV) [15] by using a voting method to generate a set of strategies from existing expert strategy and an opponent modeling scheme to choose appropriate strategies from the generated pool of possibilities. Marino et al. [11] and others constructed a new script pool by combining and replacing expert strategy scripts, using genetic algorithms to select a collection of scripts with a high winning rate.

The main drawback of rule-based methods is that they only work on a limited amount of expert-designed strategies. This led to rigid scripted tactical micro-management. By contrast, we design league training so that the trained AI can effectively explore various strategies.

2.2 Search-Based Methods

Churchill and Buro [7] proposed an Alpha-Beta Considering Durations (ABCD) [11] algorithm based on min-max search and Alpha-Beta pruning. Considering durative actions, the alpha-beta algorithm implements a tree alteration technique to address simultaneous actions. Ontañón [13] introduced NaïveMCTS, which uses a sampling strategy to exploit the tree structure of Combinatorial Multi-Armed Bandits. Yang and Ontanón [20] investigate to combine scripts into the tree policy for the convergence guarantees of MCTS. However, these tree-search based methods usually depend on an efficient forward model, which requires internal models of games to examine the outcome of different combinations. Instead, we use a model-free reinforcement learning approach.

In addition, due to real-time strategy games have a huge search space, the search algorithms described above require a lot of time. The following algorithms try to limit the search space. Puppet Search (PS) [7] defines a search space for the values of script parameters. Strategy Tactics [3] combines PS's search in the script-parameter space with a NaïveMCTS search in the original state space for the combat units. Moraes and Lelis [12] introduced a search algorithm based on combinatorial multi-armed bandits (CMABs) that used asymmetric action abstractions schemes to reduce the action space considered during search [10]. Lelis [12] proposed Stratified Strategy Selection (SSS) to micromanage units in RTS games, which divides players' units into types under the assumption that units with the same type follow the same script. Lin et al. [9] further took into account the time constraint when generating the sub-game tree and introduced the adversarial player as the opponent strategy.

In these methods, scripts are also used to guide the search. The action-abstracted space usually restricts the behaviour of agents, as such space tends to be limited to a small number of handcrafted strategies.

2.3   Learning-Based Methods

Recently, multi-agent reinforcement learning methods for specific scenarios has been studied in depth. Low et al. [10] proposed multi-agent deep deterministic policy gradient (MADDPG) [2], an adaptation of actor-critic methods that can be used in competition and cooperation scenarios. Since the proposed model requires each agent to have its own independent critic network and actor network, this model is limited to micromanagement tasks with a fixed number of input units.

In 2019, OpenAI introduced OpenAI Five [5] for playing 5v5 games in Dota 2, which can defeat a team of professional Dota 2 players. The OpenAI Five applies deep reinforcement learning via self-play. Recently, Tencent developed AI programs towards playing full 5v5 Multiplayer Online Battle Arena games (MOBA) in Honor of Kings through curriculum self-play learning, which can defeat top esports players [18]. As a sub-problem of the RTS games, there is no need to consider building base, building barracks, training different types of units and so on for MOBA games. Therefore, these models do not consider the situation that the number of agents may change with the progress of the game, i.e., new units may be produced and existing units may be eliminated. Our approach can handle the problem because we toward a full real-time strategy game instead of just micromanagement.

In 2019, DeepMind developed an AI in the full game of StarCraft II, called Alphastar, with the ability to reach the Grandmaster level for all three StarCraft races. The AI is trained through end-to-end multi-agent reinforcement learning. The main difference between our model and AlphaStar is that we explore playing games on various different scale of maps instead of playing games in specific scenarios.

3   Preliminary

3.1   Real-Time Strategy Games

The RTS game is essentially a simplified military simulation. Combat scenarios in RTS games can be described as finite zero-sum two-player games with simultaneous and durative moves defined by a tuple $(\mathcal{N}, \mathcal{S}, s_{\text{init}}, \mathcal{U}, \mathcal{A}, \mathcal{R}, \mathcal{P}, \pi)$,

- $\mathcal{N} = \{i, -i\}$: the set of players.
- $\mathcal{S} = \mathcal{D} \cup \mathcal{F}$: the set of states, where $\mathcal{D}$ is the set of non-terminal states and F is the set of terminal states.
- $s_{init} \in \mathcal{D}$: the start state of the games.
- $\mathcal{A} = \mathcal{A}_i \times \mathcal{A}_{-i}$: the set of joint actions.
- $\mathcal{A}_i(s)$: the set of legal actions that player i can perform in state s.
- $\mathcal{U} = \mathcal{U}_i \times \mathcal{U}_{-i}$: the set of joint units.
- $\mathcal{U}(s)$: the set of available units in state s.
- $\mathcal{R}_i : \mathcal{F} \to \mathbb{R}$ is a utility function with $\mathcal{R}_i(s) = -\mathcal{R}_{-i}(s)$ for any $s \in \mathcal{F}$.
- $\mathcal{P} : \delta \times \mathcal{A}_i \times \mathcal{A}_{-i} \to \mathcal{S}$ determines the successor state for a state s and a set of join actions taken at s.
- $\pi(\mathcal{A} \mid \mathcal{S}) : \mathcal{S} \times \mathcal{A} \to [0, 1]$ is the policy of a player.

3.2   Reinforcement Learning

We consider the Reinforcement Learning problem in a Markov Decision Process (MDP) denoted as $(\mathcal{S}, \mathcal{A}, \mathcal{P}, r, \gamma, T)$ [6]. Here, $r : \mathcal{S} \times \mathcal{A} \to \mathbb{R}$ is the reward function, γ is the discount factor, and T is the maximum episode length. The goal is to maximize the expected discounted return of the policy:

$$\mathbb{E}_\tau \left[\sum_{t=0}^{T-1} \gamma^t r_t \right] \tag{1}$$

where τ is the trajectory $(s_0, a_0, r_0, \ldots, s_{T-1}, a_{T-1} r_{T-1})$, $s_t \sim \mathcal{P}\left(* \mid s_{t-1}, a_{t-1}\right)$ and $a_t \sim \pi_\theta\left(* \mid s_t\right), r_t = r\left(s_t, a_t\right)$.

4   Methods

In this section, we introduce our methods from three aspects. Firstly, we introduce our neural network architecture for microRTS. Secondly, we describe the design of league learning used to generate training data. Finally, we present our multi-agent reinforcement learning algorithm.

4.1   The Architecture

We use an actor-critic network structure, where actor refers to policy network, and critic refers to value network. The network follows the paradigm of centralized training and decentralized executing, i.e., the critic network is only used for training, and the actor-network is used for training and execution. In our design, all units share a policy network and a value network. Specifically, the policy network calculates the policy for each unit independently, using the global state information and the information of an input unit, and the value network evaluates the global state value of a player. Based on the deep neural network method, we need to design a DNN-represented policy $\pi_\theta\left(a_t \mid s_t\right)$ with parameters θ. Its inputs include previous observations, current units of a player and the information of the unit types. Its output are actions and the game state value. In order to provide informative observations to the actor, we develop multi-modal features, including a comprehensive list of both scalar and spatial features. The scalar feature is consisted of units' type attributes, e.g., health point (hp), relative position, speed, action types, etc. The spatial features include convolutional channels derived from the player's global view, e.g., player resources, the type of units on each grid, and the actions of the units on each grid. To deal with the problem that different types of agents have different action spaces and the explosion of combined action spaces, we develop hierarchical action heads and discretize each head. Specifically, the actor predicts the output actions hierarchically: 1) what action to select, including move, attack, return, harvest, and produce. 2) how to act, e.g., a discretized direction to move. 3) which unit type to produce given an input unit, including worker, barrack, base, light, heavy, ranged. The network structure is shown in Fig. 1.

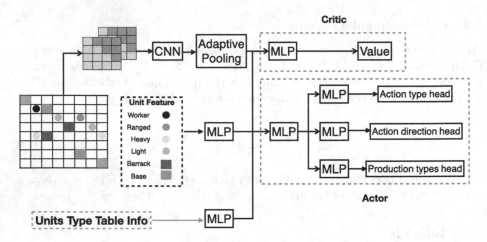

Fig. 1. Network architcture for RTS game

Internally, in order to handle different map sizes, the observations are first convolved and then passed to an adaptive pooling layer to output fixed-length vectors. The adaptive pooling layer removes the fixed size constraint of the network. Therefore, the model could process maps of different sizes. The model concatenates the encoding of the various information of the unit types and the output of the pooling layer into a vector v_1. The model concatenates the vector v_1 and the encoding unit given by a player into a vector v_2. The vector v_1 is passed to the multi-layer perceptron of the critic network to calculate the state value. The vector v_2 is passed to the multi-layer perception of the actor network to calculate the action.

4.2   League Learning

Balduzzi et al. [1] pointed out that, for the approximately transitive game, self-play generates sequences of agents of increasing strength. However, non-transitive games, such as rock-paper-scissors, may chase cycle and fail to progress. Observing this, AlphaStar computes a best response against a non-uniform mixture of opponents to avoid strategic cycles. In this paper, we use a pool of non-uniform mixed opponents as a league and then train the agent through playing against with the league. By making use of the advantages of the built-in script AI, we selected some AI to join the league according to the strength of AI. In this work, we used the following AIs as opponents:

1. Passive: A hard-coded strategy that passively attacks the nearest unit.
2. Random: A heuristic strategy that randomly selects one of the possible player-actions, but the probability of choosing an attack or harvest action is 5 times higher than the other actions.

3. WorkerRush: A hard-coded strategy that builds a barracks, and then constantly produces "worker" units to attack the nearest target (it uses one worker to mine resources).

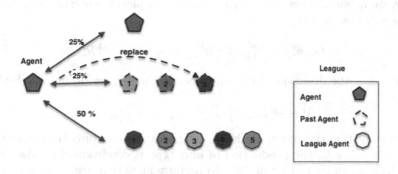

Fig. 2. League training stage three.

In the first stage, the league is set to a weaker version, including Passive and Random. The agent is trained with a proportion of 25% Self-play, 50% against Passive in the league, and an additional 25% of matches against Random in the league. This stage of training ends when the training reaches the 100 thousand timestep or the time to complete the game is stable.

In the second stage, WorkerRush was added to the league. The agent is trained with a proportion of 25% Self-play, 25% against WorkerRush in the league, 25% against Passive in the league, and an additional 25% of matches against Random in the league. This stage of training ends when the training reaches the 100 thousand timestep or the time to complete the game is stable.

In the third stage, we copy the trained agent, named Past Agent, to replace Passive. The agent is trained with a proportion of 25% Self-play, 25% against WorkerRush in the league, 25% against Passive in the league, and an additional 25% of matches against Past Agent. The training process is shown in Fig. 2.

4.3 Multi-agent Reinforcement Learning

We use the actor-critic paradigm [17] to train a value function $v_\theta(s_t)$ with a policy $\pi_\theta(a_t \mid s_t)$. We first introduce the design of reward function. Then, we present our policy gradient update algorithm.

Inspired by Wang et al. [8], we design *Semi-Advantage-Function* (SAF) as follow. Firstly, we calculate the hp of player p at time t as Eq. (2).

$$\text{HP}_t^p \triangleq \sum_i hp_t^i \tag{2}$$

where hp_t^i denotes the hp of agent i at time t. Secondly, we design a single timestep reward of player p as Eq. (3).

$$R_t^p = \left(HP_t^p - HP_{t'}^p\right) - \left(HP_t^{-p} - HP_{t'}^{-p}\right) \tag{3}$$

Thirdly, we define the action reward for unit i of player p at the beginning of the timestep t as Eq. (4).

$$r^p\left(s_t, a_t^i\right) = R_{t+1}^p + \gamma R_{t+2}^p + \cdots + \gamma^{t'-t-1} R_{t'}^p \tag{4}$$

where $t' - t$ is the duration of the unit action. Finally, we design our SAF as Eq. (5).

$$A_t^i = r^p\left(s_t, a_t^i\right) + \gamma^{t'-t} v\left(s_{t'}\right) - v\left(s_t\right) \tag{5}$$

We assume that the action heads are independent to simplify the correlations of action heads, e.g., the production of unit type is conditioned on the action type, which is similar to that of [19]. To perform an action, we select an Action Type with corresponding action parameters. Namely, a_t^i is composed of smaller actions $a_t^{\text{Action Type}|i}$, $a_t^{\text{Action Parameter}|i}$, $a_t^{\text{Produce Type}|i}$. We use Proximal Policy Optimization (PPO) [21] to train the agent. The policy gradient is updated using SAF in the following way (without considering the PPO's clipping for simplicity)

$$\sum_{t=0}^{T-1} \sum_{u \in \mathcal{U}(s_t)} A_t^u \nabla_\theta \log \pi_\theta\left(a_t^u \mid s_t\right) \tag{6}$$

$$= \sum_{t=0}^{T-1} \sum_{u \in \mathcal{U}(s_t)} A_t^u \nabla_\theta \left(\sum_{d \in D} \log \pi_\theta\left(a_t^{d|u} \mid s_t\right)\right) \tag{7}$$

$$= \sum_{t=0}^{T-1} \sum_{u \in \mathcal{U}(s_t)} A_t^u \nabla_\theta \log \left(\prod_{d \in D} \pi_\theta\left(a_t^{d|u} \mid s_t\right)\right) \tag{8}$$

where D = {Action Type, Action Parameter, Produce Type}. The detail of the proposed method is shown in Algorithm 1.

5   Experiments

5.1   Environment

We choose microRTS as the main body of the research, where microRTS is a small implementation of RTS games, aimed at AI research.

Algorithm 1. The Unified Actor-critic Learning Algorithm for RTS Games

---

1: Set initial actor network parameters θ, critic network parameters ω, buffer B, state
   $s = s_{init}$, update steps T.

2: Define the policy of league π_{league}.

3: Define the policy of actor π_{old}.

4: **for** iteration: $i = 0, 1, ..., N$ **do**

5:    **for** actor: $j = 0, 1, ..., M$ **do**

6:       Select the policy of the opponent π_{league}^{j}

7:       Choose an action for player 1 based on the actor policy $A_1 \leftarrow \pi_{\theta} old$.

8:       Choose an action for player 2 based on the critic policy $A_2 \leftarrow \pi_{league}^{j}$.

9:       Submit joint actions $A \leftarrow (A_1, A_2)$.

10:      Update states $s \leftarrow s'$.

11:      Store the actionable unit u and the unit-selected action a under state s in
    the form (s, u, a, r, s') in buffer B.

12:      **if** $|B| = T$ **then**

13:        Update parameters θ_{old} according to 6.

14:      **end if**

15:      **if** $s = s_{end}$ **then**

16:        $s \leftarrow s_{init}$

17:      **end if**

18:      $T \leftarrow T + 1$

19:    **end for**

20: **end for**

5.2 Experiment Setup

Our RL infrastructure runs on a physical computing cluster. To train our AI
model in parallel, we use 8 V100 GPUs and 128 CPUs totally. For each of the
experiments, we use 1 GPU and 16 CPUs resource to train. We have selected
three maps for experimentation (Fig. 3).

We perform two groups of experiments. One is to train an agent on each map
in a self-play manner and evaluate it with benchmarks on the three maps. The
other is to train and evaluate an agent on the three maps using the method of
league learning. The detailed description of the map is as follows. (1) On the
4×4 map, each player has 1 base and 2 resources. In addition, there are 2 public
uncollected resources. The max cycle of game is 1000 timesteps. (2) On the 6×6
map, each player has a base and 5 resources. In addition, there are 5 uncollected
resources around each player. The max cycle of game is 1500 timesteps. (3) On
the 8×8 map, each player has a base and 5 resources. In addition, there are
20 uncollected resources around each player. The max cycle of game is 2000
timesteps.

5.3 Experiment Results

We use WorkerRush, Passive, Random as the benchmarks to evaluate the performance of our trained agents. To be brief, WorkerRush is an offensive agent eager
to win, Passive is a defensive agent who wants to avoid defeat, and Random

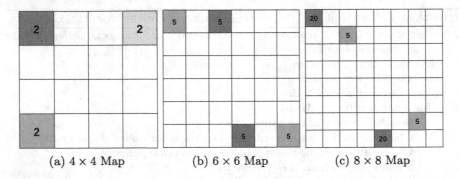

(a) 4 × 4 Map   (b) 6 × 6 Map   (c) 8 × 8 Map

Fig. 3. Training maps

wants the game to be a tie (more detailed descriptions of these three methods are shown in Sect. 4.2).

In the group of experiments, there are two agents SelfPlay6 and SelfPlay8, which represent agents trained in self-play on 6 × 6 and 8 × 8 maps, respectively. The evaluation results are shown in Fig. 4a–4b.

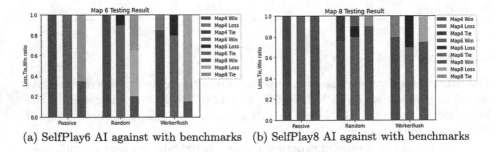

(a) SelfPlay6 AI against with benchmarks   (b) SelfPlay8 AI against with benchmarks

Fig. 4. AI against with benchmarks

Figure 4a shows the result of SelfPlay6 agent. We found that SelfPlay6 has a higher winning rate on 6 × 6 and 4 × 4 maps, with a winning rate of more than 60%. It is revealed that the agent has a certain generalization performance. However, on the 8×8 map, the winning rate is significantly reduced. By observing the game video, we found that it has obvious offensive intent on the 4 × 4 map, but not on the 8 × 8 map. We would evaluate all methods by the metric of Loss, Tie, Win ratio.

Figure 4b shows the result of SelfPlay8 agent. We found that the agent is very competitive against Passive and Random in the three maps, but has a low win rate against WorkerRush. This reveals that the learning strategy of the agent has certain effect, but the strategy is not strong enough. On the one hand, the size of these three maps is not that big, victory mainly depends on micro-management, and WorkerRush has the advantage of micromanagement. On the other hand, agents trained through self-play may chase cycles resulting in no progress.

Through the results, we find that the AI trained on the medium map has better generalization performance on the small map, while the generalization ability on the large map is weaker. But the trained AI is not strong enough via self-play, thus we examine its performance when combined with league learning.

In the second group of experiments, we train an agent LeagueAllMap, which uses league learning to train on the three maps. The result is shown in Fig. 5.

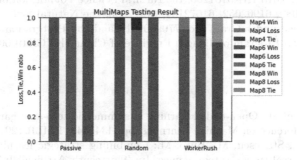

Fig. 5. LeagueAllMap AI against with benchmarks

To improve the generalization and robustness of the agent, we try to make the model play against opponents on three maps at the same time. Furthermore, we use league training to overcome chase cycles in strategy space. Through these methods, the experimental results show that the performance of the agent has been significantly improved. The agent outperforms the benchmarks on all of the three maps.

We found that multi-stage training by setting up opponents with increasing difficulty can effectively improve their abilities. It suggests that this learning method is easier to get rewards from sparse rewards than self-play. Moreover, we find that the model has better results on the three maps due to the simultaneous use of data from three different maps for training.

6 Conclusion and Future Work

In this paper, we develop a unified learning model, which aims to play a full RTS game on multiple maps through deep reinforcement learning. We explored neural networks that can be learned and trained at different scales, and then explored league training methods. To the best of our knowledge, this is the first reinforcement learning based RTS AI program that can play multiple maps and outperform the benchmark.

However, this study still has some limitations. First of all, there is a lack of theoretical proofs for the convergence of simultaneous training on multiple maps. Secondly, the maps we choose to train are not significantly different, and in general, the scale of the map is not that large. Finally, the robustness of our agent needs to be evaluated in comparison with agents with stronger and more diverse strategies. In the future, we will continue explore more diverse and complex maps for experiments.

Acknowledgement. This work is supported by the National Key R&D Program of China under grant 2018AAA0101200, the Natural Science Foundation of China under Grant No. 62102082, 61902062, 61672154, 61972086, the Natural Science Foundation of Jiangsu Province under Grant No. 7709009016 and the Postgraduate Research & Practice Innovation Program of Jiangsu Province of China (KYCX19_0089).

References

1. Balduzzi, D., et al.: Open-ended learning in symmetric zero-sum games. In: International Conference on Machine Learning, pp. 434–443. PMLR (2019)
2. Barriga, N.A., Stanescu, M., Buro, M.: Combining strategic learning with tactical search in real-time strategy games. In: Thirteenth Artificial Intelligence and Interactive Digital Entertainment Conference (2017)
3. Barriga, N.A., Stanescu, M., Buro, M.: Puppet search: Enhancing scripted behavior by look-ahead search with applications to real-time strategy games. In: Eleventh Artificial Intelligence and Interactive Digital Entertainment Conference (2015)
4. Berner, C., et al.: Dota 2 with large scale deep reinforcement learning. arXiv preprint arXiv:1912.06680 (2019)
5. Brown, N., Sandholm, T.: Superhuman AI for multiplayer poker. Science **365**(6456), 885–890 (2019)
6. Buro, M.: Real-time strategy games: a new AI research challenge. In: IJCAI, vol. 2003, pp. 1534–1535 (2003)
7. Churchill, D., Saffidine, A., Buro, M.: Fast heuristic search for RTS game combat scenarios. In: Eighth Artificial Intelligence and Interactive Digital Entertainment Conference (2012)
8. Konda, V., Tsitsiklis, J.: Actor-critic algorithms. In: Advances in Neural Information Processing Systems 12 (1999)
9. Lin, S., Anshi, Z., Bo, L., Xiaoshi, F.: HTN guided adversarial planning for RTS games. In: 2020 IEEE International Conference on Mechatronics and Automation (ICMA), pp. 1326–1331. IEEE (2020)
10. Lowe, R., Wu, Y.I., Tamar, A., Harb, J., Pieter Abbeel, O., Mordatch, I.: Multi-agent actor-critic for mixed cooperative-competitive environments. In: Advances in Neural Information Processing Systems 30 (2017)
11. Marino, J.R., Moraes, R.O., Toledo, C., Lelis, L.H.: Evolving action abstractions for real-time planning in extensive-form games. In: Proceedings of the AAAI Conference on Artificial Intelligence, vol. 33, pp. 2330–2337 (2019)
12. Moraes, R., Lelis, L.: Asymmetric action abstractions for multi-unit control in adversarial real-time games. In: Proceedings of the AAAI Conference on Artificial Intelligence, vol. 32 (2018)

13. Ontanón, S.: The combinatorial multi-armed bandit problem and its application to real-time strategy games. In: Ninth Artificial Intelligence and Interactive Digital Entertainment Conference (2013)
14. Schrittwieser, J., et al.: Mastering atari, go, chess and shogi by planning with a learned model. Nature **588**(7839), 604–609 (2020)
15. Schulman, J., Wolski, F., Dhariwal, P., Radford, A., Klimov, O.: Proximal policy optimization algorithms. arXiv preprint arXiv:1707.06347 (2017)
16. Silva, C., Moraes, R.O., Lelis, L.H., Gal, K.: Strategy generation for multiunit real-time games via voting. IEEE Trans. Games **11**(4), 426–435 (2018)
17. Sutton, R.S., Barto, A.G.: Reinforcement Learning: An Introduction. MIT Press, Cambridge (2018)
18. Vinyals, O., et al.: Grandmaster level in StarCraft II using multi-agent reinforcement learning. Nature **575**(7782), 350–354 (2019)
19. Wang, Z., Wu, W., Huang, Z.: Scalable multi-agent reinforcement learning architecture for semi-MDP real-time strategy games. In: Zhang, H., Zhang, Z., Wu, Z., Hao, T. (eds.) NCAA 2020. CCIS, vol. 1265, pp. 433–446. Springer, Singapore (2020). https://doi.org/10.1007/978-981-15-7670-6_36
20. Yang, Z., Ontanón, S.: Guiding Monte Carlo tree search by scripts in real-time strategy games. In: Proceedings of the AAAI Conference on Artificial Intelligence and Interactive Digital Entertainment, vol. 15, pp. 100–106 (2019)
21. Ye, D., et al.: Towards playing full MOBA games with deep reinforcement learning. In: Advances in Neural Information Processing Systems 33, pp. 621–632 (2020)

Research on Visual Servo Control System of Substation Insulator Washing Robot

Zhipeng Li[✉], Shouyin Lu, Zhe Jiang, Shiwei Yu, and Qiang Zhang

School of Information and Electrical Engineering, Shandong Jianzhu University,
Jinan 250101, Shandong, China
1094313106@qq.com

Abstract. This paper first describes the overall structure of the insulator washing robot, designs the visual servo control system structure of the insulator washing robot, and obtains the equipment position information through visual methods, the information is fed back to the vision servo control unit to form a closed-loop vision servo system of insulator washing robot. The binocular vision positioning principle and hand-eye relationship are analyzed, and then a visual servo strategy is proposed, and an image-based visual control method is used to realize automatic aiming and cleaning of insulator. Finally, the proposed system and method are simulated by MATLAB/Simulink platform. The experimental results proved that this strategy can be used to achieve automatic targeting and cleaning operations for insulators, and has a good control effect, and further improving the automation level and operational efficiency of the insulator washing robot, realizing high-voltage energized operations with no one involved, and guaranteeing the stable operation of substation power equipment.

Keywords: Insulator washing robot · Visual servo control system · Binocular vision positioning · Image-based

1 Introduction

As the my country's power system grows by leaps and bounds, the safe and effective operation of substation equipment has attracted more and more attention [1]. In the long-term operation of the power transmission and transformation equipment insulators in the power grid, due to the influence of the electromagnetic field environment and the harsh natural environment, the surface of the insulators in the substation is seriously polluted, which can easily cause pollution flashover, which seriously threatens the operation safety of substation equipment [2]. Therefore, for ensure the normal operation of the substation equipment, the contamination layer on the surface of the substation insulation equipment must be regularly removed.

The existing insulator water washing robot generally adopts the control method of remote operation. The robot mainly relies on manual operation and movement. The staff needs to work near the high-voltage live equipment and cannot be far away from the dangerous working environment. The operating capacity requirements of the water pump

H. Zhang et al. (Eds.): NCAA 2022, CCIS 1638, pp. 40–51, 2022.
https://doi.org/10.1007/978-981-19-6135-9_4

are relatively high, which seriously affects the efficiency of the water flushing operation [3]. So as to further promote the working efficiency of the insulator washing robot, clean the insulator more effectively and keep the operator away from the high-voltage live equipment, the servo control method can be adopted to complete the cleaning task of the target.

There are two main visual servo control methods: position-based visual servoing and image-based visual servoing. The former uses image features and through the geometric model of the given target to estimate the position of the target insulator, and then obtains the movement amount of the manipulator through the positional relationship between the end water gun and the target insulator [4], while the latter omits the step of pose estimation, The image features are directly used, and the error of the image information is used to make the robot motion [5].

This study firstly introduces the insulator washing robot, proposes a visual servo system for insulator washing robot, then analyzes the principle and method of visual positioning, then proposes a visual servo control strategy to control the chassis and robot arm of the insulator washing robot respectively, adopts the image-based visual servo method for the complex substation environment to complete the aiming of insulators. Finally, a simulation experiment of the above method was conducted using MATLAB platform, the results verified the possibility of the adopted scheme.

2 Overview of Insulator Washing Robot

This paper mainly takes the insulator washing robot as shown in Fig. 1 as the research object, which mainly includes a mobile chassis module and a robotic arm module. The mobile chassis module mainly realizes the movement of the insulator washing robot in the substation. The mobile chassis adopts a crawler structure design and is equipped with an industrial computer, an inertial measurement unit, a binocular camera, etc. It is able to pass through roads such as cable trenches and trenches, with strong flexibility and stability, and can realize the autonomous walking of the insulator washing robot and complete the task close to the work target. The robotic arm module can complete the automatic aiming of the insulator, and install a vision sensor and a water gun mechanism at the end of the water washing robot to realize the identification, positioning and cleaning of the insulator.

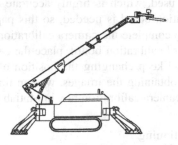

Fig. 1. The structure of the insulator washing robot

3   Visual Servo Control System

According to Fig. 2, the visual servo control system of the insulator washing robot is composed of a visual unit, a visual controller, and a robot control unit. The system uses the visual unit to analyze the acquired images, and uses the analysis results as the input and feedback of the visual servo system to form a closed-loop vision servo system of the insulator washing robot. The visual servo controller uses the feedback information from the visual unit to calculate the amount of motion of the robot, and outputs it to the robot control unit. And then controls the movement of the insulator washing robot through the control quantity, and finally realizes that the insulator washing robot aims at the insulator.

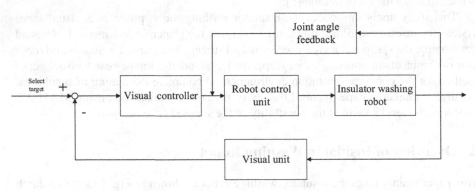

Fig. 2. The visual servo control system of the insulator washing robot

4   Visual Positioning

4.1   Camera Calibration

For obtaining the position information of the insulator through the preliminary calculation of the global camera, it is also necessary to calibrate the camera to get the internal and external parameters of the camera and eliminate the distortion of the camera itself.

Among the current camera calibration methods, the Zhengyou Zhang calibration method is the most widely used, which is highly accurate and easy to operate, and only a checkerboard calibration board is needed, so this paper adopts the Zhengyou Zhang calibration method to complete the camera calibration [6, 7]. Firstly, we select a certain size of checkerboard calibration board, place the calibration board under the camera's shooting range, and keep changing the position of the calibration board to obtain 20 photos, and after obtaining the images, we can finish the calibration of the global camera by using the camera calibration box in Matlab platform.

4.2   Binocular Vision Positioning

Binocular visual localization uses the principle of binocular parallax. The binocular Positioning model is shown in Fig. 3. The camera focal length is f. The distance of the optical center between the left camera and the right camera is B, The corresponding coordinates of the target point $P(x_c, y_c, z_c)$ on the left and right image planes are:

$$\begin{cases} P_l = (x_l, y_l) \\ P_r = (x_r, y_r) \end{cases} \tag{1}$$

The left and right camera are placed under the same horizontal line, then the image coordinates $y = y_l = y_r$, then by the triangular similarity geometric relationship is obtained.

$$\begin{cases} x_l = f \dfrac{x_c}{z_c} \\ x_r = f \dfrac{(x_c - B)}{z_c} \\ y = f \dfrac{y_c}{z_c} \end{cases} \tag{2}$$

Then $D = x_l - x_r$ is the parallax, therefore it is possible to derive the position information of the target point P under the camera coordinate system as:

$$\begin{cases} x_c = \dfrac{B * x_l}{D} \\ y_c = \dfrac{B * y}{D} \\ z_c = \dfrac{B * f}{D} \end{cases} \tag{3}$$

Therefore, a point in the image plane of the right camera can determine the 3D coordinates of this target point by finding a corresponding point in the image plane of the left camera.

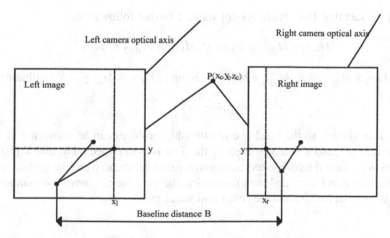

Fig. 3. Binocular stereo vision model

4.3 Hand-Eye Relationship

The hand-eye relationship can transform the position information in the global camera coordinate system into the position information in the water gun coordinate system at the end of the robot, and the insulator washing robot hand-eye relationship can be obtained with hand-eye calibration. There are two methods of hand-eye calibration, one is the eye-to-hand calibration method, the other is the eye-in-hand calibration method. This paper adopts the method of hands eye-to-hand. Figure 4 shows the diagram of the hand-eye relationship calibration. {B} indicates the base coordinate system of the robot arm, {C} represents the global camera coordinate system, {E} indicates coordinate system of the robot end water gun, {R} indicates the coordinate system of the calibration plate.

Fix the calibration plate at the robot water gun, move the robot arm, and use the global camera to take multiple sets of pictures of the calibration plate, and define the position of the robot arm at this time as i. M_{BCi} indicates the coordinate conversion relationship between the base of the robot arm and the global camera at this time. M_{RCi} indicates the coordinate transformation relationship between the calibration plate and the global camera at this time, which can be obtained directly by taking a picture of the calibration plate. M_{BEi} indicates the coordinate conversion relationship between the base of the robot arm and the end water gun at this time, and this position relationship can be obtained from the kinematic analysis of the robot arm [8, 9]. M_{REi} represents the coordinate conversion relationship between the robot end water gun and the calibration plate at this time, and this position relationship is constant during the movement of the robot.

For any moment there are:

$$M_{BC1} = M_{RC1} * M_{RE1} * M_{BE1} \tag{4}$$

The above equation can be deformed to obtain:

$$M_{RE1} = M_{RC1}^{-1} * M_{BC1} * M_{BE1}^{-1} \tag{5}$$

From the fact that M_{REi} is always constant, it further follows that:

$$M_{RC2} * M_{RC1}^{-1} * M_{BC1} = M_{BC2} * M_{BE2}^{-1} * M_{BE1} \tag{6}$$

Let $M_{RC2} * M_{RC1}^{-1} = A$, $M_{BE2}^{-1} * M_{BE1} = B$, and $M_{BC1} = M_{BC2} = X$ to obtain:

$$AX = BX \tag{7}$$

where A, B is known, so the hand-eye relationship problem can be converted to solve the above formula, and X can be solved by the Tsai two-step method to find X [10–12], which can determine the coordinate conversion relationship between the global camera and the base of robot arm, and then determine the coordinate conversion relationship between the global camera and the robot end water gun.

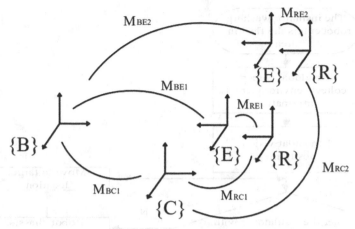

Fig. 4. Hand-eye relationship representation

5 Visual Servo Strategy

Two cameras are installed on the insulator washing robot, one is fixed on the robot motion chassis as a global camera and the other is fastened on the top of the robotic arm water gun as the local camera. As shown in Fig. 5, the visual servo flow diagram shows that the insulator washing robot initially locates the insulator by the image information obtained from the global camera and judges whether the identified insulator is within the working range of the robot arm.

When the insulator is located outside the working range of the robotic arm of the insulator washing robot, the global camera takes a large range of shots and can effectively acquire the visual information of the insulator, and is not easy to miss the insulator information. The image information is collected by the global camera and the pixel position of the target insulator are captured. The position information of the target insulator is obtained by binocular vision positioning estimation, and the position information in the global camera is converted into the position information at the end of the robot arm through the above-mentioned hand-eye relationship. The robot chassis is then controlled to approach the target insulator so that it within the shooting range of the local camera.

The local camera uses image-based vision servo to realize the aiming of insulators. Firstly, the local camera acquires the image information, compares the acquired information with the expected image and obtains the feature deviation, and the vision controller calculates the control amount of the insulator washing robot arm by the image Jacobi matrix and the feature deviation, and uses the control amount as the input of the robot controller to control the movement of the robot to further decrease the feature deviation until the end of the robot aims at the insulator.

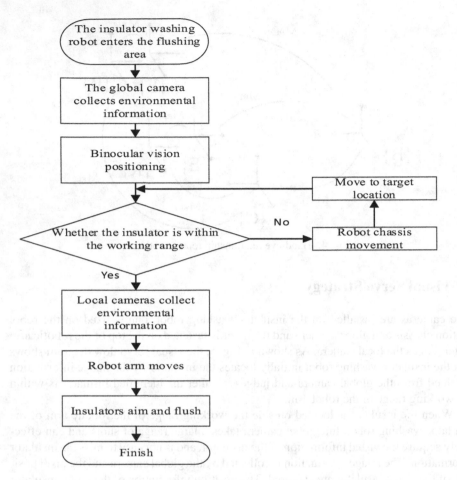

Fig. 5. Visual servo flow

6 Image-Based Visual Servo Control Method

Image-based visual servo control has the feedback signal defined in the image plane. The servo process first extracts the image features from the images captured by the local camera and compares them with the set desired image features to obtain the image feature deviations. The visual servo controller calculates the robot arm control amount from the feature deviation and image Jacobi matrix, drives the robot arm end toward the target, and completes the visual servo task.

6.1 Calculation of Jacobi Matrix

The image Jacobi matrix represents the mapping relationship between the image feature space velocity and the robot arm end motion velocity. Where P is a point in space whose coordinates under the camera coordinate system are (x_c, y_c, z_c), whose coordinates under the image coordinate system are (u, v), whose velocity of the end motion of the robotic arm is $(v_x, v_y, v_z, \omega_x, \omega_y, \omega_z)$, and f denotes the camera focal length.

From the transformation relationship between the coordinate systems of the camera model, the coordinates of the point under the camera image coordinates are:

$$\begin{pmatrix} u \\ v \end{pmatrix} = \frac{f}{z_c} \begin{pmatrix} x_c \\ y_c \end{pmatrix} \tag{8}$$

By the movements of the robot arm, the following relationship is obtained:

$$\dot{P} = -\omega * P - v \tag{9}$$

Then there are:

$$\begin{cases} \dot{x}_c = -v_x - \omega_y z_c + \omega_z y_c \\ \dot{y}_c = -v_y - \omega_z x_c + \omega_x z_c \\ \dot{z}_c = -v_z - \omega_x y_c + \omega_c x_c \end{cases} \tag{10}$$

The derivative of (8) yields:

$$\begin{cases} \dot{u} = \dfrac{\dot{x}_c}{z_c} - \dfrac{x_c \dot{z}_c}{z_c^2} \\ \dot{v} = \dfrac{\dot{y}_c}{z_c} - \dfrac{y_c \dot{z}_c}{z_c^2} \end{cases} \tag{11}$$

Bringing (10) into (11) gives:

$$\begin{pmatrix} \dot{u} \\ \dot{v} \end{pmatrix} = \begin{pmatrix} -\dfrac{f}{z_c} & 0 & \dfrac{u}{z_c} & \dfrac{uv}{f} & -\dfrac{f^2+u^2}{f} & u \\ 0 & -\dfrac{f}{z_c} & \dfrac{v}{z_c} & \dfrac{f^2+v^2}{f} & -\dfrac{uv}{f} & -v \end{pmatrix} \begin{pmatrix} v_x \\ v_y \\ v_z \\ w_x \\ w_y \\ w_z \end{pmatrix} \tag{12}$$

The above formula can be expressed as:

$$\dot{f} = J_r(r) \cdot \dot{r} \tag{13}$$

where $\dot{f} \in R^n$ denotes the rate of change of image features, $\dot{r} \in R^n$ represents the robot arm end velocity, $J_r(r)$ is the image Jacobi matrix.

The robot Jacobi matrix describes the mapping relationship between each joint of the robot arm and the end velocity. As for Eq. (14).

$$\dot{r} = J_q(q) \cdot \dot{q} \tag{14}$$

where $\dot{r} \in R^n$ denotes the end velocity of the robotic arm, $\dot{q} \in R^n$ represents the joint velocity vector of the robotic arm with n degrees of freedom, and $J_q(q)$ is the robotic Jacobi matrix, which is expressed in the form:

$$J_q(q) = \begin{bmatrix} \frac{\partial r_1}{\partial q_1} & \cdots & \frac{\partial f_1}{\partial q_n} \\ \vdots & \ddots & \vdots \\ \frac{\partial r_m}{\partial q_1} & \cdots & \frac{\partial f_m}{\partial q_n} \end{bmatrix} \tag{15}$$

From Eq. (13) and Eq. (14), the relationship between the image features and each joint angle of the robot can therefore be obtained.

$$\dot{f} = J_\theta \cdot \dot{q} = J_r(r) \cdot J_q(q)\dot{q} \tag{16}$$

where $\dot{f} \epsilon R^n$ represents the rate of change of image features, $J_r(r)$ is the image Jacobi matrix, $J_q(q)$ is the robotic Jacobi matrix, and $\dot{q} \epsilon R^n$ represents the joint velocity vector of the robotic arm with n degrees of freedom.

And $J_\theta = J_r(r) \cdot J_q(q)$, whose expression takes the form:

$$J_q(q) = \begin{bmatrix} \frac{\partial f_1}{\partial q_1} & \cdots & \frac{\partial f_1}{\partial q_n} \\ \vdots & \ddots & \vdots \\ \frac{\partial f_m}{\partial q_1} & \cdots & \frac{\partial f_m}{\partial q_n} \end{bmatrix} \tag{17}$$

Therefore, the visual servo control of the robotic arm can be achieved by image feature changes as long as the image Jacobi matrix and the robot Jacobi matrix are calculated.

In the process of image-based visual servoing, the locus between the camera and the target insulator changes from moment to moment, so the image Jacobi matrix changes at each moment, and for this problem the dynamic fitted Newton approach can be applied to achieve the estimation of the image Jacobi matrix [13–15].

6.2 Design of Vision Servo Controller

The input of the vision servo controller is the characteristic deviation of the desired image from the current image, and its output is the angular velocity of the joints of the robot, which produces movement and gradually reduces the characteristic deviation until the water gun at the end of the robot arm moves to its desired posture.

The paper uses the PID control algorithm is used to complete the design of the visual servo controller, and the general form of its control law is:

$$u = K_p e(k) + K_i \sum_{i=1}^{k} e(i) + K_d(e(k) - e(k-1)) \tag{18}$$

Assuming that the expected feature point coordinates are f_d, and the present feature point coordinates are $f(t)$, the error function is:

$$e(t) = f(t) - f_d \tag{19}$$

The paper uses proportional control and setting the proportionality factor as ?, From Eqs. (13), (18) and (19), we get:

$$u(t) = \lambda J_r^+ e(t) \tag{20}$$

where u represents the robot arm control quantity and J_r^+ represents the pseudo-inverse matrix of the image Jacobi matrix.

7 Simulation Experiments

The simulation experiments of this paper were done in MATLAB and the results of the simulation experiments are as follows. Before the simulation experiment, first define one camera model and set the coordinates of feature points under the desired image as (305, 710), (305, 310), (700, 710), (700, 310), and the coordinates of feature points under the current image as (170, 810), (200, 690), (300, 810), (310, 690). After running the simulation program, the four feature points change trajectory is displayed in Fig. 6, and all four feature points can reach the set desired position. The findings indicate that under the visual servo control system, the insulator washing robot can control the robot arm movement according to the feedback visual information, so that the feature points are constantly close to the desired feature points. As shown in Fig. 7, the robot arm posture changes during the aiming process, and it can be seen that the robot moves from the initial posture to the final posture after servo control, and the projection of the four feature points in the image space also reaches the desired position from the initial position.

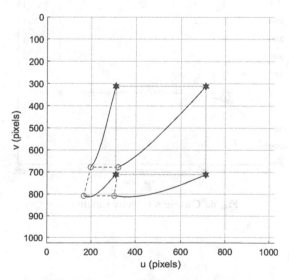

Fig. 6. Feature point change trajectory

As Fig. 8 represents the Cartesian velocity change, In the simulation process, the speed of the end of the robot arm is large at the initial time, and under the control of the visual servo system, the speed decreases continuously, and the feature points of the present image keep approaching the counterpart feature points on the desirable image, and when the image feature points reach the desired feature points, the speed also becomes zero, and the whole process runs smoothly.

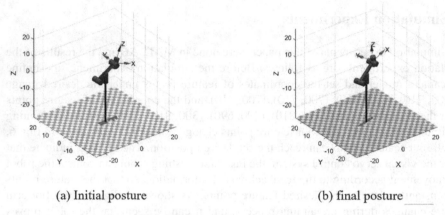

(a) Initial posture   (b) final posture

Fig. 7. Robotic arm position change

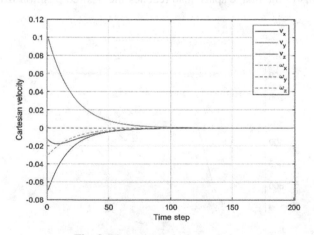

Fig. 8. Cartesian velocity variation

8 Conclusion

The paper introduces a visual servo system for insulator washing robot, which obtains equipment position information by vision method and feeds this information into the robot control system to achieve closed-loop control for insulator washing robot servo control system, and proposes a visual servo control strategy to realize the aiming and washing of insulators by using the image-based visual servo method. Through experiments, it is shown that the system can complete the aiming and flushing of insulators with no one involved, effectively reducing the operational hazards of staff and further improving the automation level of the insulator washing robot.

References

1. Zou, W., Shu, X., Tang, Q.: A survey of the application of robots in power system operation and maintenance management. In: Proceedings of Chinese Automation Congress (2019)
2. Han, S., Hao, R.: Inspection of insulators on high-voltage power transmission lines. IEEE Trans. Power Delivery 24(4), 2319–2327 (2009)
3. Jing, Z., Liao, H., Zhang, C.H.: Design of auto detection system for high-voltage insulator inspection robot. In: Proceedings of International Conference on Robotics and Automation Sciences, pp. 534–547 (2017)
4. Wilson, W.J., Williams Hulls, C.C., Bell, G.S.: Relative end-effector control using cartesian position based visual servoing. IEEE Trans. Robot. Autom. 12(5), 684–696 (1996)
5. Corke, P.I., Hutchinson, S.A.: A new partitioned approach to image-based visual servo control. IEEE Trans. Robot. Autom. 17(14), 507–515 (2001)
6. Zhang, Z.: Camera calibration with one-dimensional objects. IEEE Trans. Pattern Anal. Mach. Intell. 26(7), 892–905 (2004)
7. Liu, Z., Wu, Q., Wu, S.: Flexible and accurate camera calibration using grid spherical images. Opt. Express 25(13), 15269–15285 (2017)
8. Li, S.J., Yang, Q., Geng, M.: Structural design and kinematic analysis of moving mechanism of insulator inspection robot. Mech. Robot. Syst. 3(2), 175–192 (2016)
9. Chen, X.W., Gao, Q., Xiao, B.: Design and kinematics simulation of manipulator with five degrees of freedom based on ADAMS. Manuf. Autom. 12(4), 40–43 (2014)
10. Ma, Q.L., Li, H.Y.: New probabilistic approaches to the AX=XB hand-eye calibration without correspondence. In: 2016 IEEE International Conference on Robotics and Automation, pp. 4365–4371 (2016)
11. Tabb, A., Khalil, M.: Solving the robot world hand eye calibration problem with iterative methods. Mach. Vis. Appl. 28(5), 569–590 (2017)
12. Zeng, J., Cao, G., Chen, B.: An algorithm of hand eye calibration for arc welding robot. In: 16th International Conference on Ubiquitous Robots, pp. 1–6 (2019)
13. Piepmeier, J.A., Gumpert, A.: Uncalibrated eye-in-hand visual servoing. In: Proceedings of International Conference on Robotics and Automation, pp. 568–573 (2002)
14. Yoshimi, B.H., Allen, P.K.: Alignment using an uncalibrated camera system. Trans. Robot. Autom. 11(4), 516–521 (1995)
15. Li, H.F., Liu, J.T., Li, Y.: Visual servoing with an uncalibrated eye-in-hand camera. In: Proceedings of the 29th Chinese Control Conference, pp. 3666–3672 (2010)

A Dual-Size Convolutional Kernel CNN-Based Approach to EEG Signal Classification

Kai Zhao[✉] and Nuo Gao

Department of Information and Electrical Engineering, Shandong Jianzhu University, Jinan 250101, China
1334164473@qq.com, gaonuo@sdjzu.edu.cn

Abstract. In this paper, an algorithm based on a combination of Riemannian space and convolutional neural network is proposed for the feature extraction as well as classification of motor imagery EEG signals with four classifications (left hand, right hand, foot and tongue). In terms of motor imagery EEG signal processing and feature extraction, the covariance matrix of the symmetric positive definite matrix is chosen as the descriptor to reasonably transform the EEG signal described in the spatio-temporal domain from Euclidean space to Riemannian space, followed by a low-dimensional vector estimation of the n-dimensional symmetric positive definite matrix using a specific feature mapping to obtain a low-dimensional and efficient EEG signal feature set. The convolution operation is improved in the classifier by adding a double-size convolution kernel to capture the motor imagery EEG data features more comprehensively. Experimental testing of the method using the publicly available dataset BCI2008IV-2a competition number competition dataset shows that the algorithm can effectively identify EEG signals with high classification accuracy and good robustness.

Keywords: Motor imagery EEG signal · Brain-computer interface · Riemannian spaces · Covariance matrix · Convolutional neural network

1 Introduction

Since the beginning of the 21st century, the global ageing process has been accelerating. The advent of an ageing society is accompanied by a variety of social problems and the number of elderly people suffering from cerebrovascular disease or motor dysfunction caused by spinal cord injury is increasing. The brain of a person with motor deficits is still capable of sending motor commands, but the motor nerves, muscles or limbs of the person have been damaged and are therefore unable to execute motor commands. Brain-Computer Interface (BCI) is a technology that allows the brain to communicate and control directly between the human brain and computers or other electronic devices without the help of peripheral nerves or limbs [1, 2]. BCI devices or systems can bypass the patient's damaged neurological and muscular systems, allowing the use of electroencephalographic signals that contain a wealth of motor imagery, emotion and control commands. The use of BCI technology can help patients with physical disabilities to control external devices and improve their ability to care for themselves; it can also improve the patient's rehabilitation outcome by increasing their motivation to participate in rehabilitation therapy using real-time feedback. Research on BCI systems has become an emerging and popular area of biomedical engineering and rehabilitation engineering.

Because of the spontaneous and intrinsically natural nature of motor imagery EEG(MI-EEG) signals, MI-EEG signals are more suitable than steady-state visual evoked potentials and P300 as control signals for BCI systems [3]. When subjects perform different limb movements, neural activity in the contralateral motor-sensory areas of the brain decreases, leading to a decrease in the spectral energy of rhythms μ (8–12 Hz) and β (13–30 Hz), while neural activity in the ipsilateral motor-sensory areas of the brain increases, leading to an increase in the spectral energy of rhythms β and μ. This electrophysiological phenomenon is known as event-related desynchronisation (ERD) and event-related synchronisation (ERS) [4]. ERD/ERS energy changes based on μ and β rhythms can be used as EEG features to differentiate between different motor imagery tasks. Using the ERS/ERD phenomenon, researchers have proposed a large number of machine learning methods for decoding MI-EEG signals. Common feature extraction methods include co-space patterns [5], filter bank co-space patterns [6], constrained Boltzmann machines [7], and Riemannian geometry [8]. Since the algorithms based on Riemannian geometry have better generalization ability and robustness [9–11], the analysis method of Riemannian space provides a more efficient and novel way of signal processing for analyzing EEG signals. At present, methods based on Riemannian geometry for BCI decoding of EEG signal features are mainly in Riemannian manifolds, based on the invariance of Riemannian distance, using spatial information in the covariance matrix for direct classification, and have achieved good results.

Convolutional neural network (CNN) is a highly effective recognition method developed in recent years, especially in the fields of computer vision [12] and natural language processing [13], and have achieved a series of successful applications. The CNN not only inherits the advantages of traditional neural networks, such as adaptive, fault-tolerant and self-learning, but also has the ability to learn abstract features implicitly from the original features by means of a layered convolutional structure, thus achieving automatic feature extraction. Compared with the manual extraction of EEG features, CNN can automatically extract and select richer and more representative abstract features from the original signal without a priori knowledge, which greatly reduces the loss of information contained in the signal. At the same time, the weight-sharing feature of CNN effectively reduces the size of weights and greatly reduces the complexity of the network model [14]. It makes it more time- and effort-efficient in feature extraction, and is very suitable for feature learning and decoding of EEG signals.

The subsequent chapters of this paper are organised as follows: Sect. 2 introduces the EEG dataset and pre-processing; Sect. 3 deals with the processing of EEG signals and feature extraction based on Riemannian space; Sect. 4 introduces the classifier of this paper, a CNN with a dual-scale convolutional kernel, and the combination of feature vectors and classifier; Sect. 5 presents further experiments on the proposed EEG feature extraction method and the corresponding result analysis; Sect. 6 concludes the paper.

2   Experimental Data and Pre-processing

2.1   Experimental Test Datasets

This study was tested experimentally on the BCI Competition IV-2a public dataset. The dataset consisted of four types of MI-EEG signals from nine subjects, including imagery of left and right hand, foot and tongue movements. Each subject's dataset was divided into a 60% training set, a 20% validation set and a 20% test set. The training set was used to construct the classification model, the validation set was used to optimise the model parameters and the test set was used to evaluate the classification effectiveness of the model. Each subject underwent two sets of identical acquisition experiments in which continuous EEG signals were recorded through 22 Ag/AgCl electrodes at a sampling rate of 250 Hz. In each set of experiments, 72 motor imagery tasks were performed for each class, for a total of 288 trials. The experimental paradigm for competition data set IV2a is shown in Fig. 1. The duration of a single motor imagery task was 8 s. The first 2 s was a preparatory phase, followed immediately by a 1.25 s visual cue instructing the subject to perform the corresponding motor imagery task until the end of the 6th s. The subject then took a 2 s break to prepare for the next motor imagery task. From the continuously collected EEG data, we intercepted the 4 s after the cue as a single MI-EEG sample.

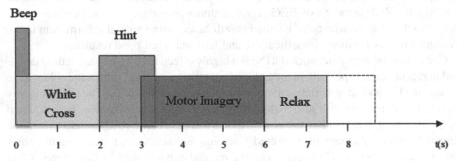

Fig. 1.   Single motion imagination task timing.

2.2   Data Pre-processing

In the pre-processing of MI-EEG signals, filtering operations are often required, the main purpose of which is to filtering out useless frequency ranges and also allows for the processing of artefacts in the EEG signal, particularly myoelectric artefacts. The frequency range of the MI-EEG signal is 8–30 Hz, so the data is band-pass filtered from 8–30 Hz. In this paper, a Butterworth filter is selected to perform a 5th order band-pass filter on the original EEG signal, which reduces the computational effort and complexity of the EEG signal processing and provides good conditions for subsequent classification and recognition.

3   Feature Extraction of EEG Signals

The MI-EEG signal is non-linear and non-smooth, and feature extraction for MI-EEG has been a challenge to deal with. Existing feature extraction of EEG signals often falls in Euclidean space, however, due to the complexity of the brain structure, the weakness of the signal and the confounding of signals emitted by neurons in the brain, it is not ideal to judge different brain activities based on the features in Euclidean space. In addition, the demands on data representation methods are increasing, requiring a compact and discriminative model to represent data containing large amounts of information, while also being as robust as possible.

Therefore, this paper adopts a Riemannian geometric approach to characterize the intrinsic relationship between EEG signals, choosing the covariance matrix as the descriptor of EEG signals, and proving by [15] that the covariance matrix is a symmetric positive definite matrix and the spatial structure of its composition is a specific manifold. This not only solves the problem of unsatisfactory judgments of different brain activities based on the features of Euclidean space, but also makes it a natural and reasonable way to represent MI-EEG signals due to the complete geometric properties and metric calculations of symmetric positive definite matrix manifolds. In this paper, the symmetric positive definite matrix manifold is used as an entry point for the feature extraction of Riemann manifold-based EEG signals.

The process of the algorithm is specified as follows:

1) Representation of the EEG signal.

From the continuously collected EEG data we intercepted the 4s after the cue as a single motion imagined EEG sample, the EEG signal data that can be collected can be expressed as:

$$X_i = [x_t, x_{t+1}, \cdots, x_{t+T_S-1}]^T \in R^{CH*T_S} \tag{1}$$

where X_i denotes the EEG signal acquired from moment t to moment $t + T_S - 1$, CH denotes the number of channels, and T_S denotes the number of signal samples.

At this point, the sample EEG signal Φ with labels can be expressed as:

$$\Phi = \{\Phi_1, \Phi_2, \ldots, \Phi_N\} = \left\{ \begin{bmatrix} X_1 \\ l_1 \end{bmatrix}, \begin{bmatrix} X_2 \\ l_2 \end{bmatrix}, \ldots, \begin{bmatrix} X_N \\ l_N \end{bmatrix} \right\} \tag{2}$$

where l_N belongs to the class label of the Nth cycle of the N motion imagined EEG signal.

2) The n-dimensional covariance matrix of the MI-EEG signal is represented as follows:

$$C = \frac{1}{N-1} \sum_{i=1}^{N} (\Phi_i - \mu)(\Phi_i - \mu)^T = \frac{1}{N-1} \Phi J J^T \Phi^T \tag{3}$$

where $\mu = 1/N \sum_{i=1}^{N} \Phi_i$ is the mean value of the N motion imagery tasks, $J = I_N - N^{-1} 1_{N \times N}$, $1_{N \times N}$ is an N-dimensional square matrix with all elements in the matrix being 1.

3) Estimation of low-dimensional vectors.

Low-dimensional vector estimation of n-dimensional covariance matrices allows the kernel for estimation to be transformed from Euclidean space to Riemannian manifold space, while inheriting the advantages of the former, resulting in a more accurate approximate representation. This allows the rich information in n-dimensional covariance matrices to be retained, while allowing processing and analysis using finite-dimensional symmetric positive definite matrix manifold geometry and related algorithms.

Low-dimensional matrix estimation of n-dimensional covariance matrices is often done using the stochastic Fourier identity method [16], but the premise that a translation-invariant kernel function can be Fourier-integrated is that it needs to satisfy Bochner's theorem [17]. Although the stochastic Fourier eigenmethod can estimate n-dimensional covariance matrices in the kernel space, not all kernel functions satisfy the conditions required by Bochner's theorem, whereas the Nyström method [18], as a data-dependent estimation method, is not subject to these conditions.

Let the n-dimensional eigenvectors $x_i \in^n$ ($i = 1, 2, \ldots, l$) be l n-dimensional eigenvectors, then a p-order approximation of a kernel matrix K of size $l \times l$ can be expressed as:

$$K \tilde{=} Z^T Z, Z = \Lambda^{1/2} U \tag{4}$$

where Λ is the diagonal matrix of the first p largest eigenvalues and U is the matrix of size $p \times l$ of the eigenvectors corresponding to these eigenvalues. Based on such a low-order approximation, a p-dimensional vector approximation representation of x_i can be computed as follows:

$$z_{Ny}(x) = \Lambda^{1/2} U (k(x, x_1), k(x, x_2), \ldots, k(x, x_l))^T \tag{5}$$

where $k(\cdot, \cdot)$ is the kernel function corresponding to the kernel mapping ϕ.

The EEG signal $\Phi = \{\Phi_1, \Phi_2, \ldots, \Phi_N\}$ is known to be an eigenmatrix composed of N $t + T_S$-dimensional eigenvectors, then an n-dimensional covariance matrix C can be computed using the Nyström method of eigenmapping z_{Ny} to calculate its p-dimensional approximate covariance matrix.

$$\hat{C} = \frac{1}{N-1} z_{Ny}(\Phi) J J^T z_{Ny}(\Phi^T) \tag{6}$$

where, $z_{Ny}(\Phi) = [z_{Ny}(\Phi_1) z_{Ny}(\Phi_1) \ldots z_{Ny}(\Phi_N)]$.

According to Eqs. (4) and (5), for t n-dimensional positive definite covariance matrices, a p-dimensional vector approximation representation of the n-dimensional positive definite covariance matrix can be calculated by the following equation:

$$z'_{Ny}(X) = \Lambda_p^{1/2} U_p(k_p(C, C_1), k_p(C, C_2), \ldots, k_p(C, C_t))^T \tag{7}$$

where $k_p(\cdot, \cdot)$ is the kernel function of the kernel mapping, and Λ and U_p are the diagonal matrices consisting of the first p largest eigenvalues and the corresponding eigenvectors obtained after eigen decomposition of the kernel matrix K_n of size $t \times t$. After the above operations, we obtain a set of p-dimensional approximation vectors Z of the n-dimensional covariance matrix, which makes it very convenient to apply the calculations in Euclidean space.

4   Classification Algorithms for EEG Signals

4.1   Classifier Selection and Improvement

In the process of EEG classification, the effectiveness of the classification also depends in large part on the choice of classification algorithm. Currently, research in the classification of EEG signals can be divided into two main categories, namely traditional machine learning-based classification algorithms and neural network-based classification algorithms. Traditional machine learning-based classification algorithms include support vector machines [19], decision trees [20], random forests [21] and K-nearest neighbours [22].

Among the neural network-based classification algorithms, including CNN, graphical neural networks, long and short-term memory neural networks and other deep neural networks, convolutional neural networks are an efficient recognition method that can directly input the original feature space, and extract and further optimise the features within the network, thus solving the classification problem of high-dimensional features again with good results. Compared to fully connected networks, CNN have two features, namely weight sharing and local awareness. Weight sharing means that the connection weights are shared between some neurons in the same layer, and local awareness means that the connections between neurons are not fully connected, but local. These two features can greatly reduce the number of connection weights and hence the complexity of the model.In addition, CNN have the capability of representational learning and are able to classify input information in a translation-invariant manner according to their own hierarchical structure. The general structure of a classical CNN model is: an input layer, alternating superimposed convolutional and pooling layers, a fully connected layer and a classifier. Similar to BP (back propagation) networks, classical CNN algorithms output the convolution of the learnable convolution kernel in the convolution layer with the input of this layer by forward propagation as the input of the next layer, and correct the network weights and biases of each layer by back propagation of errors.

Classical CNN use a single size convolutional kernel in each convolutional layer, and the size of the convolutional kernel directly affects the performance of the network. A perceptual field that is too large beyond what the convolutional kernel can express, and a perceptual field that is too small for the convolutional kernel to extract effective local features [23], are not only detrimental to the rapid convergence of the network and seriously affect network performance, but also too many network parameters leading to a model that cannot be adequately training will also lead to the degradation of network model performance. In this paper, the convolutional layer of CNN is improved, as shown in Fig. 2, that is, the convolutional kernel in the convolutional layer of CNN is changed from a single size, to a double size. A convolutional layer with a different scale of convolutional kernels is added to the original convolutional layer, turning it into a convolutional layer with dual-size convolutional kernels. When extracting data features, the convolutional layer with dual-scale convolutional kernels will perform feature extraction from a field of view of different sizes of the input data, capturing a more comprehensive view of the input data features.

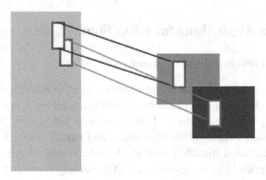

Fig. 2. Convolutional layers with double-scale convolutional kernels.

4.2 CNN Training Process

The training of the dual-scale convolutional kernel CNN is divided into two stages as in the classical CNN. The training flow is shown in Fig. 3.

The first stage is forward propagation, which is the stage where the input data is propagated from a low level to a high level.

The second stage is the back propagation of errors stage, back propagation is one of the most common and effective methods used to train artificial neural networks. Because artificial networks need to constantly calculate the deviation of the output value from the actual value to modify the network parameters during the learning process, a loss function is needed to measure the magnitude of the error between the predicted value of the training sample passing through the neural network and the actual true value. Gradient descent is one of the algorithms that minimises the loss function and is a common optimisation algorithm used in the training of CNN models. Gradient descent is an iterative process in which the weights and biases in the network are adjusted according to the error between the output value and the actual value, and the error in each layer is minimised through learning adjustments. Optimisation of the network is achieved. The specific parameters set for the network training are shown in Table 1.

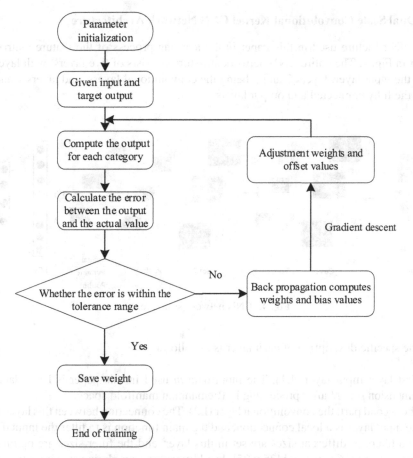

Fig. 3. CNN training process flow chart.

Table 1. CNN network main parameters.

Type	Parameter
Learning rate	0.001
Stride	1
Padding	0
Activation functions	RELU
Training iteration times	50
Sample size processed per batch	128

4.3 Dual Scale Convolutional Kernel CNN Network Architecture

The CNN structure used in this paper in the learning process of the feature matrix is shown in Fig. 4. The entire CNN network structure consists of five layers, with layer 1 being the input layer, layers 2 and 3 being the convolutional layers, and layers 4 and 5 being the fully connected and output layers.

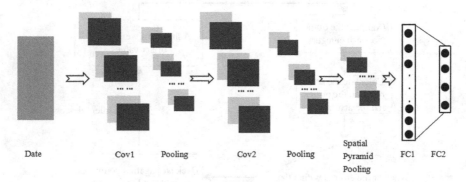

| Date | Cov1 | Pooling | Cov2 | Pooling | Spatial Pyramid Pooling | FC1 | FC2 |

Fig. 4. CNN network structure.

The specific description of each layer is as follows:

1) First layer:Input layer (L1). The input data m used in this paper is EEG data of dimension $p \times N$ after processing in Riemannian manifold space.
2) The second part: the convolutional layer (L2). The connection between this layer and the input layer is a local connection, and the main function is to filter the input data. Two filters of different sizes are set in this layer, and the filter sizes are optimally chosen to be $\{16 \times 16\}$ and $\{25 \times 25\}$. In addition, the convolution step is set to 1×1. This not only makes the derivatives better, but also allows the network to be trained faster with the ReLU activation function. The specific calculation formula is:

$$o^2 = \sigma[(m * ck_1^2 + b_1^2) + (m * ck_2^2 + b_2^2)] \tag{8}$$

where $*$ is the convolution operation, ck_1^2, ck_2^2 are the two different size filters of the layer, and b_1^2 and b_2^2 are the biases.
3) The second part: the convolutional layer (L3), the purpose of this layer is to integrate the new features of L2 and achieve the effect of dimensionality reduction, the same as L2, can be filtered by filtering the input data of the previous convolutional layer, you can obtain the size and number of features of this layer. The calculation formula is:

$$o^3 = \sigma[(o^3 * ck_1^3 + b_1^3) + (o^3 * ck_2^3 + b_2^3)] \tag{9}$$

4) The third part: this part is a fully connected layer (L4) and an output layer (L5) that first accepts new features from the computations in part 2. Since inputs of different dimensions cannot be processed in the fully connected layer, the information of the

same dimension is extracted using spatial pyramid pooling after obtaining data of different dimensions from the previous layer. After obtaining the feature vectors of the same dimensions, the number of neurons is set in the first fully-connected layer as well as the up-dimensioning of the obtained features, which serves to cooperate with the convolutional layer to form the classification of the EEG signals and to provide high-dimensional information for the classification of the final layer. All neurons in this layer are fully connected to the neurons in L3, as calculated in Eq. (10).

$$o^4 = \sigma(o^3 Y^4 + b^4) \qquad (10)$$

where Y^4 is the full connectivity matrix between L3 and L4.

The neurons in the output layer are fully connected to the neurons in L4 and are calculated as:

$$o^5 = \sigma(o^4 Y^5 + b^5) \qquad (11)$$

The number of neurons in the output layer is set to 4, where each neuron corresponds to a MI-EEG category, and the number of neurons in the layer can be determined according to the problem to be classified.

5 Experimental Results and Analysis

5.1 Evaluation Indicators

In this paper, accuracy (Acc) and Kappa coefficient are used to evaluate the classification performance of the algorithm, with accuracy being the most widely used metric to evaluate model performance.

$$Acc = \frac{TP + TN}{TP + TN + FP + FN} \qquad (12)$$

where TP (true positive): positive samples are correctly predicted as positive; FP (false positive): negative samples are incorrectly predicted as positive; TN (true negative): negative samples are correctly predicted as negative; and FN (false negative): positive samples are incorrectly predicted as negative.

The Kappa coefficient is a measure of classification consistency, characterising the ratio of classification to the reduction in error produced by a completely random classification, and is calculated as follows [24]:

$$Kappa = \frac{Acc - \overline{Acc}}{1 - \overline{Acc}} \qquad (13)$$

where Acc is the overall correct recognition rate and $\overline{Acc}$ is the correct classification rate at complete random, with $\overline{Acc}$ being 0.25 for the four classification problem.

In addition, the confusion matrix is a systematic way to account for classification accuracy in studies of four categories of motor imagery tasks (left hand, right hand, foot and tongue). The classification rates shown diagonally on the confusion matrix indicate the accuracy rate for each category and the other values indicate the percentage of misclassified samples.

6   Experimental Results and Analysis

Results of the Methodological Analysis Proposed in this Paper. In order to verify the effectiveness of the fusion algorithm of EEG signal processing based on Riemannian space and two-scale convolution kernel CNN proposed in this paper, the same network structure is used to experimentally verify 9 subjects (S1, S2,…, S9) in BCI competition IV-2a public data set.

Due to the individual variability of the EEG signal and the fact that each individual's mental state and physical condition during the extraction of the EEG signal can have an impact on the acquisition of the signal, this affects the quality of the acquired EEG signal, which in turn has an impact on the classification results. Therefore, the classification accuracy of the nine subjects in the BCI Competition IV-2a public dataset varies somewhat, provided that the same algorithm is used. The validation results can be seen in Table 2, where each subject achieved a classification accuracy of 78% or higher and a kappa coefficient of 0.717 or higher. The quality of the EEG data was most impressive in subject 6, with an accuracy of 82.21% and a kappa coefficient of 0.763. In addition, the mean classification accuracy of the nine subjects was above 80% and the kappa coefficient was above 0.74. The values of the mean square deviation were 1.107 and 0.014 respectively, which can initially verify the effectiveness and better robust performance of the method.

Table 2. Classification accuracy and kappa coefficient of the algorithm in this paper.

Subjects	The accuracy of this algorithm	Kappa coefficient
S1	80.30%	0.737
S2	82.03%	0.760
S3	80.22%	0.736
S4	78.80%	0.717
S5	79.56%	0.727
S6	82.21%	0.763
S7	81.90%	0.759
S8	80.48%	0.740
S9	80.20%	0.736
Mean ± std	80.63% ± 1.107%	0.742 ± 0.014

Analysis of Confusion Matrix Results. To further reveal that the proposed algorithm has the ability to adequately learn different kinds of MI-EEG signal patterns from the samples, the confusion matrix of the classification results is calculated in this paper to verify how well the algorithm classifies each category. As shown in Fig. 5, the horizontal axis of the confusion matrix represents the motor imagery categories predicted by the recognition method, the vertical axis represents the actual motor imagery categories, and the diagonal elements represent the rate at which each motor imagery category is correctly classified, while the non-diagonal elements represent the rate at which each motor imagery category is incorrectly classified. In addition using the classical time-frequency domain-based CNN MI-EEG signal recognition method (STFT + CNN) [25] and the Riemann space-based MI-EEG signal feature extraction with k-nearest neighbour classification method (Riemann + kNN) [26] as control groups,The recognition rates of the STFT + CNN and Riemann + kNN methods for the left hand, right hand, foot and tongue motion imagined EEG were all smaller than the accuracy of the algorithm proposed in this paper, and the misclassification of the left and right hand was the highest, higher than the misclassification rates of the other two categories, all reaching over 11% and up to 17%. The recognition rate of both the left and right hand of the proposed algorithm reached over 80%, indicating that the algorithm is effective in improving the recognition rate of each motor imagery category and can be performed to learn the unique features of EEG activity during different categories of imagery actions.

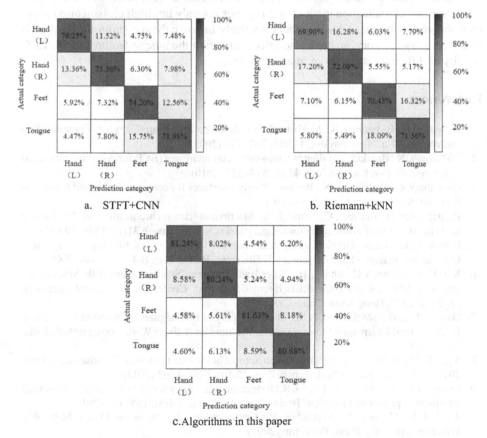

a.   STFT+CNN

b.   Riemann+kNN

c. Algorithms in this paper

Fig. 5. Confusion matrix results for three different EEG signal classifications.

7 Conclusion

This paper presents a method for feature extraction and classification based on the processing of EEG signals within Riemannian manifolds combined with a two-size convolutional kernel CNN. Firstly, the EEG signal expressed in Euclidean space is converted to Riemannian space by selecting the covariance of the symmetric positive definite matrix as a descriptor in the pre-processed raw EEG signal; then, in order to better adapt to learning classification algorithms within Euclidean space, a low-dimensional vector estimation of the n-dimensional symmetric positive definite matrix is performed using a specific feature mapping to obtain a low-dimensional efficient feature set of the EEG signal; finally, a feature set is constructed based on the Finally, a CNN network classifier is constructed based on the feature set, which uses a double-sized convolutional kernel in the convolutional layer to capture the input data features more comprehensively, and classifies them by the improved CNN network classifier.

To verify the feasibility of the method proposed in this paper, a series of experimental results conducted on the BCI2008IV-2a competition dataset show that the feature extraction and classification method in this paper not only has high classification accuracy and good robustness, but also can effectively improve the recognition rate of each motor imagery category, and can also effectively learn the unique features of the EEG activity of different categories of imagery actions.

References

1. Wolpaw, J.R., Birbaumer, N., McFarland, D.J.: Brain computer interfaces for communication and control. Clin. Neurophysiol. **113**(6), 767–791 (2002)
2. Birbaumer, N.: Breaking the silence: brain-computer interfaces (BCI) for communication and motor control. Psychophysiology **43**(6), 517–532 (2010)
3. Chaudhary, U., Birbaumer, N.: Brain-computer interfaces for communication and rehabilitation. Nat. Rev. Neurol. **12**(9), 513 (2016)
4. Pfurtscheller, G., Brunner, C., Schlogl, A.: Mu rhythm (de)synchronization and EEG single trial classification of different motor imagery tasks. Nceuro. Image **31**(1), 153–159 (2006)
5. Ramoser, H., Muller -Gerking, J., Pfurtscheller, G.: Optimal spatial filtering of single trial EEG during imagined hand movement. IEEE Trans. Rehab. Eng. **8**(4), 441–446 (2000)
6. Kai, K.A., Zhang, Y.C., Zhang, H.: Filter Bank Common Spatial Pattern (FBCSP)in brain-computer interface. In: 2008 IEEE International Joint Conference on Neural Networks, pp. 2390–2397. Hong Kong, China (2008)
7. Gao, Y., Lee, H.J., Mehmood, R.M.: Deep learninig of EEG signals for emotion recognition. In: 2015 IEEE International Conference on Multimedia & Expo Workshops, pp. 1–5. Turin, Italy (2015)
8. Yger, F., Berar, M., Lotte, F.: Riemannian approaches in brain computer interfaces: a review. 2017 IEEE Trans. Neural Syst. Rehab. Eng. **25**(10), 1753–1762 (2017)
9. Congedo, M., Barachant, A., Bhatia, R.: Riemannian geometry for EEG-based brain-computer interfaces: a primer and a review. Brain-Comput. Interfaces **4**(3), 155–174 (2017)
10. Absil, P.A., Mahony, R., Sepulchre, R.: Optimization Algorithms on Matrix Manifolds. Princeton University Press, Princeton (2009)
11. Waytowich, N.R., Lawhern, V.J., Bohannon, A.W.: Spectral transfer learning using information geometry for a user independent brain-computer interface. Front Neurosci. **10**(430), 24–39 (2016)

12. Shin, H.C., Roth, H.R., Gao, M.: Deep convolutional neural networks for computer-aided detection: CNN architectures, dataset characteristics and transfer learning. IEEE Trans. Med. Imaging 35(5), 1285–1298 (2016)
13. Al-Ayyoub, M., Nuscir, A., Alsmcarat, K.: Deep learning for Arabic NLP: a survey. J. Comput. Sci. 26, 522–531 (2018)
14. Garand, J., Gregg, D.: Low complexity multiply accumulate unit for weight-sharing convolutional neural networks. IEEE Comput. Archit. L. 16(2), 132–135 (2017)
15. Barachant, A., Bonnet, S., Congedo, M.: Multiclass brain-computer interface classification by riemannian geometry. IEEE Trans. Biomed. Eng. 59(4), 920–928 (2012)
16. Rahimi, A., Recht, B.: Random features for large-scale kernel machines. Adv. Neural Inf. Process. Syst. 07, 1177–1184 (2008)
17. Rudin, W.: Fourier Analysis on Groups. Interscience publishers, New York (1962)
18. Baker, C.T.H.: The numerical treatment of integral equations. J. Appl. Mech. 46(4), 969 (1977)
19. Chang, C., Lin, C.: LIBSVM: a library for support vector machines. ACM Trans. Intell. Syst. Technol. 2(27), 1–27 (2011)
20. Safavian, S., Landgrebe, D.: A survey of decision tree classifier methodology. IEEE Trans. Syst. Man Cybern. 21(3), 660–674 (1991)
21. Breiman, L.: Random forests. Mach. Learn. 45(1), 5–32 (2001)
22. Ye, N., Sun, Y.K., Wang, X.: Brain computer interface signal classification method based on common space pattern and k-nearest neighbor classifier. J. Northeast Univ. 30(08), 1107–1110 (2009)
23. Rui, T., Fei, J.C., Zhou, S.: Pedestrian detection based on deep convolution neural network. Comput. Eng. Appl. 52(13), 162–166 (2016)
24. Sakhavi, S., Guan, C., Yan, S.: Learning temporal information for brain-computer interface using convolutional neural networks. IEEE Trans. Neural Network Learn. 29(11), 5619–5629 (2018)
25. Hu, Z.F., Zhang, L., Huang, L.J.: Recognition method of EEG signals of motor imagination based on convolution neural network in time-frequency domain. Comput. Appl. 39(8), 2480–2483 (2019)
26. Gao, N., Gao, Z., Zhang, H.: Riemann method for feature extraction and classification of motor imagery EEG. J. Biomed. Eng. Res. 40(3), 246–251 (2021)

Research and Simulation of Fuzzy Adaptive PID Control for Upper Limb Exoskeleton Robot

Shiwei Yu$^{(\boxtimes)}$, Shouyin Lu$^{(\boxtimes)}$, Zhe Jiang, Zhipeng Li, and Qiang Zhang

School of Information and Electrical Engineering, Shandong Jianzhu University,
Jinan 250101, Shandong, China
`564875310@qq.com, sdznjqr@163.com`

Abstract. In order to promote the rehabilitation of upper limb function of patients suffering from stroke, a homogeneous and homogeneous master-slave upper limb exoskeleton rehabilitation robot was designed based on the principle of human biology, and a fuzzy adaptive PID control method strategy based on the joint position was proposed for the problems such as the robot slave arm precisely following the motion trajectory of the master arm; secondly, the kinematic model of the robot slave arm was established by using the improved D-H method, and a workspace simulation based on the Matlab robot toolbox was performed. The robot toolbox of Matlab is used for workspace simulation. And Simulink/Simscape Multibody was used to realize the simulation experiment of exoskeleton motion control to verify the effect of slave arm trajectory following using fuzzy PID control method. Experiments show that under the same conditions, the proposed control algorithm can track the main arm motion curve more quickly and accurately than the traditional PID algorithm, and the trajectory is smooth and stable without obvious jitter, and the passive rehabilitation training effect is good.

Keywords: Master-slave · Upper limb exoskeleton · Fuzzy PID control · Trajectory tracking · Modeling simulation

1 Introduction

Stroke is an accidental rupture of a cerebral blood vessel with bleeding or acute occlusion, which causes serious damage to the brain and eventually leads to unilateral upper limb numbness and muscle weakness, or even muscle atrophy and loss of muscle capacity. The traditional training mode is that the physical therapist conducts one-on-one assistance training for the patient through experience, but this training mode is not only time-consuming and labor-intensive, but also the patient's independent participation is poor, which makes it difficult to achieve the ideal treatment effect. With the continuous development of medical rehabilitation technology and the innovation of robotics, upper limb rehabilitation robots are gradually applied to the rehabilitation training of patients, which not only can effectively reduce the burden of health care workers, but also can meet the training intensity requirements of patients and achieve the recovery of upper limb function [1, 2].

H. Zhang et al. (Eds.): NCAA 2022, CCIS 1638, pp. 66–79, 2022.
https://doi.org/10.1007/978-981-19-6135-9_6

At present, the control methods of domestic upper limb exoskeleton rehabilitation robots are mainly divided into two types: passive control and active control, and for patients, at the beginning of training, the affected limb has no movement ability at all, and can only rely on the upper limb rehabilitation robot to drive passive rehabilitation training according to the planned trajectory, so as to promote the initial recovery of the affected limb. The main goal of passive rehabilitation training is to improve muscle tone, restore active movement of the affected limb, and relieve the physical injury that muscle spasm may bring to the patient. Therefore, the passive control method of the upper limb exoskeleton robot applied to the patient is particularly important at this time [3, 4]. In the passive training, how to improve the accuracy of robot motion trajectory tracking and anti-interference ability, so as to drive the patient's affected limb to be able to perform a series of periodic movements smoothly and slowly, is still a key research problem in the field of rehabilitation robotics today [5–7].

Nowadays, the more common method used by patients for passive training is PID control.PID control can adjust the parameters according to the system characteristics and can overcome the drawback of the complex model of the exoskeleton robot system to achieve a good control effect. Rahman et al. [8] used a linear PID control method to teleoperate the MARSE-4 exoskeleton robot by means of an upper limb exoskeleton master hand, which was prescribed to move along a predetermined trajectory. Validation experiments showed that MARSE-4 could effectively track the desired trajectory and passively treat the patient's wrist, elbow and forearm movements to achieve satisfactory training results. The Shenzhen Institute of Advanced Technology, Chinese Academy of Sciences, developed STAT [9], which uses a servo motor as a driving member and uses a PID algorithm to control the motor so that the servo drive system can follow a preset gait within the control computer for trajectory control, enabling the robot to follow a precise trajectory. The Lokomat [10] rehabilitation robot from Hocoma, Switzerland, uses a trajectory tracking control method that presets the trajectory curve of the gait motion of each joint of the lower limb, thus simulating human gait. YU et al. [11] used the theoretical results of the controller parameters of EXO-UL7, an exoskeleton system, to simulate and evaluate the system performance and verify the semi-global asymptotic stability of the system. However, these rehabilitation robots use traditional PID control, which achieves good trajectory tracking but is difficult to solve the problem of variable trajectory curves and susceptibility to interference of the robotic arm during passive rehabilitation training.

Based on the above deficiencies, this paper designs a master-slave upper limb exoskeleton robot based on fuzzy PID controller, which mainly includes (1) a homogeneous isomorphic master-slave symmetric upper limb exoskeleton robot is designed, in which angular displacement and angular velocity sensors are set at each joint of the master arm, so as to obtain the motion information of the patient's healthy limb and grasp its motion trajectory; (2) Modified D-H method and Lagrangian dynamics method are used to model the robot slave arm, set the linkage length and joint angle, and obtain the robot arm motion space; (3) designing an efficient and robust fuzzy logic-based control strategy, i.e., fuzzy PID control algorithm, for an upper limb exoskeleton rehabilitation robot to achieve good trajectory tracking control effect for the exoskeleton robot; (4) In order to analyze its practicality, a closed-loop simscape/MATLAB model is established for the rehabilitation robot designed in this paper, the robotic arm model in SOLIDWORKS is exported to generate a file in urdf format, which is imported into Matlab, and simulation experiments of robotic arm trajectory tracking are conducted using Simulink/Simscape Multibody The simulations are carried out by using Simulink/Simscape Multibody, and the conventional PID controller is set up to compare with the proposed controller in this paper to study the effect of trajectory tracking from the arm, so as to provide a reference for the control research algorithm of the upper limb exoskeleton robotic arm.

2   Upper Limb Exoskeleton Robot

2.1   Robotic Arm Model Design

Considering the space and existing technology, by imitating the main joints of human upper limbs and meeting the range of motion of each degree of freedom of each joint of human upper limbs, this paper designs a two-arm exoskeleton robot with five degrees of freedom, as shown in Fig. 1, whose body structure is divided into two homogeneous and homogeneous robotic arms, a cantilever arm for mounting the robotic arms and a base for connecting the two cantilever arms. The left and right arms of the upper limbs are capable of various human activities, such as shoulder abduction/adduction, large arm pitch/tilt, small arm internal/external rotation, elbow flexion/extension, and wrist abduction/adduction. Due to the semi-anthropomorphic design of the exoskeleton robot, the robot has a tight structure and small space volume, which is less likely to interfere with the human body and simpler to control.

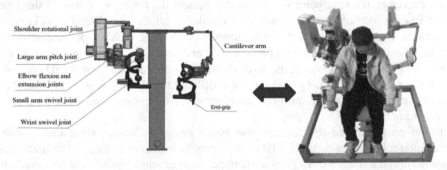

Fig. 1. Master-slave upper limb exoskeleton rehabilitation robot overall structure

2.2 Robotic Arm Control System Design

In addition, this paper designs a robotic arm control system as shown in Fig. 2, in addition to the master-slave robotic arm described above, including the host computer, the bottom controller, the angular displacement sensors and speed sensors placed at each joint of the master arm, the motor drive mechanism placed at each joint of the slave arm, i.e., the drive motor to drive the robot movement, and the communication module. The patient's healthy limb wears the master arm and the affected limb wears the slave arm, and during the movement of the healthy limb, the sensors at each joint of the master arm collect information and transmit the information to the bottom controller through the bus; the bottom controller uploads the movement information to the top computer through the communication module; the top computer processes the movement information and forms control instructions, which are transmitted to the bottom controller through the TCP/IP communication protocol; the bottom controller receives the After receiving the control command, the bottom controller drives the motor through the motor drive module, which drives the movement of the slave arm and thus the movement of the affected limb.

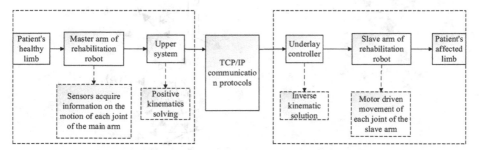

Fig. 2. Master-slave upper limb exoskeleton rehabilitation robot control system

3 Kinematic Analysis of Upper Limb Exoskeleton Robot

3.1 Kinematic Modeling

Since the robot body structure designed in this paper is composed of two homogeneous and homogeneous robotic arms, and the passive training of the patient is only studied and analyzed in this paper, i.e., the motion trajectory planning of the slave arm is analyzed, so this paper only takes the slave arm as an example for kinematic modeling. The slave arm is a five-degree-of-freedom robotic arm, and the modified D-H modeling method [12, 13] is applied to establish the kinematic joint coordinate diagram as shown in Fig. 3. According to the linkage relationship of each joint of the slave arm, that is, the cantilever arm is the base coordinate, and the point O is set as the origin of its coordinate system, and the shoulder joint, the large arm joint, the elbow joint, the small arm joint and the end of the wrist joint are taken as the origin of each joint coordinate system respectively, so as to obtain the adjacent inter-joint The transformation matrix $^{i}_{i+1}T$, $i = 1, 2 \cdots 5$, and concatenate these transformation matrices to obtain the transformation matrix $^{0}_{6}T$

of the terminal wrist with respect to the base coordinates, with O as the base coordinate system and O_6 as the end coordinate system of the wrist joint from the arm.

$$ {}^0_6T = {}^0_1T{}^1_2T{}^2_3T{}^3_4T{}^4_5T{}^5_6T = \begin{bmatrix} n_{11} & o_{12} & a_{13} & p_{14} \\ n_{21} & o_{22} & a_{23} & p_{24} \\ n_{31} & o_{32} & a_{33} & p_{34} \\ 0 & 0 & 0 & 1 \end{bmatrix} \tag{1}$$

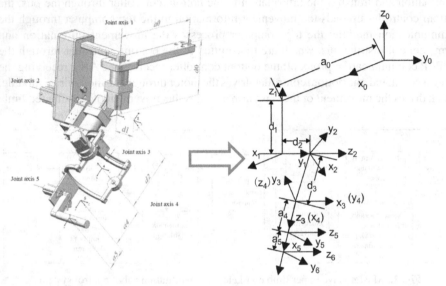

Fig. 3. Kinematic analysis of the exoskeleton robot from the arm

The values of the D-H parameters for each of these joints are shown in Table 1, where d_i denotes translation distance, α_i denotes torsion angle, a_i denotes length, and θ_i denotes joint angle.

Table 1. Upper limb exoskeleton robot from arm D-H parameters.

Linkage i	a_{i-1}/m	$\alpha_{i-1}/(rad)$	d_i/m	$\theta_i/(rad)$
1	0	0.44	-0.29	θ_1
2	-90	0	0.1	$90 + \theta_2$
3	90	0	0.27	$90 + \theta_3$
4	-90	0	0	$-90 + \theta_4$
5	-90	0.17	0	θ_5
6	0	0.1	0	0

Thus, it is concluded that

$$
\begin{cases}
n_{11} = c\theta_1 s\theta_2 s\theta_3 s\theta_4 c\theta_5 + c\theta_1 c\theta_2 c\theta_4 c\theta_5 - s\theta_1 c\theta_3 s\theta_4 c\theta_5 \\
\quad - c\theta_1 s\theta_2 c\theta_3 s\theta_5 + s\theta_1 s\theta_3 s\theta_5 \\
n_{21} = s\theta_1 s\theta_2 s\theta_3 s\theta_4 c\theta_5 + s\theta_1 c\theta_2 c\theta_4 c\theta_5 + c\theta_1 c\theta_3 s\theta_4 c\theta_5 \\
\quad - s\theta_1 s\theta_2 c\theta_3 s\theta_5 - c\theta_1 s\theta_3 c\theta_5 \\
n_{31} = c\theta_2 s\theta_3 s\theta_4 c\theta_5 - s\theta_2 c\theta_4 c\theta_5 - c\theta_2 c\theta_3 s\theta_5 \\
o_{12} = -c\theta_1 s\theta_2 s\theta_3 s\theta_4 s\theta_5 - c\theta_1 c\theta_2 c\theta_4 s\theta_5 + s\theta_1 c\theta_3 s\theta_4 s\theta_5 \\
\quad - c\theta_1 s\theta_2 c\theta_3 c\theta_5 + s\theta_1 s\theta_3 c\theta_5 \\
o_{22} = -s\theta_1 s\theta_2 s\theta_3 s\theta_4 s\theta_5 - s\theta_1 c\theta_2 c\theta_4 s\theta_5 - c\theta_1 c\theta_3 s\theta_4 s\theta_5 \\
\quad - s\theta_1 s\theta_2 c\theta_3 c\theta_5 - c\theta_1 s\theta_3 c\theta_5 \\
o_{32} = -c\theta_2 s\theta_3 s\theta_4 s\theta_5 + s\theta_2 c\theta_4 s\theta_5 - c\theta_2 c\theta_3 c\theta_5 \\
a_{13} = c\theta_1 s\theta_2 s\theta_3 c\theta_4 - c\theta_1 c\theta_2 s\theta_4 - s\theta_1 c\theta_3 c\theta_4 \\
a_{23} = s\theta_1 s\theta_2 s\theta_3 c\theta_4 - s\theta_1 c\theta_2 s\theta_4 + c\theta_1 c\theta_3 c\theta_4 \\
a_{33} = c\theta_2 s\theta_3 c\theta_4 + s\theta_2 s\theta_4 \\
p_{14} = (c\theta_1 s\theta_2 s\theta_3 + c\theta_1 c\theta_2 c\theta_4 - s\theta_1 c\theta_3 s\theta_4)(0.1c\theta_5 + 0.1) \\
\quad - 0.1 s\theta_5 (c\theta_1 s\theta_2 c\theta_3 - s\theta_1 s\theta_3) + 0.27 c\theta_1 c\theta_2 - 0.1 s\theta_1 + 0.44 \\
p_{24} = (s\theta_1 s\theta_2 s\theta_3 s\theta_4 + s\theta_1 c\theta_2 c\theta_4 + c\theta_1 c\theta_3 s\theta_4)\,(0.1c\theta_5 + 0.1) \\
\quad - 0.1 s\theta_5 (s\theta_1 s\theta_2 c\theta_3 + c\theta_1 s\theta_3) + 0.27 s\theta_1 c\theta_2 + 0.1 c\theta_1 \\
p_{34} = (c\theta_2 s\theta_3 s\theta_4 - s\theta_2 c\theta_4)(0.1c\theta_5 + 0.1) - 0.1 s\theta_3 c\theta_2 c\theta_3 - 0.27 s\theta_2 - 0.29
\end{cases}
\tag{2}
$$

where, $c\theta_i = \cos\theta_i$, $s\theta_i = \sin\theta_i$, $i = 1, 2, \cdots, 5$.

3.2 Robotic Arm Model Simulation

According to the D-H parameters of the upper limb exoskeleton from the arm set in Sect. 2.1, the linkage model is established through the Matlab toolbox toolbox, and the positive kinematic simulation can be verified, i.e., the functions provided by the toolbox are used to determine whether the chi-square transformation matrix derived from the robotic arm model is the same as the transformation matrix calculated using the positive kinematic functions by inputting the same angle, and it is verified that the It is verified that the transformation matrices obtained by both methods are the same.

Based on the established kinematic linkage model and the range of motion of each joint, and based on the results of the kinematic analysis, 30,000 discrete points were taken for each range of motion of each joint to obtain the point cloud diagram of the robot from the end of the arm as shown in Fig. 4. From this figure, it can be obtained that the robot slave arm workspace is similar to a hemisphere, which is consistent with the operating space directly in front of the human body.

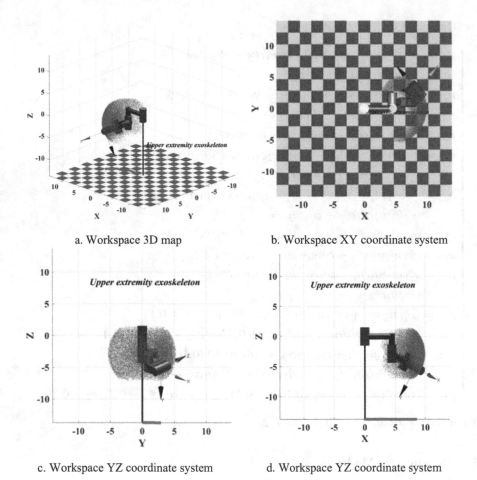

a. Workspace 3D map   b. Workspace XY coordinate system

c. Workspace YZ coordinate system   d. Workspace YZ coordinate system

Fig. 4. Workspace for the slave arm of a master-slave rehabilitation robot

4   Control Strategy

In this section, the proposed control strategy, a fuzzy PID controller for the slave arm of a master-slave upper limb rehabilitation robot, is discussed. First, the design of the conventional PID controller is discussed first. Then, the basic design of the fuzzy PID controller scheme is presented in detail.

Because the upper limb exoskeleton rehabilitation robot proposed in this paper is susceptible to external disturbances such as patient's own pressure and friction when assisting patients in training, the system is characterized by time-varying and nonlinear parameters, and although the traditional PID control can achieve good control accuracy, the control effect is hardly satisfactory [14, 15].The formula of traditional PID control algorithm is as follows.

$$u(k) = K_p e(k) + K_i \sum_{j=0}^{k} e(j) + K_d \Delta e(k) \tag{3}$$

where: $u(k)$ is the PID controller output signal; $\sum_0^k e(j)$ is the cumulative error signal; $e(k)$ is the current error signal; K_p, K_i, K_d are the proportionality coefficient, integration constant and differentiation constant of the PID controller respectively.

Fuzzy control is based on expert experience to determine the control rules and carry out fuzzy inference to obtain the control quantity [16, 17], which not only has strong robustness, but also can better adapt to the changes of the control object itself and the influence caused by external disturbances. In the paper, a parameter self-adjusting fuzzy adaptive PID control algorithm exoskeleton robot to achieve trajectory tracking control, that is, the angular displacement error e and the rate of change of error e_c of the motion trajectory of the rehabilitation robot as the input of the fuzzy controller, and ΔK_p, ΔK_i, ΔK_d as the output of the fuzzy controller, so as to complete the self-tuning of the PID parameters and enhance the adaptiveness of the system.

The PID controller parameter self-tuning formula is as follows.

$$\begin{cases} K_p = K_p(0) + \Delta K_p \\ K_i = K_i(0) + \Delta K_i \\ K_d = K_d(0) + \Delta K_d \end{cases} \tag{4}$$

where: $K_p(0)$, $K_i(0)$, $K_d(0)$ are the initial values of the PID controller parameters; ΔK_p, ΔK_i, ΔK_d are the outputs of the fuzzy controller, respectively.

The control block diagram of the upper limb exoskeleton rehabilitation robot is shown in Fig. 5. The fuzzy sets of the inputs e and e_c of the controller and the outputs ΔK_p, ΔK_i, and ΔK_d of the controller are defined as follows: The fuzzy sets of the fuzzy linguistic variables are all {negative large (NB), negative medium (NM), negative small (NS), zero (ZO), positive small (PS), positive medium (PM), and positive large (PB)}. The theoretical domains of design control error e and error rate of change e_c are both $[-3,3]$, and the theoretical domains of ΔK_p, ΔK_i, and ΔK_d are all $[-1,1]$, and the input and output variables all use the triangular affiliation function, as shown in Fig. 6.

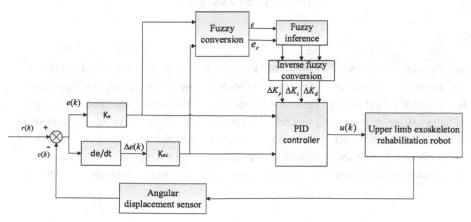

Fig. 5. Block diagram of fuzzy adaptive PID control system

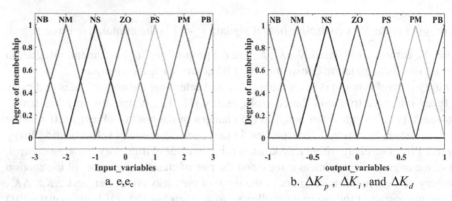

Fig. 6. Membership functions.

The affiliation function $\mu(x)$ can be used to represent a fuzzy set of input and output quantities. It indicates how close the exact quantity is to the fuzzy language, i.e., to what extent a specific value can be described in that language. A triangular affiliation function is used for both input and output variables, which are described in the following form.

$$\mu(x) = \begin{cases} \frac{x-a}{b-a}, x \in (a, b) \\ \frac{x-c}{b-c}, x \in (b, c) \end{cases} \qquad (5)$$

where: a, b, c - fuzzy interval constants.

The fuzzy controller adopts the form of two inputs and three outputs, and adopts the rule of "IF A AND B THEN C AND D AND E", combined with the principle of PID parameter adjustment, the fuzzy rules are shown in the Table 2, Fig. 7 shows the graph of the fuzzy rule surface obtained between the two considered inputs and a single output.

Table 2. Fuzzy control rules

e	e_c						
	NB	NM	NS	ZO	PS	PM	PB
NB	PB/NB/PS	PB/NB/NS	PM/NM/NB	PM/NM/NB	PS/NS/NB	ZO/ZO/NM	ZO/ZO/PS
NM	PB/NB/NS	PB/NB/NS	PM/NM/NB	PS/NS/NM	PS/NS/NM	ZO/ZO/NS	ZO/ZO/PS
NS	PB/NB/ZO	PM/NM/NS	PM/NS/NM	PS/NS/NM	ZO/ZO/NS	NS/PS/NS	NS/PS/ZO
ZO	PM/NM/ZO	PM/NM/NS	PS/NS/NS	ZO/ZO/NS	NS/PS/NS	NM/PM/PS	NM/PM/ZO
PS	PS/NM/ZO	PS/NS/ZO	ZO/ZO/ZO	NS/PS/ZO	NS/PS/ZO	NM/PM/ZO	NM/PB/ZO
PM	PS/ZO/PB	ZO/ZO/NS	NS/PS/PS	NM/PS/PS	NM/PM/PS	NM/PB/PS	NB/PB/PB
PB	ZO/ZO/PB	ZO/ZO/PM	NM/PS/PM	NM/PM/PM	NM/PM/PS	NB/PB/PS	NB/PB/PB

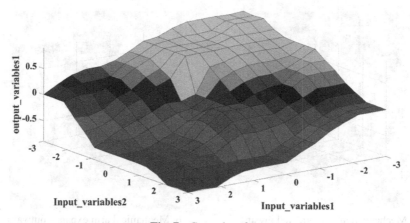

Fig. 7. Control surface

After fuzzy rule inference, the output value is a fuzzy quantity and must also be defuzzified to convert the fuzzy quantity into an exact quantity by defuzzification calculation. The selection of defuzzification method is related to the selection of inference method and affiliation function. Since the resulting set of fuzzy quantities is a collection of multiple exact quantities, this paper uses the area center of gravity method for defuzzification. Its expression are as follows.

$$u = \frac{\sum_{i=0}^{n} x_i u(x_i)}{\sum_{i=0}^{n} u(x_i)} \tag{6}$$

where: x_i is the fuzzy variable value; n is the number of subset elements, n = 7; $u(x_i)$ is the affiliation function.

5 Experimental Simulation

The kinematics as well as dynamics analysis of the exoskeleton rehabilitation robot model and the design of fuzzy PID controller have been carried out in the previous paper, and in order to verify them, Simulink/Simscape Multibody is used for simulation experiments in this paper [18, 19], based on the previously proposed exoskeleton rehabilitation robot slave arm model in Solidworks software, the Adding the coordinate system of each joint and configuring the relevant parameters of each joint to generate a file in urdf format, importing it into the Simulink environment through the plug-in of Simscape Multibody, setting the input and output of some modules by designing the joint parameters, realizing the position control of the upper limb rehabilitation robot through the fuzzy PID controller, and visualizing the robot motion As shown in Fig. 8, the left figure shows the motion pose when contracting inward from the arm, and the right figure shows the motion pose when expanding outward from the arm.

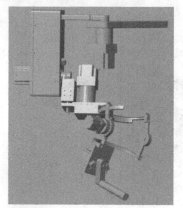

a . Mechanical arm contracted inward   b. Mechanical arm expands outward

Fig. 8. Robotic arm motion posture analysis

Taking the slave arm elbow joint as an example, in order to verify whether the joint carries out smooth motion with the master arm, the desired trajectory of the joint motion training is set as a sinusoidal trajectory as the reference trajectory input to the controller, and the tracking effect of the traditional PID and fuzzy PID algorithms is studied by comparing the output trajectory tracking through simulation. The desired trajectory of the elbow joint and the actual trajectory after different controllers are shown in Fig. 9. The tracking of the gait trajectory by the PID controller shows that although the tracking of the desired curve can be achieved by the PID algorithm, the curve does not fit completely at the curve inflection point and the trajectory tracking is not accurate enough. This is due to the fact that at the inflection point, the trend of the reference curve changes abruptly, and it is difficult for the controller to respond in time, so the tracking curve will lag behind the reference curve at the inflection point, and then the change of the reference curve tends to be stable. The tracking effect of the fuzzy PID controller on the desired trajectory of the elbow joint can be obtained from the figure, and it can be seen that using the fuzzy PID algorithm at the inflection point of the exoskeleton elbow joint curve are able to fit the desired trajectory curve better with less error, and finally fit the input curve completely, thus avoiding the problem of overshooting the tracking curve of the PID control algorithm at the inflection point position of the curve. The simulation results show that the fuzzy PID algorithm can track the desired trajectory curve more accurately and quickly than the traditional PID algorithm.

In order to verify the anti-external disturbance ability of the two control algorithms, a random external disturbance signal generated by Random number module is added to the input, and then the tracking effect of the traditional PID and fuzzy PID algorithms is observed. It can be seen from Fig. 10 that the traditional PID controller has difficulty in achieving effective tracking of the reference trajectory under external disturbance, especially at the curve inflection point where the deviation is large. In Fig. 10b, we can see that the control system based on fuzzy PID control, under the same external perturbation,

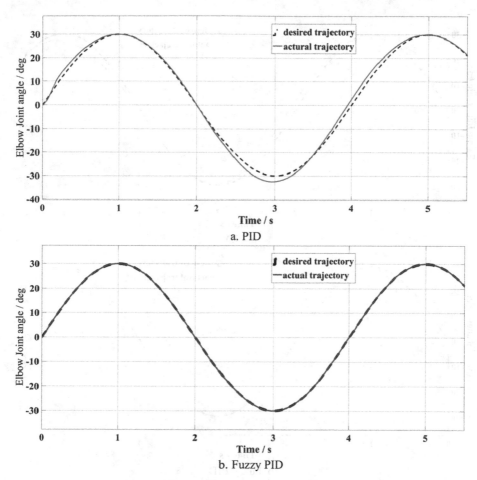

Fig. 9. Elbow joint tracking trajectory

although the actual trajectory curve of the elbow joint has some overshoot with the desired trajectory curve before 1s, the deviation is small, and the desired trajectory can still be tracked accurately after 1s, which indicates that the fuzzy PID algorithm has better suppression effect on the external perturbation than the traditional PID algorithm. The robustness of the system is significantly improved, and it has better stability and self-adaptive capability. This is because the PID parameters are fixed and it is difficult for the system to respond in time when the external disturbance is large, while the fuzzy PID can adjust the PID parameters according to the error of the tracking trajectory and the rate of change of the error, so that the target curve can still be tracked accurately even if there is disturbance from external disturbance.

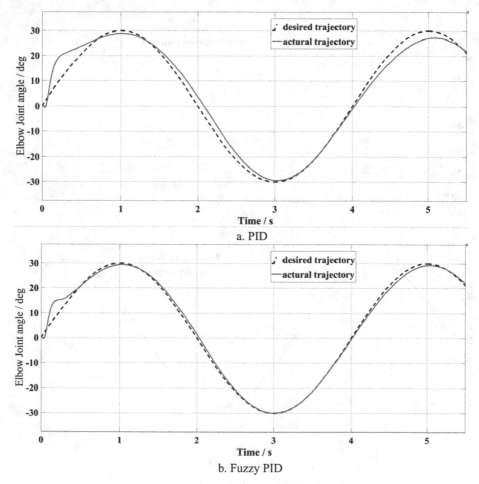

Fig. 10. Tracking trajectory after adding external perturbation to the elbow joint

6   Conclusion

In this paper, based on the rehabilitation training needs of hemiplegic patients, a five-degree-of-freedom isomorphic master-slave exoskeleton upper limb training rehabilitation robot is designed, and a fuzzy adaptive PID algorithm is proposed for it, and a fuzzy adaptive PID controller is designed. The controller organically combines fuzzy logic with classical PID control, continuously detects the error of angular displacement and error changes in the process, and online rectification of the three parameters of the PID algorithm through fuzzy inference decision, which has the advantages of high precision of classical PID control, but also brings into play the advantages of strong adaptability, fault tolerance and flexible control of fuzzy control. The simulation and experimental results show that the fuzzy adaptive PID algorithm has the advantages of high accuracy, small overshoot and excellent dynamic performance compared with the classical PID algorithm, and it can effectively suppress the joint jitter brought about by the passive training process of the exoskeleton slave arm when the system has a large error, and can follow the movement of the main arm activity trajectory more accurately with strong following and robustness to ensure that the patient achieves good It has strong following and robustness to ensure good training effect.

References

1. Zhang, X., Zhang, Y.F., J, Z.W.: Research progress of rehabilitation robot. Chin. Med. Equip. J. **41**(4), 97–102 (2020)
2. Xu, D., Xu, H., Li, Y.B.: Review on research progress of upper limb rehabilitation robot. Mod. Inf. Technol. **4**(16), 142–144 (2020)
3. Ates, S., Haarman, C.J.W., Stienen, A.H.A.: SCRIPT passive orthosis: design of interactive hand and wrist exoskeleton for rehabilitation at home after stroke. Auton. Robot. **41**(3), 711–723 (2017)
4. Liu, B., Li, N., Yu, P.: Research on the control methods of upper limb rehabilitation exoskeleton robot. J. Univ. Electron. Sci. Technol. China **49**(5), 641–651 (2020)
5. Guo, X.H., W, J., Xu, S.H.: Recent advances in functional hand rehabilitation robot research. Chin. J. Rehabil. Med. **32**(2), 235–240 (2017)
6. Hou, Z.G., Zhao, X.G., Cheng, L.: Recent advances in rehabilitation robots and intelligent assistance systems. Acta Automatica Sinica **42**(12), 1765–1779 (2017)
7. Cao, J., Xie, S.Q., Das, R.: Control strategies for effective robot assisted gait rehabilitation: the state of art and future prospects. Med. Eng. Phys. **36**(12), 1555–1566 (2014)
8. Rahman, M.H., K-Ouimet, T., Saad, M.: Tele-operation of a robotic exoskeleton for rehabilitation and passive arm movement assistance. In: IEEE International Conference on Robotics & Biomimetics, pp. 443–448. IEEE, Phuket, Thailand (2011)
9. Zhang, S.M., Wang C., Wu, X.Y.: Real Time Gait Planning for a Mobile Medical Exoskeleton with Crutche. In: IEEE International Conference on Robotics and Biomimetics, pp. 2301–2306 (2015)
10. Shi, D., Zhang, W.X., Zhang, W.: A review on lower limb rehabilitation exoskeleton robots. Chin. J. Mech. Eng. **32**(4), 2–11 (2019)
11. Yu, W., Rosen, J.: A novel linear PID controller for an upper limb exoskeleton. In: IEEE Conference on Decision and Control, pp. 3548–3553. IEEE, Atlanta, USA (2010)
12. Craig, John, J.: Introduction to Robotics: Mechanics and Control. Pearson Education (2005)
13. Vengateswaran, J.R.: Robotics. Resonance **4**(12), 76–82 (1999). https://doi.org/10.1007/BF0 2838676
14. Liu, J.K.: Intelligent Control. Publishing House of Electronics Industry (2014)
15. Zhong, G.L., Wang, C.M., Dou, W.Q.: Fuzzy adaptive PID fast terminal sliding mode controller for a redundant manipulator. Mech. Syst. Signal Process **159**, 107577 (2021)
16. Cao, G., Zhao, X., Ye, C., et al.: Fuzzy adaptive PID control method for multi-mecanum-wheeled mobile robot. J. Mech. Sci. Technol. **36**(4), 2019–2029 (2022)
17. Dettori, S., Iannino, V., Colla, V., et al.: An adaptive Fuzzy logic-based approach to PID control of steam turbines in solar applications. Appl. Energy **227**, 655–664 (2018)
18. Sharma, R., Gaur, P., Bhatt, S.: Optimal fuzzy logic-based control strategy for lower limb rehabilitation exoskeleton. Appl. Soft Comput. **105**(4), 107226 (2021)
19. Wu, Q., Wang, X., Du, F.: Fuzzy sliding mode control of an upper limb exoskeleton for robot-assisted rehabilitation. In: IEEE International Symposium on Medical Measurements and Applications, pp. 451–456. IEEE, Torino, Italy (2015)

Short-Term Wind Power Prediction Based on Convolutional Neural Network-Bidirectional Long Short-Term Memory Network

Qi Wang[✉] [iD], Zheng Xin, and Xingran Liu

Shandong Key Laboratory of Intelligent Buildings Technology, School of Information and Electrical Engineering, Shandong Jianzhu University, Jinan 250101, China
2020085122@stu.sdjzu.edu.cn

Abstract. Aiming at the nonlinear and non-stationary characteristics of wind power data, a wind power prediction method based on Convolutional Neural Network (CNN) and Bidirectional Long-Short Term Memory (BiLSTM) models is proposed. Use CNN sequence feature extraction capabilities to extract effective information, retain longer effective memory information to solve the gradient dispersion problem, and make up for the instability and gradient disappearance of the LSTM network model when facing an excessively long sequence. The proposed method is used to predict the wind power of a wind farm, and the prediction results are compared with other models. The results show that the use of CNN-BiLSTM method can significantly improve the prediction performance and reduce the wind power prediction error.

Keywords: Wind power · Short-term prediction · Bidirectional long-short term memory network · Feature extraction

1 Introduction

1.1 Background and Motivation

As a renewable clean energy, wind energy has attracted much attention in recent years. However, wind power generation is affected by meteorological and environmental factors, and there are a series of problems such as randomness and volatility, which seriously affect the utilization rate of wind energy and hinder the development of wind power generation. Accurate wind power forecasts for wind farms are helpful for reasonable wind power planning and thus increase the utilization rate of wind energy. Therefore, it has important research significance [1].

At present, the main methods of wind power prediction are physical method, statistical method and learning method [2]. The physical method mainly predicts the power based on the numerical weather forecast. It does not require historical data but lacks accuracy, and the modeling is more complicated and the calculation requirements are high. The statistical method and the learning method establish a mapping relationship between historical data and power, and use recent wind turbine data to predict the power, these two methods can make predictions, but the accuracy of each is not high. Therefore, the combined method has become a popular method for forecasting wind power.

H. Zhang et al. (Eds.): NCAA 2022, CCIS 1638, pp. 80–93, 2022.
https://doi.org/10.1007/978-981-19-6135-9_7

1.2 Literature Review

Wind power forecasting can be divided into ultra-short-term, short-term, medium-term, and long-term forecasts according to the predicted time scale and corresponding control strategies. This article focuses on the study of short-term wind power forecasting. At present, there are many methods used in short-term wind power forecasting. Bayesian model is a time series forecasting method, which has been applied in many literatures. Blonbou et al. [3] uses a combination of adaptive Bayesian learning and neural network to predict wind power generation. Zhe et al. [4] uses Bayesian model and Markov switching model for prediction. At the same time, Support Vector Machines (SVM) is also a popular algorithm. Yang et al. [5] used SVM-enhanced Markov model to predict wind power generation in consideration of wind ramps and wind farm seasonality. Huang et al. [6] established a deterministic wind power prediction model with multiple support vector regression models based on the enhanced harmony search algorithm, and the results showed that the model has good prediction accuracy. In addition, artificial neural networks are also an important method that cannot be ignored in prediction problems. Sergio Velázquez et al. [7] improved the neural network model by changing the priority period selected by the input layer parameters or adding the input layer data of the second weather station except the wind farm reference station. The results showed that when adding the data of the second weather station. When, the model performance has been improved. In addition, the prediction accuracy of the corresponding prediction model composed of some other algorithms is also very satisfactory. Shen et al. [8] proposed a prediction model that combines empirical mode decomposition and random forest algorithm (Random Forest, RF). The model uses the decomposition of wind power series into several eigenmode functions and a residual. Then use the random forest algorithm to train each component, and finally add the predicted results of each component to get the predicted value of wind power. The result shows that the model reduces the prediction error and can accurately track the change of wind power. Xu et al. [9] applied cluster analysis to wind power forecasting. He Dong et al. [10] combined the principal component analysis method with the neural network algorithm to solve the problem of wind power neural network prediction input variables and poor generalization ability. Xu Tongyu et al. [11] combined principal component analysis and genetic algorithm optimized back-propagation neural network, used principal component analysis to reduce the dimensionality of original variables, improve the quality of sample data, and perform short-term predictions of photoelectric output power, and verified Its effectiveness.

2 Analysis of Factors Affecting Wind Power

2.1 Main Influencing Factor

The principle of wind power generation is that the kinetic energy of the wind is converted into the kinetic energy of the wind turbine blades, and finally converted into electrical energy. The output power of the wind turbine can be calculated from the Eq. (1):

$$p = \frac{1}{2}\rho C_p A V^3 \qquad (1)$$

where p is the output power of wind power generation, ρ is the density of the air, C_p is the wind energy utilization coefficient of the wind turbine, A is the coverage area of the wind turbine blades, and V is the wind speed. It can be seen from Eq. (1) that C_p and A of the same type of wind turbines in the same wind farm are fixed, so the wind power is mainly determined by the wind speed and air density.

Air density refers to the mass possessed by a unit volume of air at a certain temperature and pressure. When the wind speed is constant, the kinetic energy of the wind increases with the increase of the air density, and the wind power also increases accordingly. The temperature, air pressure, humidity and other meteorological factors of the environment where the wind power generation equipment is located can not only affect the size of the air density, but also affect the operating performance of the power generation equipment, so these meteorological factors will also have a non-negligible impact on the output power of the wind power generation equipment. The wind direction mainly affects the output power of the wind turbine through the angle with the wind turbine blades and the wake effect. In summary, in order to ensure the comprehensiveness of the input factors of the wind power prediction model as much as possible and improve the prediction accuracy of the model, in addition to considering the historical power and wind speed, it is also necessary to consider the influence of other meteorological factors.

Relationship Between Wind Speed and Wind Power. The available power output by the fan will be further reduced due to the influence of various factors such as wind speed and wind direction fluctuation. Figure 1 shows the influence of wind speed and wind power generation, the output power formula of the fan can be written as:

$$p = \frac{1}{2}C_p \cdot \rho \cdot A \cdot v^3 \cdot \eta = \frac{1}{2}\rho \cdot A \cdot V^3(C_p \cdot \eta) \tag{2}$$

Set: v_i—The initial wind speed of the fan, that is, the power that the fan starts to output when the wind speed is greater than this; v_r—The rated wind speed of the fan, that is, the rated power output of the fan at this wind speed; v_o—The cut-off wind speed of the fan, that is, the fan is braked at this wind speed.

According to Eq. (2), the output power of the fan according to the wind speed can be obtained:

when $v_i < v < v_r$:

$$p = \frac{1}{2}\rho \cdot A \cdot v^3(C_p \cdot \eta)\infty v^3 \tag{3}$$

When $v_r < v < v_o$:

$$p = p_r = \frac{1}{2}\rho \cdot A \cdot v_r^3(C_p \cdot \eta) \tag{4}$$

when $v = v_o$:

$$p = 0 \tag{5}$$

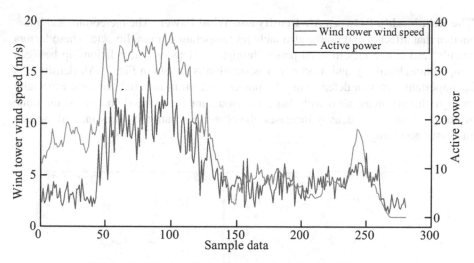

Fig. 1. The relationship between wind speed and wind power

The Relationship Between Wind Direction and Wind Power. The influence of wind direction on the power generation of wind farms is mainly reflected in two aspects:

(1) After the wind turbine extracts energy from the wind, the wind energy cannot be effectively recovered, and the wind speed decreases significantly in the long area downwind of the wind turbine, which is the wake effect [12]. Due to the influence of the wake effect of the wind turbines in the upwind direction, the wind energy obtained by the wind turbines in the downwind direction is reduced, and the output power of the corresponding wind turbines is also reduced [13]. In order to reduce the influence of the wake effect on the downstream wind turbines, a distance for wind energy recovery should be reserved between each wind turbine. Usually, the distance between two adjacent wind turbines in the vertical dominant wind direction should be set aside for wind energy recovery. Usually, the distance between two adjacent wind turbines in the vertical dominant wind direction is about 3 to 5 times the impeller diameter, and the distance in the dominant wind direction is 7 to 10 times the impeller diameter.

(2) The yaw device of the wind turbine aligns the wind turbine with the incoming wind direction according to the nacelle anemometer and the wind vane. However, the yaw device has a certain lag, and the wind turbine cannot always face the incoming wind direction. Affect the power generation of wind turbines.

In order to further quantitatively analyze the influence of wind direction on the power generated by the wind farm, the efficiency coefficient η of the wind farm is defined as

$$\eta = P_m / P_f \tag{6}$$

where p_m is the measured power generation of the wind farm at a certain wind speed and wind direction, MW; p_f is the generated power of the wind farm that is not affected by the wake under the same working conditions, MW.

The Relationship Between Air Density and Wind Power. The meteorological information that affects wind power also includes temperature, humidity, etc. These factors mainly lead to changes in wind power through air density. The relationship between temperature, humidity and wind power generation is shown in Fig. 2. Air density ρ is an important factor in determining the power generation of wind turbines, especially in high altitude areas, air density has an obvious impact on wind energy. At the same wind speed, as the air density increases, the power generation of the wind turbine also increases accordingly.

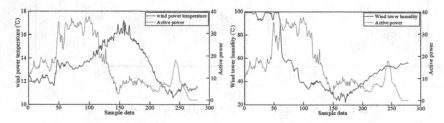

Fig. 2. The relationship between temperature, humidity and wind power

2.2 Spearman Correlation Analysis

When there are multiple variables, it is necessary to find out the variables that have a greater impact on the predicted power generation. Correlation analysis was performed using Spearman's method. The method for calculating the Spearman correlation coefficient of two n-dimensional vectors x and y is shown in Eq. (7). Where x, y represent the two variables for correlation analysis.

$$\rho_s = \frac{\sum\limits_{i=1}^{N} \left(R_i - \overline{R} \right)}{\sqrt{\sum\limits_{i=1}^{N} \left(R_i - \overline{R} \right)} \sqrt{\sum\limits_{i=1}^{N} \left(S_i - \overline{S} \right)^2}} = 1 - \frac{6 \sum\limits_{i=1}^{N} d_i^2}{N \left(N^2 - 1 \right)} \tag{7}$$

where R_i and S_i are the rank of the observed value i after sorting the vectors x and y respectively; $\overline{R}$ and $\overline{S}$ are the average rank of the vectors x and y respectively; N is the total number of i; $d_i = R_i - S_i$ means i in the two variables rank difference. The ρ_s in the Spearman correlation coefficient is a real number in $[-1, 1]$. When $\rho_s > 0$, the two variables are positively correlated, otherwise, they are negatively correlated. The larger the $|\rho_s|$, the higher the correlation between the variables x and y.

The correlation analysis method was used to calculate the wind power of the wind farm in 2019 and the correlation coefficient between the variable data of the influencing factors. The analysis is shown in Table 1. It can be seen from Table 1 that the correlation between power generation and wind speed is the largest, and the correlation coefficient is 0.9405; while the lowest correlation with wind power is air pressure, which is 0.0898. The variables are: power generation, wind speed, air temperature, and wind direction.

Table 1. Correlation analysis between wind power and impact factors

Variable	Wind speed	Wind direction	Air temperature	Air pressure	Humidity	Power
Power Generation	0.9405	−0.1663	−0.3956	0.0898	−0.1512	1

3   CNN-BiLSTM Network Structure

3.1   CNN Network

Convolutional Neural Network (CNN) is a kind of commonly used neural network model in the field of deep learning. It is widely used because it has strong feature learning ability and can greatly reduce the number of parameters in the model in image recognition and other fields.

CNN consists of input layer, convolution layer, pooling layer, fully connected layer and output layer. The convolution layer uses several different convolution kernels to convolve the input feature map, and then uses the activation function to assign nonlinear features to obtain local feature information in the input data [14]. The pooling layer performs dimensionality reduction sampling on the convolution output and extracts more critical information in the convolution output, thereby preventing the network from overfitting. The fully connected layer maps the feature map output by the pooling layer into a fixed-length column vector for subsequent classification or regression operations. Because CNN adopts convolution operation in calculation, its operation speed has been greatly improved compared with general matrix operation. The alternating use of CNN's convolution operation and pooling layer can effectively extract local features of data and reduce the dimension of local features; due to weight sharing, the number of weights can be reduced and the complexity of the model can be reduced. The feature extraction output of one-dimensional convolution for time series is:

$$Y = \sigma (W \cdot X + b) \tag{8}$$

where Y is the extracted feature; σ is the sigmoid function; W is the weight matrix; X is the time series; b is the bias vector.

3.2   BiLSTM Network

In order to effectively obtain the changing characteristics of the input time series with respect to time, a bidirectional long-short-term memory network (BiLSTM) is used to further process the output sequence of the convolutional neural network, and the memory function of the bidirectional long-term and short-term memory network is used to extract temporal characteristics [15].

BiLSTM develops a Recurrent Neural Network (RNN), which inputs the input data at each moment into the network for calculation, and the output of each hidden point is not only sent to the next layer of the network, but also to the next moment at the same time point. Its structure is shown in Fig. 3.

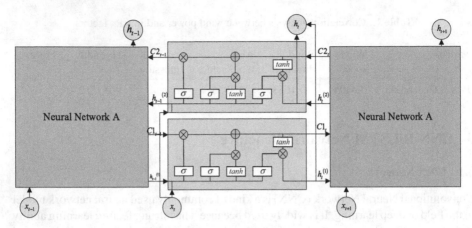

Fig. 3. BiLSTM network unit structure

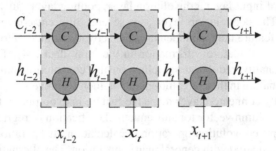

Fig. 4. BiLSTM hidden layer information transfer process diagram

The information transfer process of the hidden layer of the BiLSTM network is shown in Fig. 4.

Among them, each hidden unit contains two long short-term memory network (Long Short-Term Memory, LSTM) units, one of which is a forward LSTM, which receives the output data of the previous moment, and passes it to the next moment after calculation. In order to obtain the past information of the network, the other is backward LSTM. Relatively, it receives the output data of the next moment, and passes it to the previous moment after calculation to obtain the future information of the network. Each LSTM computing unit has four gate control units, namely input gate (i_t) , output gate (o_t), control gate (C_t) and forget gate (f_t).

The forgetting gate decides how much of the input information h_{t-1} at the previous moment is saved to the current control gate output C_t, and the output f_t and the input x_t at the current moment have the following relationship:

$$f_t = \sigma\left(\omega_f\left[h_{t-1}, x_t\right] + b_f\right) \tag{9}$$

where ω_f and b_f are the input weight and bias of the forget gate, respectively; σ is the sigmoid activation function, which is used to endow the network with nonlinear characteristics. The input gate decides how much of the input information at the current moment is retained to C_t; the intermediate variable $\widetilde{C}_t$ is used to determine whether to add i_t to the unit state. The corresponding functional relationship is as follows:

$$\widetilde{C}=\tanh\left(\omega_C \cdot [h_{t-1}, x_t] + b_i\right) \tag{10}$$

$$i_t = \sigma\left(\omega_i \cdot [h_{t-1}, x_t] + b_i\right) \tag{11}$$

where ω_c and ω_i are the weight parameters of the intermediate variable $\widetilde{C}_t$ and input i_t, respectively; b_c and b_i are the bias parameters of the intermediate variable $\widetilde{C}_t$ and input i_t, respectively; tanh is the tanh activation function. The output gate determines how much the current control gate output C_t passes to the current moment output h_t. The corresponding functional relationship is as follows:

$$C_t = C_{t-1} \cdot f_t + i_t \cdot \widetilde{C}_t \tag{12}$$

$$o_t = \sigma\left(\omega_o \cdot [h_{t-1}, x_t] + b_o\right) \tag{13}$$

$$h_t = o_t \tanh(C_t) \tag{14}$$

where ω_o and b_o are the weight parameters and bias parameters of the output gate, respectively; C_t is the control gate output; h_t is the output value of the LSTM computing unit.

The total output value of the BiLSTM calculation unit at time t is the sum of the output values of the forward LSTM unit and the backward LSTM unit it contains. The specific calculation formula is as follows:

$$h_t^{(1)} = \overrightarrow{h_t} = LSTM\left(h_{t-1}, x_t, C_{t-1}\right) \tag{15}$$

$$h_t^{(2)} = \overrightarrow{h_t} = LSTM\left(h_{t+1}, x_t, C_{t+1}\right) \tag{16}$$

$$h_t = h_t^{(1)} \oplus h_t^{(2)} \tag{17}$$

4   Wind Power Prediction Based on CNN-BiLSTM

4.1   CNN-BiLSTM Prediction Model

Aiming at the randomness and non-stationarity of wind power time series, this paper proposes a wind power prediction method based on CNN-BiLSTM. The model structure is shown in Fig. 5.

The description of each layer in the model is as follows:

1) Input layer. The historical wind power data of length n is preprocessed and used as the input of the model, expressed as

$$X = [x_1 \cdots x_{t-1}, x_t \cdots x_n]^T \tag{18}$$

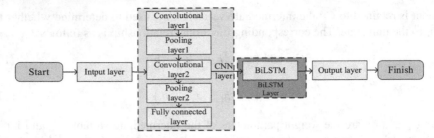

Fig. 5. Based on CNN-BiLSTM model structure

2) CNN layer. The role of the CNN layer is to extract features from historical input sequences. The CNN layer of this model is designed as 2 convolutional layers, 2 pooling layers and 1 fully connected layer. The convolutional layers are all one-dimensional convolutions, and the pooling layers are all max-pooling methods. The fully connected layer outputs the processed data, and this layer uses the sigmoid activation function. The representation of output $H_c = \left[h_{c1} \cdots h_{ct-1}, h_{ct} \cdots h_{cj} \right]^T$ of the CNN layer is as follows:

$$C_1 = Relu(X \otimes W + b_1) \tag{19}$$

$$P_1 = max(C_1) + b_2 \tag{20}$$

$$C_2 = Relu(P_1 \otimes w_2 + b_3) \tag{21}$$

$$P_2 = max(C_2) + b_4 \tag{22}$$

$$H_c = Sigmoid(P_2 \times W_3 + b_5) \tag{23}$$

where the outputs of convolutional layers 1 and 2 are C_1 and C_2 respectively; the outputs of pooling layers 1 and 2 are P_1 and P_2 respectively; W_1, W_2 and W_3 are weights; b_1, b_2, b_3, b_4 and b_5 Bias; $\otimes$ is the convolution operation.

3) BiLSTM layer. Build a BiLSTM layer to learn the regularity of the output of the CNN layer. The representation of output h_t of the BiLSTM layer is as follows:

$$h_t = BiLSTM\left(H_{C,t-1}, H_{C,t}\right), t \in [1, i] \tag{24}$$

5 Example Verification and Result Analysis

5.1 Construction of Sample Set and Training Set

According to the analysis results of the load influencing factors, the data of wind speed, wind direction, temperature, humidity and air pressure are selected to form the input feature set. The data used is from a domestic wind farm. The data from October 8, 2019 to October 8, 2020 is selected. The sampling interval is 5 min, that is, there are 288 sampling points per day, and a total of 105,409 sampling points. The data from October 8, 2019 to September 1, 2020 is used as the training set, and the data from September 2, 2020 to October 8, 2020 is used as the test set.

Real and accurate data samples are the basic guarantee for the accuracy of wind power forecasting. However, due to acquisition failures, transmission system failures, fan failures, manual operation errors, etc., there must be noise data in the historical data collected by wind farms, which will seriously reduce the application value of wind power prediction models. Therefore, this paper adopts the mean difference method to deal with abnormal data and missing data.

In order to speed up the training of the model and improve the prediction accuracy, the Min-Max algorithm is used to normalize the wind power data. The value range after processing is $[-1, 1]$, and the formula is as follows.

$$X_i' = \frac{X_i - X_{min}}{X_{max} - X_{min}} \tag{25}$$

where X_i' represents the normalized data; X_i is the original wind power data; X_{max} and X_{min} are the maximum and minimum values, respectively.

In order to make the prediction result have its physical meaning, the prediction result needs to be denormalized, the formula is as follows:

$$X_i = (X_{max} - X_{min})X_i' + X_{min} \tag{26}$$

5.2 Prediction Performance Index Selection

In this paper, the wind power generation is predicted, and the root mean square error (RMSE) and the mean absolute percentage error (MAPE) are selected as evaluation indicators. The formulas for RMSE and MAPE are as follows

$$E_{RMSE} = \sqrt{\frac{1}{m} \sum_{i=1}^{m} (\hat{y}_i - y_i)^2} \tag{27}$$

$$E_{MAPE} = \frac{1}{m} \sum_{i=1}^{m} \left| \frac{\hat{y}_i - y_i}{y_i} \right| \times 100\% \tag{28}$$

where m is the number of samples; y_i and $\hat{y}_i$ are the actual value and predicted value of the wind power of sample i, respectively.

Two evaluation metrics, RMSE and MAPE, can measure the predictive performance of the model. In wind power prediction, if the values of RMSE and MAPE are smaller, the difference between the predicted value and the actual wind power value is smaller, and the prediction accuracy is higher.

5.3 Parameter Settings

Based on the CNN-BiLSTM model, it consists of two convolutional layers, two pooling layers and a fully connected layer, using the Relu activation function. The first convolutional layer has 64 convolution kernels, the size is set to 1×4; the second convolutional layer has 32 convolution kernels, the size is set to 1 × 3, the stride is 2, and the pooling layer selects the largest. The pooling method, and then the fully connected layer converts the output. The BiLSTM network contains 2 hidden layers with 30 and 20 units respectively.

5.4 Experimental Process and Result Analysis

In order to better verify the effectiveness of the proposed model, the back-propagation neural network model, GA-BP model, LSTM model, CNN-LSTM model, and CNN-BiLSTM model were used for short-term wind power prediction, The prediction results of the four models are shown in Figs. 6, 7, 8 and 9 respectively.

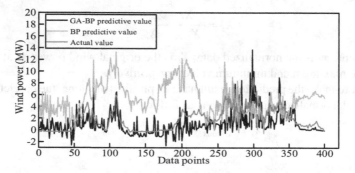

Fig. 6. Comparison of BP and GA-BP Wind Power Forecast

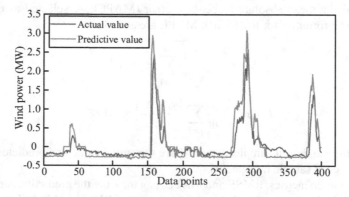

Fig. 7. Comparison of LSTM wind power prediction

It can be seen from Figs. 6, 7, 8 and 9 that in the prediction of wind power, all four models can roughly predict the trend of wind power, and the GA-BP model has the worst performance. The GA-BP model has a weaker ability to handle smoothly changing wind power data, and the LSTM model has a weaker ability to handle rapidly changing wind power data. The CNN-LSTM model combines the advantages of two single models, CNN and LSTM, so its prediction error is smaller.

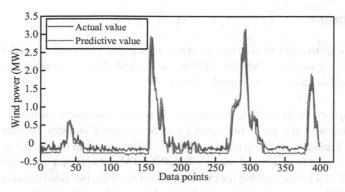

Fig. 8. Comparison of CNN-LSTM wind power prediction

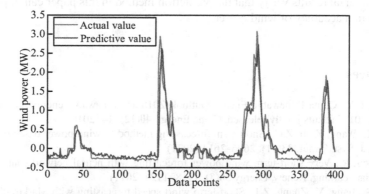

Fig. 9. Comparison of CNN-BiLSTM wind power prediction

Table 2. Comparison of forecast errors of different forecast models

Predictive model	RMSE	MAPE
GA-BP	1.3582	0.5490
LSTM	1.2321	0.4562
CNN-LSTM	0.5285	0.3352
CNN-BiLSTM	0.2665	0.1863

The experimental results are evaluated using two evaluation indicators, RMSE and MAPE. The results are shown in Table 2. Compared with other methods, the CNN-BiLSTM prediction method proposed in this paper is better than GA-BP, LSTM, and CNN-LSTM respectively 1.0917, 0.9656, 0.262. Compared with other methods on the MAPE index, it is 0.3627, 0.2699, and 0.1489 higher than GA-BP, LSTM, and CNN-LSTM, respectively. Therefore, it is verified by experiments that the CNN-BiLSTM model has better prediction performance than other models.

6 Conclusion

In order to improve the prediction accuracy, this paper proposes a wind power prediction method based on CNN-BiLSTM network, and the effectiveness is verified by experiments. The calculation example shows:

1) Using Spearman to analyze the correlation between different variables, the multivariate input of the prediction model can be screened, thereby reducing the data scale and reducing the impact of ineffective information on the model accuracy.
2) The prediction model based on the CNN-BiLSTM network not only has the advantages of CNN being suitable for extracting data features, but also includes the ability of BiLSTM to process time series, which reduces the loss of information and is more consistent with the time series and nonlinear relationship of wind power data. The experimental results verify that the prediction method in this paper can improve the prediction accuracy of wind power.

References

1. Dong, L.Y.: China Renewable Energy Outlook 2018 and renewable energy source market report 2018" results jointly released. China Energy **40**(12), 44 (2018)
2. Liu, X., Wang, Y., Ji, Z.C.: Short-term forecasting method of wind power based on random forest. J. Syst. Simul. **33**(11), 2606–2614 (2021)
3. Blonbou, R.: Very short-term wind power forecasting with neural networks and adaptive Bayesian learning. Renew. Energy **36**(3), 1118–1124 (2011)
4. Zhe, S., Jiang, Y., Zhang, Z.J.: Short-term wind speed forecasting with Markov-Switching Model. Appl. Energy (S0306–2619), **5**(130), 103–112 (2014)
5. Yang, L., He, M., Zhang, J.: Support-vector-machine enhanced Markov Model for Short-term Wind Power Forecast. IEEE Trans. Sustain. Energy **6**(3), 791–799 (2015)
6. Huang, C.M., Kuo, C.J., Huang, Y.C.: Short-term wind power forecasting and uncertainty analysis using a hybrid intelligent method. IET Renew. Power Gener. **11**(5), 678–687 (2017)
7. Medina, S.V., Ajenjo, U.P.: Performance improvement of artificial Neural Network Model in short-term forecasting of wind farm power output. J. Mod. Power Syst. Clean Energy **8**(3), 484–490 (2020)
8. Shen, W., Jiang, N., Li, N.: An EMD-RF based short-term wind power forecasting method. In: 2018 IEEE 7th Data Driven Control and Learning Systems Conference (DDCLS). Piscataway, pp. 283–288. IEEE, NJ (2018)
9. Xu, A., Yang, T., Ji, J., et al.: Application of cluster analysis in short-term Wind Power Forecasting Model. J. Eng. **9**, 5423–5426 (2019)
10. He, D., Liu, R.Y.: Ultra-short-term wind power prediction using ANN ensemble based on the principal components analysis. Power Syst. Protect. Control **41**(4), 50–54 (2013)
11. Xu, T.Y., Ma, Y.M., Cao, Y.L., et al.: Short term forecasting of photovoltaic output power based on principal component analysis and genetic optimization of BP Neural Network. Power Syst. Protect. Control **44**(22), 90–95 (2016)
12. Cao, N., Yu, Q.: Grouping method of wind turbines in grid-connected wind farms under the condition of wind speed fluctuation. Autom. Electric Power Syst. **36**(02), 42–46 (2012)
13. Jia, Y., Liu, X., Li, H., Ren, Z., Zhao, W.W.: Analysis of the influence of wake effect on the layout of wind farm units. Renew. Energy **32**(04), 429–435 (2014)

14. Yu, P., Zhao, J.S., Zhang, J.: Research on convolutional neural network image recognition based on nonlinear correction function. Sci. Technol. Eng. **15**(34), 221–225 (2015)
15. Lu, C.H.: Text classification method based on multi-channel recurrent convolutional neural network. Comput. Appl. Software **37**(08), 282–288 (2020)

Fault Arc Detection Method Based on Multi Feature Analysis and PNN

Chao Li[1]($\boxtimes$), Jiachuan Shi[1]($\boxtimes$) ⓘ, Jiankai Ma[2], Rao Fu[1] ⓘ, and Jin Zhao[3]

[1] Shandong Key Laboratory of Intelligent Buildings Technology, School of Information
and Electrical Engineering, Shandong Jianzhu University, Jinan 250101, China
2020080115@stu.sdjzu.edu.cn, {jc_shi,furao20}@sdjzu.edu.cn
[2] Shandong Taishan Pumped Storage Power Station Co. Ltd., Taian 271000, China
[3] Department Electrical and Computer Engineering, The University of Alabama,
Tuscaloosa, AL 35487, USA
jzhao26@crimson.ua.edu

Abstract. In order to accurately and timely identify the low-voltage DC fault arcs and make protection actions, this paper proposes a kind of fault arc detection method based on multi-feature analysis and PNN. Through the multi-dimensional feature analysis of the electromagnetic waves of low-voltage DC series arc fault, the effective characteristic quantity is selected to form the eigenvector group to express the fault arc information. In terms of fault detection, a low-voltage DC fault arc diagnosis algorithm based on PNN is used to identify eigenvectors. The experimental results show that the method can effectively detect DC arc fault and improve the detection accuracy, which is of great significance to protect the safe operation of low voltage DC systems.

Keywords: Series arc fault · Probabilistic neural network · Multi-dimensional feature analysis · Fault diagnosis

1 Introduction

In a photovoltaic power generation system, the DC fault arc seriously affects the normal operation of the system. According to the form of fault arcs, fault arcs can be divided into series fault arcs, parallel fault arcs, and ground fault arcs. [1] It was difficult for the arc to detect the reliability of the system effectively due to the small change in circuit current and voltage.

The current arc fault detection methods have two categories: 1) multiple current and voltage threshold metrics modeling [2] and 2) arc fault current for modeling. The threshold method mainly judges the difference between normal current and fault current sequence through the difference in current morphology, and compares it with the threshold value to determine whether arc fault occurs. In reference [3], it identified series fault arcs by calculating the zero-rest time scaling factor, and the filtered normalized correlation coefficient to compare with the empirical threshold. Although the method has strong real-time performance, it has low generalization. Modeling the current signal

H. Zhang et al. (Eds.): NCAA 2022, CCIS 1638, pp. 94–108, 2022.
https://doi.org/10.1007/978-981-19-6135-9_8

directly is a common method now. In reference [4], three different coefficients were used as combined feature quantities to train a multilayer neural network. The characteristic quantities included Fourier coefficient, wavelet feature, and MFCCs. The disadvantage of the deep learning method was that it required a large number of labeled data, and long training time, so it was not practical for arc fault detection tasks. The frequency-domain segmentation methods used in the feature analysis based on frequency domain mainly included wavelet transform [5], variational mode decomposition [6], and empirical wavelet transform [7]. The above methods have large amount of calculation and poor effect on nonlinear signal processing. A signal processing tool called Mathematical Morphology (MM) can be used as a solution to the above problems. It extracts the characteristic frequency band of the fault signal from the time domain, and the phase and amplitude of the signal will not change without time-frequency conversion. The method is simple, real-time, and less computational. It is more effective to extract features of different frequency bands for nonlinear signals. Through a mathematical morphology filtering algorithm, the influence of low-frequency current disturbance on arc fault eigenvalue was effectively filtered out, which had more obvious advantages for weak arc fault eigenvalue extraction [8]. Since the interference of the low-frequency signal was eliminated, the frequency domain feature extraction of the filtered signal could be carried out to enrich the feature dimension of arc fault detection. The accuracy of fault detection was thus improved.

In summary, a fault arc detection method based on multi-dimensional feature analysis and PNN is proposed. Firstly, the statistical analysis of the current sampling data is carried out by using the characteristic parameters such as peak value and pulse index, and then the sampling data is filtered by a mathematical morphology filter. The singular entropy of the filtered signal and the mean difference between adjacent windows are calculated. Through the above methods to describe the fault information in the current signal, and finally, determine the effective feature as the input feature parameters of the PNN classifier, and select the appropriate spread factor to improve the accuracy of the algorithm. The simulation results showed that the proposed algorithm had the ability to distinguish fault arc, high accuracy, and anti-interference ability.

2   Low Voltage DC Fault Arc Data Acquisition

2.1   Current Signal Acquisition

According to the requirements of the experiment in the literature [9, 10], the experimental platform of low-voltage DC arc was built. The schematic diagram of this experimental platform was shown in Fig. 1. The experimental platform schematic included programmable DC power supply, arc generator, load, current sensor, oscilloscope and data acquisition card.

Due to the actual operating conditions were extremely complex, this paper carried out the arc fault experiments under mixed conditions. Not only included different loads for separate arc fault experiments, but also took into account the simultaneous operation of multiple types of loads. Experiments in all types of load parameters were shown in Table 1.

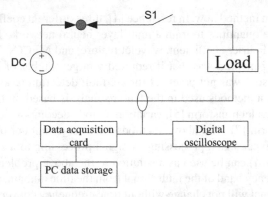

Fig. 1. Schematic diagram of experimental platform

Table 1. Load parameters information table

Serial number	Load name	Load type	Equipment parameters	Rated operating current
1	Resistors	Linear loads	200 W/8 Ω	____
2	LED lights	Non-linear load	DC12 V ~ 85 V	____
3	Electric fan	Mechanical load	DC12 V ~ 85 V	____
4	LED light + resistor	Mixed load	____	4.5 A
5	Electric fan + resistor	Mixed load	____	4.5 A
6	Electric Fan + LED Light + resistor	Mixed load	____	4.8 A

2.2   Current Waveform Analysis

By operating the experimental platform to simulate the occurrence of fault arc, the experimental data of low-voltage DC faults under different working conditions with different loads were collected. Finally, the characteristics of the collected data in various states were analyzed. The experimentally obtained current waveforms for each load were shown in Fig. 2, where Fig. 2(a) represents the resistor, Fig. 2(b) represents the LED lamp, Fig. 2(c) represents the electric fan, Fig. 2(d) represents the LED lamp + resistor, Fig. 2(e) represents the fan + resistor, and Fig. 2(f) represents the fan + LED lamp + resistor.

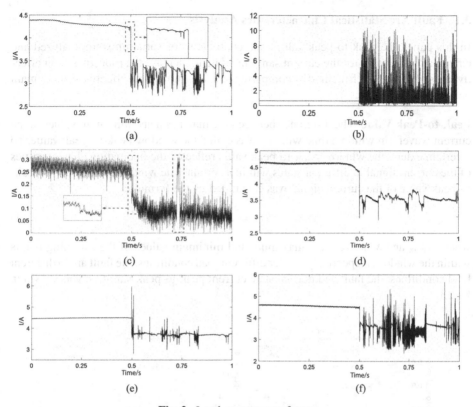

Fig. 2. Load current waveforms

By analyzing the experimental results of DC arc fault under different load conditions, it was impossible to judge the occurrence of arc fault based on the magnitude of the current amplitude under fault conditions, and it was necessary to conduct deep excavation of the current signal from multiple dimensions to select effective feature values to express the arc fault information and improved the detection accuracy of arc fault. In this paper, resistive load, fan load and LED light were selected to represent linear load, mechanical load and nonlinear load respectively, and resistive shunt electric fan was selected to represent mixed load.

3 Multi-dimensional DC Arc Fault Characterization

In this section, statistical indicators and feature analysis based on differential mathematical morphology filter were analyzed for the sampled current signal within the time window. The statistical indicators included peak-to-peak and pulse indicators. In addition the sampled signals were pre-processed by mathematical morphological filters. Thus, the information expression of low frequency components was weakened and the interference of oscillation interference and arc unstable combustion on the discrimination was reduced. The filtered signal was then subjected to adjacent window waveform difference mean and singular entropy feature extraction.

3.1 Fault Arc Statistical Characteristics Analysis

In this paper, the peak-to-peak values and coefficients of variation were analyzed and calculated separately for the current sampling data to increase the robustness and effectiveness of this research method by comparing and determining the effective time-domain characteristic indexes.

Peak-to-Peak Value The difference between the maximum and minimum values of the current waveform within a time window was called the window peak-to-peak value. To a certain extent, the window peak-to-peak value reflected the fluctuation characteristics of the current signal in different states within a certain time window. The window peak-to-peak value of the current signal was calculated by the formula:

$$\Delta X_{max-min} = X_{max} - X_{min} \tag{1}$$

where X_{max} and X_{min} were the maximum and minimum values of the sampling points within the window, respectively. Under different load conditions, the fault arc in different load conditions, the fault and normal state current peak-to-peak variation was shown in Fig. 3

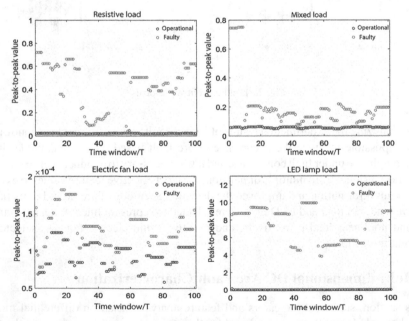

Fig. 3. Peak-to-peak value eigenvolume

As can be seen from Fig. 3, the peak-to-peak index had the ability to distinguish whether a DC series arc fault occurs in most load cases, but for mechanical loads such as electric fans, it was still necessary to improve the accuracy of detection by increasing the determination dimension.

Pulse Indicator. Through experimental analysis, the arc noise in the fault current signal was in the form of pulse signal when the arc fault occurs. Therefore, this paper used a dimensionless indicator to identify the arc fault through the pulse indicator. The expression of the pulse indicator was:

$$C_f(T) = \frac{X_{T_P}}{|\mu_T|} \tag{2}$$

where $|\mu_T|$ denoted the absolute mean value of data points within the time window; X_{T_P} denoted the maximum value of current data points within the time window. The pulse indicator was mainly used to identify whether there was a statistical indicator of shock in the signal. In the low current or weak arc state, the pulse indicator was very sensitive to the shock component in the current signal.

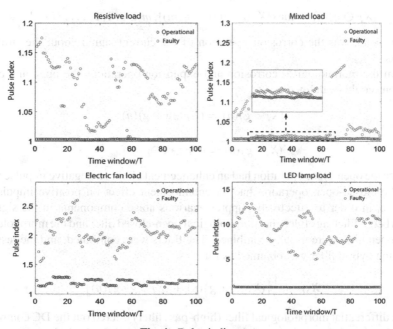

Fig. 4. Pulse indicators

Figure 4 showed the pulse indicator trends of the fault state and the normal operation state under different load conditions. Due to the unstable arc ignition phenomenon in the fault state, the pulse indicators of all kinds of loads showed a large degree of fluctuation, but the pulse indicator trend graph was always higher than that in the normal state, which had a good ability to distinguish the normal state from the fault state. Therefore, the pulse indicator could also be used as a description of whether a DC arc fault occurs.

3.2   Based on Mathematical Morphology Filter Fault Arc Characterization

Mathematical Morphology. Mathematical Morphology based filters (MM) [11, 12] were used to extract the characteristic frequency bands of the faulty signal entirely from the time domain, without further time-frequency conversion of the signal. The phase and amplitude of the signal also did not change. The method was computationally simple, real-time, lower computational effort, and more effective for feature extraction of different frequency banded for nonlinear signals.

The mathematical morphology of the corrosion and expansion calculations was the most basic calculation of the pulse signal filter [13]. Defining the element structure as $g(G)$, the expansion and corrosion operations of the arc current signal X_T with respect to the structural elements were:

$$(X_T \oplus g)(n) = \max\{X_T(n-m) + g(m)\}, m \in 0, 1, 2, \ldots, G-1 \tag{3}$$

where $n > m$, was the expansion operation of the current signal about the structural elements.

$$(X_T \ominus g)(n) = \max\{X_T(n+m) - g(m)\}, m \in 0, 1, 2, \ldots, G-1 \tag{4}$$

where $n > m$, was the corrosion operation of the current signal about the structural elements.

From the morphological corrosion and expansion operations the open and closed operations could be defined as:

$$(X_T \circ g)(n) = (X_T \ominus g \oplus g)(n) \tag{5}$$

$$(X_T \cdot g)(n) = (X_T \oplus g \ominus g)(n) \tag{6}$$

Since the open-close operation had an enhancement effect on negative impulse noise, while the closed-open operation had an enhancement effect on positive impulse by. Therefore, in order to effectively suppress various noise components in the signal to obtain a better filtering effect, the morphological open-closed filter and the morphological closed-open filter were usually combined. The filters were combined, and the resulting alternating hybrid filter was obtained as follows:

$$X_{ST} = \frac{1}{2}\Big[(X_T \cdot g \circ g)(n) + (X_T \circ g \cdot g)(n)\Big] \tag{7}$$

The differential morphological filter (high-pass filter) filtered out the DC component of the signal and the skeleton of the signal, and extracted the high-frequency signal components with the expression:

$$X_{DT} = X_T - X_{ST} \tag{8}$$

where $\cdot$, $\circ$, $\ominus$ and $\oplus$ were the operators of the corrosion, expansion, open and closed operations, respectively. The mathematical morphology tool had great advantages in detecting irregular signals such as randomly perturbed signals as well as transient pulse aspects, so it was chosen as a signal pre-processing filter. In order to reduce the influence of negative pulse interference and reduce the complexity of calculation, this paper adopted a hybrid filter based on linear structure elements with mathematical morphology. Considering the requirement of real-time and statistical width of the shock component in the current when the arc fault occurs, the length of the linear structure element was set to 3.

Adjacent Window Waveform Difference Mean. By analyzing the current waveforms of various types of loads, the components of the current waveform during the normal operation of various types of loads could be expressed by the formula:

$$I_T = I_{T-k} + \delta_r \tag{9}$$

where $T > K$; K were integers; I_T was the current waveform in the Tth window under normal operation; 4 was the disturbance factor in the current waveform in the Tth time window under normal operation. Analysis of the current signal distortion characteristics of the fault state, the fault state at each moment of the current signal paratime value could be approximated by the distortion factor should be affected, and the degree of distortion factor on the current at each moment was random. The expression could be as follows:

$$I_T^{arc} = I_{T-k} \cdot A_T + \delta_T^{arc} \tag{10}$$

where I_T^{arc} denoted the fault arc current sequence in the Tth time window; A_T denoted the aberration factor sequence in the Tth cycle, which controled the overall trend of the current, and its instantaneous value always satisfies $A_T(t) < 1$ was a changing quantity; δ_T^{arc} was the arc noise, which was usually, $\delta_T^{arc} \geq \delta_T$, more influenced by the external environment, the electrode material and the electrode gas gap. Under the above conditions, the difference between the current waveforms in the adjacent windows of the normal operation state and the fault state could be expressed as follows, respectively:

(1) Normal operation phase:

$$\Delta I_T = I_T - I_{(T-1)} = \delta_T - \delta_{(T-1)} \tag{11}$$

(2) Stable combustion stage:

$$\Delta I_T = I_{T-k} \cdot (A_T - A_{T-1}) + \delta_T^{arc} - \delta_{T-1}^{arc} \tag{12}$$

By analyzing Eq. (11), it could be seen that there are some low-frequency disturbances, and these low-frequency disturbances could affect the calculation results of ΔI_T^{arc} and made it misrepresent the arc fault information. To overcome the above problems, this paper proposed a new method based on mathematical morphological filter to pre-process the sampled signals and increased the accuracy and effectiveness of feature extraction for low-voltage DC arc faults. The mean coefficient of the difference of adjacent window waveforms was calculated as follows:

$$M(T) = \frac{\sum_{i=1}^{N} \Delta I_T(i)}{N} \tag{13}$$

Figure 5 showed the mean coefficient curve of the difference between adjacent window waveforms of the filtered signal.

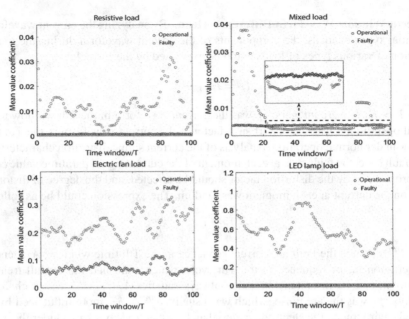

Fig. 5. Average value of waveform difference between adjacent windows under different loads

Filtered Signal Singular Entropy. Let the data signal have a sampling length of n per group and be divided into m sampled data groups, then the construction of matrix A could be performed:

$$A = \begin{pmatrix} C_1 \\ C_2 \\ \cdots \\ C_m \end{pmatrix} = \begin{bmatrix} C_{1,1} & C_{1,2} & C_{1,n-1} & C_{1,n} \\ C_{2,1} & C_{2,1} & C_{2,n-1} & C_{2,n} \\ & & & \\ C_{m,1} & C_{m,2} & C_{m,n-1} & C_{m,n} \end{bmatrix} \tag{14}$$

For matrix A the singular value decomposition was given by:

$$A = UAV^T = [u_1 \ u_2 \ \cdots \ u_n] \cdot \begin{bmatrix} \sigma_1 & & & \\ & \sigma_2 & & \\ & & \cdots & \\ & & & \sigma_n \end{bmatrix} \cdot [u_1 \ u_2 \ \cdots \ u_n]^T \tag{15}$$

$$A = \sum_{i-1}^{r} \sigma_i u_i v_i^T \tag{16}$$

where U was an orthogonal matrix of order m*n; V was an orthogonal matrix of order n*n; Λ was a diagonal diagonal matrix whose pair of elements σ_i was called the singular value of the information matrix A and was denoted as:

$$\Lambda = dig(\sigma_1, \sigma_2, \sigma_3, \cdots, \sigma_n) \tag{17}$$

Information entropy indicated the amount of information contained in the data itself, and as a measure of data distribution could effectively indicate the dispersion degree of data distribution. Therefore, the information entropy of wavelet singular value could be used as a fault feature. Taking fan inter-load harmonics as an example, the information entropy of singular value was calculated as follows:

$$S(T) = -\sum_{i=1}^{n} p_i \log_2(p_i) \tag{18}$$

where

$$p_i = \frac{\sigma_i}{\sum_{i=1}^{n} \sigma_i} \tag{19}$$

If $\sigma_i = 0$, then $p_i \log_2 p_i = 0$.

As shown in Fig. 6, the information entropy scatter plotted for different loads in normal and fault states were shown respectively. The information entropy characterized the fault information contained in the number and size of the wavelet singular values of the arc current under the influence of harmonics. From the experimental and simulation results to observe, the differential mathematical morphology filter signal singular entropy could be a good distinction between normal operation and fault arc state under different load conditions, and could be used as an effective discriminatory feature quantity.

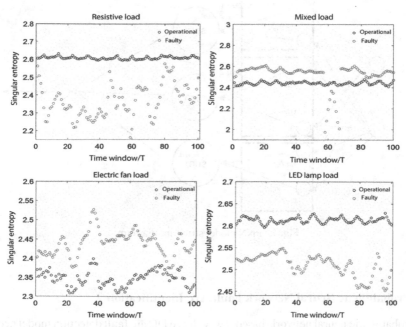

Fig. 6. Filtered signal singular entropy

4 Arc Fault Identification Based on Probabilistic Neural Network

4.1 Probabilistic Neural Networks

Probabilistic Neural Network (PNN), was a radial basis feedforward neural network proposed based on Bayesian decision theory and Parzen window probability density estimation method [14, 15]. The structure of the probabilistic neural net was not compli-cated and converges much faster than the BP neural net in the training process. Unlike the traditional BP neural network, the probabilistic neural network replaced the Sigmod activation function with the exponential activation function. It could compute nonlinear decision bounds efficiently and use linear learning algorithms to do the work of nonlinear learning algorithms. With the support of large-scale training data, the neural network did not have the problem of local extrema and had high data tolerance. Therefore, this method was more suitable for pattern classification than the traditional BP neural net-work in terms of recognition accuracy. The identification of arc fault belonged to the category of pattern classification.

PNN, as a radial basis neural network, was composed of a parallel four-layer struc-ture, as shown in Fig. 7, with the four layers of the model being: input layer, pattern layer, summation layer and output layer.

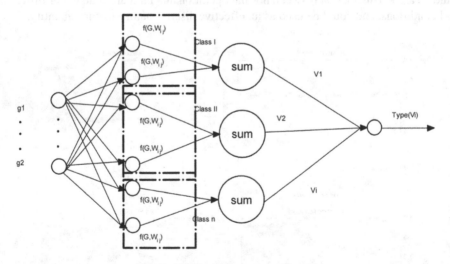

Fig. 7. PNN model structure diagram

4.2 Probabilistic Neural Network Fault Arc Identification

The probabilistic neural network-based low-voltage DC arc fault detection model needed to satisfy the categorization of a high-dimensional feature vector composed of valid feature values with current data set in different states in the historical database. The construction process of PNN-based diagnostic model mainly included: data acquisition, data processing and feature value extraction, probabilistic neural network training, and fault identification. The specific steps were: building the training data set, normalizing the eigenvalues, parameter rectification, and testing the diagnostic model.

The fault diagnosis model of low-voltage DC arc fault extracted the fault feature quantity by analyzing the current signal in time domain and frequency domain, and outputted the corresponding fault identification type after discriminating by the probabilistic neural network. Using the feature volume of length 4 to construct into the input vector, then the number of neurons in the input layer of the PNN neural network is 4. The probabilistic neural network outputs were 1 and 0 for the fault state and normal operation state, respectively. By setting the value of spread factor, the fault type belonging to the fault data was confirmed through the probabilistic neural network recognition, and finally the confirmed fault type was output to complete the fault identification process. The diagnosis model of low-voltage DC arc fault was shown in Fig. 8.

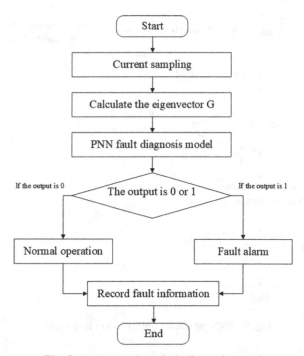

Fig. 8. PNN-based arc fault diagnosis model

5 Analysis of Results

In the sample feature vectors, a total of 1063 sets of vectors were selected for each of the four types of typical loads under normal operating conditions in this paper. Among them, 27 sets of resistive loads (resistor), including 211 sets of training samples and 60 sets of testing samples; 210 sets of mechanical loads (fan), 170 sets of training samples and

40 sets of testing samples; 290 sets of mixed loads (resistor + fan), 230 sets of training samples and 60 sets of testing samples; 292 sets of non-linear loads (LED lamp), 222 sets of training samples and 70 sets of testing samples. A total of 682 sets of vectors are selected for the fault condition, 510 sets of training samples and 172 sets of testing samples. The above test samples and training samples were basically selected according to the standard of 1:4.

In order to obtain the optimal parameters and ensure the optimization of the diagnostic model, simulation experiments were needed to observe the accuracy problem of PNN with different *spread* by using linear exhaustive method. The experimental flow was shown in Fig. 9, such that *spread* $\in$ (0.0001, 1), $\Delta spread \in 0.02$. The experimental results were shown in Fig. 10.

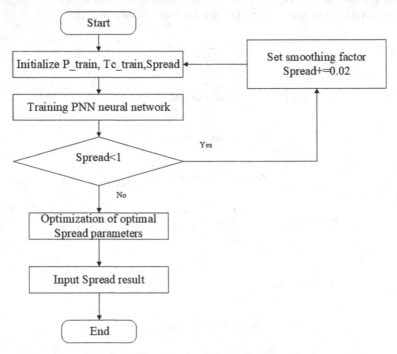

Fig. 9. PNN parameter optimization flow chart

From the simulation experimental results, the accuracy of the probabilistic neural network increased and then decreased with the increase of δ. In the process of increasing the smoothing factor from 0.0001 to 1, the accuracy of the model arc detection increased from 57.60% to 96.73%, and then decreased to 31.89%. Therefore, *spread* = 0.3 was chosen as the best in this paper. The probabilistic neural network was trained with the training data set, and the arc fault detection model was tested using the test data set.

Using the PNN diagnostic model and the traditional BP model for testing, Table 2 shows the experimental results of different classification models for arc fault detection, it can be seen that the PNN is more suitable for arc fault detection than the traditional BP neural network in terms of detection accuracy.

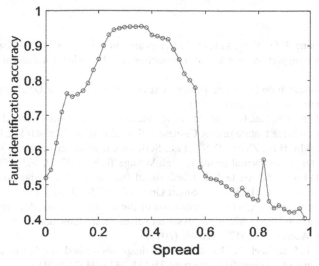

Fig. 10. Accuracy of different smoothing factors

Table 2. Comparison of test results of different classification models

Classification model name	Load type	Correctness rate
PNN	LED light	98.7%
	Electric fan	93.6%
	Mixed load	91.3%
	Resistive load	92.4%
BP	LED light	96.3%
	Electric fan	92.1%
	Mixed load	89.3%
	Resistive load	91.3%

6 Conclusion

In this paper, the low-voltage DC arc fault was taken as the research object. Through the multi-dimensional characteristic analysis of the current signal under different load conditions, the effective eigenvalues were extracted to form the feature vector to characterize the distortion characteristics of the arc current data. The PNN was used to construct the detection model *spread* Factor to improve the accuracy of the algorithm. The simulation study confirmed that the method had a high ability of arc resolution and anti-interference.

References

1. Yang, K., Zhang, R.C., Yang, J.H., et al.: Series arc fault diagnostic method based on fractal dimension and support vector machine. Transactions of China Electro-technical Society **31**(2), 70–77 (2016)
2. Liu, J.W.: Research on Fault Arc Feature Extraction and Adaptive Detection Technology. Zhejiang University, Hangzhou (2019)
3. Zhao, H.J., Qin, H.Y., Liu, K., et al.: Detection method of series fault arc based on correlation theory and zero break feature fusion. Chinese J. Scientific Instrument **41**(4), 218–228 (2020)
4. Long, G.W., Mu, H.B., Zhang, D.N., et al.: Series arc fault identification technology based on multi-feature fusion neural network. High Voltage Technol. **47**(2), 463–471 (2021)
5. Wang, Z., Ba Log, R.S.: Arc fault and flash signal analysis in DC distribution systems using wavelet transformation. IEEE Trans. Smart Grid **6**(4), 1955–1963 (2015)
6. Liu, S., Dong, L., Liao, X., et al.: Application of the variational mode decomposition-based time and time–frequency domain analysis on series dc arc fault detection of photovoltaic arrays. IEEE Access. **7**, 126177–126190 (2019)
7. Li, Z., Zhu, M., Chu, F., et al.: Mechanical fault diagnosis method based on empirical wavelet transform. Chinese J. Scientific Instrument **35**(11), 2423–2432 (2014)
8. Chun-Zhi, L.I., Rong-Jian, H.E., Tian, G.M.: Research on the application of the mathematical morphology filtering in vibration signal analysis. Computer Eng. Sci. **30**(9), 126–127 (2008)
9. Dini, D.A., Brazis, P.W., Yen, K.H.: Development of Arc-Fault Circuit-Interrupter requirements for Photovoltaic systems. In: IEEE Photovoltaic Specialists Conference, pp.1790–1794 (2011)
10. Liu, J.:Analysis of GB/T 31143—2014. Electrical & Energy Management Technology (2015)
11. Yaman, O., Yeti, H., Karakse, M.: Image processing and machine learning-based classification method for hyperspectral images. The Journal of Eng. **2**, 85-96 (2021)
12. Bouchet, A., et al.: Compensatory fuzzy mathematical morphology. Signal, Image and Video Processing (2017)
13. Mm, A., Rrp, B., Pkr, A.: A combined mathematical morphology and extreme learning machine techniques based approach to micro-grid protection. Ain Shams Eng. J. **10**(2), 307–318 (2019)
14. Pourkamali-Anaraki, F., Hariri-Ardebili, M.A.: Neural networks and imbalanced learning for data-driven scientific computing with uncertainties. IEEE Access **9**, 1533415350 (2021)
15. Chang, D.T. Probabilistic Deep Learning with Probabilistic Neural Networks and Deep Probabilistic Models. (2021)

Design of AUV Control System Based on BP Neural Network and PID

Tengguang Kong[1,2], Huanbing Gao[1,2(✉)], Shoufeng Yao[3], and Xiuxian Chen[1,2]

[1] School of Information and Electrical Engineering, Shandong Jianzhu University,
Jinan 250101, China
gaohuanbing2004@sdjzu.edu.cn
[2] Shandong Key Laboratory of Intelligent Building Technology, Jinan 250101, China
[3] Shandong Provincial Communications Planning and Design Institute Group Co., Ltd.,
Jinan, China

Abstract. Since the AUV of conventional PID controller has destitute parameter setting impact and flexibility within the commonsense operation of uncommon submerged environment location, BPNN is included on the premise of this controller to design Detect type AUV based on BPNN and PID control. Firstly, online parameter setting is realized through BPNN's automatic learning characteristic, which overcomes the complex and unsteady issue of conventional PID control for parameter setting. Furthermore, after BPNN and PID controller gets the perfect parameters, its control effectiveness amid operation is incredibly progressed, which increments the soundness of AUV in submerged environment survey. At last, the modulation process of BPNN and PID controller parameters is simulated by Matlab. The comes about appear that the control framework has great steadiness and anti-interference capacity, and essentially completes the anticipated design destinations.

Keywords: Underwater vehicle · Embedded system · PID · BPNN

1 Introduction

In later a long time, With the deepening of ocean research and development, intelligent underwater robot as an important tool for exploration and operation, its value is increasingly prominent [1], and has been widely used in offshore oil development, mineral resources development, fishing and military fields [2, 3]. Underwater robots are used to replace manual work underwater to complete underwater exploration, construction and so on. At present, the underwater robot carrying sonar detection device is used to carry out geological survey before road construction in the vast areas with karst development, and the data of underwater detection points can be collected by controlling the robot. Because of the complexity of underwater conditions, we first analyze the motion of underwater robot, according to the characteristics of the motion of underwater robot, aiming at the control circuit and control method proposed motion control research objectives [4].

The article designs a kind of the underwater robot on the strength of an embedded control system of STM32 microcontroller, so as to control the robot's advance and retreat, lateral movement, advance and retreat in the vertical plane, and snorkeling movement.

2 Underwater Vehicle Structure

The design goal of this paper is an underwater robot which can carry a sonar detection device, the robot can achieve the detection point positioning, underwater movement, horizontal hovering, remove the mud interference function, the requirements of the overall structure of the robot, motion control system hardware and software [12, 13]. The robot is required to be small in size and light in weight, with a length of 30–50 cm, and the robot has a high sealing requirement, which can not leak water under normal working conditions [5].

This robot is armed with depth transducer and attitude transducer, also can be controlled to complete forward, reverse, steering, submergence and depth through the operation instructions sent by the ground station [6]. The motor is driven by a DC brushless motor, and the main structure of the underwater robot comprises five basic components:

(1) Standby frame

The spare frame is composed of stainless steel support frame and polypropylene plate, which can be used to carry external additional sensors or external actuators such as mechanical grippers, and can also be used to carry counterweights to achieve the balance of small underwater robots in water.

(2) Propeller

The blades of the propeller generate enough thrust when rotating at high speed to drive the small underwater robot to move.

(3) Motor

Used to drive the robot to move and start the fan to remove the sludge.

(4) Main frame

The main frame is used to support, connect and fix the various parts of the small underwater robot. It is composed of stainless steel and polypropylene plate, which has strong corrosion resistance and certain strength. At the same time, threaded holes are reserved on the main frame to facilitate the loading of counterweight.

(5) Buoyancy material

Buoyancy material is a kind of composite material with low density and high strength, which has large positive buoyancy in water and can be used to trim small underwater vehicles.

Due to the serious cross-coupling in the space motion of the underwater vehicle, its accurate motion model is very complex and difficult to establish completely. The overall motion speed of the small underwater vehicle studied in this paper is relatively low in the water and in most cases it moves in the same plane, so when studying the kinematics model of the underwater vehicle, only the forward and backward and lateral motion of the underwater vehicle is considered in the horizontal plane, and only the forward and backward and snorkeling motion is considered in the vertical plane, while ignoring the motion coupling in the two planes, so as to simplify the motion model [15, 16]. On the basis of plane division, the specific motion model of the important motion forms of the underwater vehicle can be built.

3 Control System Design

The overall square chart of circuit framework structure of the underwater vehicle designed in this paper is shown in Fig. 1 below, which mainly includes the surface control platform, the microcontroller based on STMS32 F103 as the main chip, the driving module for driving the T550AUV propeller and fan, the data transmission module, the N-way voltage stabilization module, the sonar detector and the attitude sensor.

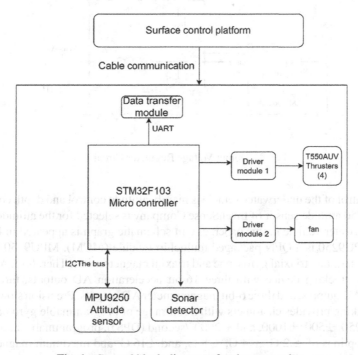

Fig. 1. General block diagram of underwater robot.

3.1 Hardware Composition of Control System

With the development of technology, the function of microcontroller is more and more powerful, and it is widely used in electrical appliances, automobiles, robots and so on. The microcontroller in this paper is STM32F103. The STM32F103 has a 32-bit ARM microcontroller with a Cortex-M3 processor using the ARMv7-M architecture [14]. The Cortex-M3 processor execute ARM order set and operates up to 72 MHz. The STM32F103 MCU has 2 DMA Control Unit, DMA1 has 7 channels and DMA2 has 5 channels. In addition, STM32F103 contains 3 SPI, 2 I2S, etc., also almost all FSMC interfaces can tolerate 5V signals.

The N-way voltage regulator module has perfect protection circuit, current limit, thermal shutdown circuit, etc. To reduce peripheral devices, form an efficient and stable voltage regulator circuit, the 3A current output step-down switching integrated voltage regulator chip produced by Texas Instruments is selected in this paper. It is capable of delivering 3A drive current with excellent linearity and load regulation. The output regulator circuit is shown in Fig. 2.

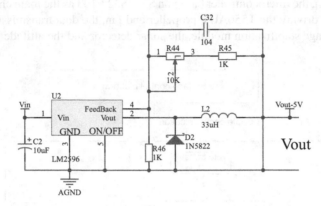

Fig. 2. Output Voltage Regulator Circuit.

The control of the underwater vehicle is mainly attitude control and depth control, so the MPU9250 attitude sensor of InvenSense Company is selected for the attitude control of the underwater vehicle in this paper, circuit schematic graph is appeared in Fig. 3.

The MPU9250 is a QFN packaged multichip module (MCM). MPU9250 includes triaxial acceleration, triaxial gyroscope and triaxial magnetometer. Therefore, As a nine-axis motion tracking device with three 16-bit acceleration AD outputs, three 16-bit gyroscope AD outputs, and three 6-bit magnetometer AD outputs. Precision slow and fast motion tracking provides customers with a full range of programmable gyro parameter options (±250, ±500, ±1000, and ± 2000°/second (DPS)), programmable acceleration parameter options of ± 2 G, ± 4 G, ± 8 G, and ±16 G, and maximum magnetometer capability of ± 480 μT. The attitude sensor is used to detect whether the underwater vehicle is in a balanced state when it is suspended.

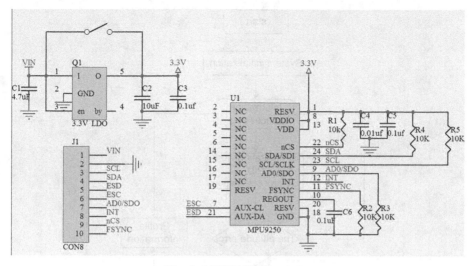

Fig. 3. MPU9250 Schematic.

The DC brushless motor is driven by the DC motor drive module, which is an H-bridge drive circuit. The L298N Double H bridge engine driver chip is selected in this paper, in which each circuit is capable of providing a current of 2A, the power supply can provide a voltage of 2. 5-48 V, and the logic part needs a voltage of 5 V. The robot drive circuit consists of two drive modules, which are respectively connected to the propeller and fan of the rov. The thruster is T550AUV thruster, which controls the robot to move forward and backward, traverse and lurk. The fan is mainly used for removing the interference of underwater silt when the underwater robot comes to the bottom of the pile. Sonar detection sensors are used to feed back data and detect karst caves.

The sonar detector adopts the sonar detection system with STM32F103 as the main control chip, which is independently designed by the laboratory. The sonar detection system detects the karst cave at the bottom of the pile through the sonar elastic wave excited in the mud and water environment. The JL-SONAR is used to transmit and collect the sonar signals, and then the PBCA signal analysis software is used to analyze and process the data, so as to realize the survey of the underground geological environment.

3.2 Control System Software Design

Computer program plan of the control framework of the underwater vehicle is an important part to ensure the better control of the motion state of the vehicle. This design mainly uses the STM32 series MCU, and its supporting software is written by KEIL MDK, including N-way voltage regulation module, drive module, data transmission module, sensor data acquisition module and controller module. The software as a whole is composed of subroutines with single functions. The software control square graph of program is appeared in Fig. 4 below:

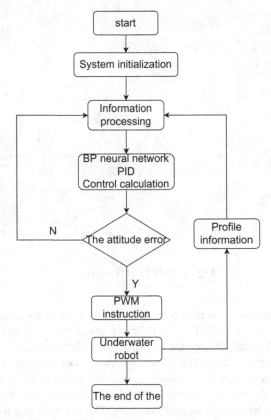

Fig. 4. Graph of underwater vehicle program.

4 Design of PID Control Unit Based on BPNN

4.1 BPNN PID Model and Parameter Determination

There are many control algorithms for underwater robots [8–10]. PID control is the foremost broadly utilized control strategy at show, which has advantages of simple algorithm and strong robustness. The advanced PID control calculation is partitioned into two types: position type and incremental type, and the incremental PID control calculation commonly utilized in engineering is utilized here. The expression is:

$$u(t) = K_p[e(t) + \frac{1}{T_i}\int_0^t e(t)dt + T_d\frac{de(t)}{dt}$$ (1)

where: Kp is the scaling factor, Ti is the amount of time, and Td is the integral time.

An incremental PID control algorithm can be obtain after that integral quantity and the differential quantity are discretize by a certain discretization method:

$$u(k) = K_p[e(k) - e(k-1)] + K_i e(k) + K_d[e(k) - 2e(k-1) + e(k-2)]$$ (2)

The principle block diagram of the PID Control Unit is appeared within the Fig. 5:

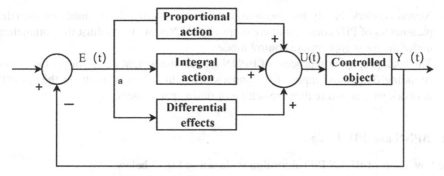

Fig. 5. Principle square graph of PID Control Unit.

Although PID controller has many advantages, there are also many limitations in practical applications. Due to the limitation of the physical characteristics of the components, the original message obtained by the PID controller deviates from the actual value, also the control function generated by the PID control unit deviates from the theoretical value. If PID control needs to attain way better control impact, it must frame the relationship of common participation and shared confinement in control quantity by altering the extent, necessarily and differential control capacities. Therefore, the PID controller of the original underwater robot is organically combined with BPNN, and the capability to learn automatically of neural network is utilized to adjust the parameters [7]. Because BPNN has the capacity of arbitrary nonlinear representation, the PID controller built on BPNN can realize the PID control unit with the most excellent combination by learning the framework performance. BPNN has the capacity to approximate any nonlinear work, its structure and learning calculation are basic and clear. The PID CU built on BPNN Kp, Ki, Kd parameters can discover the ideal control P, I, D parameters through the learning of the neural network itself. In this design, the most commonly used BP network in engineering is used to construct the BPNN PID control. Schematic diagram of PID control unit built on BPNN is appeared in Fig. 6.

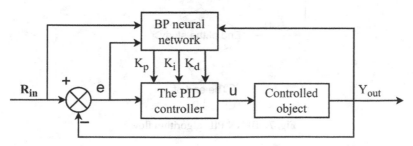

Fig. 6. Schematic diagram of PID control unit built on BPNN.

(1) PID control unit: The controlled object embraces closed-loop control, the parameter of PID Kp, Ki and Kd are balanced online.

(2) Neural network: Firstly, the operating state of the system is determined, and then the parameters of PID control unit are output by BPNN, so as to adjust the parameters under the most appropriate control mode.

(3) The back propagation process of BPNN adjusts the weight according to the corresponding output and input of the underwater vehicle, so as to obtain the optimal network weight and realize the setting of online parameters.

4.2 BPNN and PID Design

The flow chart of BPNN PID algorithm is shown in Fig. 7 below.

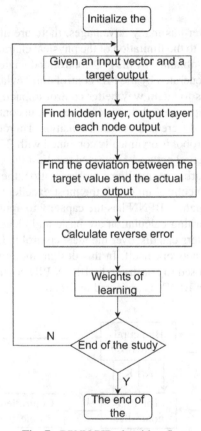

Fig. 7. BPNN PID algorithm flow.

The BPNN schematic is appeared in Fig. 8. Here, the BPNN is used, including an import level node J, a conceal level node I, and an export level node L. The import level node compares to the chosen framework working state variables, and the export level node compares to the three movable parameters Kp, Ki, Kd of the PID controller.

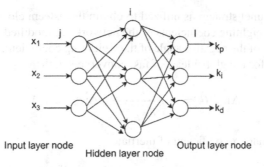

Fig. 8. BPNN architecture diagram.

Therefore, the Excitation function of the neurons in a conceal level node can be the obverse and the reverse symmetric Sigmoid function.

From BPNN architecture diagram, we can see that the input layer is:

$$O_j^{(1)} = x(j) \quad j = 1, 2, 3, 4 \tag{3}$$

where: x(j)-the import of input node; $O_j^{(1)}$-the import of input node.

Implicit node import and export are:

$$net_i^{(2)}(k) = \sum_{j=0}^{m} w_{ij}^{(2)} o_j^{(1)}, \quad i = 1, 2, 3, 4, 5 \tag{4}$$

$$o_i^{(2)}(k) = g[net_i^{(2)}(k)], \quad i = 1, 2, 3, 4, 5 \tag{5}$$

where: $w_{ij}^{(2)}$-export node weighting coefficient; g(x)-Sigmoid function, $g(x) = \tanh(x)$

The come in and go out values of the output node:

$$et_l^{(3)}(k) = \sum_{i=0}^{q} w_{li}^{(3)} o_i^{(2)}, \quad l = 1, 2, 3 \tag{6}$$

$$o_l^{(3)}(k) = f[net_l^{(3)}(k)], \quad l = 1, 2, 3 \tag{7}$$

$$o_1^{(3)}(k) = k_p \tag{8}$$

$$o_2^{(3)}(k) = k_i \tag{9}$$

$$o_3^{(3)}(k) = k_d \tag{10}$$

where: $w_{li}^{(3)}$-weighting coefficient of output node; f(x)-activation function,
$f(x) = 1/2 + [1 + \tanh(x)]$;

The objective work of the BP neural network is taken as:

$$E(k) = \frac{1}{2}[r(k) - y(k)] \tag{11}$$

The angle plummet strategy is utilized to obtain the esteem closest to the objective function, and the weighting coefficients of the network are modified. Press E(k) to look and alter the course of the negative angle of the weighting coefficient, and add an inertial term that can hunt for the global least of fast convergence, then:

$$\Delta w_{li}^{(3)}(k) = -\eta \frac{\partial J}{\partial w_{li}^{(3)}} + \alpha \Delta w_{li}^{(3)}(k-1) \tag{12}$$

where: η- earning rate; α-coefficient of inertia;

$$\frac{\partial E(k)}{\partial w_{li}^{(3)}} = \frac{\partial E(k)}{\partial y(k)} \bullet \frac{\partial y(k)}{\partial u(k)} \bullet \frac{\partial u(k)}{\partial o_l^{(3)}(k)} \bullet \frac{\partial o_l^{(3)}(k)}{\partial net_l^{(3)}(k)} \bullet \frac{\partial net_l^{(3)}(k)}{\partial w_{li}^{(3)}} \tag{13}$$

Since the partial derivative of y(k) with respect to u(k) is unknown, the approximation function is used here for calculation. To reduce the error in the calculation process, we adopt the learning rate to reduce the error. It can be obtained from the formula:

$$\frac{\partial u(k)}{\partial o_1^{(3)}(k)} = e(k) - e(k-1) \tag{14}$$

$$\frac{\partial u(k)}{\partial o_2^{(3)}(k)} = e(k) \tag{15}$$

$$\frac{\partial u(k)}{\partial o_3^{(3)}(k)} = e(k) - 2e(k-1) + e(k-2) \tag{16}$$

Therefore, according to the above method, the formula for calculating the weight of the export node of BPNN:

$$\Delta w_{li}^{(3)}(k) = \eta \delta_l^{(3)} o_i^{(2)}(k) + \alpha \Delta w_{li}^{(3)}(k-1) \tag{17}$$

$$\delta_l^{(3)} = e(k)\text{sgn}\frac{\partial y(k)}{\partial u(k)} \bullet \frac{\partial u(k)}{\partial o_l^{(3)}(k)} \bullet f'[net_l^{(3)}(k)] \tag{18}$$

In the same way, the calculation formula of the hidden layer weight of the BP neural network can be calculated as follows:

$$\Delta w_{ij}^{(2)}(k) = \eta \delta_i^{(2)} o_j^{(1)}(k) + \alpha \Delta w_{li}^{(2)}(k-1) \tag{19}$$

$$\delta_i^{(2)} = g'[net_i^{(2)}(k)] \sum_{l=1}^{3} \delta_l^{(3)} w_{li}^{(3)}(k) \tag{20}$$

where: $f(x) = f(x[1 - g(x)]); g(x) = [1 - g^2(x)]/2;$

To sum up, BPNN and PID control calculation can be summarized as [11, 17]: On the premise of the initial PID control calculation, the BPNN architecture is firstly chosen, that's, the number of J, I and L nodes is chosen, and after that K = 1 is decided by learning rate and inertia coefficient. The real come in and go out values are gotten by inspecting, and the framework blunders are calculated. The real values of the framework examined over are normalized as the input values of BPNN. The over equation is utilized to calculate the import and export values of each node of the neural organize to induce the parameters of the PID control unit, and after that calculate the esteem of the control unit u(k) and the weight of each node. At last, the k value is utilized to input the input amount and repeat the over steps, and at long last get the foremost fitting parameters.

4.3 Algorithm Verification

The PID control unit of BPNN is modeled by Matlab [15, 18], that's, the control calculation is realized by S-function programming in Simulink [11, 16], where the exchange function is: $G(s) = 1/(20s + 1)$, and the fitting result of parameters is appeared in Fig. 9.

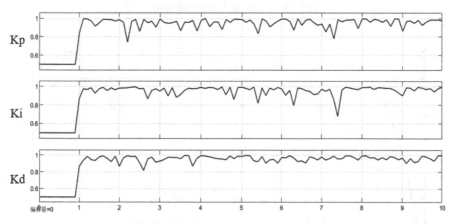

Fig. 9. Kp, K_i and K_d parameter curve.

The simulation written by S function is helpful and quick. It can be seen from Fig. 10 and Fig. 11 that the control framework simulation show of BPNN-PID control law realized by S function can make great utilize of its possess programmed learning capacity to alter PID three parameters. Agreeing to the reenactment outcpme, compared with the conventional PID control, BPNN-PID control can realize online alteration of parameters more rapidly and get the foremost appropriate parameters. PID parameters are appeared in Fig. 10. To sum up, the advantages of PID control unit built on BPNN include: (1) BPNN-PID control unit isn't like other complex control systems to set up numerical models. (2) The parameter setting of the controller is convenient. (3) Good dynamic and static characteristics.

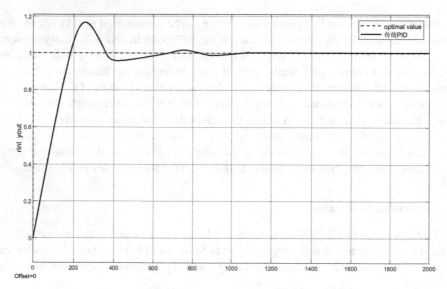

Fig. 10. Traditional step response.

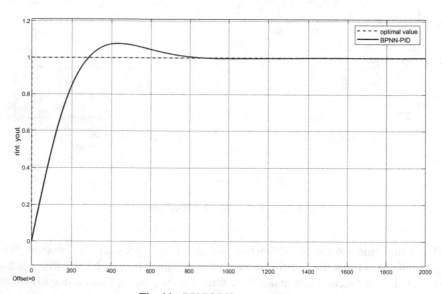

Fig. 11. BPNN-PID step response.

5 Concluding Remarks

The above handle mainly analyzes the preferences and impediments of the traditional PID controller. in arrange to overcome the impediments of the traditional PID controller which is not adaptive to uncertainty, make the controller have better adaptability. The controller itself can automatically alter the parameters of the control unit, use the capability to learn automatically of neural network to alter the arguments online. At long last, a PID controller which BPNN employments its own learning capacity to adjust the ideal parameters is shaped. It adjusts the nonlinear object well and enhance the tracking of the system, and its parameters do not need to know the parameters and digital modelling of the controlled target, so it has solid versatility to different objects.

In spite of the fact that the controller devised in this paper has great steadiness and anti-interference capacity within the confront of the obstructions of submerged environment, due to the complexity of submerged mud environment, the robot still should carry sonar discovery device in actually project. Therefore, the afterward study will assist optimize the robot gadget structure to reduce the development resistance and precision the stability and accuracy of the submerged robot.

Acknowledgments. The realization of the above content is partly bolstered by the science and innovation ventures of Shandong Hi-speed Group (Nos. KJ-2019-QLJTJT-07). We also wish to acknowledge the support of National Natural Science Foundation of China under Grant (Nos. 61903227).

References

1. Wan, L., Tang, W., Li, Y.: BP neural network S plane control for autonomous underwater vehicle. Industrial Instrumentation Automation (02), 13-17 (2019)
2. Feng, X., Liu, Y.: Present situation and trend of autonomous underwater vehicle research and development. Chinese High Technol. Letters (9), 55-59 (1999).
3. Gan, Y., Wang, L., Liu, J., Xu, Y.: The embedded basic motion control systemof autonomous underwater vehicle. Robot **26**(3), 25-29 (2004)
4. Wang, L.: Research on Motion Control of a Underwater Robot. University Of Science and Technology Of China (2018)
5. Chang, R., He, X.: Design and implementation of control system for a kind oflightweight underwater robot. J. Hefei Univ. Technol. (Natural Science) **42**(07), 935-939 (2019)
6. Zhang, H., Tian, J., Xiong, J., Zhao, Y., Shi, K.: Design and analysis on motion control system of underwater vehicle. Computer Syst. Appl. **27**(12), 83-89 (2018). https://doi.org/10.15888/j.cnki.csa.006690
7. Wang, Y., Chen, X.: Simulation of AGV speed control system based on BP neural network PID control. Manufacturing Automation **43**(06), 63–66+133 (2021)
8. Hong, L.: Research on adaptive control algorithm of underwater rescue robot. Techniques Automation Appl. **39**(11), 91-95 (2020)
9. Wang, T., Qi, X.: Application of GENETIC algorithm PID control in AUV attitude adjustment. Techniques of Automation Appl. **39**(04), 8-14 (2020)
10. Wang, J., Song, Y., Wei, G., Yuan, B.: Application of cascade PID control in pitch control system of underwater robot. J. Univ. Shanghai for Science Technol. **33**(3), 229-235 (2017). 10. 13255/j.carol carroll nki jusst.2017.03.005.
11. Han, J., Ma, J., Sang, H.: Research on obstacle avoidance of mobile robot based on bp neural network PID control. Sci. Technol. Vision (03), 34-35 (2020). 10. 19694/j.cnki.issn2095-2457.2020.03.014
12. Guo, X., Tang, H., Xu, C.: Construction and exploration of underwater robot standard system framework. Ocean Dev. Manage. **38**(01), 53-56 (2021)

13. Wang, G.: Design and analysis of underwater robot frame structure based on ANSYS. Mechanical Res. Appl. **33**(03), 163-165 (2020). https://doi.org/10.16576/j.cnki.1007-4414. 2020.03.046
14. Shi, G.: Development and application of eMMC card driver based on STM32F103. Information Technol. Informatization (04), 152-154 (2021)
15. Gu, Y.: Modeling and control of underwater biped octopus robot based on deep reinforcement learning. Hangzhou Dianzi University (2021). https://doi.org/10.27075/d.cnki.ghzdc.2021. 000550
16. Hu, L.: Modeling and simulation of underwater unmanned robot based on SIMULINK. J. Lanzhou University of Arts and Science (Natural Science Edition) **32**(03), 59–63 (2018). https://doi.org/10.13804/j.cnki.2095-6991.2018.03.011
17. Zhao, H., Wang, S., Yin, Q., Fu, T.: Research on liquid mixing system of poultry house based on NEURAL network PID algorithm control. Computer DigitalEng. **50**(03), 498-502 (2022)
18. Yang, Y., Hu, E.: BP neural network PID controller based on S function and Simulink simulation. Electronic Design Eng. **22**(04), 29–31+35 (2014). https://doi.org/10.14022/j.cnki.dzs jgc.2014.04.014

A Multi-channel Fusion Method Based on Tensor for Rolling Bearing Fault Diagnosis

Huiming Jiang[1(✉)], Song Li[1], Yunfei Shao[2], An Xu[1], Jing Yuan[1], and Qian Zhao[1]

[1] University of Shanghai for Science and Technology, 516 Jungong Rd, Shanghai 200093, China
{Hmjiang,yuanjing,qzhao}@usst.edu.cn
[2] Shanghai Aircraft Manufacturing Co., Ltd, 919 Shangfei Road, Shanghai 201324, China

Abstract. Aiming at the limitation that the single channel source of rolling bearing contains incomplete fault information and the strong correlation and coupling relationship between different channels, a multi-channel fusion diagnosis method for rolling bearing based on multi-domain tensor feature and support tensor machine is proposed. In order to fuse the bearing fault information contained in multi-channel vibration signals, firstly, multi-domain features such as time domain features, wavelet domain features, entropy domain features and trigonometric function features of each channel signal are extracted respectively, and the third-order tensor with channel, feature and time as dimensions is constructed. Then, the CP rank is determined by the kernel consistent diagnosis method, and the high-order fault state information contained in the multi-domain tensor is extracted by the low rank tensor approximation method based on CP decomposition to obtain the multi-domain tensor characteristics of multi-channel rolling bearing. Finally, the multi-domain tensor feature is input into the support tensor machine for bearing fault diagnosis. The example of bearing fault diagnosis shows that the multi-channel fusion diagnosis method based on multi-domain tensor feature and support tensor machine proposed in this paper can make full use of the high-order structural features of multi-channel information source, fuse multi-channel information to more comprehensively characterize the fault state of bearing, and obtain higher diagnosis accuracy than single channel information source.

Keywords: Rolling bearing · Multi-domain tensor characteristics · Tensor decomposition · Low rank tensor approximation · Support tensor machine

1 Introduction

Rolling bearing is not only the core component of rotating equipment, but also the main source of mechanical faults of rotating equipment. As an important basis for rolling bearing health monitoring and fault diagnosis, single channel vibration signal can not ensure the completeness of rolling bearing state information. Therefore, there is limited room to improve the fault diagnosis rate based on single channel vibration signal feature extraction and fault diagnosis method. However, each channel of the multi-channel vibration signal contains the health status information of the rolling bearing, and each channel is not independent of each other and has the characteristics of correlation and coupling.

© The Author(s), under exclusive license to Springer Nature Singapore Pte Ltd. 2022
H. Zhang et al. (Eds.): NCAA 2022, CCIS 1638, pp. 123–137, 2022.
https://doi.org/10.1007/978-981-19-6135-9_10

Therefore, the multi-channel vibration signal has higher information completeness than the single channel signal. At present, more and more researchers are committed to fault diagnosis based on multi-sensor information fusion, and explore new information fusion forms and algorithms [1–3]. Zhudanchen et al. [4] transformed the multi-channel vibration signals measured by multiple sensors into gray-scale images, and then input them into convolutional neural network for rolling bearing fault recognition; Hao et al. [5] extracted the spatial and temporal characteristics of multi-sensor measured multi-channel vibration signals by using one-dimensional convolution long-term and short-term memory network, and combined them for bearing fault diagnosis; Azamfar et al. [6] used the two-dimensional convolutional neural network to fuse the multi-channel current data measured by multiple sensors to achieve gearbox fault detection and diagnosis. However, the above methods are simply for multi-channel data fusion, and do not consider the internal correlation characteristics between channels, so the degree of information fusion is insufficient. As a powerful tool for multi-channel data, it has been widely concerned in the field of high-dimensional data signal processing and machine learning in recent years [7–9].

As a powerful tool for processing multidimensional signals, tensor decomposition theory has been widely studied and applied in recent years [10]. Sidiropoulos and Lathauwer summarized the application of tensor decomposition theory in signal processing and machine learning, and pointed out that multidimensional data fusion based on tensor decomposition is a research direction with great potential and wide application space [11]. Hu chao fan et al. [12] constructed the time domain signal, frequency and channel of rolling bearing into a third-order tensor data, reconstructed the target tensor by truncating the high-order singular value decomposition, and realized multi-channel signal fusion and noise reduction. Feng et al. [13] studied a robust tensor principal component analysis method based on tensor singular value decomposition, which can effectively extract low rank sparse components in multi-channel signals.

At the same time, as a high-order extension of vector machine learning model, tensor based machine learning model has developed rapidly in recent years. The main difference between tensor based machine learning model and vector based machine learning model lies in the data representation and operation rules, while the core idea and related models are basically similar. Theory and practice have proved that the machine learning model based on tensor representation requires fewer model parameters and is more suitable for small samples. Moreover, the data input, output and operation rules of the machine learning model based on tensor representation are based on tensor, which can avoid the damage to the structure of tensor data and information loss after vectorization. Support tensor machine (STM) is one of the successful practices.

To sum up, a multi-channel fault diagnosis method of rolling bearing based on multi-domain tensor characteristics and STM is proposed in this paper. Firstly, in order to more comprehensively characterize the fault state information in the vibration signal, based on the feature extraction in time domain, wavelet domain, entropy domain and trigonometric function domain, a high-order tensor integrating channel structure information, multi domain feature information and fault evolution information with time is constructed. Secondly, the high-order tensor feature extraction method is studied, and the CP rank is determined by kernel consistent diagnosis method, and the low rank tensor approximation method based on CP decomposition is used to extract the high-order features of multi domain tensor, the redundant information in multi domain tensor features is effectively removed, and the representation ability of multi domain tensor features to fault information is improved. Finally, combined with the support tensor machine, the multi-channel fault diagnosis of rolling bearing is realized, and the effectiveness and superiority of the proposed method are verified based on the simulated fault test data of rolling bearing.

2 Tensor Theory

As a high-dimensional generalization of scalar, vector and matrix, tensor can not only intuitively represent high-dimensional data, but also maintain the internal correlation characteristics of data to the greatest extent. It has natural advantages in processing multi-channel data. Tensor decomposition can simplify the tensor structure and extract the principal component factors of each order while maintaining the high-order structure of the data, so as to mine the information contained in the tensor [14].

One N order tensor can be expressed as $\mathcal{X} \in \mathbb{R}^{I_1 \times I_2 \times \cdots \times I_N}$, where, N it is called tensor order or modulus, which I_n represents the dimension of tensor n order, and $n = 1, 2, \cdots, N$. an element of tensor can be expressed as $\mathcal{X}_{i1i2\cdots iN}, 1 \leq i_n \leq I_n \cdot N$. N Modal vectorization should also be completed by modal matrix.

2.1 Tensor Algebra

Matrix and Vectorization of Tensors
An N order tensor can be matrixed (vectorized) into a matrix (vector) in multiple dimensions. The n modal matrix of a tensor $\mathcal{X} \in \mathbb{R}^{I_1 \times I_2 \cdots I_N}$ can be expressed as $\mathbf{X}_{(n)} \in \mathbb{R}^{I_n \times I_1 \cdots I_{n-1} I_{n+1} \cdots I_N}$. The n mode vectorization of a tensor $\mathcal{X} \in \mathbb{R}^{I_1 \times I_2 \cdots I_N}$ can be expressed as $x_{(n)} \in \mathbb{R}^{I_n I_1 \cdots I_{n-1} I_{n+1} \cdots I_N}$. N modal vectorization should also be completed by n modal matrix.

Tensor Inner Product and Norm
Given the tensor $\mathcal{X}, \mathcal{Y} \in \mathbb{R}^{I_1 \times I_2 \cdots I_N}$, the inner product (scalar product) of the two is defined as:

$$\langle \mathcal{X}, \mathcal{Y} \rangle = \sum_{i_1} \sum_{i_2} \cdots \sum_{i_N} x_{i_1 i_2 \cdots i_N} y_{i_1 i_2 \cdots i_N} \tag{1}$$

The Frobenius norm of the tensor $\mathcal{X} \in \mathbb{R}^{I_1 \times I_2 \cdots \times I_N}$ is defined as:

$$\|\mathcal{X}\|_F = \sqrt{\langle \mathcal{X}, \mathcal{X} \rangle} \tag{2}$$

Matrix Product of Tensor
Given an N order tensor $\mathcal{X} \in \mathbb{R}^{I_1 \times I_2 \cdots \times I_N}$ and a matrix $\mathbf{A} \in \mathbb{R}^{I_n \times R}$, the matrix product of $\mathcal{X}$和$\mathbf{A}$is defined as:

$$\mathcal{Y} = \mathcal{X} \times_n \mathbf{A} \in \mathbb{R}^{I_1 \times \cdots I_{n-1} \times R \times I_{n+1} \times \cdots I_N} \tag{3}$$

Both sides of the equation are tensors.

2.2 Tensor Decomposition

Tensor decomposition mainly includes CANDECOMP/PARAFAC (CP) decomposition and tucker decomposition. In CP decomposition, the tensor is decomposed into the linear sum of rank-1 tensors. Before introducing the definition of CP decomposition, the definition of rank 1 tensor is given. Suppose a tensor can be expressed as:

$$\mathcal{X} = b^1 \circ b^2 \circ \cdots b^N \tag{4}$$

where, $\mathcal{X} \in \mathbb{R}^{I_1 \times I_2 \cdots I_N}$, $b^n \in \mathbb{R}^{I_n}$, $y_{i_1,\cdots i_N} = b_{i_1}^1 \cdots b_{i_N}^N$, $\mathbb{R}$ represents the real number field. We call the tensor expressed in formula (4) as rank 1 tensor

The CP decomposition of an N order tensor $\mathcal{X} \in \mathbb{R}^{I_1 \times I_2 \times \cdots \times I_N}$ is defined as:

$$\mathcal{X} \approx \sum_{r=1}^{R} \lambda_r b_r^1 \circ b_r^2 \circ \cdots b_r^N$$
$$= \mathcal{A} \times_{1m} \mathbf{B}_1 \times_{2m} \mathbf{B}_2 \cdots \times_{Nm} \mathbf{B}_N \tag{5}$$

where, $\lambda_r = \mathcal{A}_{r,r,\ldots,r}$, $r \in [1, R]$ is the element of the diagonal core tensor $\mathcal{A} \in \mathbb{R}^{R \times R \ldots \times R}$, $\mathbf{B}_n = [b_1^n, b_2^n, \cdots b_R^n] \in \mathbb{R}^{I \times R}$ is the factor matrix, and X_n represents the principal components of each order of the tensor. Represents the - modal matrix product of tensors. For example, the CP decomposition process of the third-order tensor $\mathcal{X} \in \mathbb{R}^{I_1 \times I_2 \times I_3}$ is shown in Fig. 1.

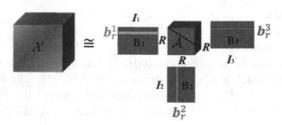

Fig. 1. CP decomposition of 3rd-order tensor

2.3 Low Rank Tensor Approximation

For a matrix, Eckart and young proved that [15, 16] can give an optimal rank - r approximation through the main factors of singular value decomposition (SVD) Similarly, taking the third-order tensor as an example, it can be expressed in the form shown in Eq. (6).

$$\mathcal{X} = \sum_{r=1}^{R} \lambda_r a_r \circ b_r \circ c_r + \mathcal{E} \Rightarrow$$
$$\mathcal{X} \approx \sum_{r=1}^{R} \lambda_r a_r \circ b_r \circ c_r \tag{6}$$

where, R is the rank of CP decomposition and ε represents the redundant components (noise, etc.) in the tensor. Therefore, the low rank tensor based on CP decomposition is approximately equivalent to finding the best rank R to minimize the following objective function:

$$\min_{\mathbf{A},\mathbf{B},\mathbf{C}} \left\| \mathcal{X} - \sum_{r=1}^{R} \lambda_r a_r \circ b_r \circ c_r \right\|_F^2 \tag{7}$$

where $\mathbf{A} = [a_1, \cdots, a_R]$, $\mathbf{B} = [b_1, \cdots, b_R]$, $\mathbf{C} = [c_1, \cdots, c_R]$, is called the factor matrix. $\| \cdot \|_F$ Representation *Frobenius* norm.

3 Multi-channel Fault Diagnosis Method of Bearing Based on Tensor Feature and STM

Based on tensor representation, it is an important prerequisite for fault diagnosis to fuse multi domain features and multi-channel high-order correlation information that can reflect the running state of rolling bearings, and complete the transformation from one-dimensional vibration signal to high-order tensor data. The tensor data based on multi-channel fault vibration signals inevitably contains a small amount of noise and other redundant information. In order to eliminate these noise and redundant information, effectively extract effective feature information from the tensor data without damaging the structure of the tensor data, low rank tensor approximation is an effective method.

In order to effectively fuse the multi-channel vibration signals of rolling bearings for fault diagnosis, a multi-channel fusion diagnosis method of rolling bearings based on multi domain tensor features and STM is proposed in this paper. First, the constructed channel × features × time of the third-order tensor, which establishes the visual expression and high-dimensional relationship of multi-channel vibration signals; Then, the rank of CP decomposition is determined by the kernel consistent diagnosis method, and the low rank tensor based on CP decomposition is used to approximately mine the bearing running state information contained in the tensor data to extract the multi domain tensor features; Finally, the multi domain tensor features are used as the training and test samples of STM model to realize the fault diagnosis based on multi-channel vibration signals of rolling bearings. The method flow is shown in Fig. 2, and the specific steps are as follows.

(1) The third-order tensor is constructed by using multi-channel signals. Firstly, the monitoring data under different channels are divided into multiple time periods by using sliding rectangular window. Then, multi domain features such as time domain, wavelet domain, entropy domain and trigonometric function domain are extracted from the vibration signals in different channels in each period, so as to construct the "channel" × features × time", thethird − ordertensorof $\mathcal{X} \in \mathbb{R}^{I_1 \times I_2 \times I_3}$.

(2) Determine the optimal CP rank R. It is NP hard to determine the CP rank of the tensor, that is, in Eq. (5), and the CP decomposition of the tensor needs to know the rank of the tensor in advance. The kernel consistent diagnosis method was proposed by Bro and Kiers [17]. In recent years, it has been used to determine the number of CP ranks in the CP decomposition model and the number of multilinear ranks in Tucker decomposition. According to the definition in reference [18], the formula of nuclear consistency diagnosis is expressed as:

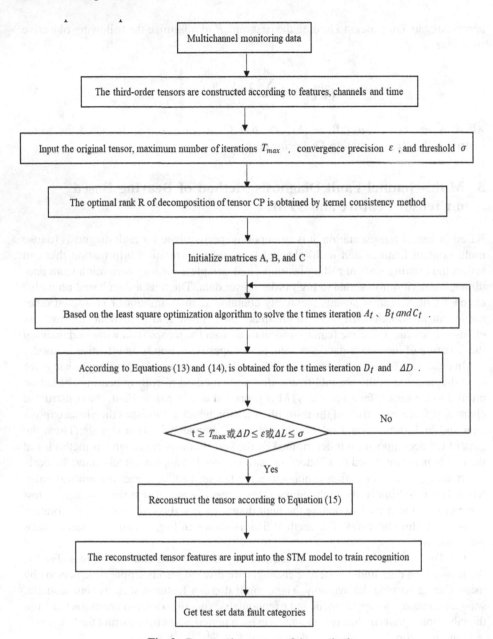

Fig. 2. Propose the process of the method

$$core(r) = \left(1 - \frac{\|\mathcal{G}_r - \mathcal{T}_r\|_F^2}{r}\right) \times 100\% \tag{8}$$

In the formula, $\mathcal{A}$ it is the kernel tensor obtained after CP decomposition in formula (5) $\mathcal{T}_r$, and $\mathcal{G}_r = (B_n \otimes \cdots \otimes B_1)^\dagger \mathcal{T}_r$, † represents the pseudo inverse, r which is the estimated CP rank. Usually, when the $r + 1$ kernel consistency value $core(r + 1)$ is less than a threshold η, the best rank r is at this time. Generally, it is considered that the value range of threshold is $70\% \leq \eta \leq 90\%$. Therefore, this paper selects the threshold $\eta = 85\%$.

(3) Set the maximum number of iterations $T_{\max}$, convergence accuracy ε, threshold σ and other parameters. After the optimal rank R is determined, the high-order feature extraction is equivalent to finding the matrix $\mathbf{A} \in \mathbb{R}^{I_1 \times R}, \mathbf{B} \in \mathbb{R}^{I_2 \times R}, \mathbf{C} \in \mathbb{R}^{I_3 \times R}$ to minimize the following objective function

$$L(\mathcal{D}, \mathbf{A}, \mathbf{B}, \mathbf{C}) = \|\mathcal{X} - \mathcal{D}_1 \times_1 \mathbf{A}_t \times_2 \mathbf{B}_t \times_3 \mathbf{C}\|_F^2 \tag{9}$$

where, diagonal core tensor $\mathcal{D} \in \mathbb{R}^{R \times R \times R}$. Equation (9) shows that the approximate CP decomposition of the whole tensor is obtained according to the tensor, and the decomposed $\mathbf{A}, \mathbf{B}, \mathbf{C}$ and $\mathcal{D}$ are used to reconstruct the whole tensor as a high-order feature. In this article, $T_{\max}$ set to 100, ε set to 10^{-8}, σ and set to 10^{-8}.

(4) Initialize the matrix $\mathbf{A}, \mathbf{B}, \mathbf{C}$. Set the matrix $\mathbf{A}, \mathbf{B}, \mathbf{C}$ as the identity matrix and record it as $\mathbf{A}_0, \mathbf{B}_0, \mathbf{C}_0$.

(5) Based on the iterative calculation $\mathbf{A}, \mathbf{B}, \mathbf{C}$ and $\mathcal{D}$ of the least squares optimization algorithm, the objective function of each parameter is as follows:

$$\mathbf{A} \leftarrow \arg\min_{\mathbf{A}} \left\|\mathbf{X}_1 - (\mathbf{C} \odot \mathbf{B})\mathbf{A}^T\right\|_F^2 \tag{10}$$

$$\mathbf{B} \leftarrow \arg\min_{\mathbf{B}} \left\|\mathbf{X}_2 - (\mathbf{C} \odot \mathbf{A})\mathbf{B}^T\right\|_F^2 \tag{11}$$

$$\mathbf{C} \leftarrow \arg\min_{\mathbf{C}} \left\|\mathbf{X}_3 - (\mathbf{B} \odot \mathbf{A})\mathbf{C}^T\right\|_F^2 \tag{12}$$

$$\mathcal{D} \leftarrow 归一化\mathbf{A}、\mathbf{B}、\mathbf{C} \tag{13}$$

where $\odot$ is Khatri Rao product operation.

(6) Judge whether the following three conditions are true. ① Reach the set number of iterations $T_{\max}$; ② The difference t times between $t - 1$ times iterations of objective function is less than σ; ③ Norm satisfaction of the difference between diagonal kernel tensors of iteration t times and $t - 1$ times

$$\Delta\mathcal{D} = \|\mathcal{D}_t - \mathcal{D}_{t-1}\|_F < \varepsilon \tag{14}$$

(7) As long as one of the above three conditions is satisfied, the iteration stops, According to the t times iteration obtained $\mathbf{A}_t, \mathbf{B}_t, \mathbf{C}_t$ and $\mathcal{D}_t$, the tensor is reconstructed based on the following formula

$$\mathcal{Y} = \mathcal{D}_t \times_1 \mathbf{A}_t \times_2 \mathbf{B}_t \times \mathbf{C}_t \tag{15}$$

(8) Identification and classification based on STM. After obtaining the multi domain tensor features, the tensor features are divided into training set and test set and input into STM training and test model, so as to complete the multi-channel vibration signal fault diagnosis of rolling bearing.

4   Experiment Verification

A typical fault simulation test-bed of gear reducer bearing is used to verify the proposed method. As shown in Fig. 3, the test bench is mainly composed of variable frequency motor, single-stage gear reducer, magnetic particle brake, tension controller and frequency converter. The input shaft is a high-speed shaft, the speed is equal to the motor speed, the output shaft is a low-speed shaft, and the transmission ratio of the gearbox is $39/28 \approx 1.393$. During the test, the bearing 3 corresponding to the sensor installation position of channel 3 is the test bearing, and the other bearings are normal bearings. The model of the test bearing is 7203 single row angular contact ball bearing, and the structural parameters are shown in Table 1. Install the four acceleration sensors on the four bearing end covers of the gear reducer to collect vibration data. The sampling frequency is 51200 Hz and the sampling time is 10 s. The vibration signals of bearings in normal state H, inner ring fault I and outer ring fault O at 750 rpm, 1000 rpm, 1250 rpm and 1500 rpm are collected as the original data.

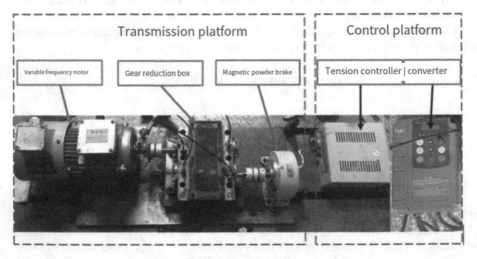

Fig. 3. Real view of the test bed

Taking the multi-channel vibration signal of inner ring fault as an example, the time domain waveform and FFT amplitude spectrum of the vibration signal under four channels are drawn respectively, as shown in Fig. 4. It can be seen from Fig. 4 that the time domain waveforms of signals in different channels reflect the existence of fault impact, but due to the difference of vibration propagation paths, the vibration signal waveforms in different channels are quite different. At the same time, from the amplitude spectrum

Table 1. Main structural parameters of rolling bearing 7203

Type	Pitch diameter D (mm)	Rolling element diameter d (mm)	Number of rolling elements Z (individual)	contact angle θ (degree)
7203	28.5	7.14	10	15

of signals in different channels, it can be seen that the four channels contain the fault impact frequency band, but the resonance frequency band distribution and amplitude are also different. It can be concluded that each channel of the multi-channel vibration signal of the rolling bearing contains the main state information of the bearing, but the degree of representation is different. The information contained in each channel is related, coupled and complementary to each other. Therefore, the multi-channel vibration signal of rolling bearing used in this paper has higher information completeness than the single channel vibration signal.

For the above bearing vibration signals, the third-order tensor is constructed according to the method described in Fig. 3, and the 30-dimensional features shown in Table 2 are extracted from the vibration signals of each channel, including 10-dimensional time domain features, 16-dimensional wavelet domain features, 2-dimensional entropy domain features and 2-dimensional trigonometric function features [18]. The length of each data point is 1024. Finally, constructed channel × features × time of the third-order tensor, the size of each tensor sample is $4 \times 30 \times 5$, and 50 tensor characteristic samples are constructed for each running state, a total of 150 samples.

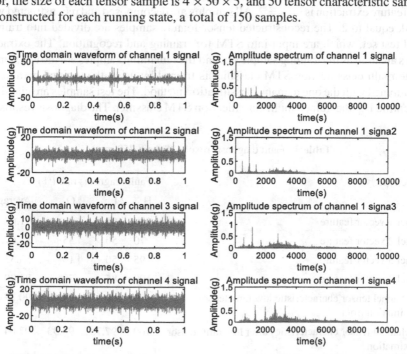

Fig. 4. Time domain waveform and FFT amplitude spectrum of 4-channel vibration signal under inner race fault

Table 2. 30-dimensional features

Feature type	Feature	
Time domain characteristics	F1: average value	F6: Waveform factor
	F2:Root mean square value	F7: Pulse factor
	F3: Peak to peak	F8: Margin factor
	F4: Skewness	F9: Kurtosis Factor
	F5: Kurtosis	F10: Skewness factor
Wavelet feature	F11-F18: Normalized energy of three-layer wavelet packet decomposition	
	F19-F26: Normalized energy entropy of three-layer wavelet packet decomposition	
Entropy domain characteristics	F27: Envelope spectral entropy	F28: Amplitude spectral entropy
Trigonometric function characteristics	F27: Inverse triangular hyperbolic cosine standard deviation (IHC_{sti})	
	F28: Inverse triangular hyperbolic sine standard deviation (IHS_{std})	

According to the kernel consistent diagnosis method, the CP rank of higher-order tensor feature extraction is rank = 2. All samples are decomposed and reconstructed with CP rank equal to 2. The reconstructed tensor feature samples are divided into training set and test set, which are input into STM for training and recognition. The extracted samples are used as the training sample set, and the remaining samples are the test sample set. The multi classification STM classifier is trained by using the optimal parameters and combined with the one to many classification method. The test samples are classified and identified by the trained multi classification STM classifier. The diagnosis results are

Table 3. Fault diagnosis results (Speed: 750 rpm)

Feature type	Fault diagnosis rate (%)			
	H	I	O	Average
Channel 1 vector feature	100	65	20	61.67
Channel 2 vector feature	70	50	35	51.67
Channel 3 vector feature	95	50	15	53.33
Channel 4 vector feature	100	85	70	85.00
Multichannel tensor characteristic low rank tensor approximation front	100	86.67	84.44	90.37
Multichannel tensor characteristics after low rank tensor approximation	100	97.78	93.33	97.04

Table 4. Fault diagnosis results (speed: 1000 rpm)

Feature type	Fault diagnosis rate (%)			
	H	I	O	Average
Channel 1 vector feature	85	75	85	81.67
Channel 2 vector feature	65	90	85	80.00
Channel 3 vector feature	95	80	65	80.00
Channel 4 vector feature	85	65	65	71.67
Multichannel tensor characteristic low rank tensor approximation front	100	97.78	84.44	94.07
Multichannel tensor characteristics after low rank tensor approximation	100	100	100	100

Table 5. Fault diagnosis results (speed: 1250 rpm)

Feature type	Fault diagnosis rate (%)			
	H	I	O	Average
Channel 1 vector feature	85	90	100	91.67
Channel 2 vector feature	95	100	90	95
Channel 3 vector feature	80	90	95	88.33
Channel 4 vector feature	85	35	65	61.67
Multichannel tensor characteristic low rank tensor approximation front	100	97.78	91.11	96.30
Multichannel tensor characteristics after low rank tensor approximation	100	100	100	100

Table 6. Fault diagnosis results (Speed: 1500 rpm)

Feature type	Fault diagnosis rate (%)			
	H	I	O	Average
Channel 1 vector feature	95	75	100	90
Channel 2 vector feature	95	75	100	90
Channel 3 vector feature	95	75	95	88.33
Channel 4 vector feature	85	40	50	58.33
Multichannel tensor characteristic low rank tensor approximation front	100	93.33	86.67	93.33
Multichannel tensor characteristics after low rank tensor approximation	100	100	100	100

shown in Tables 3, 4, 5 and 6. In order to verify the effectiveness of the proposed method, the vector features are extracted from each channel and compared with the support vector machine (SVM) classification algorithm. In Tables 3, 4, 5 and 6, the vector characteristic diagnosis rate, the multi-channel tensor characteristic low rank tensor approximation pre diagnosis rate and the multi-channel tensor characteristic low rank approximation post diagnosis rate under the four channels are listed respectively.

It can be seen from Tables 3, 4, 5 and 6 that the fault diagnosis rates of the multi-channel tensor features at the four speeds before the low rank tensor approximation are

Table 7. Fault diagnosis results (speed: 750 rpm)

Feature type	Fault diagnosis rate (%)			
	H	I	O	Average
Time domain tensor characteristics	100	48.89	86.67	78.52
Tensor characteristics in wavelet domain	95.56	60	88.89	81.48
Tensor characteristics in entropy domain	100	73.33	62.22	78.52
Characteristics of triangular tensor	100	77.78	55.56	77.78
Multi domain tensor characteristics	100	97.78	93.33	97.04

Table 8. Fault diagnosis results (speed: 1000 rpm)

Feature type	Fault diagnosis rate (%)			
	H	I	O	Average
Time domain tensor characteristics	100	40	62.22	67.41
Tensor characteristics in wavelet domain	97.78	57.78	62.22	72.59
Tensor characteristics in entropy domain	100	35.56	26.67	54.07
Characteristics of triangular tensor	100	95.56	100	98.52
Multi domain tensor characteristics	100	100	100	100

Table 9. Fault diagnosis results (speed: 1250 rpm)

Feature type	Fault diagnosis rate (%)			
	H	I	O	Average
Time domain tensor characteristics	100	100	28.89	76.30
Tensor characteristics in wavelet domain	100	97.78	64.44	87.41
Tensor characteristics in entropy domain	97.78	42.22	95.56	78.52
Characteristics of triangular tensor	100	100	100	100
Multi domain tensor characteristics	100	100	100	100

Table 10. Fault diagnosis results (speed: 1500 rpm)

Feature type	Fault diagnosis rate (%)			
	H	I	O	Average
Time domain tensor characteristics	100	60	77.78	79.26
Tensor characteristics in wavelet domain	100	71.11	93.33	88.15
Tensor characteristics in entropy domain	80	95.56	66.67	80.74
Characteristics of triangular tensor	100	100	100	100
Multi domain tensor characteristics	100	100	100	100

90.37%, 94.07%, 96.30% and 93.33% respectively, and the fault diagnosis rates after the low rank tensor approximation are 97.04%, 100%, 100% and 100% respectively. This shows that the high-order tensor feature extraction method based on the low rank tensor approximation effectively improves the representation ability of the tensor features on the bearing operation status information, The recognition accuracy of the classifier is improved. At the same time, the fault diagnosis rate of the multi-channel tensor feature is significantly higher than that of the single channel vector feature, which shows that compared with the single channel signal feature, the multi-channel tensor feature makes full use of the nonlinear relationship between the correlation coupling information of the multi-channel source and the original feature data, obtains more and more comprehensive bearing operation status information, and effectively improves the fault diagnosis rate.

In order to further illustrate the effectiveness of multi domain tensor feature fault diagnosis, time domain tensor feature, wavelet domain tensor feature, entropy domain tensor feature and triangular domain tensor feature are constructed respectively. According to the previous method, STM is used to diagnose the fault of tensor features in different domains. The diagnosis results are shown in Tables 7, 8, 9 and 10.

It can be seen from Tables 7, 8, 9 and 10 that the construction method of multi domain feature tensor is the same as that of time domain tensor feature, wavelet domain tensor feature, entropy domain tensor feature and triangle tensor feature. The former is fused from the latter. However, the diagnostic results obtained when applied to STM are different. At 1250 rpm and 1500 rpm, the triangular tensor feature and multi domain tensor feature can reach 100% and are higher than other tensor features; At 1000 rpm, the diagnostic rate of triangular tensor feature is close to that of multi domain tensor feature and higher than that of other tensor features; However, under the rotating speed of 750 rpm, the fault diagnosis rate of multi domain tensor feature is much higher than that of other tensor features. This shows that the multi domain tensor feature has a wide range of working conditions. Compared with the single domain tensor feature, the multi domain fusion tensor feature contains more effective information and can be applied to fault diagnosis under more working conditions.

5 Conclusion

The fault diagnosis method based on multi-domain tensor feature and support tensor machine can make full use of the high-order correlation information of multi-channel signals of rolling bearing, and obtain more comprehensive bearing operation state information, to the greatest extent, the omni-directional operation state information of rolling bearing is contained in the basic data of fault diagnosis to ensure the completeness of data information, and effectively improve the accuracy of fault diagnosis. The multi-channel tensor feature is used to diagnose the fault of rolling bearing, which realizes the effective fusion of multi-channel vibration signals, and can obtain better diagnosis effect than the single channel vector feature method.

The low rank tensor approximation based on CP decomposition has strong tensor data information mining ability. It can fully mine the state information in tensor features, effectively remove the redundant information of tensor features, improve the representation ability of tensor features for different operating states and improve the recognition performance of classifier without damaging the tensor data structure. It is an effective high-order feature extraction method.

Acknowledgements. This research was financially supported by the National Key R&D Program of China (No. 2019YFB2004600) and the National Natural Science Foundation of China (No. 52005335).

References

1. Li, N.P., Gebraeel, N., Lei, Y.G., et al.: Remaining useful life prediction based on a multi-sensor data fusion model. Reliab. Eng. Syst. Saf. **208**, 107249 (2020)
2. Safizadeh, M.S., Latifi, S.K.: Using multi-sensor data fusion for vibration fault diagnosis of rolling element bearings by accelerometer and load cell. Inf. Fusion **18**, 1–8 (2014)
3. Zhang, X.W., Li, H.S.: Research on transformer fault diagnosis method and calculation model by using fuzzy data fusion in multi-sensor detection system. Optik **176**, 716–723 (2019)
4. Zhu, D.C., Zhang, Y.X., Pan, Y.Y., et al.: Fault diagnosis for rolling element bearings based on multi-sensor signals and CNN. J. Vibr. Shock **39**(04), 172–178 (2020)
5. Hao, S.J., Ge, F.X., Li, Y.M., et al.: Multisensor bearing fault diagnosis based on one-dimensional convolutional long short-term memory networks. Measurement **159**, 107802 (2020)
6. Azamfar, M., Singh, J., Bravo-Imaz, I., et al.: Multisensor data fusion for gearbox fault diagnosis using 2-D convolutional neural network and motor current signature analysis. Mech. Syst. Signal Process. **144**, 106861 (2020)
7. Zhao, H.S., Zhang, W.: Fault diagnosis method for rolling bearings based on segment tensor rank-(L, L, 1) decomposition. Mech. Syst. Signal Process. **132**, 762–775 (2019)
8. Rébillat, M., Mechbal, N.: Damage localization in geometrically complex aeronautic structures using canonical polyadic decomposition of Lamb wave difference signal tensors. Struct. Health Monit. **19**, 305–321 (2019)
9. Li, Y., Gong, X.L., Zhao, Q.H.: Hyperspectral image classification based on tensor-based radial basis kernel function and support vector machine. Chin. J. Sci. Inst. **41**(12), 253–262 (2020)

10. Kolda, T.G., Bader, B.W.: Tensor decompositions and applications. SIAM Rev. **51**(3), 455–500 (2009)
11. Sidiropoulos, N.D., de Lathauwer, L., Fu, X., et al.: Tensor decomposition for signal processing and machine learning. IEEE Trans. Signal Process. **65**(13), 3551–3582 (2017)
12. Hu, C.F., Wang, Y.X.: Research on multi-channel signal denoising method for multiple faults diagnosis of rolling element bearings based on tensor factorization. J. Mech. Eng. **55**(12), 50–57 (2019)
13. Feng, L.L., Liu, Y.P., Chen, L.X., et al.: Robust block tensor principal component analysis. Signal Process. **166**, 107271 (2020)
14. Lei, Y.G., Xu, X.F., Cai, X., et al.: Research on data quality assurance for health condition monitoring of machinery. J. Mech. Eng. **57**(04), 1–9 (2021)
15. Eckart, C., Young, G.: The approximation of one matrix by another of lower rank. Psychometrika **1**(3), 211–218 (1936)
16. Bro, R., Kiers, H.A.L.: A new efficient method for determining the number of components in PARAFAC models. J. Chemom. **17**(5), 274–286 (2003)
17. Liu, K.F., So, H.C., Costa, J.P., et al.: Core consistency diagnostic aided by reconstruction error for accurate enumeration of the number of components in parafac models. In: IEEE International Conference on Acoustics, pp. 6635–6639 (2013)
18. Javed, K., Gouriveau, R., Zerhouni, N.: A feature extraction procedure based on trigonometric functions and cumulative descriptors to enhance prognostics modeling. In:2013 IEEE Conference on IEEE Prognostics and Health Management (PHM), pp. 1–7 (2013)

A Perception Method Based on Point Cloud Processing in Autonomous Driving

Qiuqin Huang[1], Jiangshuai Huang[1(✉)], Xingyue Sheng[2], and Xiaowen Yue[2]

[1] Chongqing University, 174 Shazheng Street, Chongqing, Shapingba District, China
{HuangQQ,jshuang}@cqu.edu.cn
[2] Chongqing Zhi Xiang Paving Technology Engineering CO., LTD.,
Chongqing, China

Abstract. The abstract should briefly summarize the contents In the field of autonomous driving, autonomous cars need to perceive and understand their surroundings autonomously. At present, the most commonly used sensors in the field of driverless cars are RGB-D camera and LIDAR. Therefore, how to process the environmental information collected by these sensors, and extract the features we are interested in, and then use them to guide the driving of unmanned vehicles, has become an essential research point in the field of autonomous driving. Among them, compared with 2D image information, 3D point cloud can provide 3D orientation information of objects that 2D image does not have. Based on this, how to accurately process and perceive 3D point cloud and separate objects such as obstacles, cars and roads is crucial to the safety of autonomous driving. This paper adopts a method of preprocessing point cloud data and enhancing point cloud with images. KITTI dataset is the most classic and representative dataset in the field of autonomous driving. In the experiment, KITTI dataset is used to verify that this method can get good perception effect. In addition, we use the bird's-eye view benchmark on KITTI to evaluate the performance of the original network. It is found that the improvement effect of this experiment on the original network is also very obvious.

Keywords: Autonomous driving · Point cloud and image fusion · Deep learning

1 Introduction

With the development of artificial intelligence, autonomous driving has gradually become an increasingly hot field. Automatic driving requires reducing traffic accidents caused by driver error, making full use of traffic resources, intelligent navigation and driving, and reducing manual driving time. LIDAR has the advantages of small size, high resolution and accurate measurement, and has been widely used in sweeping robots, automatic driving and other related fields. As a major form of in-formation storage by LIDAR, point cloud contains a lot of depth information, location information, 3D geometry information, etc., and is widely used in automatic driving.

The PCL point cloud library contains part of the 3D sensing algorithm, as well as the interface of LIDAR and 3D scanner, which has promoted the wide application of point cloud since its launch. Traditional point cloud processing is mainly based on the spatial structure of point cloud, such as concavity segmentation, watershed analysis, hierarchical clustering, regional growth and spectral clustering. In 2012, Princeton University established the large-scale 3D CAD model library project ModelNet [15]. With the great success of artificial intelligence technology in the field of image, the application of deep learning algorithm in 3D data such as point cloud has gradually emerged. In recent years, using deep learning to process point cloud data has become a research hotspot. In particular, the development of 3D CNN [9], PointNet++ [12], graph neural network [16] and other technologies has brought huge development opportunities to point cloud processing and played an important role in the field of autonomous driving environment perception. As 3D shape data, point cloud can extract features with deep network and be applied to target classification, model segmentation and scene semantic analysis tasks. Good results have been achieved. Point cloud processing [5] has become an increasingly popular and widely used technology.

But at the same time, point cloud processing also has the following problems and difficulties:

1) Point clouds has irregularity and disorder. Point clouds reflect the real world by discretely sampling the surface of an object. The number, location, precision and density of sampling points are arbitrary. On the other hand, point clouds are mostly generated by the original depth information collected by RGB-D camera or LiDAR and the three-dimensional point spatial coordinates generated by fusion filtering 3D reconstruction method. In these production processes, regularly arranged data can also become unordered collections. Because of these characteristics of point cloud, it is a difficult task to make use of spatial local correlation in point cloud perception using convolutional neural network.

2) Point clouds have diversity. In the environment perception of autonomous driving, the point cloud obtained by real-time scanning is sparsely distributed, and the point cloud range of road scenes is huge and contains a lot of noise. For example, it might include poles, trees, the ground, and other extraneous objects that interfere with perception.

Previous point cloud approaches rarely paid full attention to these two characteristics, only focusing on one. We use an approach that combines traditional methods of point cloud processing with deep learning. In order to reduce the diversity of point cloud, this paper preprocesses the point cloud and filters out irrelevant points and noise points. For the irregularity and disorder of point cloud in the perception task, an image enhancement method of point cloud is adopted to enhance the information of point cloud and reduce the influence of irregularity and disorder.

Based on the above methods, we first use the traditional point cloud processing method to filter the original point cloud. In this stage, RANSAC is used and proved to have good ground point filtering effect. Then, we will detect the filtered ground points. The traditional point cloud target detection network such as VoxelNe [17] only uses the original point cloud with less information. The network that partially integrates image information, such as MV3D [6], only integrates RGB information of image, sacrificing the original speed, and the improvement of information is not enough. As a result, in this stage, the image corresponding to the point cloud is semantically segmented firstly. Here, the conventional image semantic segmentation network can be used to obtain the classification category score of each pixel. After the classification score of the image is obtained, the point cloud corresponding to each pixel can be enhanced. Compared with the point cloud before enhancement, the amount of information contained in the enhanced point cloud has been significantly improved. Finally, we verify it on KITTI dataset and find that the filtered and enhanced point cloud performs better in mean average precision(mAP) than the original point cloud.

2   Methods Architecture

2.1   Point Cloud Preprocessing

The point cloud map includes a large number of ground points, which not only makes the whole map appear messy, but also brings trouble to the subsequent obstacle recognition and point cloud perception tasks, so this paper considers to remove them first.

Ground filtering can be achieved by simple point cloud segmentation. In real point cloud data, there is often some prior knowledge of objects in the scene. For objects in complex scenes, their geometric shapes can be attributed to simple geometric shapes. This brings great convenience to segmentation, because simple geometric shapes can be described by equations. In other words, in the case of determining the equation, a simple object can be described with limited parameters. The equation represents the topological abstraction of the object. Of course, for complex objects, it is necessary to describe objects through complex models such as neural networks. This description method is not used here (for the case of simple ground), because it is very likely to sacrifice a lot of calculation, but the accuracy may not be directly improved. Moreover, on the other hand, the point cloud filtering stage serves to reduce the amount of computation in the later point cloud enhancement stage. Therefore, there is no need for high precision and no need to blindly pursue high-precision filtering effect.

For point cloud data in autopilot, the ground can be approximated as a plane in 3D coordinates. That is:

$$ax + by + cz + d = 0 \tag{1}$$

where, x, y and z respectively represent the three dimensions in the X-Y-Z coordinate system.

RANSAC (Random Sample Consensus) is a fitting algorithm that repeatedly selects a group of Random subsets of data to achieve goals. In this paper, RANSCA was used to filter ground points. The assumption of RANSAC algorithm is that the sample contains data that can be described by the model, as well as noise data and abnormal data that deviate far from the normal range and cannot adapt to the mathematical model. These abnormal data may be due to incorrect measurements, incorrect assumptions, incorrect calculations, etc. RANSAC also assumes that, given a correct set of data, there is a way to calculate model parameters that conform to these data. For 3D point cloud data in autonomous driving, a large part of points are concentrated in the central position of LIDAR. The closer to the LIDAR is, the more dense the point cloud data is, and most of them are concentrated on the ground. In addition, in the research of this paper, data points such as cars and pedestrians are objects that need to be perceived, but these data points are far away from the ground and account for a small proportion of the ground, so they need to be regarded as abnormal data that cannot adapt to the mathematical model in the fitting process. Because of these characteristics of point cloud, RANSAC algorithm can better fit the ground point with the most concentrated data information.

2.2   Sense of Mission

There have been many problems in the development of 3D point cloud sensing technology. LIDAR has its own shortcomings. For example, poles and pedestrians are clearly discernible in the image, but in the radar point cloud, they look so similar that they are virtually indistinguishable. Previous experiments have shown that, for much of the KITTI [7] dataset, there is indeed recognition of the signal lamp posts as pedes-trians. So the researchers [14] considered using features from other sensors to compen-sate for LIDAR's shortcomings. Camera and LIDAR are complementary, making fusion models more efficient and popular than other sensor fusion models. As a result, there have been many techniques for enhancing point clouds with images.

Camera-LIDAR fusion is not a simple task. First, cameras record the real world by projecting it onto a plane, while point clouds preserve three-dimensional geometry. In addition, in terms of data structure, point clouds are irregular, disordered and discrete, while images are regular, orderly and continuous. These feature differences between point cloud and image lead to different feature extraction methods.

For the field of image enhancement point cloud studied in this paper, a common fusion idea is to project the RGB attribute of the image onto the point cloud, so that each point cloud not only has XYZR attribute, but also RGB attribute, which is helpful to improve the accuracy without too much calculation. Pointpainting [13] proposed a method to fuse the semantic information of pictures with point clouds. In this paper, it is considered that it is not enough to add the RGB attribute of the picture, while adding the semantic information of the picture will make the point cloud more interpretable. With semantic information added to the above KITTI dataset, pedestrians and road signs will be much easier to distinguish.

As shown in the Fig. 1. Pointpainting [13] first performs semantic segmentation on the image to obtain the category score of each pixel. Thanks to the

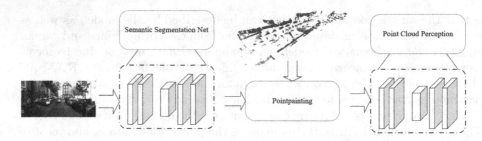

Fig. 1. Point cloud perception processing flow.

development of image semantic segmentation technology, many advanced semantic segmentation networks can be used in this stage of experiments. For example: DeepLabv3 [3], DeepLabv3+ [4] and HMAnet [11]. Table 1 shows the performance of these three semantic segmentation networks on KITTI dataset. Their mIOU can reach more than 85 points, the performance is optimistic, and the accuracy is sufficient for the method in this paper. In the structure of this paper, the input of the semantic segmentation network is the image corresponding to the point cloud image, and the output is the category score of each pixel. Each pixel has scores in four categories: background, bicycle, car and pedestrian. The matrix is expressed as $(score_a, score_b, score_c, score_d)$. Then, the output of semantic segmentation is used to enhance the point cloud. The enhancement method is the Pointpainting [13] mentioned above, that is, the image perspective is converted to the point cloud perspective, and then the dimension of each point cloud is expanded. The input of the original point cloud is (x, y, z, r), and the final point cloud after Pointpainting [13] is $(x, y, z, r, score_a, score_b, score_c, score_d)$.

Table 1. Accuracy comparison of several classical image segmentation networks.

Method	mIOU
DeepLabv3 [3]	85.7
DeepLabv3+ [4]	87.8
HMAnet [11]	87.3

The enhanced point cloud has richer world information, and only the feature size of the point cloud is modified. Therefore, any kind of neural network competing for the point cloud can be used to realize the perception function of the point cloud, such as target detection. At present, the classic point cloud target detection networks include: PointPollars, PionRCNN, PvRCNN, Second and so on. As shown in the Fig. 2. When these neural networks are used to complete the last point cloud sensing task, the input is the point cloud after pointpainting, and the output is the 3D detection frame and final category of the detected object in the point cloud.

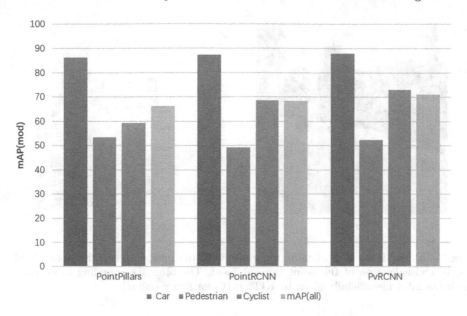

Fig. 2. Comparison diagram of several typical point cloud target detection networks.

3 Experimental Setup

In this section we will introduce the process and implementation of the whole method in detail.

3.1 Datasets

The point cloud data processed in this paper comes from KITTI dataset. KITTI dataset is jointly founded by Karlsruhe Institute of Technology and Toyota American Institute of Technology. It is the largest computer vision algorithm evaluation dataset in the autonomous driving scenario in the world at present. At the same time, KITTI dataset annotation information is very rich, easy to transplant to a variety of different networks.

KITTI officials processed the collected data sets according to different task requirements. Taking target detection as an example, the dataset includes two formats: point cloud and image, and the arrangement order of image and point cloud is one-to-one corresponding. In addition, in order to enhance the diversity of data sets, the frame data in the same folder (only containing point clouds or images) is disrupted, and different scenes (such as road width, light intensity, number of pedestrian vehicles, size of buildings, range of occluded places, etc.)

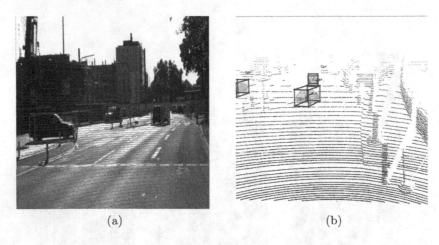

(a)   (b)

Fig. 3. (a) is a demo after aligning the image with the perspective of the point cloud; (b) is a visualization of the point cloud dataset. The blue 3D bounding boxs comes from the label file officially given by KITTI. (Color figure online)

are randomly selected into the dataset file of target detection. For the task of target detection, KITTI official provides the 3D detection frame of objects, classification categories and other target detection related annotation files in the training dataset. There are 7481 samples in the training dataset and 7518 samples in the test dataset. In the experiment, 3712 samples were randomly selected from the training dataset for training, and 3769 samples were verified and evaluated. Figure 3 is an example of a KITTI dataset in the task of target detection.

3.2 RANSCA

In the experiment, RANSAC algorithm can be verified by the following methods:

Step 1: Assume some internal points and calculate all unknown parameters of the model based on the assumed internal points.

Step 2: Use the model obtained in Step 1 to test all other point clouds. If a point fits the estimated model, it is considered as a local point.

Step 3: Then, with all the station point to estimate the assumption of the model.

Step 4: Finally, the model is evaluated by estimating the error rate between the internal points and the model.

This process needs to be repeated a fixed number of times. The model generated each time has two processing results: No.1, it is abandoned because there are too few internal points. No.2, it was chosen because it was better than existing models. In addition, some work is done on the evaluation algorithm of the local point model. For example, in the experiment, PCA principal component analysis was used to find the characteristic vector and normal vector corresponding to each point. In PCA processing, KD-tree is used to construct a search tree to facilitate searching and speed up CPU computing. After this step, the normal vector to the point is (a, b, c). Assuming that the ground point is a plane close to the X-Y plane, some point clouds parallel to the Z-axis can be removed by the normal vector:

$$\frac{\sqrt{a^2 + b^2}}{|c|} > q \tag{2}$$

where, q is the threshold value of point offset X-Y plane, which is set to 0.57 in the experiment.

Finally, you need to set the count to stop the iteration. Two conditions are also given in the experiment: No.1, reaches the maximum number of calculations 50. No. 2, if the ratio of the internal points to the total point cloud is greater than or equal to 70%, it can be considered that most ground points have been fitted, and then iteration stops.

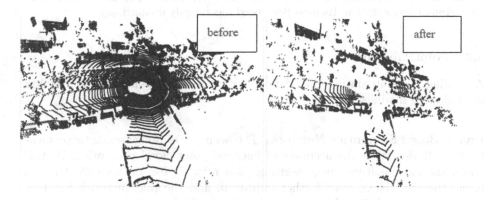

Fig. 4. Comparison experiment of ground filtration.

As can be seen from the Fig. 4, the center ground points can be well filtered by this method, and the ground points can be filtered up to 50%, which is of great help to reduce the computation amount of the subsequent neural network. In the experiment, we also tested the effect of RANSCA on the reduction of computing time, as shown in Fig. 5. The network selected in the experiment is the PointPillars network trained by the dataset processed by PointPainting. Then, we selected 1500 samples in the train dataset to test the improvement of neural network computing speed by ground segmentation.

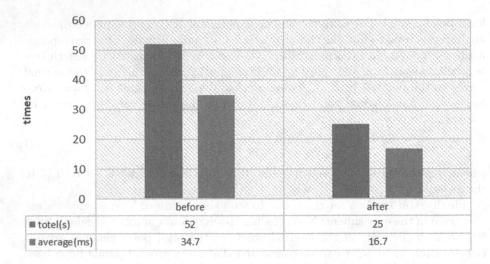

Fig. 5. The stand in effect of RANSCA for ground segmentation speed. "before" represents the original painted point cloud, and "after" represents the painted point cloud after ground segmentation. Its detection speed has roughly doubled.

3.3 PointPainting

According to Pointpainting [13], the perception part of the network mainly has the following modules.

Image Based Semantics Network. The accuracy of image semantic segmentation will also affect the accuracy of backend point cloud perception. At this stage, the more mature image segmentation network has deeplab [2]. On this basis, the researchers have further improved, and the pilot network has better performance ability. After comparing the performance of several advanced semantic segmentation networks, the experiment finally felt that the evolutionary version of DeepLabv3+ was used to complete the semantic segmentation of pictures. Based on ResNet101 [8] backbone network, the network can train the model on the data set of KITTI's semantic segmentation task, and then directly use it as the network of image input in this paper.

Point Cloud Processing and Perception. In actual measurements, the position of the LIDAR and the camera is always somewhat different. KITTI officially gave the conversion mode of LIDAR to camera, and the formula of projecting point x in LIDAR coordinates onto the color image y on the left is:

$$y = P\_rect\_n \times R0\_rect \times Tr\_velo\_to\_cam \times x \tag{3}$$

where, P_rect_n is the internal parameter matrix of the camera labeled n, which is only related to the internal parameters of the camera (such as focal length and light center position), $R0\_rect$ is the corrected rotation matrix of camera 0, and $Tr\_velo\_to\_cam$ is the external parameter matrix of the point cloud to the camera.

The transformation matrix from image to point cloud can be obtained by inverting the above matrix. By transforming matrix, the score of image semantic segmentation is extended to point cloud, and the information of point cloud with increased dimension can be obtained. Through the point cloud information processing of these images, more accurate perception effect can be obtained. The backend sensing network can use a target detection network like Pointpillars [10] for object detection. Or a network like Pointnet++ [1] can do semantic segmentation, and finally get the semantic segmentation score of point cloud.

In the experiment, two kinds of point cloud target detection networks are used to evaluate the point cloud after PointPainting, which are PointPillars and PvRCNN respectively. We evaluated the mAP of the network trained with the original KITTI dataset and the network trained with the image enhanced dataset. As shown in Table 2. Compared with the network corresponding to the original dataset, whether it is based on PointPillars or PvRCNN, the mAP value of painted network generally increases in varying degrees, but some individual patterns decreased. It can be explained that the accuracy of the same network and the same dataset will fluctuate under the same training conditions, and the decline of indicators in the table is very small, so it can be regarded as a normal range. In general, the mAP of painted network has been significantly improved. Especially for the pedestrian category, the point cloud enhancement algorithm is the most obvious to improve the detection accuracy. This is very important for improving pedestrian safety in the field of automatic driving.

Table 2. Results before and after painting on the KITTI test BEV (bird's-eye view) detection benchmark.

Method	mAP	Car			Pedestrian			Cyclist		
	Mod.	Easy	Mod.	Hard	Easy	Mod.	Hard	Easy	Mod.	Hard
PiontPillars	66.27	89.44	86.16	83.94	59.47	53.32	49.54	79.96	59.34	56.88
Painted	70.74	89.42	87.29	85.56	65.09	59.94	56.63	82.75	65.00	61.01
Delta	+4.47	−0.08	+1.13	+1.62	+5.62	+6.62	+7.09	+2.79	+5.66	+4.13
PvRCNN	71.07	90.14	87.94	87.40	59.55	52.27	48.59	85.87	73.01	70.78
Painted	75.24	89.89	87.80	87.42	72.07	65.13	62.04	87.84	72.79	70.32
Delta	+4.17	−0.25	−0.14	+0.02	+12.52	+12.86	+13.45	+1.97	−0.22	−0.46

4 Conclusion

The main work of this paper is to preprocess the point cloud based on the field of automatic driving to get a better point cloud perception effect. The pre-processing work mainly includes two aspects: Point Cloud Filtering and point cloud enhancement. Point cloud filtering is mainly aimed at the ground, and

RANSCA algorithm is adopted. This method belongs to a traditional segmentation method, which can filter out most of the ground points. The point cloud enhancement part adopts the way of image fusion to enhance. The traditional point cloud enhancement algorithm only integrates the RGB information of the image, but this is not enough in some cases. The PointPainting used in this paper integrates the semantic information of the image, so that the enhanced point cloud has richer world features. Experiments show that the accuracy of the enhanced point cloud is significantly improved compared with the original point cloud.

After the research and analysis of various point cloud-oriented processing perception, this paper finds that the following issues need to be further discussed in the future:

1) Although the computation of neural network can be significantly reduced after ground segmentation, in fact, we find that the mAP will also be reduced to varying degrees. This is most likely due to the lack of information on the ground. How to balance the calculation time and accuracy of neural network is a problem worth discussing in the future.
2) Most of the existing methods are only applicable to small-scale point clouds. In fact, the point clouds obtained by depth sensors usually have a large scale. Therefore, it is necessary to further study the perception of large-scale point cloud. In addition, the network structure that directly takes the point cloud as the input data is the current mainstream processing method. However, the attributes of points determine that they do not have explicit neighbor information. Most existing point based methods have to use the computationally expensive neighborhood point search mechanism. Future research can seek to design a relatively flexible way.

References

1. Charles, R., Li, Y., Hao, S., Leonidas, J.: Pointnet++ Deep hierarchical feature learning on point sets in a metric space. In: Proceedings of the Advances in Neural Information Processing Systems, Long Beach, CA, USA, pp. 4–9 (2017)
2. Chen, L.-C., Papandreou, G., Kokkinos, I., Murphy, K., Yuille, A.L.: Semantic image segmentation with deep convolutional nets and fully connected crfs. arXiv preprint arXiv:1412.7062 (2014)
3. Chen, L,.-C., Papandreou, G., Schroff, F., Adam, H.: Rethinking atrous convolution for semantic image segmentation. arXiv preprint arXiv:1706.05587 (2017)
4. Chen, L.-C., Zhu, Y., Papandreou, G., Schroff, F., Adam, H.: Encoder-decoder with atrous separable convolution for semantic image segmentation. In Proceedings of the European conference on computer vision (ECCV), pp. 801–818 (2018)
5. Chen, S., Liu, B., Feng, C., Vallespi-Gonzalez, C., Wellington, C.: 3D point cloud processing and learning for autonomous driving: impacting map creation, localization, and perception. IEEE Signal Process. Mag. 38(1), 68–86 (2020)
6. Chen, X., Ma, H., Wan, J., Li, B., Xia, T.: Multi-view 3D object detection network for autonomous driving. In: Proceedings of the IEEE conference on Computer Vision and Pattern Recognition, pp. 1907–1915 (2017)

7. Geiger, A., Lenz, P., Urtasun, R.: Are we ready for autonomous driving? the kitti vision benchmark suite. In: 2012 IEEE Conference on Computer Vision and Pattern Recognition, pp. 3354–3361. IEEE (2012)
8. He, K., Zhang, X., Ren, S., Sun, J.: Deep residual learning for image recognition. In: Proceedings of the IEEE Conference on Computer Vision and Pattern Recognition, pp. 770–778 (2016)
9. Ji, S., Wei, X., Yang, M., Kai, Yu.: 3D convolutional neural networks for human action recognition. IEEE Trans. Pattern Anal. Mach. Intell. $35(1)$, 221–231 (2012)
10. Lang, A.H., Vora, S., Caesar, H., Zhou, L., Yang, J., Beijbom, O.: Pointpillars: Fast encoders for object detection from point clouds. In: Proceedings of the IEEE/CVF Conference on Computer Vision and Pattern Recognition, pp. 12697–12705 (2019)
11. Niu, R., HmaNet: Hybrid multiple attention network for semantic segmentation in aerial images. arxiv 2020. arXiv preprint arXiv:2001.02870
12. Qi, C.R., Yi, L., Su, H., Guibas, L.J.: Pointnet++: Deep hierarchical feature learning on point sets in a metric space. In: Advances in neural information processing systems, vol. 30 (2017)
13. Vora, S., Lang, A.H., Helou, B., Beijbom, O.: Pointpainting: Sequential fusion for 3d object detection. In: Proceedings of the IEEE/CVF Conference on Computer Vision and Pattern Recognition, pp. 4604–4612 (2020)
14. Wang, H., Lou, X., Cai, Y., Li, Y., Chen, L.: Real-time vehicle detection algorithm based on vision and lidar point cloud fusion. J. Sensors **2019** 1–9 (2019)
15. Wu, Z.: 3D shapenets: A deep representation for volumetric shapes. In: Proceedings of the IEEE Conference on Computer Vision and Pattern Recognition, pp. 1912–1920 (2015)
16. Zhou, J., et al.: Graph neural networks: a review of methods and applications. AI Open **1**, 57–81 (2020)
17. Zhou, Y., Tuzel, O.: Voxelnet: End-to-end learning for point cloud based 3D object detection. In: Proceedings of the IEEE Conference on Computer Vision and Pattern Recognition, pp. 4490–4499 (2018)

Many-Objective Artificial Bee Colony Algorithm Based on Decomposition and Dimension Learning

Shuai Wang, Hui Wang$^{(\boxtimes)}$, Zichen Wei, Jiali Wu, Jiawen Liu, and Hai Zhang

School of Information Engineering, Nanchang Institute of Technology,
Nanchang 330099, China
huiwang@whu.edu.cn

Abstract. Artificial bee colony (ABC) has shown strong global search abilities on single-objective problems (SOPs). In order to stretch ABC to tackle many-objective optimization problems (MaOPs), a novel many-objective ABC algorithm on account of decomposition and dimension learning (called MaOABC-DDL) is proposed in this paper. By the decomposition, a MaOP is transformed to several sub-problems, which are simultaneously optimized by an improved ABC algorithm. A novel fitness function is the adoption of the ranking value of each objective. Then, an elite set is constructed according to the fitness value. Built on the elite set, a revised search strategy is designed. In addition, dimension learning is employed to amplify the search capability and Speed up convergence. To verify the performance of MaOABC-DDL, the DTLZ benchmark set is measured in the trials. The outcome shows that the proposed MaOABC-DDL obtains better results than the other five compared algorithms in two performance metrics.

Keywords: Many-objective optimization problem · Dimension learning · Artificial bee colony algorithm · Decomposition

1 Introduction

In the real world, multi-objective optimization problems generally refer to multiple conflicting objectives, and they require to be optimized simultaneously, such problems can be called multi-objective optimization problems (MOPs). In general, a MOP can be described as follows:

$$\begin{cases} \min F(X) = (f_1(X), f_2(X), \ldots, f_M(X)) \\ \text{subject to } X \in \Omega \end{cases} \tag{1}$$

where $X = (x_1, x_2, \ldots, x_D)$ is the vector of decision variables, Ω is the decision space of the problem, D is the number of decision variables, and $f_i(X)$ is the i-th objective function. For MOPs, unlike single-objective optimization problems that have a single optimal solution, MOPs can optimize all objectives simultaneously. With conflicting objectives $M > 3$, such problems are named many-objective

H. Zhang et al. (Eds.): NCAA 2022, CCIS 1638, pp. 150–161, 2022.
https://doi.org/10.1007/978-981-19-6135-9_12

optimization problems (MaOPs). All the generated solutions must be close to or equal to a point on the surface of Pareto front (PF) as much as possible [1]. MaOPs often exist in logistics scheduling, resource utilization, engine design, etc.

With the growth of objectives, the optimization capabilities of multi-objective evolutionary algorithms (MOEAs) is rapidly deteriorated. It is difficult to judge the quality of individuals grounded on Pareto dominance relationship, so that the ratio of nondominated individuals in the population increases rapidly. It significantly impairs the selection pressure and leads to difficulties in convergence. To solve MaOPs, many improved MOEAs were proposed. Deb and Jain [2] devised a method that uses non-dominated sorting to select elite solutions with better convergence, preferentially retains extreme solutions in environment selection to participate in the evolution of offspring. Finally, reference points in the vital layer are used in the solution space to guarantee the distribution of the algorithm (named NSGA-III). The decomposition-based MOEA(named MOEA/D) was suggested by Zhang and Li [4]. The results show that for MOPs, decomposition-based MOEA/D can be solved effectively. For MaOPs, some MOEA/D modifications were proposed, such as MOEA/D-DE [5] and MOEA/D_DRA [6]. Bader and Zitzler [3] proposed an indicator-based approach called Hypervolume-based fast algorithm (HypE) for solving MaOPs, where selecting candidate solutions based on hypervolume is rigorously Pareto optimal.

Artificial bee colony (ABC) is a swarm intelligence optimization algorithm that exhibits good performance on many optimization problems. [7–15]. However, most studies on ABC have concentrated on single-objective optimization problems. For MOPs, different improvements on ABC were proposed [16–19]. So far, few studies have focused on MaOPs. Because the efficiency of many ABC algorithms deteriorates with increasing of objectives. In this article, a novel many-objective artificial bee colony algorithm based on dimensional learning and decomposition (called MaOABC-DDL) is proposed. Firstly, a MaOP is converted into multiple subproblems by the decomposition, and each subproblem is optimized by an improved ABC algorithm. Secondly, a new fitness value is determined on the basis of the ranking value of each objective. An elite set is constructed according to the fitness value. On the basis of the elite set, a revised search method is designed. In addition, dimension learning is used to improve the search capability and Speed up convergence. In the final experimental results, the algorithm MaOABC-DDL obtained better results than the other five compared algorithms in two tested performance indicators.

The remaining sections of this paper are arranged as shown below. Section 2, the original ABC is briefly introduced. In Sect. 3, the proposed approach MaOABC-DDL is described. Experimental results and discussions are given in Sect. 4. Lastly, Sect. 5 summarizes the work.

2   Artificial Bee Colony Algorithm

The artificial bee colony algorithm was designed by imitating the behavior of bees, which is a swarm intelligence optimization algorithm with strong global search capability. The honey-seeking process of bees is similar to the search for the optimal value within the range of independent variables of function. In ABC, three kinds of bees cooperate with each other to seek new honey sources. Firstly, the employed bees find new honey sources through the known honey source information, and share its message with the onlooker bees. The scout bees selectively seek honey according to the transmitted honey source information, while the scout bees seek a new honey source information and conduct random search nearby.

In ABC, let the amount of employed and onlooker bees be SN. For every solution $X_i = (x_{i1}, x_{i2}, x_{i3}, \ldots, x_{iD})$ has D dimensions. For each of the X_i, a novel solution is generated as below.

$$v_{ij} = x_{ij} + \phi_{ij} \cdot (x_{ij} - x_{kj}) \tag{2}$$

where $i = 1, 2, \ldots, SN$, $j \in \{1, 2, \ldots, D\}$ and $k \in \{1, 2, \ldots, SN\}$ are random integers, in the population v_{ij} represents the j-th dimension of the i-th solution, and $k \neq i$, and ϕ_{ij} is a random value between $[-1, 1]$.

$$fit(i) = \begin{cases} \frac{1}{1 + f_i}, & f_i > 0 \\ 1 + abs(f_i), & f_i < 0 \end{cases} \tag{3}$$

Here $fit(i)$ as the fitness value of the i-th solution and f_i is the function value of the i-th solution.

An onlooker bee selects food based on its probability $prob(i)$ associated to its fitness value.

$$prob(i) = \frac{fit(i)}{\sum_{i=1}^{SN} fit(i)} \tag{4}$$

From the above equation, it is clear that the probability $prob(i)$ is positively related to the fitness value. When the food source is better, the larger chance of being selected by the roulette wheel. We use roulette to assign a good solution X_i to each onlooker bee based on the size of $prob(i)$. Then, an offspring V_i is generated by the formula (2). Finally, make a greedy choice between X_i and V_i, and a better solution is kept as the new X_i.

If a food source cannot be enhanced within the specified iteration threshold, it will be discarded. The iteration threshold is a preset value named $limit$. If the food source exceeds the iteration threshold is X_i, X_i will be initialized randomly.

3   Proposed Approach

3.1   Decomposition Technology

A multi-objective optimization algorithm based on decomposition to solve MOPs was proposed by Zhang et al. [4]. The essential idea of resolving MOP in classical mathematics is the decomposition strategy. A MOP is divided into multiple single objective problems according to the mathematical decomposition method.

Then, we can optimize it by weighting the subproblems. Each adjacent subproblem has an approximate solution, and the adjacent relationship is defined by the Euclidean distance. Each subproblem can be optimized by using the neighbor subproblem information to obtain the Pareto solution set. In MOEA/D, each weight vector $\lambda^{(i)} = \left(\lambda_1^{(i)}, \lambda_2^{(i)}, \ldots, \lambda_M^{(i)}\right)^T$ defines a subproblem. Generally speaking, there are three decomposition methods: 1) penalty based boundary intersection; 2) Tchebycheff; and 3) weighted sum. In this paper, a modified Tchebycheff method is used, and its specific descriptions are described as follows.

$$g^{ic}\left(x \mid \lambda^{(i)}, z^*\right) = \max_{1 \leq j \leq M} \left\{|f_j(x) - z_j^*|/\lambda_j^{(i)}\right\} \tag{5}$$

where $z^* = \min\{f(x) \mid x \in \Omega\}$ indicates the ideal points, z_j^* is the ideal point of the j-th objective, M represents the count of objectives.

In this paper, the decomposition is combined with a modified ABC algorithm to solve MaOPs. The MaOP is converted into Multiple subproblems, which are optimized by our ABC algorithm simultaneously. The modified Tchebycheff method is used to update and replace the neighborhood of solutions, and produce the Pareto solution set.

3.2 New Search Strategy for the Employed Bee Phase

The search strategy (Eq. (2)) of the original ABC works on single optimization problems. To solve MOPs or MaOPs, the search strategy should be modified. As mention before, the exploitation of ABC is not as intense as its exploration. So, balancing exploitation and exploration can greatly influence of the optimization performance of ABC. Inspired by PSO, Zhu and Kwong [20] suggested a *gbest-*guided ABC (GABC) algorithm. In this case, the information of *gbest* is used to lead the candidate solution search and enhance the utilization effect. The strategy of solution search in GABC is described below.

$$v_{ij} = x_{ij} + \phi_{ij} \cdot (x_{ij} - x_{kj}) + \psi_{ij} \cdot (gbest_j - x_{ij}), \tag{6}$$

The $gbest_j$ is the j-th dimension of the overall optimal solution, ψ_{ij} is a random number uniform in the range $[0, C]$, and C is a constant greater than 0. X_k is a randomly chosen solution $(k \neq i)$, and $j \in \{1, 2, \ldots, D\}$ is a random values. In [20], $C = 1.5$ is considered as the optimal setting.

For MOPs or MaOPs, the global best solution *gbest* is not unique and the Pareto front (PF) consists of a group of Pareto-optimal solutions that are not dominated by other solutions. So, the Eq. (6) is modified as follows.

$$v_{ij} = x_{ij} + \phi_{ij} \cdot (x_{ij} - x_{kj}) + \psi_{ij} \cdot (x_j^* - x_{ij}), \tag{7}$$

where x_j^* is a randomly chosen solution among the elite solution set E.

To construct the elite set E, the ranking value of each solution is calculated for each objective. For a solution X_i, $Rank_{ij}$ means the ranking value of the j-objective. Then, a new fitness function is defined as below.

$$fit(i) = \sum_{j=1}^{M} Rank_{ij} \qquad\qquad (8)$$

where $i = 1, 2, \ldots, SN$, $j = 1, 2, \ldots, M$, and $fit(i)$ is the fitness value of X_i. According to the fitness value, the top 20% solutions are selected as a member of the elite set E.

3.3   Dimension Learning for the Onlooker Bee Phase

In Sect. 3.2, the new fitness function is re-written by Eq. (8). Each solution in the population has its own fitness value. Then, the probability selection method (described in Eq. (4)) still works. Based on the probability, some excellent solutions will be selected to search again in the onlooker bee phase.

According to Eq. (2), a dimension j is randomly selected and the difference value $(x_{ij} - x_{kj})$ (called step size) is obtained. Then, the difference value of the j-th dimension is employed to update the j-th dimension of the parent solution X_i. In our previous study, we designed a different updating method called dimension learning to produce new solutions. For a solution X_i, a dimension j and another solution X_k are randomly selected. Another dimension m is randomly chosen and $m \neq j$. Then, the dimension learning used for the onlooker bee phase can be described as follows.

$$v_{ij} = x_{im} + \phi_{im} \cdot (x_{im} - x_{km}) \qquad\qquad (9)$$

where $m \in [1, D]$ is a random integer, $m \neq j$, and $\phi_{im} \in [-1, 1]$ is a random number. By learning the differences of other dimensions, ABC can prevent falling into local optimum and speed up convergence.

3.4   Resource Adjustment for the Scout Bee Phase

In the search process, each subproblem reaches the optimal solution at different speeds. Some excellent solutions use few resources to acquire the optimal solution of the sub-problem. At this time, the probability of producing better solutions is very small, so the solution update can be monitored by the scout bees. When the number of new solutions does not exceed the preset value $limit$, the scout bee will set the locking state $Trial_i$ of the solution to true. That is, the solution will be locked. The locked solution will continue to update only when a better solution is generated to replace the current one. A poor solution will be more resources to replace the random initialization solution in the scout bee phase.

3.5   Framework of MaOABC-DDL

As mentioned before, the proposed approach MaOABC-DDL mainly have three modifications: 1) the decomposition technique is used to decompose a MaOP into multiple subproblems, which are optimized by an improved ABC; 2) a new

---

Algorithm 1: Proposed Approach (MaOABC-DDL)

```
 1  Population initialization;
 2  while FEs ≤ MaxFEs do
        /* Employed bee phase                                                    */
 3      for i = 1 to SN do
 4          if IsLockᵢ == false then
 5              Generate a new solution V by Eq. (7);
 6              Update population by aggregation function value Eq. (5);
 7              Set IsLockᵢ = false and Trialᵢ++;
 8          end
 9      end
        /* Onlooker bee phase                                                    */
10      if t mod EFs = 0 then
11          Calculate the probability prob(i) by Eq. (4);
12          while num ≤ SN do
13              if rand ≤ prob(i) then
14                  Produce a new solution V by Eq. (9);
15                  Update population by aggregation function value by Eq. (5);
16                  Set IsLockᵢ = false and Trialᵢ++;
17              end
18          end
19      end
        /* Scout bee phase                                                       */
20      if max{trialᵢ} ≤ limit then
21          IsLockᵢ = true;
22          Trialᵢ = 0;
23      end
24  end
```

Table 1. Population size and stopping criterion.

Objectives (M)	Population size (SN)	Stopping criterion ($MaxEFs$)
3	91	30000
5	210	50000
8	156	80000

fitness function is defined on the basis of the ranking value of each objective and a improved search strategy is proposed; and 3) dimension learning is employed to strengthen the search ability and speed up convergence. Algorithm 1 presents the primary structure of the algorithm.

4 Experimental Results

4.1 Test Problems and Parameter Settings

To validate the performance of MaOABC-DDL, the DTLZ benchmark set is used in the below experiments. DTLZ test problems are extensively used in the performance test of MaOEAs. The number of objectives M is set to 3, 5, and 8 for each problem. To compare the performance of MaOABC-DDL, five other MaOEAs are selected. The involved algorithms are listed as follows.

- MOEA/D [4].
- MOEA/D_DRA [6].

Table 2. HV values of different algorithms on the DTLZ benchmark set.

Problem	M	MOEA/D_DRA	MOEA/D-DU	RVEA	MOEA/D-DE	MOEA/D	MaOABC-DDL
DTLZ1	3	4.9074e-1 (3.66e-1)	8.3876e-1 (2.29e-3)	8.3876e-1 (2.88e-3)	6.8933e-1 (2.38e-1)	8.4009e-1 (9.29e-4)	**8.4157e-1** **(8.38e-5)**
	5	2.2964e-1 (2.61e-1)	9.7048e-1 (4.12e-4)	9.6957e-1 (1.92e-2)	3.5851e-1 (3.22e-1)	9.6332e-1 (6.76e-3)	**9.7097e-1** **(2.44e-4)**
	8	5.7674e-1 (3.78e-1)	**9.9755e-1** **(8.93e-5)**	9.9753e-1 (7.42e-5)	8.9532e-1 (9.70e-3)	9.1436e-1 (1.78e-2)	9.3522e-1 (2.49e-1)
DTLZ2	3	5.2765e-1 (1.64e-3)	5.5941e-1 (1.14e-4)	5.5936e-1 (8.01e-5)	5.2621e-1 (1.38e-3)	**5.5947e-1** **(1.16e-4)**	5.5899e-1 (2.08e-4)
	5	6.9937e-1 (3.02e-3)	8.1088e-1 (4.47e-4)	**8.1119e-1** **(5.30e-4)**	6.8635e-1 (9.01e-3)	7.9086e-1 (1.90e-2)	8.0893e-1 (5.20e-4)
	8	5.6103e-1 (2.02e-2)	9.2312e-1 (1.31e-4)	9.2391e-1 (2.22e-4)	5.4304e-1 (1.40e-2)	8.8119e-1 (2.86e-2)	**9.2875e-1** **(8.81e-4)**
DTLZ3	3	7.8458e-2 (1.53e-1)	2.6267e-1 (2.12e-1)	1.6033e-1 (2.16e-1)	2.6203e-1 (2.42e-1)	**5.1156e-1** **(7.38e-2)**	3.8418e-1 (2.68e-1)
	5	2.9619e-2 (5.40e-2)	5.2374e-2 (1.07e-1)	2.0199e-1 (2.56e-1)	4.7612e-2 (8.21e-2)	7.2761e-1 (6.54e-2)	**8.0867e-1** **(7.67e-4)**
	8	3.0721e-1 (2.29e-1)	8.4561e-1 (2.26e-1)	9.0977e-1 (8.73e-3)	3.5199e-1 (2.21e-1)	7.9118e-1 (7.07e-2)	**9.2853e-1** **(8.28e-4)**
DTLZ4	3	5.1328e-1 (2.69e-2)	**5.5948e-1** **(1.50e-4)**	5.5939e-1 (8.60e-5)	5.1883e-1 (2.20e-2)	3.6051e-1 (1.88e-1)	3.5413e-1 (2.40e-1)
	5	6.5641e-1 (1.95e-2)	8.1110e-1 (4.17e-4)	**8.1165e-1** **(4.22e-4)**	6.6026e-1 (1.44e-2)	6.7371e-1 (1.40e-1)	8.0476e-1 (1.23e-3)
	8	6.2917e-1 (2.59e-2)	9.2784e-1 (4.71e-4)	9.2408e-1 (3.80e-4)	6.2064e-1 (1.71e-2)	7.8384e-1 (1.29e-1)	**9.2986e-1** **(1.09e-3)**
DTLZ5	3	1.9436e-1 (9.45e-5)	1.8451e-1 (3.34e-3)	1.5175e-1 (9.39e-3)	**1.9437e-1** **(7.16e-5)**	1.9049e-1 (6.23e-6)	1.9046e-1 (1.93e-5)
	5	**1.1886e-1** **(3.10e-4)**	8.6832e-2 (7.57e-3)	1.0422e-1 (3.92e-3)	1.1853e-1 (4.10e-4)	1.0049e-1 (2.52e-4)	1.0049e-1 (3.09e-4)
	8	1.0225e-1 (3.69e-4)	9.2741e-2 (1.25e-3)	9.0888e-2 (1.73e-5)	**1.0230e-1** **(4.00e-4)**	9.9646e-2 (1.57e-4)	9.9641e-2 (1.31e-4)
DTLZ6	3	**1.9484e-1** **(2.83e-5)**	1.8396e-1 (5.09e-3)	1.4751e-1 (1.07e-2)	1.9477e-1 (2.59e-5)	1.9049e-1 (7.29e-6)	1.9047e-1 (3.38e-5)
	5	**1.1918e-1** **(3.20e-4)**	2.5810e-3 (3.97e-3)	6.5074e-2 (4.51e-2)	1.1910e-1 (3.15e-4)	1.0053e-1 (2.94e-4)	1.0056e-1 (2.70e-4)
	8	**1.0251e-1** **(2.89e-4)**	9.4433e-2 (1.61e-3)	9.0984e-2 (8.89e-3)	1.0250e-1 (2.56e-4)	9.9575e-2 (2.12e-4)	9.9691e-2 (2.00e-4)
DTLZ7	3	1.7706e-1 (3.38e-2)	1.3802e-3 (5.52e-3)	**2.6312e-1** **(2.44e-3)**	2.1405e-1 (1.19e-2)	2.5139e-1 (1.09e-2)	1.8909e-1 (6.35e-2)
	5	1.7031e-2 (2.88e-2)	0.0000e+0 (0.00e+0)	**2.1964e-1** **(4.28e-3)**	5.1247e-3 (7.46e-3)	1.8844e-1 (2.45e-2)	1.3676e-1 (4.77e-2)
	8	1.9375e-4 (4.06e-4)	0.0000e+0 (0.00e+0)	1.4888e-1 (1.42e-2)	4.3703e-4 (5.45e-4)	**1.6507e-1** **(4.80e-3)**	1.5253e-1 (8.07e-3)
+/-/=		6/13/2	4/16/1	7/11/3	6/13/2	6/8/7	

- MOEA/D-DU [21].
- RVEA [22].
- MOEA/D-DE [5].
- Proposed approach MaOABC-DDL.

For the above six algorithms, the parameter settings is suggested by other existing literature. Let $D = M - 1 + k$, where D is the number of decision variables. $k = 2$, $k = 10$, and $k = 20$ are used for DTLZ1, DTLZ2-DTLZ6, and DTLZ7, respectively. We used $MaxEFs = 10000 \times M$ as the stopping criterion. Since the performance of the algorithm is influenced by the population

Table 3. GD values of different algorithms on the DTLZ benchmark set.

Problem	M	MOEA/D-DRA	MOEA/D-DU	RVEA	MOEA/D-DE	MOEA/D	MaOABC-DDL
DTLZ1	3	2.0753e-1 (3.38e-1)	2.8623e-4 (1.31e-4)	3.2196e-4 (1.72e-4)	4.9391e-2 (1.17e-1)	2.2428e-4 (3.40e-5)	**1.9143e-4 (6.61e-6)**
	5	3.2423e+0 (8.93e-1)	1.6376e-3 (4.32e-5)	**1.0775e-3 (7.95e-5)**	2.0494e+0 (9.49e-1)	1.8547e-2 (6.77e-2)	1.5935e-3 (4.42e-5)
	8	3.6600e-1 (5.85e-1)	**3.6496e-3 (7.51e-5)**	3.9764e-3 (1.10e-4)	2.2231e-2 (5.64e-2)	9.3476e-3 (4.28e-3)	1.1297e-2 (3.18e-2)
DTLZ2	3	8.1137e-4 (6.59e-5)	**4.9397e-4 (1.38e-5)**	5.0816e-4 (7.23e-6)	8.5923e-4 (8.50e-5)	4.9674e-4 (9.75e-6)	4.9771e-4 (2.13e-5)
	5	4.7571e-3 (1.93e-4)	3.1867e-3 (3.69e-5)	3.3962e-3 (1.31e-5)	5.7233e-3 (4.20e-4)	3.0780e-3 (7.71e-5)	**2.9760e-3 (7.79e-5)**
	8	1.9805e-2 (1.15e-3)	1.2346e-2 (2.34e-4)	1.4012e-2 (3.80e-5)	1.9840e-2 (1.23e-3)	1.0957e-2 (5.93e-4)	**9.8988e-3 (3.54e-4)**
DTLZ3	3	6.7060e+0 (8.26e-1)	1.0413e-1 (1.19e-1)	2.0972e-1 (1.92e-1)	1.5203e+0 (2.84e+0)	**6.6330e-3 (1.81e-2)**	3.2787e-1 (5.02e-1)
	5	9.0392e+0 (1.86e+0)	2.2949e-1 (1.06e-1)	1.3433e-1 (1.03e-1)	7.7853e+0 (2.88e+0)	1.2038e-2 (3.60e-2)	**2.9923e-3 (6.18e-5)**
	8	1.3514e+0 (2.37e+0)	1.3431e-2 (1.83e-2)	1.2402e-2 (4.87e-4)	1.3944e+0 (2.44e+0)	8.2244e-2 (1.98e-1)	**9.4475e-3 (3.09e-4)**
DTLZ4	3	9.7684e-4 (3.79e-4)	4.8552e-4 (1.74e-5)	5.0713e-4 (4.61e-6)	1.1540e-3 (4.92e-4)	3.3293e-4 (2.04e-4)	**2.6976e-4 (2.46e-4)**
	5	3.5509e-3 (4.77e-4)	3.1656e-3 (3.87e-5)	3.3660e-3 (2.20e-5)	3.7206e-3 (2.65e-4)	2.8856e-3 (2.23e-4)	**2.8228e-3 (6.57e-5)**
	8	1.2491e-2 (1.87e-3)	1.1515e-2 (1.38e-4)	1.3544e-2 (3.97e-4)	1.1470e-2 (1.34e-3)	9.4467e-3 (1.73e-3)	**8.2539e-3 (3.58e-4)**
DTLZ5	3	1.1654e-4 (4.13e-5)	7.6242e-4 (1.15e-3)	4.4686e-2 (2.51e-2)	8.5558e-5 (2.54e-5)	5.5178e-6 (5.02e-7)	**4.3903e-6 (5.46e-7)**
	5	1.2889e-1 (1.29e-2)	2.7325e-1 (1.27e-2)	1.7114e-1 (5.19e-3)	1.2675e-1 (1.52e-2)	9.1692e-2 (4.09e-3)	**8.8523e-2 (1.70e-3)**
	8	1.3075e-1 (2.91e-2)	1.6491e-1 (8.55e-3)	2.9250e-1 (1.79e-2)	1.0461e-1 (2.54e-2)	**8.8096e-2 (1.79e-2)**	9.0142e-2 (3.67e-3)
DTLZ6	3	4.7634e-6 (1.65e-7)	7.3209e-3 (1.63e-2)	7.4117e-2 (1.18e-1)	4.7386e-6 (3.40e-7)	4.3388e-6 (3.25e-7)	**4.0111e-6 (5.84e-7)**
	5	3.5795e-1 (5.95e-2)	1.2489e+0 (2.30e-1)	**2.6861e-1 (3.03e-2)**	3.6034e-1 (1.32e-2)	3.2792e-1 (4.47e-2)	3.2816e-1 (2.45e-4)
	8	3.2594e-1 (4.88e-3)	6.4633e-1 (6.11e-2)	3.7071e-1 (3.12e-2)	4.1261e-1 (6.85e-2)	3.2654e-1 (1.08e-2)	**2.6337e-1 (4.09e-2)**
DTLZ7	3	4.3336e-2 (4.08e-2)	2.7267e+0 (1.80e+0)	6.5849e-3 (1.26e-3)	8.8769e-3 (1.13e-2)	8.4037e-4 (5.29e-5)	**4.3365e-4 (2.71e-4)**
	5	2.6173e-1 (1.53e-1)	9.4260e+0 (1.49e+0)	1.9205e-2 (1.34e-3)	1.9053e-1 (8.30e-2)	4.9110e-3 (1.21e-3)	**2.0853e-3 (2.06e-3)**
	8	7.4359e-1 (1.94e-1)	3.0833e+0 (1.00e+0)	4.3608e-2 (8.02e-3)	3.6764e-1 (7.11e-2)	2.2146e-2 (3.96e-3)	**1.9507e-2 (7.25e-3)**
+/−/=		0/21/0	1/18/2	3/17/1	0/21/0	2/11/8	

size, the different SN values are set for different objectives. For different M, the corresponding population size (SN) and stopping criterion ($MaxEFs$) are given in Table 1. The experimental results are averaged over 20 runs, for all comparison algorithms.

The popular Tchebycheff approach is selected as the aggregation method in MOEA/D. In MOEA/D the parents are selected from the neighborhood of the subproblem, and for MOEA/D-DE there is a 0.9 probability of selecting the parent from the population. The initial crossover rate (CR_0) and amplification factor (F_0) of differential evolution (DE) are set to 1 and 0.5. The neighborhood size of MOEA/D, MOEA/D_DRA, and MOEA/D-DE is set to $N/10$.

The generation of weight vector has a certain impact on the algorithm based on decomposition. So this paper uses a unified weight vector generation method. Through the m-dimensional space sampling processing, we can get C_{H+m-1}^{m-1} evenly distributed weight vectors, $H > 0$ means the number of samples in each objective direction and the sampling step is $1/H$.

4.2 Performance Indicators

Considering the balance of the algorithm, The hypervolume (HV) is selected to assess the comprehensive performance of convergence and diversity. In addition, the generation distance (GD) is used to evaluate the speed of the algorithm converging to the real frontier.

a) HV indicator: The quality of the algorithm needs to be constant through a variety of indicators in multi-objective optimization algorithms. The HV index used in this paper is a comprehensive evaluation index and does not require real Pareto front to compare multiple algorithms. HV indicator can be defined as:

$$\text{HV}(X, P) = \bigcup_{x \in X}^{X} v(x, P) \tag{10}$$

where P denotes the reference points and X denotes the non-dominated solution set. The HV index is the hypervolume of the space between the solution collection X and the reference points P. The solution with higher hypervolume is generally considered to have better performance.

b) GD indicator: GD represents the distance between the solution set obtained by the algorithm and the real PF, which is generally employed to assess the convergence of the algorithm. The GD indicators can be defined as below.

$$GD\,(P, P^*) = \frac{\sqrt{\sum_{y \in P} \min_{x \in P^*} \text{dis}(x, y)^2}}{|P|} \tag{11}$$

In the above equation, the final population obtained by the algorithm is P. The uniformly distributed set of reference points P^* is sampled from PF, and $dis(x, y)$ denotes the Euclidean distance between the point y in the solution set P and the point x in the reference set P^*.

4.3 Experiments Analysis

In this paper, six MaOEAs including MOEA/D_DRA, MOEA/D-DU, RVEA, MOEA/D-DE, MOEA/D, and MaOABC-DDL are tested on the DTLZ benchmark set. Table 2 presents the HV outcomes of six algorithms, where "+/ − / =" in the last row of the table shows the overall comparison results. The symbols "+", "−", and "=" indicates that the proposed approach MaOABC-DDL is worse, better, and similar than/to the compared algorithm, respectively.

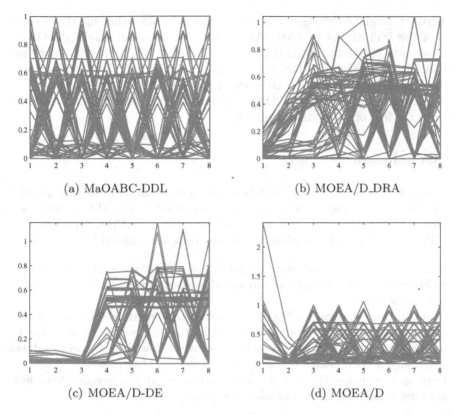

(a) MaOABC-DDL   (b) MOEA/D_DRA

(c) MOEA/D-DE   (d) MOEA/D

Fig. 1. The solution of MaOABC-DDL, MOEA/D_DRA, MOEA/D-DE, and MOEA/D on DTLZ1 with 8 objectives. (a) MaOABC-DDL. (b) MOEA/D_DRA. (c) MOEA/D-DE. (d) MOEA/D.

From the results, MaOABC-DDL is better than most algorithms on DTLZ1-DTLZ4. MaOABC-DDL is worse than MOEA/D_DRA on five test instances, while it achieves better results on 13 test instances. Compared with MOEA/D-DU, MaOABC-DDL obtains better results on 16 test instances. For the rest of five problems, MOEA/D-DU outperforms MaOABC-DDL on four problems. MaOABC-DDL is superior to RVEA on 11 problems, while RVEA performs better than MaOABC-DDL on seven problems. MOEA/D and MOEA/D-DE are better than MaOABC-DDL on six problems, while MaOABC-DDL surpasses MOEA/D and MOEA/D-DE on 8 and 13 problems, respectively.

Table 3 displays the GD results of six algorithms. From the results, MaOABC-DDL still outperforms other algorithms in the majority of the tested instances. With the increase of objectives, the performance of MaOABC-DDL is not seriously affected. On all 21 test instances, MaOABC-DDL surpasses MOEA/D_DRA and MOEA/D-DE. MOEA/D-DU outperforms MaOABC-DDL on a single test instance, while MaOABC-DDL obtains better performance on 18 test instances. Compared with RVEA and MOEA/D, MaOABC-DDL achieves better results on 17 and 11 problems, respectively.

Figure 1 displays the results of the comparison algorithm for the 8 objectives of the DTLZ1 test problem. As can be seen from the graphs, MaOABC-DDL achieves the best solution distributions among the comparison algorithms. MOEA/D_DRA, MOEA/D-DE, and MOEA/D cannot find the upper boundary of the Pareto front. MaOABC-DDL can approach the Pareto front very well, while MOEA/D fails. MaOABC-DDL, MOEA/D_DRA, and MOEA/D-DE suffer from the loss of population diversity.

5 Conclusion

The original ABC and its most modifications focus on single objective optimization problems. This paper introduces a new many-objective ABC algorithm on the basis of decomposition and dimension learning (MaOABC-DDL) to solve MaOPs. The significant contributions of our work are outlined below.

- A MaOP is transformed into several sub-problems through decomposition technology. These sub-problems are simultaneously optimized by an improved ABC algorithm.
- A new fitness function for MaOP is constructed based on the ranking value of each objective. All solutions are sorted by their fitness values and some top solutions are selected to build an elite solution. Based on the elite solution set, design a novel search method for the employed bee phase.
- In order to enhance the search ability and speed up convergence, dimension learning is adopted in the onlooker bee phase.

All comparison algorithms tested on the DTLZ benchmark set of 3, 5, and 8 objectives. According to HV and GD indicators, MaOABC-DDL achieve promising convergence and diversity, outperforming the competition on most test instances. From the graphs of the final solution set, MaOABC-DDL can approach the Pareto front very well.

In the experiments, the maximum value of objectives is 8. When the count of objectives increases, our approach will face severe challenges. So, the effectiveness of MaOABC-DDL should be further studied in the coming work.

Acknowledgment. This work was supported by National Natural Science Foundation of China (No. 62166027), and Jiangxi Provincial Natural Science Foundation (Nos. 20212ACB212004, 20212BAB202023, and 20212BAB202022).

References

1. Tanabe, R., Ishibuchi, H.: A review of evolutionary multimodal multiobjective optimization. IEEE Trans. Evol. Comput. **24**(1), 193–200 (2020)
2. Deb, K., Jain, H.: An evolutionary many-objective optimization algorithm using reference-point-based nondominated sorting approach, part I: solving problems with box constraints. IEEE Trans. Evol. Comput. **18**(4), 577–601 (2014)

3. Bader, J., Zitzler, E.: HypE: an algorithm for fast hypervolume based many-objective optimization. Evol. Comput. **19**(1), 45–76 (2011)
4. Zhang, Q., Li, H.: MOEA/D: A multiobjective evolutionary algorithm based on decomposition. IEEE Tran. Evol. Comput. **11**(6), 712–731 (2007)
5. Li, H., Zhang, Q.: Multiobjective optimization problems with complicated Pareto sets, MOEA/D and NSGA-II. IEEE Trans. Evol. Comput. **13**(2), 284–302 (2009)
6. Zhang, Q., Liu, W., Li, H.: The performance of a new version of MOEA/D on CEC09 unconstrained MOP test instances. In: Proceedings of the IEEE Congress on Evolutionary Computation, pp. 203–208 (2009)
7. Akay, B., Karaboga, D.: A modified artificial bee colony algorithm for realparameter optimization. Inf. Sci. **192**, 120–142 (2012)
8. Karaboga, D., Gorkemli, B.: A quick artificial bee colony (qABC) algorithm and its performance on optimization problems. Appl. Soft Comput. **23**(1), 227–238 (2014)
9. Wang, H., Wang, W.J., Xiao, S.Y., Cui, Z.H., Xu, M.Y., Zhou, X.Y.: Improving ar tififficial Bee colony algorithm using a new neighborhood selection mechanism. Inf. Sci. **527**, 227–240 (2020)
10. Xiao, S., Wang, H., Wang, W., Huang, Z., Zhou, X., Xu, M.: Artificial bee colony algorithm based on adaptive neighborhood search and Gaussian perturbation. Appl. Soft Comput. **100**, 106955 (2021)
11. Ye, T.Y., Zeng, T., Zhang, L.Q., Xu, M., Wang, H., Hu, M.: Artificial bee colony algorithm with an adaptive search manner. In: Neural Computing for Advanced Applications, pp. 486–497. Springer Singapore, Singapore (2021). https://doi.org/10.1007/s00521-022-06981-4
12. Zeng, T., Ye, T., Zhang, L., Xu, M., Wang, H., Hu, M.: Population diversity guided dimension perturbation for artificial bee colony algorithm. In: Zhang, H., Yang, Z., Zhang, Z., Wu, Z., Hao, T. (eds.) NCAA 2021. CCIS, vol. 1449, pp. 473–485. Springer, Singapore (2021). https://doi.org/10.1007/978-981-16-5188-5_34
13. Zeng, T., et al.: Artificial bee colony based on adaptive search strategy and random grouping mechanism. Expert Syst. Appl. **192**, 116332 (2022)
14. Ye, T.Y., et al.: Artificial bee colony algorithm with efficient search strategy based on random neighborhood structure. Knowl.-Based Syst. **241**, 108306 (2022)
15. Ye, T.Y., Wang, H., Wang, W.J., Zeng, T., Zhang, L.Q., Huang, Z.K.: Artificial bee colony algorithm with an adaptive search manner and dimension perturbation. Neural Computing and Applications (2022)
16. Zou, W., Zhu, Y., Chen, H., Zhang, B.: Solving multiobjective optimization problems using artificial bee colony algorithm. Discrete Dyn. Nature Soc. **11**(2), 1–37 (2011)
17. Akbari, R., Hedayatzadeh, R., Ziarati, K., Hassanizadeh, B.: A multi-objective artificial bee colony algorithm. Swarm Evol. Comput. **2**, 39–52 (2012)
18. B. Akay.: Synchronous and asynchronous Pareto-based multi-objective artificial bee colony algorithms. J. Global Opt. **57**(2), 415–445 (2013)
19. Xiang, Y., Zhou, Y., Liu, H.: An elitism based multi-objective artificial bee colony algorithm. Europ. J. Oper. Res. **245**(1), 168–193 (2015)
20. Zhu, G., Kwong, S.: Gbest-guided artificial bee colony algorithm for numerical function optimization. Appl. Math. Comput. **217**(7), 3166–3173 (2010)
21. Yuan, Y., Xu, H., Wang, B., Zhang, B., Yao, X.: Balancing convergence and diversity in decomposition-based many-objective optimizers. IEEE Trans. Evol. Comput. **20**(2), 180–198 (2016)
22. Cheng, R., Jin, Y., Olhofer, M., Sendhoff, B.: A reference vector guided evolutionary algorithm for many-objective optimization. IEEE Trans. Evol. Comput. **20**(5), 773–791 (2016)

LCSW: A Novel Indoor Localization System Based on CNN-SVM Model with WKNN in Wi-Fi Environments

Xuhui Wang[1], Xin Deng[1], Hao Zhang[1(✉)], Kai Liu[2], and Penglin Dai[3]

[1] College of Computer Science and Technology, Chongqing University of Posts and Telecommunications, Chongqing 400065, China
S200231028@stu.cqupt.edu.cn, {dengxin,zhanghao}@cqupt.edu.cn
[2] College of Computer Science, Chongqing University, Chongqing 400044, China
[3] School of Computing and Artificial Intelligence, Southwest Jiaotong University, Chengdu 611756, Sichuan, China

Abstract. In this paper, we propose a novel indoor localization system, which fuses convolutional neural network (CNN) and support vector machine (SVM) model with an upgraded weighted K-nearest neighbor (WKNN) algorithm, called LCSW, to enhance the localization accuracy and robustness of the system. To this end, we propose a two-layer localization scheme. Specifically, in the first locating layer, we primarily partition the whole environment into certain subareas, then continuously collect the sequence data of RSSI in different time and reshape the data format as square matrix to serve as input of the modified CNN-SVM model to locate the target to a subarea. Then, in the second locating layer, the improved WKNN is used to calculate the precise location of the target in the corresponding subarea, which adopts the variance characteristic of Wi-Fi signal to assist the calculation of weights in measuring the distance and the cosine similarity to assist the assignment of weights in computing the coordinate, respectively. Finally, extensive real-world experiments are conducted to demonstrate the effectiveness of the proposed methods.

Keywords: Indoor localization · Convolutional neural network · Support vector machine · Wi-Fi fingerprinting

1 Introduction

In recent years, with the great development of the Internet of Things and mobile Internet, the demands for accurate and universal location-based services (LBS) are increasing. In the past decades, the Global Navigation Satellite System (GNSS) has achieved great success and brought significant commercial value. However, the inability of satellite signals to reach indoors has rendered GNSS useless in indoor environments. On account of the vast development space and huge application prospects [1,2], there have been a lot of indoor localization technologies, such as Wi-Fi [3–6], RFID [7,8], Bluetooth [9,10], Ultra-Wide Band

H. Zhang et al. (Eds.): NCAA 2022, CCIS 1638, pp. 162–176, 2022.
https://doi.org/10.1007/978-981-19-6135-9_13

(UWB) [11,12], etc. Among them, Wi-Fi has been the most frequently used technology due to its high availability and coverage. In many indoor environments (e.g., shopping mall, airport, hospital, etc.), Wi-Fi is accessible through the intelligent mobile terminal, which boosts the development of LBS-related applications in indoor localization filed. In this context, on account of easy availability, Received Signal Strength Indication (RSSI) is used to locate in this paper.

Taking advantage of RSSI, the methods available today can be divided into two categories: distance-based methods and fingerprinting-based methods. The idea of the distance-based methods are to build a model of signal strength versus distance, convert the measured signal strength into the distance between the location to be measured and the AP, and calculate the localization to be measured by the distance of multiple APs. Its performance depends on the accuracy of the signal propagation model, which is susceptible to the multipath effects of the received signal. [13]. In addition, the trilateral localization method in the distance-based methods rely heavily on the distance measurement between target and the access points(APs) to calculate the final uncertainty points. In sharp contrast, the fingerprinting-based methods treat the physically measurable attributes of Wi-Fi as fingerprints of each discrete spatial point to distinguish locations, which makes RSSI fingerprinting-based localization methods more resistant to multipath effects. Besides, the fingerprinting-based localization method does not require the addition of other sensors in the experimental environment. However, due to various interferences, the RSSI value is unstable, which downgrades localization accuracy.

Fingerprinting-based methods generally include offline training phase and online localization phases. In offline training phase, some reference points (RPs) are set in the experimental environment, and RSSI fingerprints with location are collected by mobile terminal devices (e.g., smartphone, pad, etc.) to form fingerprint database (radio map) automatically. In the online localization phase, the real-time RSSI measurements are collected to match with the offline fingerprint database and figure out the location of target by corresponding algorithm. The matching process can be achieved by using machine learning techniques such as K-nearest Neighbor (KNN) [14], weighted K-nearest Neighbor (WKNN) [15], cosine similarity (COS) [16], etc. The conventional WKNN uses the inverse of the Euclidean distance between the reference point and the test point to weight the coordinates of the reference point. In order to obtain more accurate localization results, [17] proposes a new weighting algorithm based on the physical distance of RSSI, which achieves good results. [18] also proposes a new weighting algorithm based on physical distance and cosine similarity, which breaks through the new weighting method.

In addition, noise signals have a certain influence on the accuracy of localization, and Kalman filtering is a frequently used method [19], which can effectively eliminate the influence of noise signals and improve the localization accuracy. The Wi-Fi fingerprint localization method alone is prone to fluctuations in localization accuracy, so many scholars have studied the combination of Wi-Fi fingerprint localization with other sensor localization [20] to improve the localization accuracy. In addition, high-precision indoor fingerprint localization algorithms are also highly dependent on the accuracy of the database, and in view of this situation, some scholars have proposed the idea of zonal localization based on spatial characteristics [21] to solve the problem that a single fingerprint database cannot achieve full coverage. With successful applications in imaging, deep neural networks [22] have been applied in the field of indoor localization. In view of the above studies, to better balance localization accuracy and real-time performance, we propose an indoor localization method based on a novel CNN-SVM model and an improved WKNN algorithm called LCSW. In this work, we increase the dimensionality of the image input by collecting fingerprint information at different time periods in the offline phase. In addition, a support vector machine (SVM) is added after the CNN training process in order to obtain more accurate classification results in the testing phase. To better harvest the signal features, we partition the localization region into several sub-regions and train the CNN-SVM model in each sub-region. The trained parameters are stored in an offline database and used to locate the coarse-grained position of the target. In the online phase, considering the real-time problem of the system, we collect the RSSI values of all APs only once and fill the same values into the 2D image input. The classification result is the number of sub-regions. To obtain fine-grained locations, we further use an improved WKNN algorithm to localize the corresponding sub-regions by CNN-SVM model, which saves a significant amount of matching time compared to the global retracing time and avoids sharp jumps due to signal fluctuations or acquisition failures.The main contributions of this paper can be summarized as follows:

1) We propose a localization system called LCSW and implement the prototype, which contains a coarse-gained location through the proposed CNN-SVM model in indoor localization to determine a general area and on this basis to calculate a fine-grained location by an improved WKNN algorithm.

2) In order to better extract the signal feature of RSSI, we propose to use CNN model in the offline phase. Different from the CNN model used in traditional indoor localization, we use SVM to replace the Softmax layer in CNN model for a better classification result in terms of small sample. Besides, we apply this model in the first layer localization to acquire a coarse-gained location by dividing the environment into subareas. Through this way, the computational overhead can be much reduced.

3) We use a modified WKNN in the second layer of localization to calculate the exact position of the target in the corresponding subregion. The variance of the Wi-Fi signal is used as an auxiliary weight in the distance calculation, reducing the effect of unstable signals in this way. The advantages of Euclidean distance and cosine similarity are fully combined by using Euclidean distance to select K pending points in coordinate calculation and using cosine similarity to assist in weight assignment.

4) We embed the proposed model into the system prototype and carry real-world experiment for performance evaluation. We test different sizes of input format by using different numbers of APs in different environments. The com-

prehensive experiment results conclusively demonstrate that the proposed methods can effectively improve the localization accuracy, reduce the offline cost and maintain a good robustness.

The rest of this paper is organized as follows: section II simply reviews the related works. The details of the system architecture are elaborated in section III. Section IV gives a comprehensive analysis of the proposed solutions. Section V evaluates the proposed methods in real-world. Finally, Section VI concludes this paper and discusses pertinent future research directions.

2 Related Work

CNN first achieves success in the image field, but with the continuous progress of training methods and graphics processing units, CNNs have started to venture into other fields as well. Mittal et al. [23] propose an indoor localization method based on CNN. The method focuses on organizing the RSSI values in the database into a time-frequency matrix similar to that of an image. As a feature of localization, it models localization as a classification problem, and uses a five-layer CNN model consisting of three convolution layers and two fully connected layers to solve the localization problem. Tasaki et al. [24] propose a 3D-CNN model, which can take advantage of temporal fluctuations that cannot be captured by fingerprints to handle the 3D spatio-temporal structure of RSSI datasets and improve localization accuracy. Other scholars also consider the influence of time series issues on the CNN training model, Li et al. [25] design a framework based on deep learning to eliminate the influence of signal variation and device heterogeneity contained in the collected RSSI fingerprints, and extract device independent and anti-dynamic features.

In addition to short time series RSSI wave features, some researchers have tried to extract more useful features from other aspects. Qin et al. [26] design a new network model combining CDEA and CNN. In this model, the CDAE network is used to reduce the dimensionality of data, and the training process of adding noise and performing noise reduction is used to force the network to learn more robust invariant features and obtain more efficient input representation. Chen et al. [27] extend convolutional Neural network (D-CNN) and support vector regression (SVR) models are used for localization. Only one RSSI value is collected for each localization, and more useful features are extracted by expanding the receiving domain through extended convolution. With the development of other network structures, Qian et al. [28] propose a mixed density model (MDN) combining CNN and RNN. The role of CNN sub-model is to detect the features of high-dimensional input. The role of RNN sub-model is to capture the time dependence, while the MDN sub-model predicts the final output, but this leads to high time complexity and time-consuming localization. The above mentioned paper does a good job in extracting more features, but the same image dimensions used for training and testing do not sufficiently consider the real-time problem and do not realize real-time localization of the system, but use the test set for online localization verification. In this paper, high-dimensional data with different time series (i.e., temporal and spatial information of RSSI) are used as training data for CNN models, and single-dimensional data (i.e., spatial information of RSSI) are used for real-time localization in online localization, in addition, SVM is used to improve the accuracy of localized subregions. After obtaining the subregions of the target location, the final localization is performed using the improved WKNN. Due to the limitation of subregions, not only the drift problem of the original WKNN method is reduced, but also the consumption of position matching is reduced, thus improving the localization effect.

3   Architecture

The whole system architecture is based on Fingerprinting localization method, which includes two phases: offline training phase and online localization phase. In the offline training phase, we first partition the whole localization area into some subareas and collect RSSI data in the central point (CP) of each subarea. Then, a padding process is conducted to fill up the missing data in part dimensions. After that, we normalize these data and assemble them as matrix format base on the temporal property. Put these data as input of the proposed CNN-SVM model to train the related parameters. The trained parameters construct the first database. Then, build second database by collecting RSSI data in each preset reference point within each subarea via the traditional method. In the online localization phase, the online measured RSSI data are primarily used to located the corresponding subarea via the proposed CNN-SVM model and then compute the specific localization with the second database under an improved WKNN method.

The whole system architecture is shown in Fig. 1. Suppose there are N APs in the environment. One set recording in the CP of one subarea can be presented as $< RSSI_1, RSSI_2, ..., RSSI_N >$, and its label is the serial number of corresponding subareas. To enable CNN model, we collect certain recordings in every CP for a period of time and reshape the raw data as the format of

$$RI = \begin{matrix} RSSI_{11} & RSSI_{12} & ... & RSSI_{1N} \\ RSSI_{21} & RSSI_{22} & ... & RSSI_{2N} \\ ... & ... & ... & ... \\ RSSI_{N1} & RSSI_{N2} & ... & RSSI_{NN} \end{matrix} \tag{1}$$

which can be viewed as a $N \times N$ pixel image. Take these matrixes as input of the proposed CNN-SVM model after a normalization process to train involved parameters. The trained parameters form the first database. Then, we recollect RSSI in every preset RP within each subarea and denote ith recording as $< RSSI_i, x_i, y_i >$. After a padding process, these data form the second database. In the online phase, the measured real-time data cannot be used as the input of CNN-SVM model directly, so the system repeats the collected RSSI data enough times to form a $N \times N$ format data to determine which subarea the target belongs to. Then, the final location of the target is calculated by the proposed improved WKNN method with the initial real-time RSSI data.

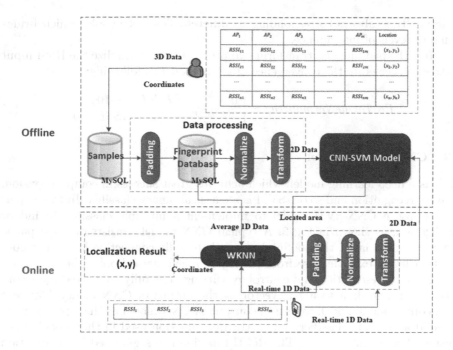

Fig. 1. System architecture

4   Proposed Model

4.1   Data Preprocessing

Due to the noise of the environment, the outliers are unavoidable in general scenes. We delete the maximum and minimum values of each set data and pad them with the average value of the corresponding rest data. Compared with local average value, the average of rest value of each set data is more stable and not susceptible to other contiguous RSSI values. Besides, there exists absence of the collected RSSI vector in certain dimensions, which also leads the raw data cannot be used to train the neural network or locate directly. Similarly, we pad the missing data with average value of the rest data.

The collected RSSI are one-dimensional vector. However, the convolution kernel in CNN is based on the assumption that adjacent elements have a certain spatial relationship. Therefore, the feature representation ability of one-dimensional RSSI vector as the input of network model is limited and cannot promote the extraction of regional features. To enable CNN model, we reshape the raw RSSI data as square matrix format by utilizing its temporal property. Through continuous sampling in each CP, we collect the same number data with the dimension of RSSI vector, which guarantees the input of CNN model is a square matrix. Besides, to contain more environment information, we collect RSSI data in different time (i.e., morning, noon, afternoon, evening) of one day.

Different time always corresponding to different flows of people, which brings different levels of interference.

After the padding and transform two processes, we normalize the RSSI input matrix to accelerate the convergence of CNN model as following:

$$EroToOneNormalized_i = \begin{cases} 0 & RSSI_i = -100 \\ \frac{RSSI_i - min}{-min} & -99 \leq RSSI_i \leq 0 \end{cases} \tag{2}$$

4.2   CNN-SVM Model

CNN is a deep learning model, which achieves great success in computer vision, activity recognition and so on. SVM is a classical binary classifier. In this paper, we propose a CNN-SVM model to implement a multiple classifier in indoor localization, as shown in Fig. 2. A typical CNN model consists of three parts: the convolution layer, the pooling layer and the fully connection layer. The convolution layers extra feature from the input image data. Pooling layers reduce the size of representation and control overfitting. The fully connected layer converts the image data to a 1-D vector as the output of CNN model. Different from conventional model, we use the output of CNN model as the input of SVM to build a multiple classifier. The activation function used in this model is the rectified linear unit (ReLU). The ReLU function is a segmented linear function commonly used in neural network models. The unilateral inhibition function, which makes the neurons in the neural network sparsely activated, improves the nonlinear relationship between the layers of the neural network and maintains the convergence rate in the steady state model, thus making better use of the correlation features of the RSSI values. The CNN uses a model with two convolutional layers, two pooling layers, and one connection layer, and the exact number of convolutional layers is presented in the experimental section. The number of convolutional layers to obtain the optimal localization accuracy is different for different datasets. To prevent overfitting, a dropout layer is added after the fully connected layer. After fully training the temporal and spatial information of RSSI using the CNN model, the one-dimensional data tiled by the fully connected layer is used as the input of the SVM model for classification, and the final output classification result is the subregion we divide. Considering that the Softmax function is more accurate for classifying samples with large amount of data, and the advantage of SVM is accurate for classifying small data samples, while we need to consume a lot of human effort to collect data in real environment, and it is not realistic to create very large data samples, so we choose SVM to replace the Softmax function for the final sub-region classification, and then use WKNN algorithm in the sub-region. The final localization result, i.e., the location coordinates, is obtained.

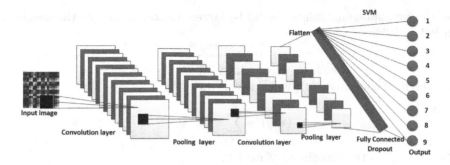

Fig. 2. The structure of CNN-SVM model

4.3 Improved WKNN Algorithm

After the subarea is determined, an improved WKNN algorithm is proposed to calculated the localization of the target, which contains two improvements: the measurement of Euclidean distance and the definition of weight.

In each RP within the subarea, we collect RSSI data certain times and store the average value in each dimension as the final fingerprint. Besides, we calculate the variance to evaluate the quality of RSSI value in each dimension. Bigger variance means the RSSI is more unstable, which should be assigned a smaller weight in calculating the distance. Suppose the fingerprint and variance of ith RP are $rp_i = (\overline{rssl_{l1}}, \overline{rssl_{l2}}, ..., \overline{rssl_{lN}})$ and $\sigma_i = (\sigma_{i1}, \sigma_{i2}, ..., \sigma_{iN})$. The measured real-time test point is $tp = (rssi_1, rssi_2, ..., rssi_N)$.Based on above ideas, we measure the weighted Euclidean distance between test point and RP as follows:

$$D_i = \sqrt{\sum_{j=1}^{N} (\overline{rssl_{lJ}} - rssi_j) \cdot \omega_{ij}} \qquad (3)$$

where the ω_{ij} is the assigned weight in the jth dimension of rp_i and which is calculate as follows:

$$\omega_{ij} = \frac{\frac{1}{\sigma_{ij}+1}}{\sum_{k=1}^{N} \frac{1}{\sigma_{ik}+1}} \qquad (4)$$

Through this way, the influence caused by fluctuation of RSSI signal can be alleviated and localization accuracy can be improved. Then, we choose K nearest RPs to the test point with above weighted Euclidean distance. Based on the chosen K RPs, here we use cosine similarity to assign weights to each RP. First, calculate the cosine similarity between the test point and RP as follows:

$$\cos \theta_i = \frac{\sum_{j=1}^{N} \overline{rssl_{lJ}} \cdot rssi_j}{\sqrt{\sum_{k=1}^{N} (\overline{rssl_{lk}})^2} \cdot \sqrt{\sum_{l=1}^{N} (\overline{rssi_l})^2}} \qquad (5)$$

The K cosine similarity values are presented as $C = (\cos \theta_1, \cos \theta_2, ..., \cos \theta_K)$. Larger cosine similarity means more closer of the TP and RP in signal space, and the corresponding weight should be larger. Hence, we denote the weight of jth RP as follows:

$$\eta_j = \frac{\cos\theta_j}{\sum_{i=1}^{K}\cos\theta_i} \tag{6}$$

And the final coordinate of the TP is calculated by:

$$(\hat{x},\hat{y}) = \sum_{i=1}^{K}\eta_i(x_i,y_i) \tag{7}$$

where (x_i, y_i) is the coordinate of ith RP.

5   Experimental Evaluation

5.1   Setting

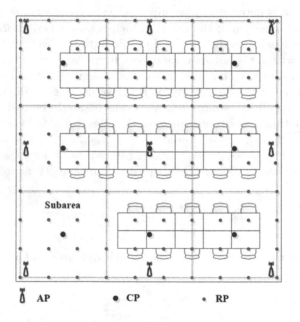

Fig. 3. The experimental configuration

The experiment is conducted in the laboratory, and the floorplan of which is shown in Fig. 3. We partition the whole zone into 9 subareas, which contains 9 CP and 100 RPs (part RP are commonly used with other adjacent subareas). The CP is presented as blue dot and RP is denoted by red dot. Neighboring RPs are separated by one meter. Besides, 9 APs are deployed in the environment. In the offline phase, data collected in each CP and RP are the input of CNN-SVM to train involved parameters and locate the target, respectively. In the online phase, we collect data at anywhere of this laboratory. We pad the missing value with −100 dBm.

5.2   Model Training

Table 1. Network structure configuration

#	Network structure
1	Conv (64-1)+max-1+ Conv (32-1)+max-1
2	Conv (64-3)+max-2+ Conv (32-3)+max-2
3	Conv (64-5)+max-2+ Conv (32-4)+max-2
4	Conv (64-4)+max-3+ Conv (32-4)+max-2
5	Conv (64-3)+max-2+ Conv (32-5)+max-2
6	Conv (64-1)+max-2+ Conv (32-3)+max-2

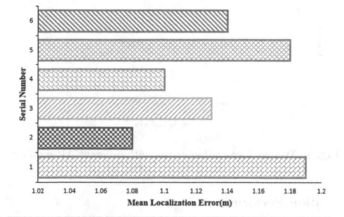

Fig. 4. The comparison of different network structures

Considering the network structure have a big influence on the localization performance, we carry a series of real world experiments to evaluate different network configurations of CNN-SVM model. As Table 1 shows, we compare 6 configurations, which contains different convolution layers and pooling layers. The corresponding localization performance are presented in Fig. 4. As we can see that the second configuration achieves the best localization performance with 1.08m average localization error. The network structure has two convolutional layers and two pooling layers. The first convolutional layer has 64 output filters with convolutional kernel size 3, and the second convolutional layer has 32 output filters with convolutional kernel size 3 and the pooling layer has a maximum pooling window size of 2.

AP number is another important factor that affects RSSI feature directly. In the experiments, we change the number of AP from 4 to 9 to analyze its influence on localization accuracy, and the results are shown in Fig. 5.

From Fig. 5 we can see that the average localization error decreases and the downward trend is gradually smoothing out along with the increasing of AP number. More AP brings richer RSSI feature, however, the cost and overhead are also increased. Keep increasing the number of AP may further reduce localization error, but low cost effective. In the following experiments, we fix the AP number at 9.

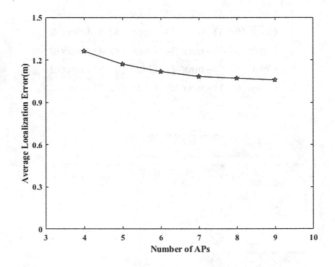

Fig. 5. Average localization errors with different AP number

5.3 Localization Performance

To evaluate the superiority of the proposed system. We compare the proposed system with other methods (i.e., KNN, WKNN, CNN) in localization performance with the same experimental configuration. In the online phase, we test 100 points. The CDF of localization errors of different methods are presented in Fig. 6. From the Fig. 6 we can see that the proposed method has the best locating performance, which achieves 95% localization error within 2m and improves 36%, 12% and 22% comparing with other three methods respectively. The average localization error of the proposed method is 1.05m, which surpass other three methods 35%, 20% and 24% respectively.

Furthermore, we assess the robustness of the proposed method in different times of one day (i.e., morning, noon, afternoon, evening, respectively). As the Fig. 7 shows, the locating performance of the four times are similar, which shows the proposed method have a good robustness in the same space.

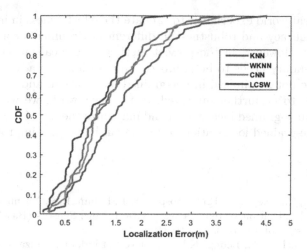

Fig. 6. The CDF curves of different methods

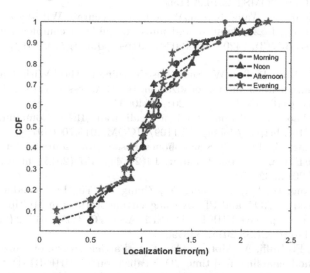

Fig. 7. The CDF curves of different times

6   Conclusion

In this paper, we propose a new localization system based on a CNN-SVM model
and an improved WKNN method. Through combining the superiority of CNN
and SVM, the CNN-SVM model extracts feature based on the sequential and
spatial characteristics of RSSI signal and locate the target in a subarea. Then,
by analyzing the fluctuation of RSSI and adopting its variance in each dimension
to assist the calculation of weight in measuring the distance when retrieving RPs
from the database. After that the cosine similarity is utilized to assigned weight
rather than Euclidean distance in computing the coordinate of target. Finally,

we carry out real-world experiment to validate the effectiveness in improving the localization accuracy and robustness in different environments comparing with other methods. However, in the present method, the localization accuracy at the region boundary needs to be improved. In addition, the advantages of area localization are not maximized in fine-grained localization, and the localization accuracy needs to be further improved. In the next work, we will explore the accuracy of coarse-grained localization and maximize the use of area localization so that the fine-grained localization accuracy can be further improved.

References

1. Pascacio, P., Casteleyn, S., Torres-Sospedra, J., Lohan, E.S., Nurmi, J.: Collaborative indoor localization systems: a systematic review. Sensors (Basel) **21**(3), 1002 (2021). https://doi.org/10.3390/s21031002
2. Zafari, F., Gkelias, A., Leung, K.K.: A survey of indoor localization systems and technologies. IEEE Commun. Surv. Tutorials **21**(3), 2568–2599 (2019). https://doi.org/10.1109/COMST.2019.2911558
3. Zhu, Q., Xiong, Q., Wang, K., Lu, W., Liu, T.: Accurate WiFi-based indoor localization by using fuzzy classifier and mlps ensemble in complex environment. J. Franklin Inst. **357**(3), 1420–1436 (2020). https://doi.org/10.1016/j.jfranklin.2019.10.028
4. Qwn, C.-M., Hou, J., Tao, W.: Signal fuse learning method With dual bands WiFi signal measurements in indoor localization. IEEE Access **7**, 131805–131817 (2019). https://doi.org/10.1109/ACCESS.2019.2940054
5. Yang, C., Shao, H.: WiFi-based indoor localization. IEEE Commun. Mag. **53**(3), 150–157 (2015). https://doi.org/10.1109/MCOM.2015.7060497
6. Wu, C., Yang, Z., Liu, Y.: Smartphones based crowdsourcing for indoor localization. IEEE Trans. Mobile Comput. **14**(2), 444–457 (2015). https://doi.org/10.1109/TMC.2014.2320254
7. Ma, Y., Wang, B., Pei, S., Zhang, Y., Zhang, S., Yu, J.: An indoor localization method based on AOA and PDOA using virtual stations in Multipath and NLOS environments for passive UHF RFID. IEEE Access **6**, 31772–31782 (2018). https://doi.org/10.1109/CCESS.2018.2838590
8. Bernardini, F., Buffi, A., Motroni, A., et al.: Particle swarm optimization in SAR-based method enabling real-time 3D localization of UHF-RFID tags. IEEE J. Radio Frequency Identif. **4**(4), 300–313 (2020). https://doi.org/10.1109/JRFID.2020.3005351
9. Kotrotsios, K., Orphanoudakis, T.: Accurate gridless indoor localization based on multiple bluetooth beacons and machine learning. In: 2021 7th International Conference on Automation, Robotics and Applications (ICARA), pp. 190–194 (2021). https://doi.org/10.1109/ICARA51699.2021.9376476
10. Obreja, S.G., Vulpe, A.: Evaluation of an indoor localization solution based on bluetooth low energy Beacons. In: 2020 13th International Conference on Communications (COMM), pp. 227–231 (2020). https://doi.org/10.1109/COMM48946.2020.9141987
11. Poulose, A., Han, D.S.: Feature-based deep LSTM network for indoor localization using UWB measurements. In: 2021 International Conference on Artificial Intelligence in Information and Communication (ICAIIC), pp. 298–301 (2021). https://doi.org/10.1109/ICAIIC51459.2021.9415277

12. Zhang, S., Han, R., Huang, W., Wang, S., Hao, Q.: Linear bayesian filter based low-cost UWB systems for indoor mobile robot localization. In: 2018 IEEE SENSORS, pp. 1–4 (2018). https://doi.org/10.1109/ICSENS.2018.8589829
13. You, Y., Wu, C.: Indoor localization system with cellular network assistance based on received signal strength indication of Beacon. IEEE Access 8, 6691–6703 (2020). https://doi.org/10.1109/ACCESS.2019.2963099
14. Xue, W., Hua, X., Li, Q., Qiu, W., Peng, X.; Improved clustering algorithm of neighboring reference points based on KNN for indoor localization. In: 2018 Ubiquitous Localization, Indoor Navigation and Location-Based Services (UPINLBS), pp. 1–4 (2018). https://doi.org/10.1109/UPINLBS.2018.8559874
15. Liu, W., Fu, X., Deng, Z., Xu, L., Jiao, J.: Smallest enclosing circle-based fingerprint clustering and modified-WKNN matching algorithm for indoor localization. In: 2016 International Conference on Indoor Localization and Indoor Navigation (IPIN), pp. 1–6 (2016). https://doi.org/10.1109/IPIN.2016.7743694
16. Han, S., Zhao, C., Meng, W., Li C.: Cosine similarity based fingerprinting algorithm in WLAN indoor localization against device diversity. In: 2015 IEEE International Conference on Communications (ICC), pp. 2710–2714 (2015). https://doi.org/10.1109/ICC.2015.7248735
17. Xue, W., et al.: A new weighted algorithm based on the Uneven spatial resolution of RSSI for Indoor localization. IEEE Access 6, 26588–26595 (2018). https://doi.org/10.1109/ACCESS.2018.2837018
18. Han, X., Yang, G., Qu, S., Zhang, G., Chi, M.: A weighted algorithm based on physical distance and cosine similarity for Indoor localization. In: 2019 14th IEEE Conference on Industrial Electronics and Applications (ICIEA), pp. 179–183 (2019). https://doi.org/10.1109/ICIEA.2019.8833982
19. Yang, B., Jia, X., Yang, F.: Variational bayesian adaptive unscented kalman filter for RSSI-based Indoor localization. Int. J. Control Autom. Syst. 19, 1183–1193 (2021). https://doi.org/10.1007/s12555-019-0973-9
20. Guo, S., Niu, G., Wang, Z., Pun, M.-O., Yang, K.: An Indoor knowledge graph framework for efficient pedestrian localization. IEEE Sensors J. 21(4), 5151–5163 (2021). https://doi.org/10.1109/JSEN.2020.3029098
21. Yang, H., Zhang, Y., Huang, Y., Fu, H., Wang, Z.: WKNN indoor location algorithm based on zone partition by spatial features and restriction of former location. Pervas. Mobile Comput. 60, 1192–1574 (2019). https://doi.org/10.1016/j.pmcj.2019.101085
22. Chen, S., Zhu, Q., Li, Z., Long, Y.: Deep neural network based on feature fusion for Indoor wireless localization. In: 2018 International Conference on Microwave and Millimeter Wave Technology (ICMMT), pp. 1–3 (2018). https://doi.org/10.1109/ICMMT.2018.8563629
23. Mittal, A., Tiku, S., Pasricha, S.: Adapting Convolutional Neural Networks for Indoor Localization with Smart Mobile Devices, pp. 117–122. Association for Computing Machinery (2018). https://doi.org/10.1145/3194554.3194594
24. Tasaki, K., Takahashi, T., Ibi, S., Sampei, S.: 3D convolutional neural network-aided Indoor localization based on fingerprints of BLE RSSI. In: 2020 Asia-Pacific Signal and Information Processing Association Annual Summit and Conference (APSIPA ASC), pp. 1483–1489 (2020)
25. Li, D., Xu, J., Yang, Z., Lu, Y., Zhang, Q., Zhang, X.: Train once, locate anytime for anyone: adversarial Learning based wireless localization. In: IEEE INFOCOM 2021 - IEEE Conference on Computer Communicatlons, pp. 1–10 (2021). https://doi.org/10.1109/INFOCOM42981.2021.9488693

26. Qin, F., Zuo, T., Wang, X.: CCpos: WiFi fingerprint Indoor localization system based on CDAE-CNN. Sensors **21**(4), 1114 (2021). https://doi.org/10.3390/s21041114
27. Chen, H., Wang, B., Pei, Y., Zhang, L.: A WiFi Indoor localization method based on dilated CNN and support vector regression. In: 2020 Chinese Automation Congress (CAC), pp. 165–170 (2020) https://doi.org/10.1109/CAC51589.2020.9327326
28. Qian, W., Lauri, F., Gechter, F.: Convolutional Mixture Density Recurrent Neural Network for Predicting User Location with WiFi Fingerprints. arXiv2019 (2019)

Large Parallax Image Stitching via Structure Preservation and Multi-matching

Yuanyuan Chen, Wanli Xue[✉], and Shengyong Chen

School of Computer Science and Engineering, Tianjin University of Technology,
Tianjin 300384, China
onachan@stud.tjut.edu.cn, xuewanli@email.tjut.edu.cn

Abstract. In image stitching tasks, parallax ghosts and composite ghosts widely exist in stitching results whose input images usually have complex moving objects (especially multiple pedestrians) and large parallax scenes, which bring greater challenges to image stitching. In order to efficiently remove these ghosts and maintain visually natural stitching results, we propose a novel image stitching method via structure preservation and multi-matching (SPMM). Specifically, the method is composed of the structure preservation module based on grid constraints that eliminates parallax ghosts by ensuring the alignment of overlapping regions and keeping object structures in non-overlapping regions; and the composite ghost removal module based on Multi-Matching, which effectively eliminates composite ghosts caused by multiple pedestrians and obtains natural stitching results. The extensive experiments on five different image stitching frameworks and real datasets demonstrate the effectiveness and advantages of the proposed method.

Keywords: Image stitching · Structure preservation · Multi-matching

1 Introduction

Image stitching is the process of combining two or more images with overlapping regions into a high-resolution, natural and wide-field panorama [1]. It is an important technique in computer vision, widely used in many tasks: such as panoramic image synthesis, remote sensing, unmanned aerial vehicle (UAV) photographyand virtual reality, etc.

When the camera baseline is too large or the captured scene is not a planar, the input images will present parallax problems, where the relative position of the motionless objects changes, resulting in the parallax ghost [2] and structural distortion in the stitching results. In addition, when there are moving multiple pedestrians in the input images, the relative position of the pedestrians and the background changes due to the movement. The moving pedestrians may be cut or duplicated in the stitching results, leading to the composite ghost [3]. Therefore, both large parallax scenes and complex multiple pedestrians will bring great challenges to image stitching tasks.

H. Zhang et al. (Eds.): NCAA 2022, CCIS 1638, pp. 177–191, 2022.
https://doi.org/10.1007/978-981-19-6135-9_14

Fig. 1. Comparisons of image stitching methods [4,5]. Evident ghosts and deformations are highlighted in red bounding boxes and our method does not have these unpleasant effects as shown by the green box. (Color figure online)

Early,the global warps [6,7] strive to minimize the alignment error between pixels in overlapping regions through a global homography. Works in [4,8–10] mainly focus on the spatially varying warps by dividing the images into multiple grids to compute the local homographies for better alignment optimization. However, when facing with the challenge of large parallax, a group of adjacent pixels in one image may not have the corresponding adjacent pixels in the other image, local homographies will produce inconsistencies, causing serious parallax ghosts. In addition, the border areas of images are stretched easily, which damage the structure and produce deformation. As shown in Fig. 1(a), APAP [4] fails to align in overlapping areas, resulting in obvious parallax artifacts, while bending the vertical relationship between the stone pillars and the ground.

In addition, when facing difficulties with complex multiple pedestrians, the seam-driven warps [11–14] are dominated by finding the optimal seam, which is searched in the overlapping region by optimizing the pixel value dependent penalty function. OGIE [5] proposes a motion-aware ghost identification and elimination strategy to ensure that pedestrians in the stitching results will not be distorted, cut and duplicated. However, the seam-driven methods are prone to fall into local optima because of misdirection by complex multiple pedestrians in overlapping areas, resulting in failure to find an optimal seam. Work in [5] cannot identify composite ghosts well when the complex multiple pedestrians are too small or excessive deformation, pedestrians are cut and the corresponding pedestrians appear twice in the final stitching result (see Fig. 1(b)).

In this paper, an image stitching algorithm is put forward to solve the challenges brought by large parallax scenes and complex multiple pedestrians to the stitching tasks, so as to ensure the naturalness of the entire stitching results. As shown in Fig. 1(c), our method not only effectively removes parallax ghosts, preserves the entire structure, but also successfully removes all composite ghosts and maintains a natural stitching result. Specifically:

- In order to minimize the parallax ghost and maintain the overall structure of our stitching result, structure preservation based on grid constraints is designed by adding constraints to the energy term (see Sect. 3.2).
- In order to effectively eliminate the composite ghost, ensure that the pedestrians in the stitching result satisfy non-distortion, uniqueness and integrity, the

composite ghost removal based on Multi-Matching is proposed to dedicate to a natural panorama (see Sect. 3.3).

2 Related Work

This paper is dedicated to solving the challenges brought by large parallax and complex multiple pedestrians to the stitching tasks, so this section reviews previous work from the two aspects.

2.1 Large Parallax Image Stitching Method

Most of the existing image stitching methods focus on solving the parallax ghosts and structural distortion problems caused by large parallax in static scenes.

The traditional methods usually estimate an optimal global warp (such as affine, similarity or homography) on the input images, and use an average blending method to blend the two images. In 2007, AutoStitch [7] proposed by Brown M et al. also aligned images through a global homography. But these methods are only suitable for images with translation and small rotation.

Therefore, some methods no longer use a global warp and try to use multiple homographies to align different areas of the image. Gao et al. proposed a dual-homography (DH) warp [8], which used two homographies to align the background and foreground for further enhancing the aligned performance.

Zaragoza J et al. [4] proposed the as-projective-as-possible (APAP) algorithm in 2013, which divided the images into several uniform grids and computed multiple local homographies to align images. Advanced the use of APAP, Lin et al. [10] proposed an adaptive method to adjust the Gaussian weights in local homographies. However, these methods are still not flexible enough to solve parallax ghosts and structural distortion in large parallax scenes.

Inspired by APAP, experts try to use grid optimization to improve the ability of image stitching. Global Similarity Prior (GSP) [15] used APAP to initialize the grid, which improved the natural appearance by minimizing an energy function composed of aligned, local and global similar terms. Recently, SPW [16] proposed by Liao et al. simultaneously emphasized different features under single-perspective warps, including alignment, distortion, and saliency. The work in [14] integrated point-line features as alignment items and used the pixel value difference evaluation model to iteratively compute and update the smoothing term for the purpose of finding the most suitable seam. Jia et al. [17] proposed the characteristic number to match co-planar local sub-regions for input images, and introduced global collinear structures into an objective function. Though those methods alleviate parallax artifacts to a certain extent, they still cannot avoid the distortion of image boundary regions.

2.2 Image Stitching with Complex Multiple Pedestrians

Currently, few studies have been done on removing composite ghosts caused by moving complex multiple pedestrians.

James Davis *et al.* [18] segmented the image into disjoint regions and sampled the pixels in each region, using Dijkstra's algorithm to find seams for avoiding blur caused by pedestrian areas. The work in [19] focused on removing pedestrians from Google Street View images, based on Liebe [20] to extract pedestrian bounding boxes, using a seam-driven approach to remove artifacts. Pulli *et al.* [21] were dedicated to creating high-resolution panoramic images, using seam-finding to eliminate ghosting effects caused by moving pedestrians. Boutellier *et al.* [22] detected multiple pedestrians in binary difference images, but the detection method was limited by image quality and scene. In addition, the work in [23] found that blur or ghost become prominent when pedestrians move in multiple images, and proposed to minimize artifacts caused by multiple pedestrians based on a video stitching framework. An object-aware ghost removal method [5] was proposed, which used SSD method to detect moving pedestrians in the scenes.

Most of the existing stitching methods for multiple pedestrians use seam-driven methods to remove composite ghosts, but when the moving pedestrians become more complex, seam-driven methods may not find optimal seam, causing pedestrians to be cut, stretched or copied.

In summary, existing methods struggle to address all the challenges posed by large parallax scenes and complex multiple pedestrians.

3   Proposed Method

3.1   Overview

Two input images are given: the reference image I_1 and the target image I_2 respectively, which has an overlapping area Ω with I_1. Image stitching is to blend I_1 and I_2 into a panorama whose overlapping area has no repeated pedestrians, maintaining a natural result.

For large parallax images with complex multiple pedestrians, the current challenges are mainly in two aspects: on the one hand, large parallax and wide baseline problems will exist due to excessively rotated angles and different photographing positions, causing parallax ghosts and structural distortion.

On the other hand, it is difficult to detect and match complex multiple pedestrians when they are too many, too small or large deformed, which increases the difficulty of removing composite ghosts. Therefore, we comprehensively design an image stitching algorithm, which has higher requirements for structure preservation and ghost removal:

- First, a global pre-warped homography $\hat{H}_*$ is obtained through the dual features of points and lines;
- Then, based on $\hat{H}_*$, the optimal grid warp $\hat{V}$ is solved by adding constraints to the energy terms, then I_1 and I_2 are turned into $\hat{I}_1$ and the warped target image $\hat{I}_2$ by transforming into the same coordinates;
- Finally, potential composite ghosts will be searched and matched on $\hat{I}_1$ and $\hat{I}_2$ based on the idea of Multi-Matching, then eliminated during the seamless image blending process to obtain the final stitching result I_S.

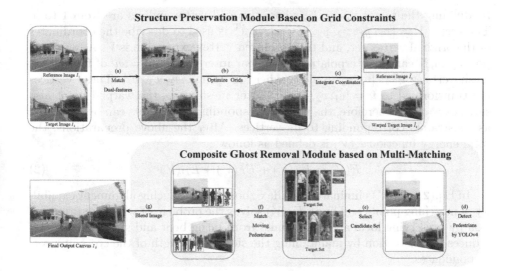

Fig. 2. An outline of our proposed method.(a) Match Dual-features (b) Optimize Grids (c) Integrate Coordinates (d) Detect Pedestrians by YOLOv4 (e)Select Candidate Set (f)Match Moving Pedestrians (g)Blend image.

The outline of our method is shown in Fig. 2 (a)–(c) presents the structure preservation based on grid constraints and (d)–(g) demonstrates the composite ghost removal based on Multi-Matching.

3.2   Structure Preservation Based on Grid Constraints

Pre-alignment: Let $\{(p_i, p_i')\}_{i=1,2,...,N}$ and $\{(l_j, l_j')\}_{j=1,2,...,L}$ be the matched point pairs and the set of matching line pairs respectively, where N and L are their numbers. The line l_j can be represented by its two endpoints $q_j^{0,1}$. In order to achieve a better alignment, the Euclidean distance between matched points and lines is minimized after warp. Therefore, a global homography can be jointly estimated by dual features:

$$\hat{H}_* = \arg\min_H \left(\sum_{i=1}^{N} \|p_i' \times Hp_i\|^2 + \sum_{j=1}^{L} \left\| l_j'^T H q_j^{0,1} \right\|^2 \right) \tag{1}$$

where $\hat{H}_* \in R^{3 \times 3}$ is the desired global homography and $\hat{H}_*$ can easily be minimized by SVD.

Preparation for Mesh Warp: The global homography $\hat{H}_*$ estimated by pre-alignment only provides an approximate warp, there are still inevitable parallax ghosts and structural distortion. In order to optimize local adjustment, a grid-based method is further adopted. In our work, regular grids were obtain through by dividing the target image I_2, where the grid vertex indices are from 1 to n. The vector $V = [x_1 \ y_1 \ x_2 \ y_2 \ \cdots \ x_n \ y_n]^T$ is used to describe the coordinates of the original vertex set, and the optimized vertex expression is $\hat{V}$. Any sample point p in I_2 can be interpolated by bilinear interpolation $p = \omega\nu$ of four closed mesh vertices $\nu = [v_1, v_2, v_3, v_4]^T$, where $\omega = [w_1, w_2, w_3, w_4]$. These vertices are transformed to $\hat{\nu} = [\hat{v}_1, \hat{v}_2, \hat{v}_3, \hat{v}_4]^T$ after mesh warp, the warped point also satisfies $\hat{p} = \omega\hat{\nu}$. Therefore, the point corresponding constraints can be expressed as constraints corresponding to its vertices. After the above preparations, the total energy function $E(\hat{V})$ is defined as follows:

$$E(\hat{V}) = E_A(\hat{V}) + E_L(\hat{V}) + E_D(\hat{V}) \tag{2}$$

In Eq.(2), $E_A(\hat{V})$ eliminates parallax ghosts by enhancing alignment capability of matching points and lines, $E_L(\hat{V})$ solves the problem of structure preservation by preserving local and global lines from being bent and stretched, $E_D(\hat{V})$ reduces the distortion by maintaining the slope and length of the cross-line correspondences.

Alignment Item $E_A(\hat{V})$: In order to further ensure the alignment in the overlapping area, the point-line feature matching is constrained. The alignment term $E_A(\hat{V})$ is divided into point-aligned item $E_p(\hat{V})$ and line-aligned item $E_l(\hat{V})$.

$E_p(\hat{V})$ aligns the matching point features as much as possible, where the matching points $p'_i \in I_1$, $p_i \in I_2$, the warped point can be represented by $\hat{p}_i = \omega_i\hat{\nu}_i$, which ensures the alignment of the overlapping area. $E_l(\hat{V})$ constrains the distance between matching lines to be minimized. The transformed line $\hat{l}_j$ can be represented by $\omega_j^{0,1}\hat{\nu}_j^{0,1}$, the line $l'_j \in I_1$ can be represented by $a_jx + b_jy + c_j = 0$ where $\sqrt{a_j^2 + b_j^2} = 1$. $E_l(\hat{V})$ not only ensures that the corresponding lines are well aligned, but also maintains the straightness of the straight line structure. Defined as follows:

$$E_p(\hat{V}) = \sum_{i=1}^{N} \|\omega_i\hat{\nu}_i - p'_i\|^2, \ E_l(\hat{V}) = \sum_{j=1}^{L} \left\| l'_j \cdot \left(\omega_j^{0,1}\right)\hat{\nu}_j^{0,1}/\sqrt{a^2 + b^2} \right\|^2$$

$$E_A(\hat{V}) = \lambda_p E_p(\hat{V}) + \lambda_l E_l(\hat{V}) \tag{3}$$

where λ_p and λ_l denote the weights of each term.

Line Preservation Item $E_L(\hat{V})$: It is especially necessary to make higher requirements on the linear structure by reason that curved lines especially affect humans' visual experience. Therefore, according to the set of original lines S_l given by LSD [24], we give the merging and division strategy of global lines for any two lines l_m and l_n from S_l, and then optimize the local and global lines at the same time. The merger conditions are as follows specifically:

1. The slopes of the two lines $slope(l_m)$ and $slope(l_n)$ should be as similar as possible, i.e. $\theta = \arctan\left|\frac{k(l_m)-k(l_n)}{1+k(l_m)k(l_n)}\right| < \beta_1$;

2. The distances from the endpoints of one to the other line should be closer, i.e. $dis(p_{l_m}^0, l_n) - dis(p_{l_m}^1, l_n) < \beta_2$;
3. The distance between the adjacent endpoints of l_m and l_n should also be smaller, i.e. $dis\left(p_{l_m}^0, p_{l_n}^1\right) < \beta_3 dis\left(p_{l_m}^1, p_{l_n}^0\right)$.

If the above three conditions are met, l_m and l_n can be merged into a long line segment into S_l. According to the length of each line, S_l can be divided into local lines S_{lq} and global lines S_{lg} by threshold χ. In most cases, local lines are obtained by the original lines and global lines are originated from the merging process. Local and global lines are uniformly sampled with points, denoted as $\{p_k^q\}_{k=1,\ldots,M_q}^{q=1,\ldots,Q}$ and $\{p_b^g\}_{b=1,\ldots,M_g}^{g=1,\ldots,G}$, where Q and G are the number of local and global lines, M_q and M_g are the number of sampling points on each line. In order to minimize the distance between adjacent sampling points, $E_L(\hat{V})$ is divided into local preserving term $E_{lq}(\hat{V})$ and global preserving term $E_{lg}(\hat{V})$, giving the definition:

$$
\begin{aligned}
E_{lq}(\hat{V}) &= \sum_{q=1}^{Q} \sum_{k=1}^{M_q-1} \left\| \left\langle \omega_{k+1}^q \hat{v}_{k+1}^q - \omega_k^q \hat{v}_k^q, \boldsymbol{n}_q \right\rangle \right\|^2, \\
E_{lg}(\hat{V}) &= \sum_{g=1}^{G} \sum_{b=1}^{M_g-1} \left\| \left\langle \omega_{b+1}^g \hat{v}_{b+1}^g - \omega_b^g \hat{v}_b^g, \boldsymbol{n}_g \right\rangle \right\|^2, \\
E_L(\hat{V}) &= \lambda_{lq} E_{lq}(\hat{V}) + \lambda_{lg} E_{lg}(\hat{V})
\end{aligned}
\tag{4}
$$

λ_{lq} and λ_{lg} are the weights of $E_{lq}(\hat{V})$ and $E_{lg}(\hat{V})$ respectively.

Distortion Control Item $E_D(\hat{V})$: It is easy to generate distortion and shape deformation when complex multiple pedestrians exist in the input images. Inspired by the quasi-homography warp [25], optimizing the slope of the cross-lines can effectively reduce distortion. Suppose that l_u is represented as a horizontal line that retains its slope after warp, which keeps a vertical relationship with l_v. Given the collection of cross-line features: $\{(l_u^i, l_u'^i)\}$ and $\{(l_v^j, l_v'^j)\}$, where l_u^i and l_v^j are parallel to l_u and l_v, $l_u'^i$ and $l_v'^j$ represents the warped lines. We uniformly sample L_i and K_j points (i.e. $\{p_k^{u,i}\}$, $\{p_k^{v,j}\}$) on cross-lines, in order to effectively reduce the distortion, $E_D(\hat{V})$ is divided into the global distortion control term $E_{ds}(\hat{V})$ and the non-overlapping area distortion control term $E_{dn}(\hat{V})$:

$$
\begin{aligned}
E_{ds}(\hat{V}) &= \sum_{i=1}^{S} \sum_{k=1}^{L_i-1} \left| \left\langle \omega_{k+1}^{u,i} \hat{v}_{k+1}^{u,i} - \omega_k^{u,i} \hat{v}_k^{u,i}, \boldsymbol{n}_i^u \right\rangle \right|^2 \\
&+ \sum_{j=1}^{T} \sum_{k=1}^{K_j-1} \left| \left\langle \omega_{k+1}^{v,j} \hat{v}_{k+1}^{v,j} - \omega_k^{v,j} \hat{v}_k^{v,j}, \boldsymbol{n}_j^v \right\rangle \right|^2 \\
&+ \sum_{j=1}^{T} \sum_{k=1}^{K_j-2} \left\| \omega_k^{v,j} \hat{v}_k^{v,j} + \omega_{k+2}^{v,j} \hat{v}_{k+2}^{v,j} - 2\omega_{k+1}^{v,j} \hat{v}_{k+1}^{v,j} \right\|^2
\end{aligned}
\tag{5}
$$

$$E_{dn}(\hat{V}) = \sum_{i=1}^{S} \sum_{k=1}^{|p_k^{u,i} \subset \pi| - 2} \left\| \omega_k^{u,i} \hat{\nu}_k^{u,i} + \omega_{k+2}^{u,i} \hat{\nu}_{k+2}^{u,i} - 2\omega_{k+1}^{u,i} \hat{\nu}_{k+1}^{u,i} \right\|^2 \qquad (6)$$

$$E_D(\hat{V}) = \lambda_{ds} E_{ds}(\hat{V}) + \lambda_{dn} E_{dn}(\hat{V}) \qquad (7)$$

where λ_{ds} and λ_{dn} represent the weights of $E_{ds}(\hat{V})$ and $E_{dn}(\hat{V})$ respectively. In Eq.(5), S and T are the numbers of cross-line features separately, $\boldsymbol{n}_i^u$ and $\boldsymbol{n}_j^v$ are the normal vectors of l_u^i and l_v^j.

In summary, $E(\hat{V})$ can be reformulated and minimized with a sparse linear solver since all constraints are quadratic.

3.3   Composite Ghost Removal Based on Multi-matching

Because of complex multiple pedestrians in input images, the final stitching results are particularly prone to exist composite ghosts. The existing challenges mainly have the following two aspects: First, the pedestrians and their entire contour are difficult to be recognized completely when they are too small or severely deformed. Second, when multiple pedestrians are relatively similar, it is difficult to accurately match the corresponding pedestrians, which brings greater challenges to the removal of composite ghosts. Therefore, committed to effective and comprehensive removal of synthetic ghosts, we comprehensively design a composite ghost removal method based on Multi-Matching.

Foreground Pedestrians Detection: To begin with, we use $yolov4$ [26] to detect the pedestrian coordinate information in $\hat{I}_1$ and $\hat{I}_2$, denoted as $\{O_i^1\}_{i=1}^{T_0}$ and $\{O_j^2\}_{j=1}^{S_0}$, where $O_i^1 = [y_{i1}, y_{i2}, x_{i1}, x_{i2}]$ stores the coordinate of the pedestrian. $yolov4$ is the current advanced target detection method. Pedestrians are detected by the pre-trained model on MS COCO dataset [27] that basically meet all the situations we have for real scenes. Experiments show that $yolov4$ can detect pedestrians in warped images well.

Candidate Set Construction: If the overlap rate of areas from $\{O_i^1\}_{i=1}^{T_0}$ is too large, it is likely to detect that there are sub objects in the same pedestrian, such as backpacks or bicycles. In order to reduce the redundancy, a region merging strategy is designed based on the idea of IoU, the formula is as follows:

$$IoU(O_{i_1}^1, O_{i_2}^1) = \frac{O_{i_1}^1 \cap O_{i_2}^1}{O_{i_1}^1 \cup O_{i_2}^1}, \ i_1 = 1, ..., T_0 - 1, i_2 = i_1 + 1, ..., T_0$$

$$O_{i_1}^1 = O_{i_1}^1 \cup O_{i_2}^1, \ O_{i_2}^1 = [\ ], if \ IoU(O_{i_1}^1, O_{i_2}^1) > \xi_1 \qquad (8)$$

when $IoU(O_{i_1}^1, O_{i_2}^1) > \xi_1$, it indicates that $O_{i_1}^1$ and $O_{i_2}^1$ have large overlapping area. $O_{i_1}^1$ is assigned their maximum range, and $O_{i_2}^1$ is deleted at the same time. $\{O_j^2\}_{i=1}^{S_0}$ use the same method to update. After the above operations, the number of pedestrians will be updated to T_1 and S_1.

Then the sub-images in $\hat{I}_1$ and $\hat{I}_2$ are cropped to construct candidate set, i.e. *Target Set PED^1* and *Search Set PED^2*, containing T_1 and S_1 sub-pictures respectively, in which pedestrians Ped_i^1 and Ped_j^2 correspond to coordinates O_i^1 and O_j^2.

Pedestrians Multi-matching: It is also a particularly important task to establish matching relationship for the corresponding pedestrians in candidate set. Inspired by *ReID* [28], First the depth features of images from candidate set are extracted by the best model trained by *Resnet*18 on the *Market* $-$ 1501 dataset [29]. Then the features extracted from PED^1 and PED^2 will be stacked, denoted as $\mathbf{A} \in \mathbb{R}^{T_1 \times W}$ and $\mathbf{B} \in \mathbb{R}^{S_1 \times W}$, where $W = 512$. The similarity of images is measured by calculating the Euclidean distance between features, which will be stored in the distance matrix $C \in \mathbb{R}^{T_1 \times S_1}$. Finally, use the Hungarian algorithm [30] to calculate the minimum distance matching from *Target Set* to *Search Set*, and obtain the corresponding index values T_{id}, S_{id}.

However, not all pedestrians in PED^1 can find the corresponding matching pedestrians in PED^2, when the feature distance between the matching pedestrians is more than the threshold ξ_2, it is determined that Ped_i^1 does not have the matching pedestrian in PED^2. Further, set matched id M_{id}^1 and M_{id}^2 into $\{O_i^1\}$ and $\{O_j^2\}$, defined as follows:

$$T_{id} \ , \ S_{id} = Hungarian(C),$$

$$C_{i,j} = \sqrt{\sum_{k=1}^{W}(A_{ik} - B_{jk})^2}, \ T_{id} = [1, \ 2, \ ...,i, \ ..., \ T_1] \tag{9}$$

$$\begin{cases} M_{id_i}^1 = 0, M_{id_{S_{id}[i]}}^2 = 0, if \ C_{i,S_{id}[i]} > \xi_2 \\ M_{id_i}^1 = i, M_{id_{S_{id}[i]}}^2 = S_{id}[i], \qquad else \end{cases} \tag{10}$$

In Eq. (9), $C_{i,j}$ donates the distance on the feature level between the i-th image in *Target Set* and the j-th image in *Search Set*, $S_{id}[i]$ represents the index value of matching pedestrian in PED^2, which corresponds to the i-th image in PED^1. In Eq. (10), $M_{id} = 0$ donates that there are no matching pedestrians.

Reserved Area τ Selection: Intuitively, composite ghosts are usually generated by moving pedestrians in or at the boundaries of overlapping regions. It is necessary to judge the location of pedestrians based on the overlapping area Ω of $\hat{I}_1$ and $\hat{I}_2$ calculated by the optimal warp and add Ω_{id}^1 and Ω_{id}^2 into $\{O_i^1\}$ and $\{O_j^2\}$. Furthermore, the composite ghost will not generate if the corresponding matching pedestrians are both in the overlapping area and not in motion.

At present, all potential moving pedestrians have been identified and matched. In order to ensure that the final result is sufficiently natural, a reserved area selection strategy is designed. Specifically, Because the pedestrians in $\hat{I}_2$ inevitably have distortion and deformation after warp, the pedestrians in $\hat{I}_1$ should be selected as reserved area τ as much as possible. However, when there is a pedestrian on the boundary of Ω or out of Ω, we must choose it as reserved area τ. According to the three rules of reserved area: non-distortion, integrity and uniqueness rule, we store the pedestrian areas that needs to be preserved into τ.

Seamless Image Blending: The seamless image blending method is designed for the purpose of eliminating composite ghosts to generate final stitching results. In the image blending process, first, obtain an initial stitched image I_{init} by naively blending $\hat{I}_1$ and $\hat{I}_2$ on average, in which their overlapping area is taken as the average blending area. Then, use the gradient in the initial stitching result I_{init} as the guide gradient, minimize the difference between the reserved area τ and I_{init} on the fusion boundary through poisson reconstruction, smoothly transition to the blended image block and blend it seamlessly into the final result I_S:

$$I_{init} = AverageImageBlending(\hat{I}_1, \hat{I}_2)$$
$$I_S = SeamlessCloning(I_{init}, \tau)$$

(11)

4   Experimental Evaluation

In order to assess the capability and validate the effectiveness of our method in solving complex multiple pedestrians and large parallax challenges, experiments are conducted on a series of real images, which cover different angles and parallax, while contain challenging multiple pedestrians.

We compare the proposed method with five state-of-the-art methods: APAP [4], AANAP [10], ELA [20], SPW [16], OGIE [31], and use the source codes provided by authors of the papers to obtain comparison results. In our implementation, the size of input images are resized to 3000×4000 pixels. In structure preservation, the grid size is set to 100×100, the threshold χ for dividing local and global line segments is set to three times the length of the mesh diagonal. In $E(\hat{V})$, $\lambda_p = 1, \lambda_l = 5, \lambda_{lq} = 50, \lambda_{lg} = 150, \lambda_{ds} = 50, \lambda_{dn} = 100$. In composite ghost removal, ξ_1 is set to 0.2 to judge whether to merge overlapping objects, ξ_2 is set to 15 to determine whether there is a corresponding matching pedestrian, which are relatively stable in our experiment.

We verify the effectiveness of the proposed method from three aspects: qualitative evaluation, quantitative comparison and ablation experiment.

4.1   Qualitative Comparison

In order to testify the effectiveness of the proposed method, we compare the performance of ours with APAP [4], AANAP [10], ELA [32], SPW [16] and OGIE [31]. We select five groups of representative images for presentation, which are shown in Fig. 5(a)–(e). It can be observed from Fig. 5(a)–(e) that these five methods all have different degrees of parallax ghosts and composite ghosts, pedestrians appear twice and are stretched in scenes, which visually leads to particularly unnatural stitching results. Generally speaking, there are moving pedestrians that are too small in *Case* 1 shown in Fig. 5(a), APAP and OGIE exist obvious parallax artifacts and stretch the shape of the dustbin. In Fig. 5(d), the vertical relationship between the stone pillar and the ground is bent in APAP, AANAP and SPW. In Fig. 5(e), AANAP and ELA bent the rectilinear structure of the yellow mesh line. As can be seen in the Fig. 5, our method can not only remove parallax ghosts, keep the linear structure well, but also effectively remove all composite ghosts, ensuring that each pedestrian appears only once in the final stitching result, and pedestrians are not segmented.

4.2 Quantitative Comparison

In the study of image stitching, Root Mean Squared Error (RMSE), Structural Similarity (SSIM) and Peak Signal to Noise Ratio (PSNR) are widely used in quantitative matrics. Among them, RMSE is to evaluate the composite accuracy and SSIM is employed to measure the structure-preserving quality. PSNR is good at evaluating the image quality and naturalness of stitching results, which has been widely used in recent image stitching works [10,33]. To obtain convincing experimental results, we will focus on the naturalness of stitching and the ability to preserve structure in overlapping regions by PSNR and SSIM, since it is readily available and often performance-sensitive. PSNR is defined as follows:

$$PSN = 10 \cdot \lg \frac{(MAX)^2}{MSE}, MSE = \frac{1}{mn} \sum_{i=0}^{m-1} \sum_{j=0}^{n-1} [G_{I'}(i,j) - G_I(i,j)]^2 \quad (12)$$

where MAX is the maximum value of pixel color, MSE represents mean square error, $G_{I'}$ is the overlapping area cropped in the final stitching result, G_I is the same regions cropped in the original images i.e. I_1, since we usually keep the pedestrians in the reference image.

We compared our method with APAP, AANAP, ELA, SPW and OGIE and the results are shown in Table 1, where bold values represent biggest values, it is obvious to see that the PSNR and SSIM values of our method outperform the other methods.

Table 1. Comparison on PSNR and SSIM

Experimental Case	PSNR						SSIM					
	APAP	AANAP	ELA	SPW	OGIE	Ours	APAP	AANAP	ELA	SPW	OGIE	Ours
Case 1	18.061	17.541	17.553	17.515	17.733	**18.260**	0.567	0.498	0.553	0.548	0.554	**0.579**
Case 2	22.073	21.658	21.145	21.339	21.790	**22.162**	0.654	0.582	0.632	0.634	0.626	**0.656**
Case 3	22.179	20.498	21.858	20.995	22.270	**22.366**	0.910	0.668	0.768	0.737	0.758	**0.780**
Case 4	24.264	24.921	23.283	25.062	24.962	**25.435**	0.807	0.808	0.799	0.814	0.817	**0.820**
Case 5	23.831	17.691	17.829	24.734	23.379	**24.981**	0.780	0.491	0.636	0.825	0.782	**0.789**

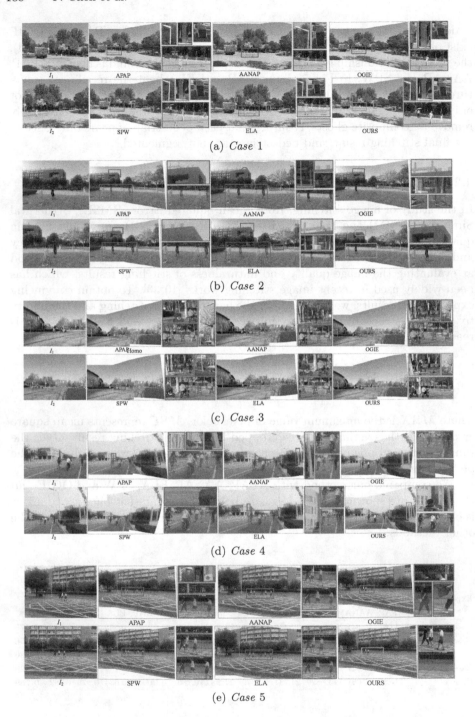

Fig. 3. Comparison of visual results on five image pairs with other five methods. Some details are highlighted in right. The red boxes show repeated pedestrians and curved structure, and the green boxes show satisfactory stitching results. (Color figure online)

4.3 Ablation Study

Structure Preservation: This module reduces parallax ghosts and effectively suppresses the structural distortion by using grid constraints to optimize the energy items. Our structure preservation module is substituted with a global homography matrix. The results are shown in Fig. 3, there are two stitching instances, the enlarged areas are shown on the right of each result. It can be seen that there are obvious ghosts and deformation in red boxes, while our structure preservation module produces natural stitching results in green boxes.

Composite Ghost Removal: This module successfully solves the unnatural problem caused by composite ghosts due to the presence of complex pedestrians on the stitching result. The composite ghost removal module is removed with other parts unchanged. As shown in Fig. 4, case1(a)–3(a) without the module produces severe synthetic ghosts, with the same pedestrian appearing twice in the scene. Our method does not have these unpleasant results, removes composite ghosts well in Fig. 4 case1(b)–3(b).

Fig. 4. Ablation study for structure preservation module.

Fig. 5. Ablation study for composite ghost removal module. Highlight regions are enlarged to the right of the result, the red boxes have severe composite ghosts while the green boxes present a natural look. (Color figure online)

5   Conclusions

In this paper, we design an image stitching method for complex multiple pedestrians and large parallax scenes. Our two main contributions are mainly in two aspects: first, the structure preservation based on grid constraints module removes parallax ghosts well and maintains the linear structure. Then we design the composite ghost removal based on Multi-Matching module, which effectively eliminates composite ghosts and ensures that each pedestrian can be unique and not cut even when there are complex multiple pedestrians in the scene. The comprehensive results on real challenging datasets demonstrated the effectiveness and advantages of our method over the state-of-the-art methods.

Acknowledgements. This work was supported in part by the National Natural Science Foundation of China (Grant number 61906135, 62020106004, 92048301); in part by the Tianjin Science and Technology Plan Project under Grant 20JCQNJC01350; and in part by the Tianjin Postgraduate Research and Innovation Project under Grant 2021YJSB244. W. Xue is the corresponding author of this paper.

References

1. Camera, T.M.: Image alignment and stitching: A tutorial. Fnd (2016)
2. Zhu, C., Sun, M.-T., Lei, J., Hou, C.: Shuai: depth coding based on depth-texture motion and structure similarities. IEEE Trans. Circuits Syst. Video Technol. **25**, 275–286 (2015)
3. Uyttendaele, M., Eden, A., Szeliski, R.: Eliminating ghosting and exposure artifacts in image mosaics. In: IEEE Computer Society Conference on Computer Vision & Pattern Recognition (2001)
4. Zaragoza, J., Chin, T.J., Brown, M.S., Suter, D.: As-projective-as-possible image stitching with moving dlt. IEEE Trans. Pattern Anal. Mach. Intell. **2013**, 2339–2346 (2013)
5. Zhang, Z., Ren, X., Wanli, X., Zhang, C., Guo, Q., Chen, S.: Object-aware ghost identification and elimination for dynamic scene mosaic. IEEE Trans. Circuits Syst. Video Technol. **32**, 2025–2034 (2021)
6. Szeliski, R.: Creating full view panoramic image mosaics and environment maps. In: Proceedings of Siggraph (1997)
7. Brown, M., Lowe, D.G.: Automatic panoramic image stitching using invariant features. Int. J. Comput. Vision **74**(1), 59–73 (2007)
8. Gao, J., Kim, S.J., Brown, M.S.: Constructing image panoramas using dual-homography warping. In: 2011 IEEE Conference on Computer Vision and Pattern Recognition (CVPR) (2011)
9. Chang, C.H., Sato, Y., Chuang, Y.Y.: Shape-preserving half-projective warps for image stitching. In: 2014 IEEE Conference on Computer Vision and Pattern Recognition (CVPR) (2014)
10. Lin, C.C., Pankanti, S.U., Ramamurthy, K.N., Aravkin, A.Y.: Adaptive as-natural-as-possible image stitching. In: Computer Vision & Pattern Recognition (2015)
11. Gao, J., Li, Y., Chin, T.J., Brown, M.: Seam-driven image stitching. In: Eurographics (2013)
12. Lin, K., Jiang, N., Cheong, L.F., Do, M., Lu, J.: Seagull: Seam-guided local alignment for parallax-tolerant image stitching. In: European Conference on Computer Vision (2016)

13. Nan, L., Liao, T., Chao, W.: Perception-based seam cutting for image stitching. SIViP **12**(3), 1–8 (2018)
14. Xue, W., Xie, W., Zhang, Y., Chen, S.: Stable linear structures and seam measurements for parallax image stitching. IEEE Trans. Circuits Syst. Video Technol. **32**, 253–261 (2021)
15. Chen, Y.-S., Chuang, Y.-Y.: Natural image stitching with the global similarity prior. In: Leibe, B., Matas, J., Sebe, N., Welling, M. (eds.) ECCV 2016. LNCS, vol. 9909, pp. 186–201. Springer, Cham (2016). https://doi.org/10.1007/978-3-319-46454-1_12
16. Liao, T., Li, N.: Single-perspective warps in natural image stitching. IEEE Trans. Image Process. **29**, 724–735 (2018)
17. Jia, Q., ZhengJun Li, X.F., Haotian Zhao, S.T., Latecki, L.J.: Leveraging line-point consistence to preserve structures for wide parallax image stitching. In: 2021 IEEE Conference on Computer Vision and Pattern Recognition (CVPR) (2021)
18. DaVis, J.: Mosaics of scenes with moving objects. In: IEEE Computer Society Conference on Computer Vision & Pattern Recognition (1998)
19. Flores, A., Belongie, S.: Removing pedestrians from google street view images. In: Computer Vision & Pattern Recognition Workshops (2010)
20. Leibe, B., Leonardis, A., Schiele, B.: Robust object detection with interleaved categorization and segmentation. Int. J. Comput. Vis. **77**, 259–289 (2008)
21. Pulli, K., Tico, M., Xiong, Y., Wang, X., Liang, C.K.: Panoramic imaging system for camera phones. In: International Conference on Consumer Electronics (ICCE), 2010 Digest of Technical Papers (2010)
22. Boutellier, J.J., López, M., Silvén, O., Tico, M., Vehviläinen, M.: Creating panoramas on mobile phones. In: Spie Electronic Imaging (2007)
23. Kakli, M.U., Cho, Y., Seo, J.: Minimization of parallax artifacts in video stitching for moving foregrounds. IEEE Access **6**, 57763–57777 (2018)
24. Gioi, R.G.V., Jakubowicz, J., Morel, J.M., Randall, G.: Lsd: A line segment detector. Image Process. On Line **2**(4), 35–55 (2012)
25. Nan, L., Xu, Y., Chao, W.: Quasi-homography warps in image stitching. IEEE Trans. Multimedia **20**, 1365–1375 (2018)
26. Bochkovskiy, A., Wang, C.Y., Liao, H.: Yolov4: optimal speed and accuracy of object detection. In: IEEE Conference on Computer Vision and Pattern Recognition (CVPR) (2020)
27. Lin, T.-Y., et al.: Microsoft COCO: Common objects in context. In: Fleet, D., Pajdla, T., Schiele, B., Tuytelaars, T. (eds.) ECCV 2014. LNCS, vol. 8693, pp. 740–755. Springer, Cham (2014). https://doi.org/10.1007/978-3-319-10602-1_48
28. Gong, S., Cristani, M., Yan, S., Chen, C.L.: Person re-identification. Adv. Comput. Vis. Pattern Recogn. **42**(7), 301–313 (2014)
29. Zheng, L., Shen, L., Lu, T., Wang, S., Qi, T.: Scalable person re-identification: a benchmark. In: IEEE International Conference on Computer Vision (ICCV) (2015)
30. Mills-Tettey, A., Stent, A., Dias, M.B.: The dynamic hungarian algorithm for the assignment problem with changing costs. carnegie mellon university (2007)
31. Xue, W., Zhang, Z., Chen, S.: Ghost elimination via multi-component collaboration for unmanned aerial vehicle remote sensing image stitching. Remote Sensing **13**, 1388 (2021)
32. Li, J., Wang, Z., Lai, S., Zhai, Y., Zhang, M.: Parallax-tolerant image stitching based on robust elastic warping. IEEE Trans. Multimedia **20**, 1672–1687 (2018)
33. Xiang, T.Z., Xia, G.S., Xiang, B., Zhang, L.: Image stitching by line-guided local warping with global similarity constraint. Pattern Recogn. **83**, 481–497 (2017)

Traffic Congestion Event Mining Based on Trajectory Data

Yanfei Li[1], Nianbo Hu[1], Chen Wang[2], and Rui Zhang[1(✉)]

[1] Wuhan University of Technology, Wuhan 430070, China
contact.work@qq.com, zhangrui@whut.edu.cn
[2] Huazhong University of Science and Technology, Wuhan 430074, China
chenwang@hust.edu.cn

Abstract. Traffic congestion, the most common event affecting traffic, is becoming serious. In this paper, the traffic congestion event (TCE) is analyzed and applied to trajectory data. TCE can be defined in terms of time, location, average speed, and degree of congestion. Then methods are designed to extract TCE from massive trajectory data. Specifically, an efficient method is developed that combines NumPy's customized universal function to calculate the instantaneous velocity of each trajectory point. Using the results, the average speed of the road is calculated. Finally, sliding windows are employed to identify the road congestion state and compare it with the official news released by Chengdu Traffic Police. This is to prove the effectiveness of the proposed method. Therefore, the recognized TCE will lay the groundwork for further activities.

Keywords: Trajectory data · Traffic congestion · Event mining

1 Introduction

GPS-equipped taxis generate and collect extensive trajectory data like "moving blood cells of the city". Therefore, by mining and analyzing trajectory data, we can understand the real external performance and internal relations of urban traffic information. Moreover, traffic congestion, as an inevitable problem in big cities, has affected people's daily lives in many ways such as time waste and a higher risk of traffic accidents. If passengers and drivers are able to get timely traffic congestion information, then a lot of trouble can be avoided. Therefore, it is of great practical significance to study traffic congestion events. The contributions of this paper are as follows:

- In most cases, people are only able to perceive traffic congestion manually. By defining the Traffic Congestion Event by several key elements, this paper accurately depicts traffic congestion and makes it easy to conduct quantitative analysis.
- An algorithm for identifying traffic congestion state based on sliding window voting method is proposed, and the effectiveness of the algorithm is verified by comparing it with Traffic Police's official information

H. Zhang et al. (Eds.): NCAA 2022, CCIS 1638, pp. 192–204, 2022.
https://doi.org/10.1007/978-981-19-6135-9_15

2   Related Work

There have been many studies and applications on the evaluation of traffic conditions and mining effective traffic information from data. Wang D. et al. [1] use social media information to warn of traffic conditions. Based on the characteristics of social media, they propose a method to continuously learn new information, which can adjust and improve the model. Since trajectory data can be generated from different traffic tools with different characteristics and their features of speed are also diverse, Zhang R. et al. [2] use deep learning method to classify. By taking the road network as a directed graph, Zhang Y. et al. [3] use GPS trajectory data to build and update digital maps. Cheng W et al. [4] propose a method for protecting the private information in trajectory data while maintaining the useful information it contains. Jiang H. et al. [5] use some metrics derived from trajectory data of private cars to uncover people's travel patterns and enhance destination prediction. In the field of event modeling, Xu Z. et al. [6] propose a 5W ("What, Where, When, Who, and Why") traffic congestion model to perform event mining. Chen X. et al. [7] not only summarize some existing event mining techniques but also re-classify them from a new perspective. For traffic congestion, D'Este G. M et al. [8] point out that traffic congestion is a key factor for evaluating traffic condition by using travel speed and duration. Liu Z. et al. [9] use position encoding to find abnormal state of traffic based on the influence on traffic congestion brought by accidents.

However, studies on event modeling mentioned above do not combine the characteristics of traffic congestion. This paper focuses on traffic congestion events modeling and finds factors that can accurately and effectively describe traffic congestion. Meanwhile, compared to the above high-demand methods, this paper is based simply on the principle that the lower the average speed of the road, the worse the congestion.

3   Method

Traffic congestion is one of the most common events in the transportation field. A comprehensive, accurate, and structured description of a traffic congestion event is of great significance for storage, query, data analysis, and application. Therefore, a traffic congestion framework is designed and algorithms are developed to complete this framework from data collected in daily life. Data sources are diverse, including social media data, driving log information, road sensor data, traffic cameras, etc. This paper uses trajectory data to build traffic congestion events. The architecture diagram is shown in Fig. 1.

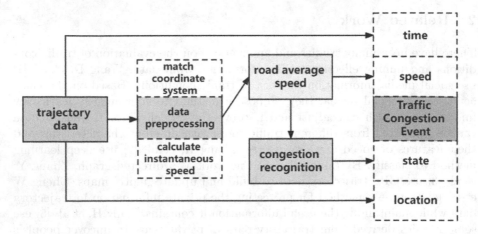

Fig. 1. Architecture diagram

3.1   Traffic Congestion Event Definition

A traffic congestion event is defined as Definition 1 after considering the key and representative factors.

Definition 1. *TCE={time, location, speed, state}*

where *TCE* represents traffic congestion event, *time* represents the time period when the traffic congestion event occurs, *location* represents where the traffic congestion event happens, *speed* represents the average speed of the road, and *state* represents whether a road is congested.

The following steps introduce how to discover and extract TCE from trajectory data.

3.2   Calculate the Instantaneous Velocity of Each Trajectory Point

The attribute "speed" in *TCE* defined in Definition 1 is an average speed of a road. However, only when the instantaneous speed of each trajectory point is obtained can the average speed of the road be calculated. There is no instantaneous velocity attribute in the original trajectory data. Fortunately, the instantaneous velocity can be approximately estimated from the latitude, longitude and timestamp of two adjacent sample points. Due to the huge amount of trajectory data, if traverse each row of trajectory data to calculate the instantaneous speed, the time consumption in this process is very long. Therefore, an efficient method is proposed to calculate the instantaneous speed of trajectory data based on NumPy's customized universal function, which greatly improves the operation's efficiency.

The surface distance between the two points is calculated according to the latitude and longitude coordinates of the two sampling points using Eq. 1.

$$D_i D_j = R \cdot arcos \left(\cos\left(Y_i\right) \cdot \cos\left(Y_j\right) \cdot \cos\left(X_i - X_j\right) + \sin\left(Y_i\right) \cdot \sin\left(Y_j\right) \right) \quad (1)$$

where $D_i D_j$ represents the distance between point i and point j, the coordinate of point i is (X_i, Y_i), and the coordinate of point j is (X_j, Y_j), R represents the radius of the earth [10].

Because the sampling interval of the original trajectory data set is mostly within 2–4 s, the instantaneous speed of a point can be approximated by the average speed of two adjacent points using Eq. 2.

$$V_i = \left| \frac{D_i D_{i-1}}{t_i - t_{i-1}} \right| \quad (2)$$

where i and $(i-1)$ are two adjacent points, V_i represents the instantaneous speed of point i, $D_i D_{i-1}$ is calculated by Eq. 1, t_i represents the timestamp at point i, and t_{i-1} represents the timestamp at point $(i-1)$. All unix timestamps are in seconds.

Yu Q et al. [11] extract origin and destination points from trajectory data by transforming adjacent load states in one line and comparing them, which gives an inspiration to process original trajectory data. Thus, a method using Python's NumPy package with customized universal function is proposed, which can optimizes the process of obtaining the instantaneous speed. Some fields of original trajectory data containing TIME1(timestamp), LON1(longitude), and LAT1(latitude) are stored in DataFrame(a data structure of package Pandas in Python)[1] shown in Fig. 2 and we call it DataFrame1.

Move the three columns TIME1, LON1, and LAT1 of the original trajectory data in DataFrame1 down by one row as a whole, and get three new fields of TIME2, LON2, and LAT2, which converts into a new DataFrame called DataFrame2. In this way, calculating the speed of the trajectory point does not need to traverse each row of data. The TIME1, LON1, LAT1, TIME2, LON2, and LAT2 columns in the DataFrame2 are regarded as vectors respectively. "$frompyfunc()$" is a function in NumPy, which can transform ordinary numerical calculations into a new function that performs computations on each element in the vector[2]. "$frompyfunc()$" transforms Eq. 1 into universal function and is applied to computing instantaneous speed based on DataFrame2. The algorithm is shown in Algorithm 1.

For the convenience of subsequent use, the instantaneous velocity of each trajectory point is taken as a new field of the original trajectory dataset called "InstantSpeed".

3.3 Match Trajectory Data with Roads' Information

Select one road as research object to be the attribute "location" in $TCE = \{time, location, speed, state\}$. Find all the trajectory sampling points passing

---

[1] https://pandas.pydata.org/pandas-docs/stable/reference/frame.html.
[2] https://numpy.org/devdocs/reference/generated/numpy.frompyfunc.html.

	DataFrame1					DataFrame2					
index	TIME1	LON1	LAT1		index	TIME1	LON1	LAT1	TIME2	LON2	LAT2
0	t_0	x_0	y_0		0	t_0	x_0	y_0	NaN	NaN	NaN
1	t_1	x_1	y_1		1	t_1	x_1	y_1	t_0	x_0	y_0
2	t_2	x_2	y_2		2	t_2	x_2	y_2	t_1	x_1	y_1
3	t_3	x_3	y_3		3	t_3	x_3	y_3	t_2	x_2	y_2
4	t_4	x_4	y_4		4	t_4	x_4	y_4	t_3	x_3	y_3
5	t_5	x_5	y_5	→	5	t_5	x_5	y_5	t_4	x_4	y_4
...	...	...	...		...	...	...	...	...	...	...
i-5	t_{i-5}	x_{i-5}	y_{i-5}		i-5	t_{i-5}	x_{i-5}	y_{i-5}	t_{i-6}	x_{i-6}	y_{i-6}
i-4	t_{i-4}	x_{i-4}	y_{i-4}		i-4	t_{i-4}	x_{i-4}	y_{i-4}	t_{i-5}	x_{i-5}	y_{i-5}
i-3	t_{i-3}	x_{i-3}	y_{i-3}		i-3	t_{i-3}	x_{i-3}	y_{i-3}	t_{i-4}	x_{i-4}	y_{i-4}
i-2	t_{i-2}	x_{i-2}	y_{i-2}		i-2	t_{i-2}	x_{i-2}	y_{i-2}	t_{i-3}	x_{i-3}	y_{i-3}
i-1	t_{i-1}	x_{i-1}	y_{i-1}		i-1	t_{i-1}	x_{i-1}	y_{i-1}	t_{i-2}	x_{i-2}	y_{i-2}
i	t_i	x_i	y_i		i	t_i	x_i	y_i	t_{i-1}	x_{i-1}	y_{i-1}

Fig. 2. Conversion to format that can perform NumPy's customized universal function

Algorithm 1. Calculate instantaneous speed by NumPy's customized universal function

$T_1 \leftarrow DataFrame1$
$T_2[TIME1] \leftarrow T_1[TIME1]$
$T_2[LON1] \leftarrow T_1[LON1]$
$T_2[LAT1] \leftarrow T_1[LAT1]$
$T_2[TIME2] \leftarrow T_1[TIME1].shift()$
$T_2[LON2] \leftarrow T_1[LON1].shift()$
$T_2[LAT2] \leftarrow T_1[LAT1].shift()$
$point1 \leftarrow (time1, lon1, lat1)$
$point2 \leftarrow (time2, lon2, lat2)$
function $GetAdjacentDistance$:
$\quad distance \leftarrow$ result from putting $(lon1, lat1, lon2, lat2)$ in Eq. 1
$\quad$ **return** distance
$UniversalFunction \leftarrow numpy.frompyfunc(GetAdjacentDistance, 4, 1)$
$T_2[\Delta distance] \leftarrow UniversalFunction(T_2[LON1], T_2[LAT1], T_2[LON2], T_2[LAT2])$
$T_2[\Delta time] \leftarrow T_2[TIME1] - T_2[TIME2]$
$T_2[instantSpeed] \leftarrow T_2[\Delta distance]/T_2[\Delta time]$

through this road. The core idea of matching the trajectory data and the road is to take the intersection of the geographic information of the two parts.

In general, road geographic data is stored in Shapefile. The "geometry" field in Shapefile stores the geographic information of the road in the form of "LineString". "Point" and "LineString" are geometric object data types in the Python plane geometry library "Shapely"[3]. "LineString" is internally composed of a series of "Point", and the coordinates of each "Point" are composed with latitude and longitude.

---

[3] https://shapely.readthedocs.io/en/stable/manual.html.

The LineString data type represents a line without an area, but the road in reality is not a line, instead the road has an area. So a buffer with a radius should be added to the LineString of the road, which can achieve the purpose of converting from a line to an area that simulates an actual road. Next, convert the latitude and longitude coordinates of each trajectory point to the data type of "Point". In order to filter out all the trajectory data points passing through the selected road, take the intersection of the coordinate points of the trajectory data and the area obtained by adding the buffer zone to the road.

3.4 Calculate the Average Speed of Road

On the basis of instantaneous speed calculation results, calculate the average speed of the road. Specific steps are as follows.

After obtaining the instantaneous speed, Eq. 3 can be used to calculate average speed of a certain road during certain time period as the attribute "speed" in $TCE = \{time, location, speed, state\}$

$$\bar{V}_t = \frac{1}{n} \sum_{i=1}^{n} V_i \qquad (3)$$

where $\bar{V}_t$ represents the average speed in time period t, V_i represents the speed of the i-th GPS trajectory sampling point, n represents the number of GPS trajectory sampling points on a certain road segment in the t-th time period [10].

3.5 Traffic Congestion State Recognition

To obtain the attribute "state" in TCE defined in Definition 1, a fine-grained traffic congestion status identification method is proposed based on sliding window. First, the time period is divided into hours with coarse granularity, so the 24-h day is divided into 24 sections, after calculating the average speed per hour using the Eq. 3, the value near the lowest point of the average speed in one day is regarded as the threshold value for judging whether there is congestion [12]. Then the time period is divided into 2-min fine-grained units. Calculate the average speed per 2-min unit using the Eq. 3 as well. Every three fine-grained time periods compose a sliding window. According to the relationship that the lower average speed of the road and the more serious traffic congestion, vote in a sliding window based on the rule that if the number of average speed in the same window whose value is less than or equal to the threshold is greater than or equal to 2, it is considered that the time periods in the window are in a traffic congestion state. It is shown in Algorithm 2, where "state=0" means the road traffic is normal, "state=1" means congested. This method can not only accurately identify the fine-grained time period of congestion, but also has strong robustness to noisy data.

Algorithm 2. Sliding window voting method

$windowSize \leftarrow 3$
$TCE[state] \leftarrow 0$
$i \leftarrow 0$
while $i < (TCE.length - windowSize)$ **do**
  $window \leftarrow TCE[i, i + windowSize]$
  if $(window[speed] \leq threshold).length \geq 2$ **then**
    $TCE[state][i, i + windowSize] \leftarrow 1$
  end if
  $i \leftarrow i + 1$
end while

4 Experiments

This paper mines and constructs the traffic congestion event on the real trajectory data. During this process, record the time required to calculate the instantaneous speed of massive trajectory data using Algorithm 1, and compare the required time using the traditional traversal method to prove the advantage of Algorithm 1. The traffic congestion state identified by the sliding window is compared with the traffic news released by the Chengdu Traffic Police to verify the effectiveness of the traffic congestion identification algorithm.

4.1 Dataset Description

The trajectory data in this paper comes from Didi Chuxing's "Gaia" data open plan[4], containing trajectory data generated in some areas of Chengdu from November 1 to November 30, 2016. The sampling area of initial trajectory data is a quadrilateral ABCD. Table 1 shows the longitude and latitude range of trajectory data. The sampling point interval of the trajectory data is 2–4 s. Since there may be noise or abnormal points when collecting data, which will affect subsequent research, it is necessary to remove these data. The fields of the trajectory dataset contains Driver ID, Order ID, timestamp, longitude, latitude of each sampling point. This trajectory dataset's coordinate system is GCJ-02. The data for the map comes from OpenStreetMap[5] using the WGS-84 coordinate system.

Table 1. Longitude and latitude range of trajectory data.

A (30. 727818, 104. 043333)	B (30. 726490, 104. 129076)
C (30. 655191, 104. 129591)	D (30. 652828, 104. 042102)

---

[4] https://gaia.didichuxing.com.
[5] https://www.openstreetmap.org/.

In this experiment, when discovering and mining traffic congestion events from trajectory data, the geographic space is limited to Shuncheng Street, and the time is limited to November 11, 2016.

4.2 Data Preprocessing

Match Coordinate System. The coordinate systems of the map and trajectory data do not match, which may cause errors. Therefore, the latitude and longitude coordinates of the trajectory data are first mapped from the original GCJ-02 coordinates to the WGS-84 coordinates.

Calculate the Instantaneous Velocity of Trajectory Points. Use the algorithm in Algorithm 1 to calculate the instantaneous speed of each point. For the reason that the trajectory dataset contains the trajectory data of different vehicles, it is considered that the Driver ID can represent different vehicles. When calculating the instantaneous speed, the information of the two adjacent sampling points should be under the condition of the same Driver ID. Through experiment, it is found that when calculating the instantaneous speed of trajectory points on 2016-11-11 using Algorithm 1, it only takes about 51 s, however, if use traversal method, it takes about 1414 s, which proves the efficiency of Algorithm 1. The instantaneous velocity of each trajectory point calculated by above steps is takens as a new field in the original trajectory dataset.

4.3 Road Average Speed Calculation

Match Trajectory Data with Roads' Information. Shuncheng Street is selected as research object to be the attribute "location" in TCE. Find relevant data of Shuncheng Street in Chengdu geographic data and result is in Table 2.

The complete Shuncheng Street LineString geographic information called $SC\_LINESTRING$ is obtained by taking the union of $L1$, $L2$, $L3$ shown in Eq. 4.

$$SC\_LINESTRING = L1 \cup L2 \cup L3 \tag{4}$$

Table 2. Shuncheng Street geographical information.

osm_id	Name	Highway	z_order	Geometry
28620975	Shuncheng Street	Primary	7	L1
423484278	Shuncheng Street	Tertiary	4	L2
576653470	Shuncheng Street	Primary	7	L3

L1 : LINESTRING (104.06639 30.65477, 104.06648 30.6...)
L2 : LINESTRING (104.06622 30.65466, 104.06639 30.6...)
L3 : LINESTRING (104.06625 30.65486, 104.06635 30.6...)
Each pair of coordinates in LINESTRING is a Point

Add a buffer with a radius of 15 to $SC\_LINESTRING$ and in this way, SC_buffer is generated. To filter out the trajectory points passing through Shuncheng Street on November 11th which are called T_{SC}, take the intersection of SC_buffer and coordinates of all the trajectory points on November 11th which are called T_{all}, just like Eq. 5. T_{SC} is visualized in Fig. 3

$$T_{SC} = SC\_buffer \cap T_{all} \tag{5}$$

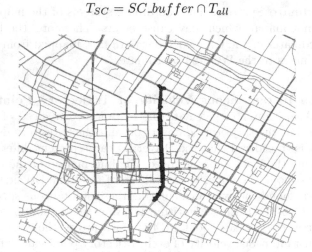

Fig. 3. Distribution of T_{SC} on the Map

Calculate the Average Speed of Road. The average speed of the road is calculated to be the component "speed" in $TCE = \{time, location, speed, state\}$. Trajectory points on November 11 of Shuncheng Street(T_{SC}) have obtained in Eq. 5. The day of November 11, 2016 is divided into 720 segments with 2 min as the basic unit, and calculate the average speed of Shuncheng Street road in each time period using Eq. 3. Part of calculation results are shown in "speed" field of Table 3.

4.4 Identify Congestion State

In this part, sliding window is used to get the attribute "state" in $TCE = \{time, location, speed, state\}$ of Shuncheng Street on November 11th, 2016.

Calculate the Threshold Speed Value. When want to judge the traffic state, the threshold for dividing the state must be known at first.

Divide the day into 24 time periods in hours, and calculate the average speed of Shuncheng Street in each time period using Eq. 3. The result is shown in Fig. 4. From Fig. 4 and the calculation results, it is found that the average speed near 18:00 is 8.148390 km/h, which is the lowest average speeds of Shuncheng Street on November 11th, so 8.148390 km/h is approximately used as the threshold for dividing the congestion state [12].

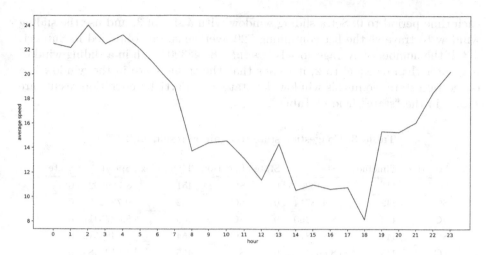

Fig. 4. Average speed in different hours.

Sliding Window Voting Method to Identify Congestion State. According to Fig 5, it can be seen that the variance of instantaneous speed from 0:00 to 8:00 and 22:00 to 24:00 are large, so data in these two periods is highly volatile and they cannot represent the overall road state. For above reason, only the trajectory data from 8:00 to 22:00 is selected for research in one day.

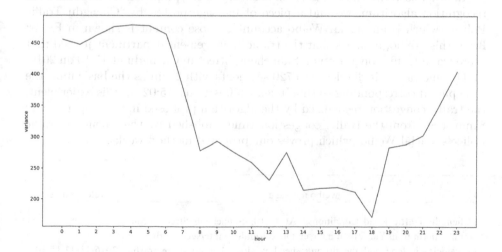

Fig. 5. Variance of speed in different hours.

Use timeIndex to represent each time period and add the "state" attribute for each time period. "state = 0" means the road status is normal, and "state = 1" means the road status is congested. At the beginning, set the state of each time period to 0. Set a sliding window with a size of 3, and use the sliding window to traverse the list containing 720 average speeds obtained in Subsect. 4.3. If the number of average speeds less than 8.148390 km/h in a sliding window is greater than or equal to 2, it means that the traffic state in the window is a congestion state. So in this window, let state = 1. Partial recognition results are shown in the "state" field of Table 3.

Table 3. Congestion state recognition results of TCE.

Location	TimeIndex	Speed	State	Location	TimeIndex	Speed	State
SC	441	10.482615	0	SC	451	8.599802	0
SC	442	9.438714	0	SC	452	9.732238	0
SC	443	8.972460	0	SC	453	8.647541	0
SC	444	9.682873	0	SC	454	10.486512	0
SC	445	8.602741	0	SC	455	12.121484	0
SC	446	8.982265	0	SC	456	12.183429	0
SC	447	8.475458	1	SC	457	10.410194	0
SC	448	7.217363	1	SC	458	10.441968	0
SC	449	6.737524	1	SC	459	12.637348	0
SC	450	8.645791	1	SC	460	10.501866	0

SC represents Shuncheng Street

In order to verify the validity and accuracy of the proposed traffic congestion recognition algorithm, we find a piece of message posted by 'Chengdu Traffic Police', which is an official Weibo account , whose content is shown in Fig. 6[6] From this Weibo, it shows that the traffic management department judged that there was a traffic congestion on Shuncheng Street in Chengdu at 15:10 on 2016-11-11. Since a day is divided into 720 segments with 2 min as the basic unit, the time period corresponding to timeIndex=450 is 15:00–15:02. In this experiment, the traffic congestion recognized by the algorithm proposed in this paper is only 8 min away from the traffic congestion time published by the Chengdu Traffic Police's official Weibo, which proves our proposed method works.

weibo message	release time
#Chengdu Traffic Road Conditions# At 15:10, Shuncheng Street: From south to north from Yanshikou intersection to Rendong intersection, the line of vehicles is long and the driving speed is slow. For real-time traffic information, please pay attention to listening to provincial and municipal traffic broadcasts FM101.7 and FM91.4.	2016-11-11 15:10

Fig. 6. Official Weibo posted by Chengdu traffic police

---

[6] https://weibo.cn/chengdujiaojing.

5   Conclusion

At the beginning, we give a definition to Traffic Congestion Event as $TCE = \{time, location, speed, state\}$. Then, we use some methods to build TCE, which not only introduce how to get the attributes of TCE from trajectory data, including time, location, speed, road state, but also introduce how to obtain them efficiently and accurately. Finally, we perform experiments to prove the proposed method work. In this paper, TCE is only extracted from trajectory data. However, data sources are diverse, which contain TCE as well. If extract TCE from different sources, the fusion of multi-source information can facilitate the application of TCE. Now, the size of the sliding window judging congestion is set to 3, and the voting number threshold whether congested or not is 2. In the future work, the setting of sliding windows' sizes and the threshold of voting results can be explored to meet practical needs. What's more, traffic congestion can be classified according to the number of votes, which means the higher the proportion of speeds less than the threshold, the more serious the congestion is. Last but not least, machine learning method can be used to predict future TCE, which can provide reference for people's traveling.

Acknowledgements. This work is partially support by the National Natural Science Foundation of China under Grants 52031009; by the Natural Science Foundation of Hubei Province, China, under Grant No.2021CFA001. Partial data source comes from Didi Chuxing.

References

1. Wang, D., Al-Rubaie, A., Clarke, S.S., et al.: Real-time traffic event detection from social media. ACM Trans. Internet Technol. (TOIT) **18**(1), 1–23 (2017)
2. Zhang, R., Xie, P., Wang, C., et al.: Classifying transportation mode and speed from trajectory data via deep multi-scale learning. Comput. Netw. **162**, 106861 (2019)
3. Zhang, Y., Liu, J., Qian, X., et al.: An automatic road network construction method using massive GPS trajectory data. Int. J. Geo-Inf. **6**(12), 400 (2017)
4. Cheng, W., Wen, R., Huang, H., et al.: OPTDP: towards optimal personalized trajectory differential privacy for trajectory data publishing. Neurocomputing **472**, 201–211 (2022)
5. Jiang, H., Zhang, Y., Xiao, Z., et al.: An empirical study of travel behavior using private car trajectory data. IEEE Trans. Netw. Sci. Eng. **8**(1), 53–64 (2020)
6. Xu, Z., Liu, Y., Yen, N.Y., et al.: Crowdsourcing based description of urban emergency events using social media big data. IEEE Trans. Cloud Comput. **8**(2), 387–397 (2016)
7. Chen, X., Li, Q.: Event modeling and mining: a long journey toward explainable events. VLDB J. **29**(1), 459–482 (2020)
8. D'Este, G.M., Zito, R., Taylor, M.A.P.: Using GPS to measure traffic system performance. Comput. Aided Civil Infr. Eng. **14**(4), 255–265 (1999)
9. Liu, Z., Zhang, R., Wang, C., et al.: Spatial-temporal Conv-sequence learning with accident encoding for traffic flow prediction. IEEE Trans. Netw. Sci. Eng. **9**, 1765–1775 (2022)

10. Yao, B., Chen, C., Cao, Q., et al.: Short-term traffic speed prediction for an urban corridor. Comput. Aided Civil Infras. Eng. **32**(2), 154–169 (2017)
11. Yu, Q., Yuan, J.: TransBigData: a Python package for transportation spatio-temporal big data processing, analysis and visualization. J. Open Source Softw. **7**(71), 4021 (2022)
12. Liu, X., Ban, Y.: Uncovering spatio-temporal cluster patterns using massive floating car data. ISPRS Int. J. Geo Inf. **2**(2), 371–384 (2013)

Many-Objective Evolutionary Algorithm Based on Dominance and Objective Space Decomposition

Zichen Wei, Hui Wang[✉], Tingyu Ye, Shuai Wang, Jiali Wu, Jiawen Liu, and Hai Zhang

School of Information Engineering, Nanchang Institute of Technology, Nanchang 330099, China
huiwang@whu.edu.cn

Abstract. Many-objective optimization problems (MaOPs) refer to multi-objective optimization problems (MOPs) containing large number of objectives, typically more than four. The complexity of MOPs grows rapidly in size with the number of objectives, making the problem quickly intractable. Though various many-objective evolutionary algorithms (MaOEAs) have been proposed to solve MaOPs, it is difficult to balance the convergence and population diversity. In order to address the above issues, this paper proposes a many-objective evolutionary algorithm based on dominance and objective space decomposition (called DDMaOEA), in which fast non-dominated sorting cooperates with objective space decomposition to maintain the population diversity and convergence during the environment selection. To verify the performance of DDMaOEA, 13 benchmark problems with 3, 5, 8, and 10 objectives are used. Experimental results show that DDMaOEA achieves better performance when compared with five other Many-objective optimization algorithms.

Keywords: Many-objective optimization · Many-objective evolutionary algorithms · Evolutionary algorithm · Objective space decomposition · Dominance

1 Introduction

Many real-world problems consist of many conflicting objective functions [1,2]. Different objective functions can conflict with each other. Usually, a specific optimal solution for such problems does not exist, but a non-inferior set of solutions exists. Solving a multi-objective optimization problems (MOPs) is more challenging than solving a single-objective optimization problem. For a single-objective optimization problem, there is one and only one optimal solution in the search space. Yet, MOPs can have many non-inferior solutions in the search space at the same time. These non-inferior solutions form the real Pareto frontier (PF). In general, the many-objective optimization algorithm has to find the set of non-dominated solutions that are closest to the true PF in the objective space.

H. Zhang et al. (Eds.): NCAA 2022, CCIS 1638, pp. 205–218, 2022.
https://doi.org/10.1007/978-981-19-6135-9_16

However, in practical applications, there exist many problems having more than three objectives. Those problems are called many-objective optimization problems (MaOPs). With increasing of objectives, the convergence and diversity of many multi-objective evolutionary algorithms (MOEAs) are seriously degraded [3]. First, many solutions become non-dominated. It results in slowing down the search process. Second, crowding distance and clustering operator are two common operators, and they will become computationally expensive when dealing with MaOPs. So, the diversity maintenance is a challenging work.

In this paper, a many-objective evolutionary algorithm based on dominance and objective space decomposition (called DDMaOEA) is proposed. Firstly, fast non-dominated sorting is introduced to divide the entire population into several layers. Solutions in the first layer dominates all other layers. Secondly, an elite strategy is introduced to merge the parent and child populations. Non-dominated solutions are selected from the merged population to form the next population. It ensures that better parent solutions are preserved and the convergence rate can be improved. Finally, a novel subspace selection method is used to maintain population diversity. In the experiments, thirteen benchmark problems with 3, 5, 8, and 10 objectives are used to test the performance of the proposed DDMaOEA. Results show that DDMaOEA achieves a good performance in balancing the convergence and population diversity.

The rest of the paper is organized as follows. The related work about MaOPs are given in Sect. 2. In Sect. 3, the proposed approach DDMaOEA is described. Simulation experiments and results are presented in Sect. 4. Finally, the work is summarized in Sect. 5.

2   Background

A general MOP can be defined as follows:

$$
\begin{aligned}
\min F(x) &= [f_1(x), f_2(x), \cdots, f_m(x)], \\
\text{s.t.} \quad & X \in \Omega,
\end{aligned}
\tag{1}
$$

where $\Omega \subseteq R^n$ denote the decision space and $X = (X_1, X_2, \ldots, X_n)$ the decision vector. F is a mapping from R^n to R^m, i.e., from the decision space to the objective space, and $f_i(X)$ denotes the i-th objective. Generally, many-objective optimization problem (MaOP) is one in which the number of objective functions in the MOP exceeds 3. Since optimizing one objective leads to the deterioration of at least other objectives, we expect to find a set of equilibrium solutions called Pareto-optimal solution set [4]. A Pareto-optimal solution, defined in that none of the objectives can be improved without lowering the other objective values, so the objective for a multi-objective optimization problem is to find the Pareto optimal solution set. For example, if X_1 dominates X_2 (expressed as $X_1 \prec X_2$), then if and only if $F_i(X_1) \geq F_i(X_2)$, here $i \in \{1, \cdots, m\}$, and $F_i(X_1) < F_i(X_2)$ holds. The set of all Pareto solutions is called Pareto-optimal set (PS) [6]. The set of all Pareto-optimal objective values is called Pareto front (PF) [4].

In order to solve MOPs and MaOPs, Deb et al. [4] proposed the non-dominated sorting genetic algorithm (NSGA). In [5], an enhanced version of NSGA called NSGA-II was designed. In NSGA-II, the population is divided into several layers by dominance relationship. The first layer is the set of non-dominant solutions in the population. The crowding degree sorting for environmental selection can maintain good diversity, but it has a high computational complexity. To solve MaOPs, Deb and Jain [7] designed a non-dominated sorting algorithm based on preference points (NSGA-III). Zhang et al. [8] proposed a new multi-objective evolutionary algorithm (MOEA/D) based on the decomposition. The multi-objective problem is decomposed into multiple single-objective subproblem to find the real Pareto front. However, MOEA/D needs to set the parameters, and the convergence rate performs poorly [11]. To solve MaOPs problem, some improved MOEA/D algorithms have been proposed successively, such as MOEA/D with a DE operator and polynomial mutation (MOEA/D-DE) [9] and MOEA/D with dynamical resource (MOEADDRA) [10]. Through the above discussions, it can be found that sacrificing partial convergence can enhance the population diversity, and sacrificing partial population diversity can strengthen the convergence. It is challenging to make a balance between convergence and population diversity.

3   Proposed Approach

3.1   Motivation

Pareto-based MOEAs [13] usually divide the whole group into different levels. Groups are divided into different hierarchical relationships based on their dominance. Individuals closer to the Pareto frontier should be retained, followed by the use of indicators of diversity to maintain the diversity of the population. So, the final population will have good convergence and diversity. For the fast non-dominant sorting [5], it divides the whole population into several levels. According to the hierarchy, n solution are chosen from small to large. As long as choose $N - n$ solutions to ensure the diversity of the solution set, we can get the expected convergence and diversity, where N represents the population size.

Diversity refers to the distribution of individuals in the population in the objective space. If the distribution of individuals in the population is more uniform, the diversity of the population is better. There are many methods to ensure the diversity, such as crowding degree operator. The crowding operator table represents the crowding distance between each individual. So picking solutions with large crowding can maximize the crowding between individuals. In MOEA/D, the objective space is decomposed into multiple subspaces. Individuals in each subspace will be linked to the weight of that space. A selected individual is closely related to the weight of another individual. So, the diversity of the population can also be guaranteed as long as the distribution of the weights is uniform enough. The population diversity and the distribution of individuals in the objective space is interconnected. The objective space is evenly decomposed into several subspaces. According to the number of individuals to be selected in the subspace, fewer individuals in the subspace are selected to enter the next population, worse diversity is obtained. As long as the subspace distribution is uniform and large enough, the population diversity will maintain a good level.

Based on the above analysis, this paper proposes a new environment selection strategy based on domination and objective space decomposition.

3.2 Tournament Selection Based on Objective Space Decomposition

In MOEAD, the objective space is decomposed by adjusting the weight vector of each objective function. The relationship between the individual and the subspace is determined according to the angle between the individual and the weight vector. In [12,14], a different decomposition method called objective space decomposition was proposed. Each subspace contains a set of solutions that can treat each subspace as a subproblem. The whole population is divided into multiple subpopulations. Through collaboration between subpopulations, the whole population diversity is to maintained.

Definition 1 (subspace): First, we predefine N uniformly distributed unit vectors: $w^1, w^2, \ldots, w^N$, and divide the objective space into N subspaces $S^1, S^2, \ldots, S^N$, where $S^i = \{v | \langle v, w^i \rangle \leq \langle v, w^j \rangle, \forall j \in \{1, 2, \ldots, N\}\}$ and $\langle v, w^i \rangle$ indicates the sharp angle between v and w^i. Then, the subspace is associated according to the sharp angle between the solution and the unit vector. According to the definition of the subspace, a vector v belongs to S^i if and only if v angles w^i from other unit vectors. During the optimization process, each solution is associated with a unit vector and a subspace.

Definition 2 (neighborhood subspace): Let K be the size of the neighborhood space. For the subspace S^i, $(i = 1, 2, \ldots, N)$, its neighborhood subspace is defined as $NS_i = \{S^{i_1}, \ldots, S^{i_K}\}$, where $\forall S^j \in NS_i, \forall S^k \notin NS_i, \langle w^j, w^i \rangle \leq \langle w^k, w^i \rangle$. In other words, the neighborhood subspace of the subspace S^i is a union of the subspace whose unit vector has the first K smallest sharp angles in all the unit vectors. In this paper, the number of neighborhood subspaces is set to M.

The standard tournament selection randomly selects k individuals from the current population size N [15]. Then, the individual with the best fitness is chosen from the k individuals. The selection process of competition selection consists of two steps: sampling and selection. The usual size of the sampling is 2, 4, and 7. The size of the sampling in this paper is set 2. At present, the tournament selection is different, such as the crowding operator used in NSGA-II and ABC [16] uses the fitness value to choose. In this paper, based on the number of individuals in the neighborhood subspace [14], the more the number of individuals in the neighborhood subspace, the less the probability of being selected. For the population of the multi-objective optimization problem, the less the number of individuals in the subspace where the individual is located, the worse the diversity in the subspace. So individuals in the next generation will necessarily hope that more solutions can be generated in the poor diversity subspace. Therefore, selecting individuals with small numbers of individuals in their neighborhood subspace to produce the next generation can somewhat

---

Algorithm 1: Environmental selection

```
1  for i = 1 to | Q | do
        /* The solution with good convergence is selected into the next generation first.
        */
2      if Q_i in the front MaxFNo-1 layer then
3          | P = P ∪ Q_i;
4      end
5  end
6  The angular cosine value (PWcos) between the individual and the weight W in P was
   calculated;
7  S = max(PWcos);
        /* With the largest angular cosine value between the individual P_i and the weight W_j,
           then the individual P_i belongs to the subspace S^j.                              */
8  for i = 1 to N − | P | do
        /* Individuals were selected based on the objective space decomposition    */
9      for j = 1 to | S | do
            /* The number of individuals in the neighborhood subspace was calculated.    */
10         if j = 1 then
11             | S_j^{sum} = S_1^{num} + S_2^{num} + S_{|S|}^{num};
12         end
13         else if j = | S | then
14             | S_j^{sum} = S_1^{num} + S_{|S|}^{num} + S_{|S|-1}^{num};
15         end
16         else
17             | S_j^{sum} = S_{j-1}^{num} + S_j^{num} + S_{j+1}^{num};
18         end
19     end
20     Find the individual P^{min} with the lowest number of individuals in the subspace;
21     if | P^{min} | > 1 then
22         Find the individual p with the lowest number of individuals in the neighborhood
           subspace in P^{min};
23     end
24     else
25         | p = P^{min};
26     end
27     q = p_i  ∀i ∈ (1, 2, ..., | p |);
28     P = P ∪ q;
29     update S;
30 end
```

---

guarantee that the diversity of the resulting next-generation population is not very poor. Before ensuring diversity, the convergence must be determined, so we should first judge the individual layer in the non-dominant ranking, and first choose the individual with a small number of layers. If the number of two individual layers is the same, then choose according to the number of individuals in the neighborhood subspace where they are located.

3.3 Environmental Selection Based on the the Objective Space Decomposition

In Algorithm 1, an environmental selection strategy based on the objective space decomposition is described. The objective of this function is to select suitable individuals from the combined population Q to enter the next generation.

Here $MaxFNo$ represents that the total number of individuals in the former $MaxFNo$ layer after non-dominated sorting is greater than or equal to the

---

Algorithm 2: Framework of DDMaOEA

---

1 Population initialization;
2 **while** $FEs < MFEs$ **do**
3     $MatingPool = TournamentSelection(P)$;
    /* Tournament competition by selecting the right individuals to produce individuals
    */
4     $P' = Offspring(MatingPool)$;
5     $Q = P \bigcup P'$;
6     $FastNonDominatedSort(Q)$;
7     $P = EnvironmentalSelection(Q, N, P_{t+1}, W)$;
8 **end**

---

number of populations. The smaller the number of layers the individual is in after non-dominated sorting, the better the convergence of the individual. Therefore, individuals whose layers are less than $MaxFNo$ are put into the next population P, and the convergence of the individuals in the offspring population is enhanced (lines 1–6). S of the individual in the population P is calculated (lines 7–8). Next, suitable individuals are selected from the $MaxFNo$ layer and stored into the offspring population P (lines 9–33). The number of selected individuals S^{sum} in the neighborhood subspace is calculated based on the number of selected individuals in the subspace (lines 9–22). Then, the individual P^{min} with the smallest S^{num} in the subspace of the individual in layer $MaxFNo$ is found. If the number of individuals in P^{min} is greater than 1, the individual p with the lowest number of individuals and S^{sum} selected in the neighborhood subspace is found in P^{min}. Finally, an individual is randomly selected from p to enter the next population P. The less the number of selected individuals S_i^{num} in the subspace S_i where the individual Q_i is located in the $MaxFNo$ layer, the worse the diversity of this subspace S_i, Then we need to select the individual P^{min} with the least selected individuals in the subspace S_j from the $MaxFNo$ layer to enter the next population P first. If there are more than one individuals in P^{min}, we should select the one with the smallest number S^{sum} in the neighborhood subspace from P.

3.4   Framework of DDMaOEA

The proposed DDMaOEA consists of three main modifications: 1) an elite selection strategy is used to retain good solutions from the parent to the offspring; 2) fast non-dominated sorting is used to find non-dominated solutions; 3) the objective space decomposition is used to enhance the diversity of the population. The framework of the proposed DDMaOEA is shown in Algorithm 2. At the initialization stage, the initial population, the appropriate weight vector based on dimension M are generated. Each weight vector represents a subspace. The appropriate individual $MatingPool$ is selected with tournaments based on the objective space decomposition to generate offspring P' (line 4). Then, the offspring P' and the parent P are merged into Q. Finally, N individuals are selected in Q as the next population using environmental selection (line 7).

4   Experimental Study

4.1   Benchmark Problems and Parameter Settings

In order to verify the performance of our approach DDMaOEA, 13 different kinds of problems (DTLZ1-DTLZ4, WFG1-WFG3, and MaF1-MaF6) with 3, 5, 8, and 10 objectives are tested in the following experiments [12,22]. So, there are 52 test instances. These problems contain various properties, such as, multi-modal, irregular PF, deceptiveness, etc. These different properties can comprehensively evaluate of the performance of the DDMaOEA.

In the experiments, DDMaOEA is compared with five other many-objective algorithms. All testing algorithms are implemented in the PlatEMO [21] platform. The proposed DDMaOEA is also encoded with MATLAB and embedded in the PlatEMO. The involved algorithms are listed as follows.

- MMOPSO [17].
- MPSOD [18].
- NSGAIIARSBX [19].
- MOPSO [20].
- MOEADM2M [14].
- DDMaOEA.

For decomposition-based algorithms, the number of populations depends on the total number of reference points. For problems with $M > 8$, the two-layer vector generation policy can be generated not only on the outer boundary, but also on the reference (or weight) vectors on the inner layer of the Pareto front. As a result, the dimensions of the test problem are set to 3, 5, 8, and 10. And the corresponding populations are set to 91, 212, 156, and 275, respectively. The maximum number of function evaluations is set to $10000 \times M$. All algorithms are tested 30 times for each problem and the average results are reported.

In this paper, two performance indicators: generation distance (GD) [23] and inverse generation distance (IGD) [16] are utilized. The GD values represent the average minimum distance between the individual in the population to the reference population. A smaller value of GD indicates a better convergence of the population. The IGD is an indicator used to evaluate the diversity and distribution of the algorithms. IGD has become a commonly used evaluation indicator and is accepted by its peers. The values of IGD represent the mean of the distance from each reference point to the nearest solution. The smaller the IGD value, the better the comprehensive performance of the algorithm.

4.2   Experimental Results

For all 52 test instances, the GD values are calculated and the results are shown in Tables 1 and 2. The symbols "-, +, =" in the last row of table indicate that the GD values corresponding to the comparison algorithms are significantly lower, better, or similar to DDMaOEA. From Table 1 and Table 2, we can see that the convergence of DDMaOEA algorithm is better than the four algorithms

Table 1. GD values of different algorithms on the benchmark set.

Problem	M	MMOPSO	MPSOD	NSGAIIARSBX	MOPSO	MOEADM2M	DDMaOEA
DTLZ1	3	5.0376e-2	5.4259e-1	2.7748e-2	4.8651e+0	4.2755e+0	**7.9570e-4**
		(5.71e-2)	(1.75e-1)	(4.09e-2)	(9.77e-1)	(5.90e-1)	**(4.94e-4)**
	5	1.2378e+0	5.8662e-1	5.4550e-1	3.0354e+0	5.9487e+0	**2.0731e-1**
		(1.18e+0)	(4.02e-2)	(5.94e-1)	(2.47e-1)	(6.67e-1)	**(2.99e-1)**
	8	2.8987e+0	**2.6948e-1**	2.5853e+0	6.8944e+0	6.9155e+0	6.9122e-1
		(9.00e-1)	**(4.59e-2)**	(1.87e-1)	(1.84e+0)	(1.03e+0)	(4.52e-1)
	10	1.0241e+1	**3.4357e+0**	1.4921e+1	5.9168e+0	6.3297e+0	3.6556e+0
		(2.49e+0)	**(3.78e-1)**	(1.10e+0)	(1.86e+0)	(3.58e-1)	(4.86e-1)
DTLZ2	3	1.2662e-3	**5.2057e-4**	1.7845e-3	1.7244e-2	1.4336e-1	7.1297e-4
		(2.99e-4)	**(2.69e-5)**	(1.59e-4)	(4.89e-3)	(6.58e-3)	(7.11e-5)
	5	1.9459e-2	**3.3708e-3**	6.2108e-2	9.2152e-3	1.4658e-1	4.3769e-3
		(4.59e-3)	**(6.40e-5)**	(2.64e-2)	(4.70e-3)	(4.92e-3)	(2.29e-4)
	8	1.3112e-1	**1.2786e-2**	1.5064e-1	1.9721e-2	1.4039e-1	1.5818e-2
		(8.72e-3)	**(9.79e-5)**	(5.54e-3)	(1.97e-3)	(1.73e-3)	(1.05e-3)
	10	1.3308e-1	3.4889e-2	1.3996e-1	**1.4485e-2**	1.4676e-1	3.5899e-2
		(1.76e-3)	(2.57e-3)	(1.62e-3)	**(1.02e-3)**	(5.84e-4)	(2.01e-3)
DTLZ3	3	2.1608e+0	1.5611e+1	2.0525e+0	2.7347e+1	2.5862e+1	**2.0870e-1**
		(1.30e+0)	(2.53e+0)	(1.34e+0)	(4.93e+0)	(6.50e+0)	**(1.92e-1)**
	5	1.3079e+1	1.0535e+1	1.7600e+1	2.5260e+1	2.9253e+1	**8.7381e-1**
		(6.10e+0)	(3.61e-1)	(3.70e+0)	(6.39e+0)	(2.88e+0)	**(5.95e-1)**
	8	9.9913e+1	**9.0285e+0**	7.7964e+1	3.7666e+1	3.2336e+1	1.7659e+1
		(1.05e+1)	**(7.08e-1)**	(5.72e+0)	(4.70e+0)	(3.81e+0)	(1.78e+0)
	10	7.2353e+1	3.2458e+1	9.4953e+1	3.7264e+1	**3.1471e+1**	4.3342e+1
		(7.72e+0)	(1.20e+0)	(1.41e+0)	(6.75e+0)	**(2.26e+0)**	(2.79e+0)
DTLZ4	3	1.3138e-3	**5.4259e-4**	1.6942e-3	2.9741e-2	4.1501e-2	5.9984e-4
		(2.18e-4)	**(1.30e-5)**	(1.99e-4)	(1.60e-2)	(6.83e-3)	(1.44e-4)
	5	1.5339e-2	4.7934e-3	1.4343e-2	8.4403e-2	1.2445e-1	**3.1764e-3**
		(2.41e-3)	(1.75e-3)	(2.35e-3)	(1.85e-2)	(5.67e-3)	**(1.94e-4)**
	8	1.6029e-1	9.6244e-3	1.4312e-1	1.2544e-1	1.4398e-1	**7.3617e-3**
		(8.01e-3)	(2.91e-4)	(3.79e-3)	(8.64e-3)	(2.75e-3)	**(8.91e-4)**
	10	1.1878e-1	**4.3049e-2**	1.1455e-1	1.2310e-1	1.4779e-1	4.8065e-2
		(4.26e-3)	**(1.77e-3)**	(2.85e-3)	(2.00e-2)	(2.75e-3)	(2.18e-3)
WFG1	3	3.0476e-2	2.1964e-1	2.2545e-2	1.4170e-1	1.3330e-1	**1.5030e-2**
		(1.01e-2)	(1.46e-2)	(5.17e-3)	(1.05e-2)	(1.21e-2)	**(6.10e-3)**
	5	6.5142e-2	2.3510e-1	**5.9713e-2**	1.3749e-1	1.3668e-1	6.0028e-2
		(7.55e-3)	(2.09e-2)	**(5.78e-3)**	(6.06e-3)	(3.42e-3)	(9.32e-3)
	8	1.0015e-1	3.0681e-1	**9.6892e-2**	1.7558e-1	1.6590e-1	1.2525e-1
		(1.32e-2)	(3.26e-2)	**(5.02e-3)**	(2.67e-3)	(1.66e-3)	(1.92e-2)
	10	**1.1544e-1**	3.3368e-1	1.1734e-1	1.9849e-1	1.8900e-1	1.5247e-1
		(9.03e-3)	(3.18e-2)	(7.48e-3)	(2.08e-3)	(1.05e-3)	(1.34e-2)
WFG2	3	2.8299e-2	**7.3784e-3**	3.1092e-2	2.9901e-2	3.9401e-2	8.6299e-3
		(9.99e-3)	**(1.20e-3)**	(1.23e-2)	(6.66e-3)	(1.08e-2)	(6.62e-3)
	5	6.6662e-2	**1.0571e-2**	7.6304e-2	3.1902e-2	5.7759e-2	1.6301e-2
		(1.14e-2)	**(6.73e-4)**	(5.97e-3)	(6.81e-3)	(9.11e-3)	(2.04e-3)
	8	1.3333e-1	**2.7976e-2**	1.1941e-1	7.0400e-2	7.2405e-2	4.7172e-2
		(6.96e-3)	**(9.12e-4)**	(9.39e-3)	(1.09e-2)	(1.29e-2)	(9.30e-3)
	10	1.4405e-1	**2.2824e-2**	1.5187e-1	1.1659e-1	1.0480e-1	6.0523e-2
		(8.15e-3)	**(4.52e-4)**	(1.07e-2)	(2.12e-2)	(1.45e-2)	(5.08e-3)
WFG3	3	9.3342e-2	1.0205e-1	**8.6430e-2**	8.9079e-2	1.3085e-1	1.0693e-1
		(2.33e-3)	(3.08e-3)	**(4.01e-3)**	(2.91e-3)	(5.39e-3)	(7.70e-3)
	5	2.0237e-1	1.6322e-1	2.0212e-1	**1.2883e-1**	1.9106e-1	1.4556e-1
		(5.40e-2)	(7.38e-3)	(2.74e-3)	**(2.69e-2)**	(6.08e-3)	(6.90e-3)
	8	4.9570e-1	3.3864e-1	4.0808e-1	**1.2309e-1**	3.7594e-1	3.1442e-1
		(7.24e-3)	(1.20e-2)	(8.34e-3)	**(4.24e-2)**	(3.35e-3)	(1.98e-2)
	10	5.5805e-1	4.9043e-1	5.4488e-1	**2.2481e-1**	4.9782e-1	3.2718e-1
		(5.59e-3)	(8.45e-3)	(9.15e-3)	**(1.44e-1)**	(3.77e-3)	(1.75e-2)

Table 2. GD values of different algorithms on the benchmark set.

Problem	M	MMOPSO	MPSOD	NSGAIIARSBX	MOPSO	MOEADM2M	DDMaOEA
MaF1	3	1.6862e-3 (2.76e-4)	3.7712e-2 (2.23e-2)	2.3705e-3 (2.10e-4)	1.3961e-2 (1.58e-3)	2.0944e-2 (6.26e-3)	**6.4302e-4 (1.75e-4)**
	5	3.7795e-3 (2.36e-4)	7.5392e-2 (1.65e-2)	6.1747e-3 (7.73e-4)	1.3020e-2 (6.30e-3)	8.4263e-2 (1.11e-2)	**2.6443e-3 (6.75e-5)**
	8	1.0937e-2 (1.17e-3)	2.0933e-1 (6.97e-2)	1.3050e-2 (9.36e-4)	1.2257e-2 (5.77e-3)	1.8098e-1 (1.52e-2)	**8.5313e-3 (2.84e-4)**
	10	1.2039e-2 (9.46e-4)	2.3862e-1 (6.25e-2)	1.6774e-2 (1.09e-3)	1.1576e-2 (6.14e-3)	2.2251e-1 (1.59e-2)	**7.8152e-3 (2.80e-4)**
MaF2	3	4.7261e-3 (1.06e-3)	1.9558e-3 (4.51e-4)	5.5131e-3 (1.25e-3)	6.9551e-3 (4.77e-4)	1.0226e-2 (9.97e-4)	**8.5312e-4 (2.36e-4)**
	5	9.4375e-3 (1.44e-4)	6.5219e-3 (1.04e-4)	9.3821e-3 (2.32e-4)	**5.4961e-3 (2.37e-4)**	1.0975e-2 (1.81e-4)	6.7376e-3 (2.79e-4)
	8	9.3064e-3 (3.17e-4)	9.8178e-3 (2.95e-4)	**7.4664e-3 (2.01e-4)**	1.0487e-2 (6.56e-4)	7.8707e-3 (1.20e-4)	8.7800e-3 (3.79e-4)
	10	7.4621e-3 (1.31e-4)	8.2643e-3 (8.12e-5)	**7.3529e-3 (9.59e-5)**	1.2674e-2 (5.04e-4)	8.4375e-3 (9.36e-5)	9.7349e-3 (3.54e-4)
MaF3	3	1.2660e+5 (2.29e+5)	6.5725e+3 (4.45e+3)	9.9836e+4 (2.47e+5)	3.3988e+8 (2.86e+8)	3.8355e+8 (3.68e+8)	**3.7525e+2 (1.19e+3)**
	5	6.4458e+6 (9.11e+6)	**1.8769e+4 (8.21e+3)**	4.1865e+8 (6.56e+8)	3.4209e+8 (3.16e+8)	5.0353e+9 (2.42e+9)	2.2791e+6 (4.27e+6)
	8	5.3745e+11 (9.13e+10)	**7.7168e+3 (2.42e+3)**	1.6777e+11 (2.43e+10)	3.4511e+10 (1.55e+10)	1.9306e+10 (1.16e+10)	1.0362e+9 (6.80e+8)
	10	6.7324e+11 (4.61e+10)	**8.9103e+3 (1.29e+3)**	6.2565e+11 (6.72e+10)	6.8955e+10 (1.21e+10)	9.1584e+9 (5.02e+9)	5.0102e+9 (1.53e+9)
MaF4	3	1.3230e+1 (1.15e+1)	3.8638e+2 (1.36e+2)	1.5266e+1 (1.24e+1)	2.6108e+2 (6.68e+1)	6.8006e+1 (3.71e+1)	**4.1631e+0 (7.94e+0)**
	5	4.1596e+1 (2.37e+1)	2.6630e+3 (5.77e+2)	1.1677e+2 (4.81e+1)	6.3736e+2 (2.40e+2)	4.9591e+2 (1.57e+2)	**1.2519e+0 (1.18e+0)**
	8	3.1171e+2 (2.56e+2)	3.5063e+4 (8.39e+3)	4.9806e+2 (3.84e+2)	4.3625e+3 (8.13e+2)	3.7721e+3 (8.17e+2)	**1.2000e+1 (1.54e+1)**
	10	1.6953e+3 (4.43e+2)	1.4928e+5 (2.12e+4)	2.8514e+3 (1.68e+3)	1.9763e+4 (6.20e+3)	1.5575e+4 (2.89e+3)	**1.9345e+2 (1.68e+2)**
MaF5	3	4.4303e-3 (1.71e-3)	3.1336e-3 (2.33e-4)	6.4749e-3 (1.20e-3)	2.2783e-1 (1.70e-1)	1.5995e-1 (3.00e-2)	**2.5971e-3 (6.08e-4)**
	5	8.6633e-2 (2.20e-2)	3.3838e-1 (2.44e-1)	9.4799e-2 (1.04e-2)	1.6393e+0 (3.98e-1)	1.2622e+0 (1.22e-1)	**3.8841e-2 (3.51e-3)**
	8	1.5199e+1 (1.65e+0)	1.5822e+0 (9.28e-1)	1.2813e+1 (5.05e-1)	9.9749e+0 (3.96e+0)	5.4293e+0 (4.59e-1)	**5.3791e-1 (6.53e-2)**
	10	4.6697e+1 (1.58e+0)	1.0360e+1 (2.59e+0)	4.5275e+1 (2.07e+0)	3.4344e+1 (1.74e+1)	1.8278e+1 (2.51e+0)	**6.1453e+0 (1.36e+0)**
MaF6	3	**2.7721e-5 (2.33e-5)**	3.6753e+0 (4.64e+0)	6.0633e-5 (1.78e-5)	1.5391e-2 (1.06e-2)	1.4097e+1 (3.32e+0)	5.5178e-4 (1.36e-3)
	5	**8.1194e-6 (3.54e-6)**	8.5581e+0 (7.68e+0)	2.8023e-5 (4.81e-6)	1.6431e-3 (1.63e-3)	1.3916e+1 (7.86e-1)	8.2807e-5 (1.96e-4)
	8	1.2552e+1 (7.97e+0)	1.3543e+1 (2.93e+0)	9.8728e+0 (6.82e+0)	**8.7397e-4 (7.13e-4)**	1.7489e+1 (6.94e-1)	5.7445e+0 (6.11e+0)
	10	1.4674e+1 (1.89e-1)	1.4364e+1 (8.67e+0)	1.4253e+1 (1.87e-1)	7.1130e+0 (4.08e+0)	1.8452e+1 (3.54e-1)	**6.4105e+0 (2.34e+0)**
+/-/=		6/44/2	14/30/8	5/43/4	5/42/5	3/49/0	

except MPSOD for DTLZ and WFG function clusters. For the MaF1, MaF4 and MaF5 test problems, the DDMaOEA algorithm has better convergence than the other four algorithms in different dimensions. Overall, for these 52 test problems, the DDMaOEA algorithm outperforms MMOPSO, MPSOD, NSGAIIARSBX, MOPSO, and MOEADM2M by 44, 30, 43, 42, and 49, respectively.

It can be seen from Table 1 and Table 2 that DDMaOEA is better than other algorithms in both low-dimensional and high-dimensional convergence on the

MaF function. On DTLZ and WFG, although the performance of DDMaOEA is better than most algorithms, it is still not better than all other algorithms.This result shows that individuals selected by fast non-dominated sorting can effectively enhance the convergence of the population.

Table 3. IGD values of different algorithms on the benchmark set.

Problem	M	MMOPSO	MPSOD	NSGAIIARSBX	MOPSO	MOEADM2M	DDMaOEA
DTLZ1	3	3.7691e-1 (3.87e-1)	1.1137e+0 (3.78e-1)	2.1948e-1 (3.03e-1)	8.4130e+0 (3.03e+0)	1.6785e+0 (1.14e+0)	**3.4198e-2** (**4.84e-3**)
	5	1.0899e+1 (1.17e+1)	2.3062e+0 (6.04e-1)	4.4613e+0 (5.06e+0)	7.0859e+0 (3.65e+0)	3.3695e+0 (2.32e+0)	**1.7020e-1** (**1.09e-1**)
	8	1.1620e+1 (6.53e+0)	**8.9835e-1** (**2.41e-1**)	**1.3176e+1** (**3.37e+0**)	1.9846e+1 (8.98e+0)	2.1244e+0 (1.79e+0)	9.2698e-1 (5.30e-1)
	10	2.7705e+1 (9.84e+0)	**1.0527e+1** (**2.80e+0**)	3.0636e+1 (7.90e+0)	2.2325e+1 (7.52e+0)	**2.0171e+0** (**1.17e+0**)	6.8389e+0 (2.44e+0)
DTLZ2	3	7.0877e-2 (2.34e-3)	**5.4943e-2** (**1.36e-4**)	6.9030e-2 (2.29e-3)	1.2997e-1 (1.55e-2)	1.5044e-1 (7.39e-3)	9.9214e-2 (1.20e-2)
	5	2.2816e-1 (7.20e-3)	**1.6539e-1** (**3.87e-4**)	3.8628e-1 (8.02e-2)	5.5514e-1 (8.97e-2)	4.4954e-1 (5.43e-3)	2.4976e-1 (1.63e-2)
	8	8.0862e-1 (1.05e-1)	**3.2125e-1** (**8.55e-4**)	**1.7700e+0** (**4.53e-1**)	9.0247e-1 (1.10e-2)	8.0921e-1 (8.26e-3)	4.8785e-1 (3.10e-2)
	10	1.4202e+0 (1.22e-1)	**5.5546e-1** (**1.50e-2**)	**1.3217e+0** (**7.38e-2**)	**9.5176e-1** (**1.18e-2**)	8.1956e-1 (6.94e-3)	7.4560e-1 (2.76e-2)
DTLZ3	3	1.5576e+1 (9.29e+0)	6.1645e+1 (1.31e+1)	1.5119e+1 (9.40e+0)	5.3770e+1 (4.38e+1)	4.2527e+1 (1.76e+1)	**1.6543e+0** (**1.58e+0**)
	5	9.6212e+1 (6.47e+1)	6.9750e+1 (5.47e+0)	1.0698e+2 (3.10e+1)	1.5901e+2 (4.85e+1)	6.0097e+1 (1.33e+1)	**2.7604e+0** (**1.55e+0**)
	8	1.7355e+2 (2.22e+1)	**3.9456e+1** (**9.16e+0**)	**2.9847e+2** (**6.53e+1**)	1.6744e+2 (4.57e+1)	5.3565e+1 (1.78e+1)	5.3485e+1 (1.69e+1)
	10	2.1573e+2 (1.16e+1)	1.7626e+2 (3.23e+1)	7.0317e+2 (1.89e+2)	1.9390e+2 (3.06e+1)	**6.1531e+1** (**7.71e+0**)	1.9874e+2 (5.21e+1)
DTLZ4	3	6.9984e-2 (2.92e-3)	**5.4920e-2** (**2.25e-4**)	6.8481e-2 (2.36e-3)	3.5010e-1 (2.12e-1)	1.0374e-1 (6.50e-3)	1.8308e-1 (1.90e-1)
	5	2.1081e-1 (3.98e-3)	**1.7784e-1** (**4.86e-3**)	2.2776e-1 (6.23e-3)	7.1502e-1 (2.80e-1)	4.8002e-1 (1.90e-2)	**2.2351e-1** (**1.44e-2**)
	8	9.0548e-1 (1.70e-1)	**3.4802e-1** (**1.56e-3**)	**1.6392e+0** (**2.34e-1**)	2.1238e+0 (3.18e-1)	8.1841e-1 (3.80e-2)	**5.1177e-1** (**3.98e-2**)
	10	1.2328e+0 (6.94e-2)	**6.4572e-1** (**1.36e-2**)	**1.2421e+0** (**5.02e-2**)	2.2810e+0 (3.62e-1)	8.5229e-1 (2.83e-2)	1.0098e+0 (6.58e-2)
WFG1	3	4.2069e-1 (7.23e-2)	1.5109e+0 (7.44e-2)	**3.5711e-1** (**4.48e-2**)	**1.7623e+0** (**1.55e-1**)	1.2785e+0 (8.48e-2)	**4.8932e-1** (**1.05e-1**)
	5	1.0609e+0 (7.59e-2)	2.0605e+0 (3.71e-2)	**1.0142e+0** (**9.46e-2**)	**2.3113e+0** (**1.39e-1**)	1.9375e+0 (4.35e-2)	1.0933e+0 (1.00e-1)
	8	**1.4818e+0** (**1.26e-1**)	2.5356e+0 (7.16e-2)	**1.6068e+0** (**6.95e-2**)	2.9933e+0 (2.56e-1)	2.4795e+0 (6.76e-2)	1.7774e+0 (1.66e-1)
	10	**1.6343e+0** (**1.01e-1**)	2.8581e+0 (4.83e-2)	1.7782e+0 (7.88e-2)	3.4317e+0 (2.52e-1)	2.8137e+0 (7.78e-2)	2.3870e+0 (1.74e-1)
WFG2	3	2.3294e-1 (1.55e-2)	**2.1139e-1** (**1.03e-2**)	**2.1889e-1** (**1.01e-2**)	3.1608e-1 (4.00e-2)	2.9245e-1 (1.15e-2)	2.3937e-1 (2.54e-2)
	5	6.5954e-1 (3.93e-2)	**5.3000e-1** (**2.33e-2**)	7.7367e-1 (6.37e-2)	2.8628e+0 (1.08e+0)	7.0674e-1 (3.99e-2)	**4.6663e-1** (**1.92e-2**)
	8	1.4060e+0 (6.20e-2)	**1.2191e+0** (**2.66e-2**)	1.3619e+0 (5.61e-2)	7.2610e+0 (1.38e+0)	1.2028e+0 (4.56e-2)	**1.0658e+0** (**5.25e-2**)
	10	1.5468e+0 (6.58e-2)	**1.3964e+0** (**1.98e-2**)	1.6171e+0 (6.65e-2)	9.4073e+0 (2.21e+0)	1.3161e+0 (4.12e-2)	**1.1940e+0** (**5.59e-2**)
WFG3	3	1.0101e-1 (2.56e-2)	2.7487e-1 (1.41e-2)	**9.4880e-2** (**1.40e-2**)	**4.8033e-1** (**1.03e-1**)	3.2152e-1 (3.26e-2)	1.1656e+0 (2.37e-1)
	5	3.8966e-1 (6.46e-2)	6.9137e-1 (4.18e-2)	**3.5588e-1** (**6.33e-2**)	**2.7758e+0** (**1.12e+0**)	6.9752e-1 (6.81e-2)	1.3691e+0 (1.25e-1)
	8	1.0432e+0 (1.91e-1)	1.2056e+0 (4.83e-2)	**9.7013e-1** (**1.80e-1**)	**3.4405e+0** (**1.53e+0**)	1.6500e+0 (2.56e-1)	2.7258e+0 (5.63e-1)
	10	1.5211e+0 (2.69e-1)	1.5696e+0 (6.97e-2)	**1.3687e+0** (**1.82e-1**)	**4.7313e+0** (**2.76e+0**)	2.4340e+0 (5.73e-1)	2.6723e+0 (3.24e-1)

Results of IGD values on these questions are shown in Tables 3 and 4. From the results, we can see that DDMaOEA is better than the other four algorithms in comprehensive performance, but it is still worse than individual algorithms in some specific test problems. For example, MPSOD algorithm is to outperforms all other algorithms on DTLZ2. The overall performance of DDMaOEA in some

Table 4. IGD values of different algorithms on the benchmark set.

Problem	M	MMOPSO	MPSOD	NSGAIIARSBX	MOPSO	MOEADM2M	DDMaOEA
MaF1	3	6.1538e-2	7.6640e-2	**5.9473e-2**	9.6636e-2	1.4154e-1	**7.7736e-2**
		(1.90e-3)	(1.58e-3)	**(2.18e-3)**	**(7.73e-3)**	(9.29e-3)	**(7.69e-3)**
	5	**1.3785e-1**	2.3648e-1	1.5063e-1	3.4495e-1	3.0359e-1	1.4599e-1
		(3.75e-3)	(1.17e-2)	(6.38e-3)	(4.30e-2)	(1.16e-2)	(1.91e-2)
	8	2.5044e-1	5.1244e-1	2.5119e-1	4.6661e-1	3.7881e-1	**2.4610e-1**
		(7.46e-3)	(3.65e-2)	(8.47e-3)	(3.08e-2)	(7.65e-3)	**(1.33e-2)**
	10	2.6748e-1	5.8788e-1	2.9401e-1	5.0603e-1	3.6389e-1	**2.5561e-1**
		(6.01e-3)	(3.01e-2)	(9.14e-3)	(1.45e-2)	(3.39e-3)	**(1.57e-2)**
MaF2	3	5.0908e-2	**3.5768e-2**	4.9667e-2	6.1106e-2	1.2452e-1	**5.2064e-2**
		(2.31e-3)	**(1.45e-4)**	**(4.17e-3)**	(3.39e-3)	(9.07e-3)	**(3.24e-3)**
	5	1.2717e-1	**9.3123e-2**	1.3098e-1	2.0284e-1	1.8987e-1	1.0002e-1
		(4.35e-3)	**(1.63e-3)**	(4.25e-3)	(3.54e-2)	(7.10e-3)	(2.89e-3)
	8	1.7028e-1	2.2069e-1	**1.5253e-1**	4.7519e-1	1.8835e-1	1.7787e-1
		(2.90e-3)	(3.75e-3)	**(2.55e-3)**	**(7.24e-2)**	(1.11e-3)	(8.23e-3)
	10	1.6815e-1	1.8782e-1	**1.6378e-1**	5.3313e-1	1.9622e-1	1.9955e-1
		(2.63e-3)	(1.65e-3)	**(1.93e-3)**	**(1.59e-2)**	(2.13e-3)	(1.44e-2)
MaF3	3	9.5878e+2	3.6008e+3	6.8104e+2	3.2796e+4	5.3158e+3	**1.3340e-1**
		(1.17e+3)	(1.58e+3)	(9.73e+2)	(1.40e+4)	(2.44e+3)	**(9.23e-2)**
	5	3.0233e+3	**7.1814e+3**	2.2498e+4	5.1096e+4	1.0385e+4	**8.2811e+0**
		(2.77e+3)	**(1.63e+3)**	(1.28e+4)	(7.04e+3)	(6.37e+3)	**(1.13e+1)**
	8	1.2635e+5	**3.7875e+3**	4.8535e+5	4.6510e+5	6.6850e+3	3.8027e+4
		(2.43e+5)	**(1.17e+3)**	**(3.37e+5)**	(2.60e+5)	(4.22e+3)	(1.02e+5)
	10	4.7934e+4	**4.4783e+3**	1.3598e+6	4.0483e+8	5.6931e+3	3.4318e+4
		(1.05e+4)	**(1.43e+3)**	**(1.38e+6)**	(9.35e+8)	(3.63e+3)	(1.22e+4)
MaF4	3	5.6152e+1	4.1032e+2	6.6264e+1	4.4589e+2	1.4977e+2	**2.1176e+0**
		(5.50e+1)	(9.06e+1)	(5.63e+1)	(2.43e+2)	(1.09e+2)	**(2.39e+0)**
	5	1.7714e+2	2.3326e+3	5.7464e+2	1.1294e+3	6.7501e+2	**6.6647e+0**
		(1.17e+2)	(2.13e+2)	(2.54e+2)	(1.03e+3)	(4.00e+2)	**(4.53e+0)**
	8	6.5015e+2	1.2696e+4	1.4042e+3	8.1109e+3	4.8058e+3	**7.5805e+1**
		(6.84e+2)	(2.20e+3)	(1.23e+3)	(6.97e+3)	(2.48e+3)	**(8.20e+1)**
	10	1.9580e+3	6.0260e+4	7.9680e+3	3.4612e+4	2.0353e+4	**1.1489e+3**
		(8.70e+2)	(8.19e+3)	(6.65e+3)	(3.07e+4)	(9.03e+3)	**(1.05e+3)**
MaF5	3	7.7132e-1	3.1747e-1	**3.0790e-1**	9.2515e-1	5.3470e-1	9.0340e-1
		(1.45e+0)	(5.00e-3)	**(8.63e-3)**	**(1.14e-1)**	(8.23e-2)	(6.50e-1)
	5	**1.9858e+0**	4.3735e+0	2.0948e+0	6.5007e+0	6.1595e+0	2.1891e+0
		(5.58e-2)	(2.06e-1)	(5.25e-2)	(1.82e+0)	(5.20e-1)	(1.27e-1)
	8	3.4652e+1	7.3678e+1	3.4414e+1	5.4393e+1	3.8309e+1	**2.1700e+1**
		(4.46e+0)	(9.00e+0)	(2.72e+0)	(1.79e+1)	(3.00e+0)	**(5.13e+0)**
	10	9.7303e+1	2.4600e+2	1.0352e+2	1.8083e+2	1.6011e+2	**8.9673e+1**
		(6.43e+0)	(3.69e+1)	(8.95e+0)	(4.41e+1)	(1.24e+1)	**(6.20e+0)**
MaF6	3	**6.2085e-3**	5.6342e-1	**5.1778e-3**	1.5485e-1	3.2453e-2	1.8342e-2
		(4.80e-4)	(1.20e-1)	**(1.52e-4)**	(1.06e-1)	(3.62e-3)	(3.65e-3)
	5	**3.1025e-3**	6.9401e-1	**2.7159e-3**	2.5302e-2	5.1246e-2	1.5190e-2
		(3.29e-4)	(1.36e-1)	**(9.98e-5)**	(2.29e-2)	(2.72e-2)	(8.49e-3)
	8	1.1149e-1	5.4304e-1	6.7120e-1	**1.4897e-2**	2.0762e-1	1.0606e+0
		(8.72e-2)	(7.20e-2)	(5.59e-1)	**(1.06e-2)**	**(9.10e-2)**	(1.17e+0)
	10	3.8127e-1	7.2635e-1	1.5537e+0	4.1491e+1	**1.8015e-1**	2.7750e+1
		(1.37e-1)	(2.13e-1)	(5.33e-1)	(4.78e+1)	**(2.64e-2)**	(2.28e+1)
+/-/=		18/23/11	20/27/5	17/27/8	2/44/6	8/36/8	

test functions is still better than other algorithms, such as the IGD value on MaF4 is better than other algorithms.

Figure 1 shows the Pareto fronts of the six algorithms on DTLZ3 with 5 objectives. It can be seen that the convergence and diversity of the Pareto fronts of MMOPSO, NSGAIIARSBX, MOPSO, and MOEADM2M are not as good as DDMaOEA. Although MPSOD outperforms DDMaOEA in diversity, DDMaOEA outperforms MPSOD in convergence. After the above comparison, it can be found that the performance of DDMaOEA on IGD is still better than the other five comparison algorithms.

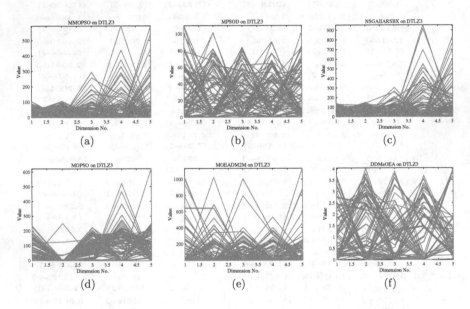

Fig. 1. The Parato fronts of six different MaOEAs on DTLZ3 with 5 objectives.

DDMaOEA is not as good as GD on IGD. However, the number of benchmark questions in which DDMaOEA outperforms other test algorithms is still more than half of the total number of benchmark questions. DDMaOEA did not perform as expected on IGD, probably due to excessive selection pressure for fast non-dominant. Therefore, the next step to improve the average direction is to reduce the selection pressure to enhance the diversity at the expense of partial convergence.

From the above analysis, DDMaOEA is better than other algorithms in terms of GD value, indicating that the convergence of DDMaOEA has a good guarantee, and the IGD value is also better than most algorithms. However, it does not exceed other algorithms too much more, indicating that the method of DDMaOEA still has room for improvement.

5   Conclusion

This paper proposes a new many-objective evolutionary algorithm based on dominance and objective space decomposition. First, an elite selection strategy is used to retain good solutions from the parent to the offspring. Then, fast non-dominated sorting is used to find non-dominated solutions. Finally, the objective space decomposition is used to enhance the diversity of the population. Experimental study on 13 benchmark problems with 3, 5, 8, 10 objectives show that the proposed approach DDMaOEA achieves better performance than five other algorithms.

However, DDMaOEA is not good enough in diversity because the selection of the neighborhood subspace in the algorithm is not detailed enough. The population may be evenly distributed in some areas. However, there are no individuals in some areas, resulting in no individuals. The regional diversity that exists is impoverished. DDMaOEA combines the objective space decomposition with fast non-dominant sorting to obtain good results in convergence and diversity. The future improvement direction focuses on an adaptive neighborhood subspace adjustment method, which can automatically adjust the size of the neighborhood subspace according to the distribution of the population. In addition, the conditions of non-dominated sorting are stringent, and a relax domination method may be suitable for MaOPs. These will be investigated in the future work.

Acknowledgment. This work was supported by National Natural Science Foundation of China (No. 62166027), and Jiangxi Provincial Natural Science Foundation (Nos. 20212ACB21 2004, 20212BAB202023, and 20212BAB202022).

References

1. Zitzler, E., Thiele, L.: Multiobjective evolutionary algorithms: a comparative case study and the strength Pareto approach. IEEE Trans. Evol. Comput. **3**(4), 257–271 (1999)
2. Wanger, A., Mills, K., Nelson, P.W., Rex, J.H.: Comparison of Etest and national committee for clinical laboratory standards broth macrodilution method for antifungal susceptibility testing: enhanced ability to detect amphotericin B-resistant Candida isolates. Antimicrob Agents Chemother. **39**(11), 2520–2522 (1995)
3. Zhan, Z.H., Li, J.J., Cao, J.N., Zhang, J., Chung, H.H., Shi, Y.H.: Multiple populations for multiple objectives: a coevolutionary technique for solving multiobjective optimization problems. IEEE Trans. Cybern. **43**(2), 445–463 (2013)
4. Srinivas, N., Deb, K.: Multiobjective optimization using nondominated sorting in genetic algorithms. Evol. Comput. **2**(3), 221–248 (1994)
5. Deb, K., Pratap, A., Agarwal, S., Meyarivan, T.: A fast and elitist multiobjective genetic algorithm: NSGA-II. IEEE Trans. Evol. Comput. **6**(2), 182–197 (2002)
6. Knowles, J.D., Corne, D.W.: Approximating the nondominated front using the pareto archived evolution strategy. Evol. Comput. **8**(2), 149–172 (2014)
7. Deb, K., Jain, H.: An evolutionary many-objective optimization algorithm using reference-point-based nondominated sorting approach, part I: solving problems with box constraints. IEEE Trans. Evol. Comput. **18**(4), 577–601 (2014)

8. Zhang, Q., Li, H.: MOEA/D: a multiobjective evolutionary algorithm based on decomposition. IEEE Trans. Evol. Comput. **11**(6), 712–731 (2007)
9. Li, H., Zhang, Q.: Multiobjective optimization problems with complicated Pareto sets, MOEA/D and NSGA-II. In: IEEE Transactions on Evolutionary Computation, **13**(2), 284–302 (2009)
10. Zhang, Q., Liu, W., Li, H.: The performance of a new version of MOEA/D on CEC09 unconstrained MOP test instances. In: Proceedings of the IEEE Congress on Evolutionary Computation, pp. 203–208 (2009)
11. Chen, B., Zeng, W., Lin, Y.: A new local search-based multiobjective optimization algorithm. IEEE Trans. Evol. Comput. **19**(1), 50–73 (2015)
12. Wq, A., Jz, A., Gw, B.: Evolutionary many-Objective algorithm based on fractional dominance relation and improved objective space decomposition strategy - ScienceDirect. Swarm and Evolutionary Computation. 60 (2020)
13. Reed, P.M., Hadka, D., Herman, J.D.: Evolutionary multi-objective optimization in water resources: The past, present, and future. In: Advances in Water Resources. 51(JAN.), 438–456 (2013)
14. Liu, H., Gu, F., Zhang, Q.: Decomposition of a multiobjective optimization problem into a number of simple multiobjective subproblems. IEEE Trans. Evol. Comput. **18**(3), 450–455 (2014)
15. Miller, B.L., Goldberg, D.E.: Genetic algorithms, tournament selection, and the effects of noise. Complex Systems. **9**(3), 193–212 (1995)
16. Zeng, T., et al.: Artificial bee colony based on adaptive search strategy and random grouping mechanism. Expert Syst. Appl. **192**, 116332 (2022)
17. Lin, Q., Li, J., Du, Z., Chen, J., Ming, Z.: A novel multi-objective particle swarm optimization with multiple search strategies. Europ. J. Oper. Res. **247**(3), 732–744 (2015)
18. Dai, C., Wang, Y., Ye, M.: A new multi-objective particle swarm optimization algorithm based on decomposition. Inf. Sci. **325**, 541–557 (2015)
19. Pan, L., Xu, W., Li, L., He, C., Cheng, R.: Adaptive simulated binary crossover for rotated multi-objective optimization. Swarm Evol. Comput. **60**, 100759 (2021)
20. Coello, C.A., Lechuga, M.S.: MOPSO: A proposal for multiple objective particle swarm optimization. In: Proceedings of the IEEE Congress on Evolutionary Computation. 1051–1056 (2002)
21. Tian, Y., Cheng, R., Zhang, X.Y., Jin, Y.C.: PlatEMO: a MATLAB platform for evolutionary multi-objective optimization. IEEE Comput. Intell. Mag. **12**(4), 73–87 (2017)
22. Pierro, F.D., Khu, S.T., Savic, D.A.: An investigation on preference order ranking scheme for multiobjective evolutionary optimization. IEEE Trans. Evol. Comput. **11**(1), 17–45 (2007)
23. Karafotias, G., Hoogendoorn, M., Eiben, A.E.: Parameter Control Evol. Algorithms. Evol. Comput. IEEE Trans. **19**(2), 167–187 (2007)

A New Unified Control Approach for Finite-/Fixed-Time Synchronisation of Multi-weighted Dynamical Networks

Jinyao Shi[1], Shuiming Cai[1(✉)], and Qiang Jia[2]

[1] School of Mathematical Sciences, Jiangsu University, Zhenjiang 212013, China
shuimingcai@ujs.edu.cn
[2] Institute of Applied System Analysis, Jiangsu University, Zhenjiang 212013, China
qiangjia@ujs.edu.cn

Abstract. This paper aims to propose a unified control scheme to explore the problems of finite-/fixed-time (FTT/FXT) synchronisation in multi-weighted dynamical networks (MWDNs). Firstly, a new unified FTT/FXT stability results is assessed for nonlinear dynamical systems, wherein a more accurate approximation of the settling time is acquired. Secondly, a novel feedback controller is designed for a class of MWDNs and a unified sufficient condition is obtained for FTT/FXT synchronisation of the considered MWDNs. It is shown that the conversion of FTT/FXT synchronisation can be achieved by adjusting only one control parameter, indicating the superiority of the control protocol in practical applications. Moreover, the designed unified control scheme excludes signum function, which can avoid the chattering phenomena in the synchronisation process. Finally, a numerical example is presented to authenticate the theoretical results.

Keywords: Multi-weighted dynamical networks · Finite-time synchronisation · Fixed-time synchronisation · Unified control framework

1 Introduction

During the past few decades, the dynamic behaviors of complex dynamical networks have been considerably researched because they are effective modeling tools for describing various real-world complex systems, such as biological networks, power grids, multi-agent networks, and communication networks. Considering the coupling forms between dynamical nodes in many realistic networks are multiple, the recent attention has converged on the study of multi-weighted dynamical networks (MWDNs) [1–3]. In fact, numerous practical networks can be represented by MWDNs. Typical examples include social networks, biological neural networks, and transportation networks [1,3,4]. In transportation networks, due to the existing of different transportation ways between two locations, the nodes are linked by multiple weights [3]. In neural networks, neurons have multiple connectivity and there exist diverse links corresponding to different neural structures [4].

H. Zhang et al. (Eds.): NCAA 2022, CCIS 1638, pp. 219–233, 2022.
https://doi.org/10.1007/978-981-19-6135-9_17

Synchronous control, a hot research issue in the realm of dynamical networks, has sparked a surge of interest owing to its potential applications in the engineering field [5–11]. Currently, there is a vast literature on synchronisation of dynamical networks, for instance, asymptotic synchronisation, exponential synchronisation, finite-time (FTT) synchronisation, fixed-time (FXT) synchronisation. Among them, FTT synchronisation and FXT one have been brought into focus gradually in view of the fact that they have superiority in convergence time and possess the properties of better robustness or resisting disturbance, which is more in line with the practical requirements [12–21]. Compared with the FTT synchronisation where the settling time (ST, a time required for reaching synchronisation) depending on the original states of the dynamical networks, FXT synchronisation means that the ST is bounded by a static constant independence of the original states, which thus facilitates its real-world applications. In [15], the authors investigated the FTT/FXT cluster synchronisation problems by designing some distributed protocols containing or excluding pinning control. In [18], the problem of FTT/FXT synchronisation in discontinuous complex networks was explored. In [19], authors studied the FTT/FXT synchronisation problem on complex networks with stochastic perturbations. In [20], the issue of FTT/FXT synchronisation was discussed on delayed multi-layer networks.

It is noted that, in order to guarantee dynamical networks can realize FTT/FXT synchronisation, signum function is indispensable for the designed controllers in most of the existent researches [15–21]. Unfortunately, the signum function would switch discontinuously, which often leads to high-frequency chattering and therefore damages apparatus and incurs unexpected results [22–27]. Considerable efforts have been devoted recently to the elimination of this undesirable phenomenon. In [22], a new discontinuous controller without signum function was developed to investigate FTT synchronisation for coupled networks with impulsive effects and Markovian topology. In [23], an innovative continuous controller was first designed and then applied to fixed-timely synchronize impulsive dynamical networks. Later, the control strategy given in [23] was adopted to analyze the FXT synchronisation of dynamical networks with stochastic perturbations [24]. In [25] and [26], based on the idea in [23], a more simpler chatter-free control scheme excluding the linear feedback term, was developed to drive dynamical networks to achieve FXT synchronisation and FTT cluster one, respectively. More recently, Hu *et al.* put forward a novel saturation continuous controller to fixed-timely or preassigned-timely stabilize chaotic systems [27]. So far as we know, however, using a common control scheme without signum function to synchronize MWDNs in finite time and fixed time simultaneously, has not been considered in the literature.

Motivated by the above discussion, this paper aims to propose a unified control scheme to research the issues of FTT/FXT synchronisation in MWDNs simultaneously. Firstly, a new unified FTT/FXT stability results is attained for nonlinear dynamical systems, wherein a more accurate approximation of the ST is acquired. Secondly, for a class of MWDNs, a novel feedback controller is designed and a unified sufficient condition is obtained for FTT/FXT synchronisation of the considered MWDNs. It is illustrated that the conversion of FTT synchronisation and FXT one can be realized by adjusting only one control parameter, indicating the superiority of the control protocol in practical applications. Moreover, the designed unified control scheme excludes signum function that can avoid the chattering phenomena in the synchronisation process. Finally, a numerical example is presented to support the theoretical results.

2 Preliminaries and Problem Formulations

Consider a general MWDN comprised of N nodes described by

$$\dot{x}_i(t) = f(x_i(t)) + \sum_{k=1}^{m}\sum_{j=1}^{N} b_{ij}^{(k)} \Gamma_k x_j(t), \quad i = 1,\ldots,N, \tag{1}$$

where $x_i(t) = \big(x_{i1}(t),\ldots,x_{in}(t)\big)^\top \in \mathbb{R}^n$ represents the state of the ith node, $f(x_i(t)) \in \mathbb{R}^n$ is a continuous function, characterizing the intrinsic dynamics of an uncoupled node, $\Gamma_k = \mathrm{diag}\big(\gamma_1^{(k)},\ldots,\gamma_n^{(k)}\big) > 0$ $(k = 1,2,\ldots,m)$ denotes the inner-coupling matrix of the kth coupling way, and $B^{(k)} = (b_{ij}^{(k)})$ $(k = 1,\ldots,m)$ is the coupling weight matrix, reflecting the network topology structure in the kth coupling way, in which $b_{ij}^{(k)} > 0$ if node j has a directed link to node i $(i \neq j)$; otherwise, $b_{ij}^{(k)} = 0$. In addition, for the matrix $B^{(k)}$, its diagonal elements are set equal to $\sum_{j=1,j\neq i}^{N} b_{ij}^{(k)} = -b_{ii}^{(k)}$, $i = 1,\ldots,N$. The initial values of system (1) are $x_i(0) = \varphi_i \in \mathbb{R}^n$, $i = 1,\ldots,N$.

System (1) is noticed as the drive system and its corresponding response system follows:

$$\dot{y}_i(t) = f(y_i(t)) + \sum_{k=1}^{m}\sum_{j=1}^{N} b_{ij}^{(k)} \Gamma_k y_j(t) + u_i(t), \quad i = 1,\ldots,N, \tag{2}$$

where $y_i(t) = \big(y_{i1}(t),\ldots,y_{in}(t)\big)^\top \in \mathbb{R}^n$, and $u_i(t) \in \mathbb{R}^n$ stands for the controller to be designed, which satisfies $u_i(t) = 0$ when $y_i(t) = x_i(t)$. The initial values of system (2) are $y_i(0) = \psi_i \in \mathbb{R}^n$, $i = 1,\ldots,N$.

Defining the error of the ith node as $e_i(t) = y_i(t) - x_i(t)$, the error system can be written as

$$\dot{e}_i(t) = \tilde{f}(e_i(t)) + \sum_{k=1}^{m}\sum_{j=1}^{N} b_{ij}^{(k)} \Gamma_k e_j(t) + u_i(t), \quad i = 1,\ldots,N, \tag{3}$$

where $\tilde{f}(e_i(t)) = f\big(y_i(t)\big) - f(x_i(t))$.

Definition 1. *Systems (1) and (2) are said to be FTT synchronisation if there exists a $T > 0$ which is dependent of the initial states, such that for any φ_i, ψ_i, $i = 1,\ldots,N$,*

$$\lim_{t \to T} \|e_i(t)\| = 0 \text{ and } e_i(t) = 0, \ t \geq T.$$

Definition 2. *Systems (1) and (2) are said to be FXT synchronisation if there exists a $T > 0$ bounded by a constant T_{max} which is independent of the initial states, such that for any φ_i, ψ_i, $i = 1, ..., N$,*

$$\lim_{t \to T} \|e_i(t)\| = 0 \text{ and } e_i(t) = 0, \ t \geq T_{max}.$$

Assumption 1 ([10]). *There exists a non-negative constant $\eta \geq 0$ such that*

$$(z_1 - z_2)^\top \left(f(z_1) - f(z_2) \right) \leq \eta (z_1 - z_2)^\top (z_1 - z_2) \tag{4}$$

holds for any $z_1, z_2 \in \mathbb{R}^n$.

Lemma 1 ([18]). *Set $\zeta_i \geq 0$, $i = 1, 2, ..., n$, $0 < r \leq 1$, $s \geq 1$, then*

$$\sum_{i=1}^{n} \zeta_i^r \geq \left(\sum_{i=1}^{n} \zeta_i \right)^r, \quad \sum_{i=1}^{n} \zeta_i^s \geq n^{1-s} \left(\sum_{i=1}^{n} \zeta_i \right)^s.$$

Lemma 2 ([18]). *Suppose that $V(e(t))$ is a continuous, radially unbounded positive-definite function on $\mathbb{R}^{nN}$, where $e(t) = \left(e_1^\top(t), e_2^\top(t), \ldots, e_N^\top(t) \right)^\top$. For system (3), if the following inequality holds*

$$\frac{\mathrm{d}}{\mathrm{d}t} V(e(t)) \leq -k_1 - k_3 V^q(e(t)),$$

where k_1, $k_3 > 0$, and $q \geq 0$. Then the statements below are true.

1) *When $0 \leq q \leq 1$, the zero solution of system (3) is FTT stable and the ST $T(e(0))$ is estimated by*

$$T(e(0)) \leq \begin{cases} \tilde{T}_1 \triangleq \dfrac{1}{k_1 + k_3} V(e(0)), & q = 0, \\[2mm] \tilde{T}_2 \triangleq \dfrac{1}{1-q} \left(\dfrac{k_1^{1-q}}{k_3} \right)^{\frac{1}{q}} \left(\left(1 + \left(\dfrac{k_3}{k_1} \right)^{\frac{1}{q}} V(e(0)) \right)^{1-q} - 1 \right), & 0 < q < 1, \\[2mm] \tilde{T}_3 \triangleq \dfrac{1}{k_3} \ln \dfrac{k_1 + k_3 V(e(0)}{k_1}, & q = 1. \end{cases}$$

2) *When $q > 1$, the zero solution of system (3) is FXT stable and the ST $T(e(0))$ is estimated by*

$$T(e(0)) \leq \tilde{T}_4 \triangleq \frac{1}{k_1} \left(\frac{k_1}{k_3} \right)^{\frac{1}{q}} \left(1 + \frac{1}{q-1} \right).$$

Lemma 3 ([20]). *For system (3), if $V(e(t))$ is a continuous, radially unbounded positive-definite function on $\mathbb{R}^{nN}$ and satisfies*

$$\frac{\mathrm{d}}{\mathrm{d}t} V(e(t)) \leq -k_1 - k_2 V(e(t)) - k_3 V^q(e(t)),$$

where $e(t) = \left(e_1^\top(t), e_2^\top(t), \ldots, e_N^\top(t) \right)^\top \in \mathbb{R}^{nN}$, k_1, k_2, $k_3 > 0$, and $q \geq 0$. Then the following conclusions are true.

1) When $0 \leq q \leq 1$, the zero solution of system (3) is FTT stable and the ST $T(e(0))$ is estimated by

$$T(e(0)) \leq \begin{cases} \hat{T}_1 \triangleq \dfrac{1}{k_2} \ln \dfrac{k_1 + k_3 + k_2 V(e(0))}{k_1 + k_3}, & q = 0, \\[3mm] \hat{T}_2 \triangleq \dfrac{1}{k_2} \ln \dfrac{k_1 + k_2 V(e(0))}{k_1}, & 0 < q < 1, \\[3mm] \hat{T}_3 \triangleq \dfrac{1}{k_2 + k_3} \ln \dfrac{k_1 + (k_2 + k_3) V(e(0))}{k_1}, & q = 1. \end{cases}$$

2) When $q > 1$, the zero solution of system (3) is FXT stable and the ST $T(e(0))$ is estimated by

$$T(e(0)) \leq \hat{T}_4 \triangleq \frac{1}{k_2} \ln \frac{k_1 + k_2}{k_1} + \frac{1}{k_3(q - 1)}.$$

Lemma 4. *For system (3), if there exists a continuous, radially unbounded positive-definite function $V(e(t))$ satisfying the inequality below*

$$\frac{\mathrm{d}}{\mathrm{dt}} V(e(t)) \leq -k_1 - k_2 V(e(t)) - k_3 V^q(e(t)), \tag{5}$$

where $e(t) = \left(e_1^\top(t), e_2^\top(t), \ldots, e_N^\top(t)\right)^\top \in \mathbb{R}^{nN}$, $k_1, k_2, k_3 > 0$, and $q \geq 0$. Then the following results hold.

1) *When $q = 0$, the zero solution of system (3) is FTT stable and the ST $T(e(0))$ is estimated by*

$$T(e(0)) \leq T_1^* \triangleq \frac{1}{k_2} \ln \frac{k_1 + k_3 + k_2 V(e(0))}{k_1 + k_3}.$$

2) *When $0 < q < 1$, the zero solution of system (3) is FTT stable and the ST $T(e(0))$ is estimated as follows:*
① *If $k_1 < q^{\frac{q}{1-q}} k_2^{\frac{q}{q-1}} k_3^{\frac{1}{1-q}}$, then*

$$T(e(0)) \leq T_2^* \triangleq \begin{cases} \dfrac{1}{1-q} \left(\dfrac{k_1^{1-q}}{k_3}\right)^{\frac{1}{q}} \left(\left(1 + \left(\dfrac{k_3}{k_1}\right)^{\frac{1}{q}} V(e(0))\right)^{1-q} - 1\right), & V(e(0)) \leq \zeta^*, \\[4mm] \dfrac{1}{1-q} \left(\dfrac{k_1^{1-q}}{k_3}\right)^{\frac{1}{q}} \left(\left(1 + \left(\dfrac{k_3}{k_1}\right)^{\frac{1}{q}} \zeta^*\right)^{1-q} - 1\right) \\[4mm] \quad + \dfrac{1}{k_2} \ln \dfrac{k_1 + k_3 (\zeta^*)^q + k_2 V(e(0))}{k_1 + k_3 (\zeta^*)^q + k_2 \zeta^*}, & V(e(0)) > \zeta^*. \end{cases}$$

where $\zeta^* > 0$ is the unique root of the equation $\left(k_1^{\frac{1}{q}} + k_3^{\frac{1}{q}} s\right)^q - k_1 - k_2 s = 0$.
② *If $k_1 \geq q^{\frac{q}{1-q}} k_2^{\frac{q}{q-1}} k_3^{\frac{1}{1-q}}$, then*

$$T(e(0)) \leq T_2^* \triangleq \frac{1}{k_2} \ln \frac{k_1 + k_2 V(e(0))}{k_1}.$$

3) *When $q = 1$, the zero solution of system (3) is FTT stable and the ST $T(e(0))$ is estimated by*

$$T(e(0)) \leq T_3^* \triangleq \frac{1}{k_2 + k_3} \ln \frac{k_1 + (k_2 + k_3)V(e(0))}{k_1}.$$

4) *When $q > 1$, the zero solution of system (3) is FXT stable and the ST $T(e(0))$ is estimated by*

$$T(e(0)) \leq T_4^* \triangleq \frac{q}{k_2(q-1)} \ln \left(1 + \frac{k_2}{k_1} \left(\frac{k_1}{k_3} \right)^{\frac{1}{q}} \right).$$

Remark 1. When $k_1 \geq q^{\frac{q}{1-q}} k_2^{\frac{q}{q-1}} k_3^{\frac{1}{1-q}}$, one has $T_2^* = \hat{T}_2$ and

$$T_2^* - \tilde{T}_2 = \frac{1}{k_2} \ln \frac{k_1 + k_2 V(e(0))}{k_1} - \frac{1}{1-q} \left(\frac{k_1}{k_3} \right)^{\frac{1}{q}} \frac{1}{k_1} \left(\left(1 + \left(\frac{k_3}{k_1} \right)^{\frac{1}{q}} V(e(0)) \right)^{1-q} - 1 \right).$$

Let $\mathcal{P}(z) = \frac{1}{k_2} \ln \frac{k_1 + k_2 z}{k_1} - \frac{1}{1-q} \left(\frac{k_1}{k_3} \right)^{\frac{1}{q}} \frac{1}{k_1} \left(\left(1 + \left(\frac{k_3}{k_1} \right)^{\frac{1}{q}} z \right)^{1-q} - 1 \right)$, $z \geq 0$, then one can deduce

$$\frac{\mathrm{d}}{\mathrm{d}z} \mathcal{P}(z) = \frac{1}{k_1 + k_2 z} - \frac{1}{\left(k_1^{\frac{1}{q}} + k_3^{\frac{1}{q}} z \right)^q}.$$

Clearly, when $k_1 \geq q^{\frac{q}{1-q}} k_2^{\frac{q}{q-1}} k_3^{\frac{1}{1-q}}$, $\left(k_1^{\frac{1}{q}} + k_3^{\frac{1}{q}} z \right)^q < k_1 + k_2 z$ for all $z > 0$, we have $\frac{\mathrm{d}}{\mathrm{d}z} \mathcal{P}(z) < 0$ for all $z > 0$. Hence, $\mathcal{P}(z) < \mathcal{P}(0) = 0$, i.e., $T_2^* < \tilde{T}_2$.

When $k_1 < q^{\frac{q}{1-q}} k_2^{\frac{q}{q-1}} k_3^{\frac{1}{1-q}}$, one has $\left(k_1^{\frac{1}{q}} + k_3^{\frac{1}{q}} s \right)^q > (k_1 + k_2 s)$ for all $s \in (0, \zeta^*)$ and $\left(k_1^{\frac{1}{q}} + k_3^{\frac{1}{q}} s \right)^q < (k_1 + k_2 s)$ for all $s \in (\zeta^*, +\infty)$. It follows that $\frac{\mathrm{d}}{\mathrm{d}z} \mathcal{P}(z) > 0$ for all $z \in (0, \zeta^*)$ and $\frac{\mathrm{d}}{\mathrm{d}z} \mathcal{P}(z) < 0$ for all $z > \zeta^*$, which implies that $\mathcal{P}(z) > \mathcal{P}(0) = 0$ for all $z \in (0, \zeta^*)$ and $\mathcal{P}(z) < \mathcal{P}(\zeta^*)$ for all $z > \zeta^*$. Therefore, if $V(e(0)) \leq \zeta^*$, then $T_2^* = \tilde{T}_2$ and $T_2^* < \tilde{T}_2$. On the other hand, if $V(e(0)) > \zeta^*$, then

$$T_2^* - \tilde{T}_2 = \frac{1}{(1-q)k_3^{\frac{1}{q}}} \left(k_1^{\frac{1}{q}} + k_3^{\frac{1}{q}} \zeta^* \right)^{1-q} + \frac{1}{k_2} \ln \frac{k_1 + k_3(\zeta^*)^q + k_2 V(e(0))}{k_1 + k_3(\zeta^*)^q + k_2 \zeta^*}$$
$$- \frac{1}{(1-q)k_3^{\frac{1}{q}}} \left(k_1^{\frac{1}{q}} + k_3^{\frac{1}{q}} V(e(0)) \right)^{1-q},$$

and

$$T_2^* - \hat{T}_2 = \frac{1}{1-q} \left(\frac{k_1^{1-q}}{k_3} \right)^{\frac{1}{q}} \left(\left(1 + \left(\frac{k_3}{k_1} \right)^{\frac{1}{q}} \zeta^* \right)^{1-q} - 1 \right)$$
$$+ \frac{1}{k_2} \ln \frac{k_1 + k_3(\zeta^*)^q + k_2 V(e(0))}{k_1 + k_3(\zeta^*)^q + k_2 \zeta^*} - \frac{1}{k_2} \ln \frac{k_1 + k_2 V(e(0))}{k_1}.$$

Let $\mathcal{Q}_1(z) = \frac{1}{(1-q)k_3^{\frac{1}{q}}}\left(k_1^{\frac{1}{q}} + k_3^{\frac{1}{q}}\zeta^*\right)^{1-q} + \frac{1}{k_2}\ln\frac{k_1+k_3(\zeta^*)^q+k_2 z}{k_1+k_3(\zeta^*)^q+k_2\zeta^*} - \frac{1}{(1-q)k_3^{\frac{1}{q}}}\left(k_1^{\frac{1}{q}} + \right.$

$\left. k_3^{\frac{1}{q}}z\right)^{1-q}$, $z \geq \zeta^*$, and $\mathcal{Q}_2(z) = \frac{1}{1-q}\left(\frac{k_1^{1-q}}{k_3}\right)^{\frac{1}{q}}\left(\left(1 + \left(\frac{k_3}{k_1}\right)^{\frac{1}{q}}\zeta^*\right)^{1-q} - 1\right) +$

$\frac{1}{k_2}\ln\frac{k_1+k_3(\zeta^*)^q+k_2 z}{k_1+k_3(\zeta^*)^q+k_2\zeta^*} - \frac{1}{k_2}\ln\frac{k_1+k_2 z}{k_1}$, $z \geq \zeta^*$, then

$$\frac{d}{dz}\mathcal{Q}_1(z) < 0, \quad z > \zeta^*,$$

since $\left(k_1^{\frac{1}{q}} + k_3^{\frac{1}{q}}s\right)^q < (k_1+k_2 s)$ for all $s > \zeta^*$. Hence, $\mathcal{Q}_1(V(e(0))) < \mathcal{Q}_1(\zeta^*) = 0$, i.e., $T_2^* < \tilde{T}_2$.

Additionally, noting that $\mathcal{P}(\zeta^*) > 0$, i.e., $\frac{1}{k_2}\ln\frac{k_1+k_2\zeta^*}{k_1} > \frac{1}{1-q}\left(\frac{k_1}{k_3}\right)^{\frac{1}{q}}\frac{1}{k_1}\left((1 + \right.$

$\left.\left(\frac{k_3}{k_1}\right)^{\frac{1}{q}}\zeta^*\right)^{1-q} - 1\right)$, we have $\mathcal{Q}_2(\zeta^*) < 0$. Therefore, $\mathcal{Q}_2(V(e(0))) < \mathcal{Q}_2(\zeta^*) < 0$ owing to the fact that $\frac{d}{dz}\mathcal{Q}_2(z) < 0$ for all $z \geq \zeta^*$. This means that $T_2^* < \hat{T}_2$.

From the analysis above, it can be concluded that $T_2^* \leq \tilde{T}_2$ and $T_2^* \leq \hat{T}_2$.

Remark 2. Let $\epsilon = \frac{1}{k_1}\left(\frac{k_1}{k_3}\right)^{\frac{1}{q}}$, then one has

$$T_4^* - \tilde{T}_4 = \frac{q}{k_2(q-1)}\ln\left(1 + \frac{k_2}{k_1}\left(\frac{k_1}{k_3}\right)^{\frac{1}{q}}\right) - \frac{1}{k_1}\left(\frac{k_1}{k_3}\right)^{\frac{1}{q}}\left(1 + \frac{1}{q-1}\right)$$

$$= \frac{q}{k_2(q-1)}\ln\left(1 + k_2\epsilon\right) - \frac{q}{q-1}\epsilon.$$

Set $\Psi(z) = \frac{q}{k_2(q-1)}\ln\left(1 + k_2 z\right) - \frac{q}{q-1}z$, $z > 0$, then a simple calculation shows

$$\frac{d}{dz}\Psi(z) = \frac{q}{q-1}\left(\frac{1}{1+k_2 z} - 1\right) < 0, \quad \text{for all } z > 0.$$

This means that $\Psi(z)$ is monotonic decreasing and $\Psi(z) < \Psi(0) = 0$ for all $z > 0$. Hence, $T_4^* < \tilde{T}_4$. Based on the theory of optimum values, it is easy to verify that the function $H(z) = \frac{1}{k_2}\ln\left(1 + \frac{k_2}{k_1}z\right) + \frac{1}{k_2(q-1)}\ln\frac{k_3+k_2 z^{1-q}}{k_3}$, $z > 0$ reaches its smallest value at $z^* = \left(\frac{k_1}{k_3}\right)^{\frac{1}{q}}$, which yields

$$T_4^* = H(z^*) \leq H(1) = \frac{1}{k_2}\ln\left(\frac{k_1+k_2}{k_1}\right) + \frac{1}{k_2(q-1)}\ln\left(1 + \frac{k_2}{k_3}\right).$$

Since $\ln(1+z) < z$ for all $z > 0$, one has

$$T_4^* < \frac{1}{k_2}\ln\left(\frac{k_1+k_2}{k_1}\right) + \frac{1}{k_2(q-1)}\frac{k_2}{k_3} = \frac{1}{k_2}\ln\left(\frac{k_1+k_2}{k_1}\right) + \frac{1}{k_3(q-1)} = \hat{T}_4.$$

Based on the foregoing analysis, one can conclude that $T_4^* \leq \tilde{T}_4$ and $T_4^* \leq \hat{T}_4$.

Remark 3. According to the discussions in Remarks 1 and 2, one can see that the estimation of the ST in our Lemma (Lemma 4) is less conservative than those given in [18] and [20]. In practice, it usually requires that the gap between the estimate and the real synchronous time should be as small as possible. Therefore, a high-precision estimation of the ST is a prerequisite for its practical applications. Here, by disposing of the corresponding proper integral or improper one skillfully, a more accurate approximation of the ST is acquired in Lemma 4, which indicates the advantage of Lemma 4 over Lemmas 2 and 3.

3   Main Results

In this section, a general control scheme is developed under which the FTT synchronisation and FXT one between system (1) and (2) can be investigated simultaneously. With the help of Lemma 4, a unified sufficient condition for FTT/FXT synchronisation will be obtained.

Theorem 1. *Under Assumption 1, if the controller in the response system (2) is given as*

$$u_i(t) = -d_i e_i(t) - \alpha e_i(t) \Upsilon(e_i(t)) - \beta e_i(t) \Upsilon^\theta(e_i(t)), \tag{6}$$

and

$$\left(\eta I_N - D\right) \otimes I_n + \sum_{k=1}^{m} \left(\tilde{B}^{(k)} \otimes \Gamma_k\right) < 0, \tag{7}$$

in which $d_i > 0$, $\alpha > 0$, $\beta > 0$, $D = \mathrm{diag}(d_1, d_2, \ldots, d_N)$, $\tilde{B}^{(k)} = \frac{1}{2}\left(B^{(k)^\top} + B^{(k)}\right)$, and $\Upsilon(e_i(t)) = 0$, if $\|e_i(t)\| = 0$; $\Upsilon(e_i(t)) = \left(e_i^\top(t)e_i(t)\right)^{-1}$, otherwise, then the results below are true.

1) *When $\theta = 1$, the drive-response system (1) and (2) is FTT synchronized, and the ST T_5 is estimated by*

$$T_5 \leq T_5^* \triangleq \frac{1}{\Theta_2} \ln \frac{\Theta_1 + 2\beta + \Theta_2 V(e(0))}{\Theta_1 + 2\beta}.$$

2) *When $0 < \theta < 1$, the drive-response system (1) and (2) is FTT synchronized, and the ST T_6 is estimated as follows:*

① *If $\Theta_1 < (1-\theta)^{\frac{1-\theta}{\theta}} (\Theta_2)^{\frac{\theta-1}{\theta}} (2\beta)^{\frac{1}{\theta}}$, then*

$$T_6 \leq T_6^* \triangleq \begin{cases} \dfrac{1}{\theta}\left(\dfrac{(\Theta_1)^\theta}{2\beta}\right)^{\frac{1}{1-\theta}} \left(\left(1 + \left(\dfrac{2\beta}{\Theta_1}\right)^{\frac{1}{1-\theta}} V(e(0))\right)^\theta - 1\right), & V(e(0)) \leq \zeta^*, \\[4ex] \dfrac{1}{\theta}\left(\dfrac{(\Theta_1)^\theta}{2\beta}\right)^{\frac{1}{1-\theta}} \left(\left(1 + \left(\dfrac{2\beta}{\Theta_1}\right)^{\frac{1}{1-\theta}} \zeta^*\right)^\theta - 1\right) \\[2ex] \quad + \dfrac{1}{\Theta_2} \ln \dfrac{\Theta_1 + 2\beta(\zeta^*)^{1-\theta} + \Theta_2 V(e(0))}{\Theta_1 + 2\beta(\zeta^*)^{1-\theta} + \Theta_2 \zeta^*}, & V(e(0)) > \zeta^*. \end{cases}$$

② If $\Theta_1 \geq (1-\theta)^{\frac{1-\theta}{\theta}}(\Theta_2)^{\frac{\theta-1}{\theta}}(2\beta)^{\frac{1}{\theta}}$, then

$$T_6 \leq T_6^* \triangleq \frac{1}{\Theta_2}\ln\frac{\Theta_1 + \Theta_2 V(e(0))}{\Theta_1}.$$

3) When $\theta = 0$, the drive-response system (1) and (2) is FTT synchronized, and the ST T_7 is estimated by

$$T_7 \leq T_7^* \triangleq \frac{1}{\Theta_2 + 2\beta}\ln\frac{\Theta_1 + (\Theta_2 + 2\beta)V(e(0))}{\Theta_1}.$$

4) When $\theta < 0$, the drive-response system (1) and (2) is FXT synchronized, and the ST T_8 is estimated by

$$T_8 \leq T_8^* \triangleq \frac{\theta - 1}{\Theta_2\theta}\ln\left(1 + \frac{\Theta_2}{\Theta_1}\left(\frac{\Theta_1}{2\beta N^\theta}\right)^{\frac{1}{1-\theta}}\right).$$

where $\zeta^* > 0$ is the unique root of the equation $\left((\Theta_1)^{\frac{1}{1-\theta}} + (2\beta)^{\frac{1}{1-\theta}}s\right)^{1-\theta} - \Theta_1 - \Theta_2 s = 0$, $\Theta_1 = 2\alpha N$, $\Theta_2 = -2\lambda_{\max}\left((\eta I_N - D)\otimes I_n + \sum_{k=1}^{m}\left(\tilde{B}^{(k)}\otimes\Gamma_k\right)\right)$, and $V(e(0)) = \sum_{i=1}^{N}e_i^{\top}(0)e_i(0)$.

Proof. Let $e(t) = \left(e_1^{\top}(t), e_2^{\top}(t), \ldots, e_N^{\top}(t)\right)^{\top}$ and construct a positive definite quadratic function as follows:

$$V(e(t)) = e^{\top}(t)e(t) = \sum_{i=1}^{N}e_i^{\top}(t)e_i(t). \tag{8}$$

By using Assumption 1, we have

$$\frac{\mathrm{d}}{\mathrm{d}t}V(e(t)) \leq 2\sum_{i=1}^{N}(\eta - d_i)e_i^{\top}(t)e_i(t) + 2\sum_{k=1}^{m}e^{\top}(t)(B^{(k)}\otimes\Gamma_k)e(t)$$
$$-2\alpha N - 2\beta\sum_{i=1}^{N}\left(e_i^{\top}(t)e_i(t)\right)^{1-\theta}$$
$$\leq -\Theta_1 - \Theta_2 V(e(t)) - 2\beta\sum_{i=1}^{N}\left(e_i^{\top}(t)e_i(t)\right)^{1-\theta}. \tag{9}$$

Case 1 ($\theta = 1$): It yields that

$$\frac{\mathrm{d}}{\mathrm{d}t}V(e(t)) \leq -\Theta_1 - \Theta_2 V(e(t)) - 2\beta. \tag{10}$$

According to Lemma 4–1), the drive-response system (1) and (2) is FTT synchronized, and the ST T_5 satisfies

$$T_5 \leq T_5^* \triangleq \frac{1}{\Theta_2}\ln\frac{\Theta_1 + 2\beta + \Theta_2 V(e(0))}{\Theta_1 + 2\beta}.$$

Case 2 $(0 < \theta < 1)$: One can deduce from Lemma 1 that

$$\frac{\mathrm{d}}{\mathrm{d}t}V\big(e(t)\big) \leq -\Theta_1 - \Theta_2 V\big(e(t)\big) - 2\beta \bigg(\sum_{i=1}^{N} e_i^\top(t)e_i(t)\bigg)^{1-\theta}$$

$$= -\Theta_1 - \Theta_2 V\big(e(t)\big) - 2\beta V^{1-\theta}\big(e(t)\big). \tag{11}$$

From Lemma 4–2), it is clear that the drive-response system (1) and (2) is FTT synchronized, and when $\Theta_1 < (1-\theta)^{\frac{1-\theta}{\theta}}(\Theta_2)^{\frac{\theta-1}{\theta}}(2\beta)^{\frac{1}{\theta}}$, the ST T_6 is estimated by

$$T_6 \leq T_6^* \triangleq \begin{cases} \frac{1}{\theta}\left(\frac{(\Theta_1)^\theta}{2\beta}\right)^{\frac{1}{1-\theta}}\left(\left(1+\left(\frac{2\beta}{\Theta_1}\right)^{\frac{1}{1-\theta}}V(e(0))\right)^\theta - 1\right), & V(e(0)) \leq \zeta^*, \\ \frac{1}{\theta}\left(\frac{(\Theta_1)^\theta}{2\beta}\right)^{\frac{1}{1-\theta}}\left(\left(1+\left(\frac{2\beta}{\Theta_1}\right)^{\frac{1}{1-\theta}}\zeta^*\right)^\theta - 1\right) \\ \quad + \frac{1}{\Theta_2}\ln\frac{\Theta_1 + 2\beta(\zeta^*)^{1-\theta} + \Theta_2 V(e(0))}{\Theta_1 + 2\beta(\zeta^*)^{1-\theta} + \Theta_2\zeta^*}, & V(e(0)) > \zeta^*. \end{cases}$$

where $\zeta^* > 0$ is the unique positive of the equation $\left((\Theta_1)^{\frac{1}{1-\theta}} + (2\beta)^{\frac{1}{1-\theta}}s\right)^{1-\theta} - \Theta_1 - \Theta_2 s = 0$. Additionally, when $\Theta_1 \geq (1-\theta)^{\frac{1-\theta}{\theta}}(\Theta_2)^{\frac{\theta-1}{\theta}}(2\beta)^{\frac{1}{\theta}}$, the ST T_6 is estimated by

$$T_6 \leq T_6^* \triangleq \frac{1}{\Theta_2}\ln\frac{\Theta_1 + \Theta_2 V(e(0))}{\Theta_1}.$$

Case 3 $(\theta = 0)$: In this case, one has

$$\frac{\mathrm{d}}{\mathrm{d}t}V\big(e(t)\big) \leq -\Theta_1 - \Theta_2 V\big(e(t)\big) - 2\beta V\big(e(t)\big). \tag{12}$$

It follows from Lemma 4–3) that the drive-response system (1) and (2) is FTT synchronized and the upper bound of the ST T_7 is determined by

$$T_7 \leq T_7^* \triangleq \frac{1}{\Theta_2 + 2\beta}\ln\frac{\Theta_1 + (\Theta_2 + 2\beta)V(e(0))}{\Theta_1}.$$

Case 4 $(\theta < 0)$: Employing Lemma 1, one has

$$\frac{\mathrm{d}}{\mathrm{d}t}V\big(e(t)\big) \leq -\Theta_1 - \Theta_2 V\big(e(t)\big) - 2\beta N^\theta V^{1-\theta}\big(e(t)\big). \tag{13}$$

Based on Lemma 4–4), the drive-response system (1) and (2) is FXT synchronized, and the upper bound of the ST T_8 is given by

$$T_8 \leq T_8^* \triangleq \frac{\theta - 1}{\Theta_2\theta}\ln\left(1 + \frac{\Theta_2}{\Theta_1}\left(\frac{\Theta_1}{2\beta N^\theta}\right)^{\frac{1}{1-\theta}}\right).$$

The proof of Theorem 1 is completed. □

Remark 4. Dissimilar form many existing works with controllers containing the signum function $sign(\cdot)$ which would switch discontinuously [15–21], this paper excludes it. As the state and control signals of the networks do not suffer from unexpected chattering arising from the signum function, our control method have great superiority in engineering applications. Furthermore, from Theorem 1, one can observe that when the control parameter θ belongs to the closed interval $[0, 1]$, the FTT synchronisation can be realized, while the FXT synchronisation can be achieved with $\theta \in (-\infty, 0)$. Hence, by changing only the control parameter θ from $[0, 1]$ to $(-\infty, 0)$, the conversion of FTT/FXT synchronisation can be accomplished, indicating the flexibility of our control protocol in practical applications.

Remark 5. Based on the inequality (11) presented in Theorem 1, if utilizing Lemma 2–2) or Lemma 3–2) to assess the ST, one can obtain the following two different upper bounds of the ST for the FTT synchronisation:

$$\tilde{T}_6 \triangleq \frac{1}{\theta}\left(\frac{\Theta_1^\theta}{2\beta}\right)^{\frac{1}{1-\theta}}\left(\left(1 + \left(\frac{2\beta}{\Theta_1}\right)^{\frac{1}{1-\theta}}V(e(0))\right)^\theta - 1\right),$$

$$\hat{T}_6 = \frac{1}{\Theta_2}\ln\frac{\Theta_1 + \Theta_2 V(e(0))}{\Theta_1}.$$

Analogously, assessing the ST of the FXT synchronisation utilizing the inequality (13) in Theorem 1 by Lemma 2–4) or Lemma 3–4), we can obtain:

$$\tilde{T}_8 \triangleq \frac{1}{\Theta_1}\left(\frac{\Theta_1}{2\beta N^\theta}\right)^{\frac{1}{1-\theta}}\left(1 - \frac{1}{\theta}\right),$$

$$\hat{T}_8 = \frac{1}{\Theta_2}\ln\frac{\Theta_1 + \Theta_2}{\Theta_1} - \frac{1}{2\beta\theta N^\theta}.$$

Similar to Remarks 1 and 2, it is easy to prove that $T_6^* \leq \tilde{T}_6$, $T_6^* \leq \hat{T}_6$, $T_8^* \leq \tilde{T}_8$ and $T_8^* \leq \hat{T}_8$, which indicates that the assessed ST in Theorem 1 is less conservative than the existent works [18] and [20].

4   Numerical Simulations

In this section, a numerical simulation is illustrated to support the synchronisation criterion given above. Consider a network consisting 15 nodes with linear diffusive couplings described below:

$$\dot{x}_i(t) = f(x_i(t)) + \sum_{j=1}^{15}b_{ij}^{(1)}\Gamma_1 x_j(t) + \sum_{j=1}^{15}b_{ij}^{(2)}\Gamma_2 x_j(t), \quad i = 1, \dots, 15, \qquad (14)$$

where $x_i(t) = (x_i^1(t), x_i^2(t), x_i^3(t))^\top \in \mathbb{R}^3$ and $\Gamma_k = I_{3\times3}$. The topological structure of the network is presented in Fig. 1.

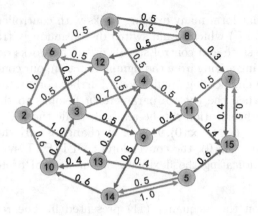

Fig. 1. A network contains 15 nodes and two weights.

Choose the node dynamics as the following Chua circuit [28]:

$$\begin{cases} \dot{x}_{i1} = 10\big((x_{i2} - x_{i1}) - \varphi(x_{i1})\big) \\ \dot{x}_{i2} = x_{i1} - x_{i2} + x_{i3} \\ \dot{x}_{i3} = -14.87x_{i2} \end{cases} \tag{15}$$

where $\varphi(x_{i1}) = -0.68x_{i1} - 0.295(|x_{i1} + 1| - |x_{i1} - 1|)$, $i = 1, ..., 15$.
The response system is described as:

$$\dot{y}_i(t) = f(y_i(t)) + \sum_{j=1}^{15} b_{ij}^{(1)} \Gamma_1 y_j(t) + \sum_{j=1}^{15} b_{ij}^{(2)} \Gamma_2 y_j(t) + u_i(t), \quad i = 1, \dots, 15, \tag{16}$$

with the following controller:

$$u_i(t) = -d_i e_i(t) - \alpha e_i(t) \Upsilon(e_i(t)) - \beta e_i(t) \Upsilon^\theta(e_i(t)). \tag{17}$$

Firstly, we consider the FTT synchronisation between systems (14) and (16) under controller (17).

By simple calculation, one has $\eta = 9.0620$. Choose $d_1 = \cdots = d_{15} = 12.7978$, $\alpha = 1$, $\beta = 4.4$, $\theta = 0.4$, then condition (7) holds. According to Theorem 1, the system (14) is FTT synchronized with (16) and the assessed ST is computed by $T_6^* = 1.6452$. In addition, based on Remark 5, two different estimates for the ST by using the methods in [18] and [20] are $\tilde{T}_6 = 2.6263$ and $\hat{T}_6 = 2.1730$. Figure 2 shows the time evolutions of the errors $e_i(t)$ $(i = 1, 2, \dots, 15)$, where the initial values of systems (14) and (16) are chosen on the interval $[0, 2]$ randomly, respectively. It can seen from Fig. 2 that the FTT synchronisation is achieved and our estimate (T_6^*) is less conservation than those $(\tilde{T}_6$ and $\hat{T}_6)$ calculated using the methods in [18] and [20].

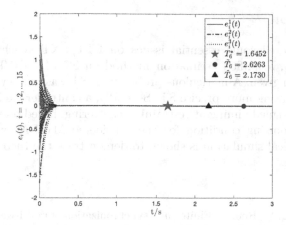

Fig. 2. FTT synchronisation between systems (14) and (16).

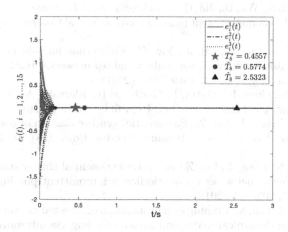

Fig. 3. FXT synchronisation between systems (14) and (16).

Then, we consider the FXT synchronisation between systems (14) and (16). Under controller (17), choose $\alpha = 1$, $\beta = 3$, $\theta = -1$, $d_i = 12.7978$ ($i = 1, 2, \ldots, 15$), based on Theorem 1, the system (14) is FXT synchronized with (16) and the estimated ST calculated by $T_8^* = 0.4557$. Additionally, following Remark 5, two different estimates for the ST by utilizing the approaches in [18] and [20] are $\tilde{T}_8 = 0.5774$ and $\hat{T}_8 = 2.5323$. The time evolutions of the errors $e_i(t)$ ($i = 1, 2, \ldots, 15$) are shown in Fig. 3, where the initial values are the same as above. Figure 3 illustrates clearly that the FXT synchronisation is realized and our estimate (T_8^*) is more precise than those ($\tilde{T}_8$ and $\hat{T}_8$) obtained utilizing the approaches in [18] and [20].

5   Conclusion

This paper explored some essential issues on FTT/FXT synchronisation for MWDNs. Firstly, a unified estimation method on ST of FTT/FXT synchronisation has been given. A meticulous probe about ST was made with the intention of approximating more precise ST. Secondly, a controller excluding signum functions was designed aiming at restraining chattering, based on which, a novel sufficient synchronizing condition for the considered MWDNs was presented. Finally, a numerical simulation is shown to demonstrate the theoretical results.

References

1. Qiu, S., Huang, Y., Ren, S.: Finite-time synchronization of multi-weighted complex dynamical networks with and without coupling delay. Neurocomputing **275**, 1250–1260 (2018)
2. Mwanandiye, E.S., Wu, B., Jia, Q.: Synchronization of delayed dynamical networks with multi-links via intermittent pinning control. Neural Comput. Appl. **32**, 11277–11284 (2020)
3. Xu, Y., Wu, X., Mao, B., Lü, J., Xie, C.: Finite-time intra-layer and inter-layer quasi-synchronization of two-layer multi-weighted networks. IEEE Trans. Circuits Syst. I: Regular Papers **4**(68), 1589–1598 (2021)
4. Katchinskiy, N., Goez, H., Dutta, I., Godbout, R., Elezzabi, A.: Novel method for neuronal nanosurgical connection. Sci. Rep. **6**(1), 20529 (2016)
5. Cai, S., He, Q., Hao, J., Liu, Z.: Exponential synchronization of complex networks with nonidentical time-delayed dynamical nodes. Phys. Lett. A **374**, 2539–2550 (2010)
6. Zhou, P., Cai, S., Shen, J., Liu, Z.: Adaptive exponential cluster synchronization in colored community networks via aperiodically intermittent pinning control. Nonlinear Dyn. **92**, 905–921 (2018)
7. Zhou, P., Shi, J., Cai, S.: Pinning synchronization of directed networks with delayed complex-valued dynamical nodes and mixed coupling via intermittent control. J. Frankl. Inst. **357**, 12840–12869 (2020)
8. Shen, Y., Shi, J., Cai, S.: Exponential synchronization of directed bipartite networks with node delays and hybrid coupling via impulsive pinning control. Neurocomputing **453**, 209–222 (2021)
9. Cai, S., Hou, M.: Quasi-synchronization of fractional-order heterogeneous dynamical networks via aperiodic intermittent pinning control. Chaos Solitons Fract. **146**, 110901 (2021)
10. Jia, Q., Bram, A.K., Han, Z.: Synchronization of drive-response networks with event-based pinning control. Neural Comput. Appl. **33**, 8649–8658 (2021)
11. Sun, M., Lyu, D., Jia, Q.: Event-triggered leader-following synchronization of delayed dynamical networks with intermittent coupling. Neural Comput. Appl. **34**, 6163–6170 (2022)
12. Haimo, V.T.: Finite time controllers. SIAM J. Control. Optim. **24**(4), 760–770 (1986)
13. Polyakov, A.: Nonlinear feedback design for fixed-time stabilization of linear control systems. IEEE Trans. Autom. Control **57**(8), 2106–2110 (2012)
14. Lu, W., Liu, X., Chen, T.: A note on finite-time and fixed-time stability. Neural Netw. **81**, 11–15 (2016)

15. Liu, X., Chen, T.: Finite-time and fixed-time cluster synchronization with or without pinning control. IEEE Trans. Cybern. **48**(1), 240–252 (2018)
16. Cai, S., Zhou, F., He. Q.: Fixed-time cluster lag synchronization in directed heterogeneous community networks, Phys. A **525**, 128–142 (2019)
17. Xu, Y., Li, Y., Li, W.: Adaptive finite-time synchronization control for fractional-order complex-valued dynamical networks with multiple weights. Commun. Nonlinear Sci. Numer. Simulat. **85**, 105239 (2020)
18. Ji, G., Hu, C., Yu, J., Jiang, H.: Finite-time and fixed-time synchronization of discontinuous complex networks: a unified control framework design. J. Frankl. Inst. **355**, 4665–4685 (2018)
19. Liu, X., Ho, D., Song, Q., Xu, W.: Finite/fixed-time pinning synchronization of complex networks with stochastic disturbances. IEEE Trans. Cybern. **49**(6), 2398–2403 (2019)
20. Xu, Y., Wu, X., Mao, B., Xie, C.: A unified finite-/fixed-time synchronization approach to multi-layer networks. IEEE Trans. Circuits Syst. II Exp. Briefs **1**(68), 311–315 (2021)
21. Hu, C., He, H., Jiang, H.: Fixed/preassigned-time synchronization of complex networks via improving fixed-time stability. IEEE Trans. Cybern. **51**(6), 2882–2892 (2021)
22. Yang, X., Lu, J.: Finite-time synchronization of coupled networks with markovian topology and impulsive effects. IEEE Trans. Autom. Control **61**(8), 2256–2261 (2016)
23. Yang, X., Lam, J., Ho, D., Feng, Z.: Fixed-time synchronization of complex networks with impulsive effects via nonchattering control. IEEE Trans. Autom. Control **62**(11), 5511–5521 (2017)
24. Zhang, W., Yang, X., Li, C.: Fixed-time stochastic synchronization of complex networks via continuous control. IEEE Trans. Cybern. **49**(8), 3099–3104 (2018)
25. Xu, Y., Wu, X., Li, N., Liu, L., Xie, C., Li, C.: Fixed-time synchronization of complex networks with a simpler nonchattering controller. IEEE Trans. Circuits Syst. II Express Briefs **67**(4), 700–704 (2020)
26. Jiang, S., Qi, Y., Cai, S., Lu, X.: Light fixed-time control for cluster synchronization of complex networks. Neurocomputing **424**, 63–70 (2021)
27. Hu, C., Jiang, H.: Special functions-based fixed-time estimation and stabilization for dynamic systems. IEEE Trans. Syst. Man Cybern. Syst. **52**(5), 3251–3262 (2022)
28. Cai, S., Zhou, P., Liu, Z.: Pinning synchronization of hybrid-coupled directed delayed dynamical network via intermittent control. Choas **24**, 033102 (2014)

Aperiodic Sampling Based Event-Triggering for Synchronization of Nonlinear Systems

Qiang Jia[1] and Jingyuan Wang[2]

[1] Institute of Applied System Analysis, Jiangsu Unviersity, Zhenjiang, China
qiangjia@ujs.edu.cn
[2] College of Civil and Transportation Engineering, Shenzhen University, Shenzhen, China
jywang@szu.edu.cn

Abstract. This work investigates the master-slave synchronization of nonlinear systems with aperiodic data-sampling, where the controller updates intermittently according to certain state-dependent triggering condition. Some sufficient criteria for reaching synchronization are established with the aid of an extended Hanalay's inequality, and the largest admissible sampling period is also estimated. The study is also extended to some typical issues with either dynamic triggering condition or sampling periods frequently beyond the estimated bound, providing more flexibility in practice. An interesting discovery is that the average inter-event time can be enlarged under certain specific sampling setting. Some numerical examples are given to demonstrate the validity of our theoretical results.

Keywords: Synchronization · Nonlinear system · Data-sampling · Event-triggered control

1 Introduction

The nonlinearity involved in dynamical systems often causes complex behavior, such as multi-stability [16], chaos [21], and chimera state [14]. In their pioneering paper in 1990 s, Pecora and Carroll achieved the synchronization between two coupled chaotic systems for the first time [15], which inspired tremendous interest in the realization and application of the synchronization issue of nonlinear systems, covering the voltage regulation in electrical circuits, frequency control in power grids, and synchronization in biological networks (e.g. memristive neural networks) [22], and secure design in communication systems [19]. Amongst the vast literature, a seminal technique for realizing synchronization is the master-slave (M-S) framework, where the master system has no control input, while the slave system receives some external control signal so that the difference between their states eventually vanishes, i.e., the slave system can finally track the trajectory of the master system. Such a technique was initially

H. Zhang et al. (Eds.): NCAA 2022, CCIS 1638, pp. 234–246, 2022.
https://doi.org/10.1007/978-981-19-6135-9_18

used to synchronize nonlinear systems, such as the delay-free systems [17] and systems with delays [20], and recently found many substantial applications in synchronization between large-scale networks, such as the outer synchronization for dynamical networks [11], and two delayed neural networks [8].

In most early studies, the control was designed as continuous for simplicity, i.e., the controller updates continuously with latest data. In practice, digital controllers relying on sampled data are used, i.e., the control signal discretely updates, and remains constant via a zero-order-hold device before the new sampled data arrive. For the ease of implementation, the data-sampling is often configured as periodic, and the controller updates with the periodic sampled-data accordingly. To deal with continuous systems with sampled-data based control, some useful techniques have been adopted. For instance, in [2], a delay-to-input method was used to convert a system with sampled data into one with delayed input, which can thus be tackled by using stability of time-delay systems [3], where the Lyapunov-Krosovskii functional method is usually utilized. Another powerful tool for analyzing delay systems is the Hanalay's inequality, i.e., $\dot{w}(t) \leq -aw(t) + b \sup_{-\tau \leq s \leq 0} w(t+s)$, and some generalizations have been made in recent studies, such as the cases with either time-varying coefficients [13] or external disturbances [5]. Compared with the conventional Lyapunov theory, exponential stability can be guaranteed, which is more preferable in practice [12,23].

In data-sampling based control, the control is usually designed to update when the latest sampled data is available, which inevitably leads to redundant execution of the control. In order to reduce the redundant execution, an event-triggered control approach was proposed in [1], where the controller updates according to certain predesigned condition. It was later used in various environments, such as phase locking in coupled Kuramoto oscillators [9] and consensus in networked chaotic circuits [10]. Recent studies revealed that such a technique can also be used to data-sampling based control. For instance, in [4], an event-triggered transmission strategy was designed for reaching consensus of multi-agent system by using sampled-data, and synchronization criteria in terms of linear matrix inequalities were constructed. It was later shown in [18] that the event-triggered mechanism is applicable in achieving master-slave system even when only some of the sampled-data outputs of the master system is available for control design. By designing some time- and state-dependent triggering conditions, a network of coupled nonlinear systems can synchronize under event-triggered control and periodic sampling [5]. It is worth to point out that existing works on this issue used periodic sampling with certain fixed period. However, in practice, the sampling period may vary over time due to either certain intentional setting or intermittent faults of the sampler, which cannot be tacked by using these existing results, and remains a challenging problem.

Motivated by the above discussions, this work aims to investigate the synchronization issue of nonlinear systems with event-triggered control and aperiodic sampling. The organization of the remaining content is given as below. Section 2 gives some preliminary knowledge, addresses the master-slave synchronization problem in concern, and proposes an aperiodic sampling based event-triggered control scheme. Section 3 presents the main theoretial results, and establishes the theorems to guarantee synchronization under different scenarios, respectively. Section 4 demonstrates the feasibility and validity of our design and criteria by using some numerical examples, and Sect. 5 finally gives the conclusion.

2   Problem Formulation

This section presents some useful preliminary knowledge, formulates the master-slave synchronization problem in concern, and proposes the event-triggered control scheme based on aperiodic sampled-data to guarantee synchronization between two coupled nonlinear systems.

In order to facilitate the later analysis, the below lemma is firstly given, which can be readily obtained from the Lemma 1 in [7].

Lemma 1. *Assume that* $w(t) > 0$ *is a scalar function, satisfying*

$$\dot{w}(t) \leq -aw(t) + b(t) \sup_{-\tau \leq s < 0} w(t+s), \tag{1}$$

where $a > 0$ *is a fixed constant,* $b(\cdot) : [0, \infty) \mapsto \mathbb{R}^+$ *is a periodic function with period* $\mathbb{T} > 0$, *and* $\tau > 0$ *is the time delay. If the time average of* $b(t)$ *over interval* $[t, t + \mathbb{T})$, *i.e.,* $\bar{b} \triangleq \frac{1}{\mathbb{T}} \int_t^{t+\mathbb{T}} b(s)\mathrm{d}s$, *fulfills*

$$a - \bar{b} > (e^{\tau\mu} - 1)b_{\max}, \tag{2}$$

where $\mu = \sup_{t \geq t_0} |b(t) - \bar{b}|$ *and* $b_{\max} = \sup_t |b(t)|$, *then,* $w(t)$ *converges exponentially for any delay subject to* $\tau < \frac{1}{\mu} \ln \left(1 + \frac{a - \bar{b}}{b_{\max}}\right)$.

Remark 1. when $b(t) \equiv b > 0$ is constant, Lemma 1 degenerates to the conventional Halanay's inequality, i.e., $w(t)$ converges exponentially if $a > b > 0$.

Consider the below nonlinear system

$$\dot{x}(t) = f(t, x(t)) \tag{3}$$

where $x(t) \in \mathbb{R}^n$ is the system state, and $f(\cdot, \cdot) : [t_0, \infty) \times \mathbb{R}^n \mapsto \mathbb{R}^n$ is a continuous nonlinear function, characterizing how the system evolves over time.

Following the master-slave framework [15], by letting (3) be the master system, one may obtain a slave system under control as below

$$\dot{y}(t) = f(t, y(t)) + u(t) \tag{4}$$

where $y(t) \in \mathbb{R}^n$, and the term $u(t) \in \mathbb{R}^n$ is the event-triggered controller to be designed.

Assumption 1. *For the nonlinear function* $f(t, \cdot)$, *there exist some constants* $\rho, \rho_0 > 0$ *such that the below inequalities*

$$(z_1 - z_2)^\top (f(t, z_1) - f(t, z_2)) \leq \rho \|z_1 - z_2\|^2,$$
$$\|f(t, z_1) - f(t, z_2)\| \leq \rho_0 \|z_1 - z_2\|. \tag{5}$$

hold for any $z_1, z_2 \in \mathbb{R}^n$ *and* $t \geq 0$.

Remark 2. Assumption 1 is very general and covers many well-known chaotic systems, such as the Lorenz system and Chua's chaotic oscillator, and ρ_0 is the Lipschitz constant.

Definition 1. *The master system and the slave system are said to achieve synchronization if their solutions satisfy* $\lim_{t \to +\infty} \|y(t) - x(t)\| = 0$.

Our objective herein is to propose an efficient event-triggered control based on aperiodic sampling, and establish some sufficient criteria to guarantee synchronization.

In our design, the instants at which the sampling occurs follow an increasing time sequence $\{t_k, k \in \mathbb{N}\}$. Assume that there exists a fixed value $\varsigma > 0$ such that $t_{k+1} - t_k > \varsigma$ holds for any $k \in \mathbb{N}$, which is reasonable as the sampler cannot sample infinitely fast. For simplicity, assume that $t_1 = 0$. By letting $e(t) = y(t) - x(t)$ be the error between the states of the two systems at instant t, we hereby propose the below event-triggered control under aperiodic sampling

$$u(t) = ce(t_{k_r}) \tag{6}$$

where $c > 0$ stands for the control gain, and t_{k_r} is a subsequence of the sampling instants, characterizing the time instants at which the control updates.

It is assumed that the first trigger happens at the initial instant $t = t_1$. Given a triggering instant t_{k_r}, one may define the measurement error $\epsilon(t) = e(t) - e(t_{k_r})$, and $\epsilon(t_k)$ for $k > k_r$ can be computed by using the sampled data. Then, the next triggering instant $t_{k_{r+1}}$ is determined by

$$t_{k_{r+1}} = \min\{t_k| \ t_k > t_{k_r}, \|\epsilon(t_k)\| \geq \sigma \|e(t_k)\|\} \tag{7}$$

where $\sigma > 0$ is a parameter to be designed.

Following controller (6) with (7), one may detect the validity of condition (7) and decide its triggering instants $t_{k_r}, k_r \in \mathbb{N}$. Then, by substituting the controller (6), the evolution of the synchronization error follows

$$\dot{e}(t) = f(t, y(t)) - f(t, x(t)) + ce(t_{k_r}). \tag{8}$$

Noting that $e(t_k) = \epsilon(t_k) + e(t_{k_r})$ for $t_{k_r} \leq t_k < t_{k_r+1}$, one further obtains that

$$\dot{e}(t) = f(t, y(t)) - f(t, x(t)) - c(e(t_k) + c\epsilon(t_k)).$$

In order to derive some sufficient criteria for master-slave synchronization, the delay-to-input approach [2] is hereby used to convert the error system with sampled-data into one with time-varying delay. It thus follows that, for $t \in [t_k, t_{k+1})$, a delay $\tau_k(t) = t - t_k$ is defined, where $0 \leq \tau(t) < h$, and $h > 0$ is the largest admissible sampling period. Consequently, the error dynamics for $t \in [t_k, t_{k+1})$ satisfies

$$\dot{e}(t) = f(t, y(t)) - f(t, x(t)) - ce(t - \tau_k(t)) + c\epsilon(t_k). \tag{9}$$

Apparently, the master-slave synchronization problem is equivalent to the stability issue of the delay differential system (9), and our objective herein is to establish some stability condition for delay system (9) so as to guarantee master-slave synchronization.

3   Main Theorems

This section presents the main theorem to guarantee master-slave synchronization, and further investigates two typical problems with different configurations.

3.1   Event-Triggered Synchronization Under Aperiodic Sampling

Now, we are in a position to present the main theorem.

Theorem 1. *For the master system* (3) *and slave system* (4) *under aperiodic sampling, denote* $h = \sup_k \{t_{k+1} - t_k\}$. *If the below condition*

$$1 - \sigma > \frac{\rho}{c} + h(\rho_0 + c + c\sigma) \tag{10}$$

holds, synchronization is guaranteed under control (6) *with triggering condition* (7). *In addition, the controlled system exhibits no Zeno phenomenon.*

Proof. First consider the below quadratic form function

$$V(t) = \frac{1}{2} e(t)^\top e(t). \tag{11}$$

For $t \in [t_k, t_{k+1})$, the change rate of $V(t)$ along the trajectory of the delay system (9) can be estimated as

$$\dot{V}(t) = e(t)^\top \big(f(t, y(t)) - f(t, x(t)) \big) \\ - c e(t)^\top e(t - \tau_k(t) + c e(t)^\top \epsilon(t_k). \tag{12}$$

By using the condition in Assumption 1, one may deduce that

$$e(t)^\top \big(f(t, y(t)) - f(t, x(t)) \big) \le \rho \|e(t)\|^2. \tag{13}$$

Meanwhile, the below inequality always holds,

$$e(t)^\top \int_{t-\tau_k(t)}^t \dot{e}(s) \, ds$$

$$\le \|e(t)\| \int_{t-\tau_k(t)}^t \|f(s, y(s)) - f(s, x(s))\| \, ds$$

$$+ c\|e(t)\| \int_{t-\tau_k(t)}^t (\|e(t_k)\| + \|\epsilon(t_k)\|) \, ds$$

$$\le h(\rho_0 + c)\|e(t)\| \sup_{-h \le s \le 0} \|e(t+s)\| + hc\|e(t)\| \|\epsilon(t_k)\|.$$

Recalling the triggering condition (7), one has

$$\|\epsilon(t_k)\| \le \sigma\|e(t_k)\| \le \sigma \sup_{-h \le s \le 0} \|e(t+s)\|. \tag{14}$$

By substituting the above inequalities into (12), it yields that

$$\dot{V}(t) \leq 2(\rho - c)V(t) + hc(\rho_0 + c)\|e(t)\| \sup_{-h \leq s \leq 0} \|e(t + s)\|$$
$$+ (c^2 \sigma h + c\sigma)\|e(t)\|\|e(t - \tau_k(t))\|. \tag{15}$$

Further taking the below Lyapunov function, $W(t) = \sqrt{V(t)}$, it gives $\|e(t)\| = \sqrt{2}W(t)$ and $\|e(t + s)\| = \sqrt{2}W(t + s)$. Thus, by using (15), $\dot{W}(t)$ when $V(t) \neq 0$ can be estimated as below

$$\dot{W}(t) = \frac{1}{2W(t)}\dot{V}(t)$$
$$\leq -aW(t) + b \sup_{-h \leq s \leq 0} W(t + s) \tag{16}$$

with $a = c - \rho$ and $b = hc(\rho_0 + c + c\sigma) + c\sigma$.

In order to guarantee the convergence of $W(t)$, it follows from the conventional Halanay's inequality that the synchronization condition should satisfy $a > b > 0$. Thus, $W(t)$ converges to zero exponentially if (10) holds, implying that the two systems (3) and (4) can achieve synchronization with an exponential rate.

Zeno phenomenon is naturally excluded since the inter-event time is no less than the smallest sampling period. It thus completes the proof.

Remark 3. In the design of event-triggered controllers, it is of significance to eliminate the Zeno phenomenon to circumvent the infinite triggering in finite time. Some recent studies on state-dependent triggering conditions introduced an artificial inter-event time to exclude Zeno behavior [6]. In our design, the triggers only occur at the multiple of the sampling period, and the inter-event time is thus lower bounded by $\varsigma > 0$ as $t_{k+1} - t_k > \varsigma$, indicating that Zeno phenomenon is naturally excluded.

Remark 4. When the system is linear, i.e., $\dot{x}(t) = Ax(t)$ with $A \in \mathbb{R}^{n \times n}$, Assumption 1 is fulfilled by simply letting $\rho = \rho_0 = \|A\|$, and the synchronization condition (10) is still applicable.

In addition, when $\rho = \rho_0 > 0$, the condition (10) can be further simplified, and some algebraic manipulations render the below corollary.

Corollary 1. *If there exist two constants, η, δ, subject to $0 < \delta < \eta < 1$, such that*

$$\frac{\delta(1 - \eta)}{1 + \eta} > h\rho, \tag{17}$$

where $h > 0$ is the largest sampling period, the control scheme (6) with (7) under parameters $c = \frac{\rho}{\delta}$ and $\sigma = \eta - \delta$ suffices to guarantee synchronization.

3.2   Applications to Two Typical Problems

1. *Reaching synchronization via dynamic triggering condition*
   Theorem 1 reveals the crucial role of the control parameters, and manifests that a larger value of σ in (7) helps to reduce the triggering frequency, and consequently cut down the control cost. Inspired by this observation, we propose a dynamic triggering condition with a tunable parameter, and the triggering frequency can thus be adjustable.
   An extended version for (7) with a tunable parameter is given as below

$$t_{k_{r+1}} = \min\{t_k|\ t_k > t_{k_r}, \|\epsilon(t_k)\| \geq \sigma(t_k)\|e(t_k)\|\} \tag{18}$$

where $\sigma(\cdot) : [t_0, \infty) \mapsto \mathcal{M}$ is a piecewise constant function, and $\mathcal{M} = \{\sigma_i > 0, i = 1, 2, \cdots, m\}$ is a finite set.
For simplicity, it is assumed that the parameter σ switches between two fixed values periodically, i.e., $m = 2$, and fulfills

$$\sigma(t) = \begin{cases} \sigma_1, [mT, (m+\eta)T), \\ \sigma_2, [(m+\eta)T, (m+1)T) \end{cases} \tag{19}$$

where $\eta \in (0, 1)$ is a constant, and $T > 0$ is the period of the switching rule. For scenarios with $m > 2$ and an aperiodically switched $\sigma(t)$, the below analysis can be readily extended.
In fact, one may still consider the Lyapunov function $W(t) = \sqrt{V(t)}$. By following the steps used in Sect. 3.1, one may obtain that

$$\dot{W}(t) \leq -aW(t) + b(t) \sup_{-h \leq s \leq 0} W(t+s) \tag{20}$$

with $a = c - \rho$ and $b(t) = hc(\rho_0 + c + c\sigma(t)) + c\sigma(t)$.
For brevity, denote $\bar{\sigma} = \eta\sigma_1 + (1-\eta)\sigma_2$, $\tilde{\sigma} = \max_i |\sigma_i - \bar{\sigma}|$, $\bar{b} = \frac{1}{T}\int_{(m-1)T}^{mT} b(s)\mathrm{d}s = hc(\rho_0 + c + \bar{\sigma}c) + c\bar{\sigma}$ for any $m \in \mathbb{N}$, $b_{\max} = hc(\rho_0 + c + c\sigma_{\max}) + c\sigma_{\max}$, and $\mu = c(hc+1)\tilde{\sigma}$. In light of Lemma 1, another theorem for synchronization of master-slave system under dynamic triggering condition can be derived. The proof can be simply done by following the steps in Theorem 1, and is thus omitted due to space limitation.

Theorem 2. *For the master system* (3) *and slave system* (4) *under triggering condition* (18), *if the sampling period satisfies*

$$a - \bar{b} > b_{\max}(e^{\mu h} - 1) \tag{21}$$

the control scheme (6) *together with* (18) *suffices to guarantee synchronization. Also, Zeno phenomenon is successfully excluded.*

Remark 5. In Sect. 3.1, the parameter in triggering condition (7) is set as fixed, i.e., the triggering condition is referred as static. In contrast, (18) admits a tunable parameter, and thus brings more flexibility as the frequency of the triggers can be tuned by using (19) with certain appropriate value of T and μ, which is demonstrated in Sect. 4.

2. *Reaching synchronization with switched sampling periods*

This part aims to derive the synchronization criterion under the assumption that the sampling periods frequently exceed the threshold value h given in Theorem 1.

Given the controller (6) with triggering condition (7), the upper bound for the admissible sampling period is estimated as $h > 0$. It is of interest to consider situations with certain sampling period larger than h. Assume that the sampling period is changeable and switches between two possible positive values h_1 and h_2 with period, satisfying $0 < h_1 < h < h_2$. For simplicity, denote $t_{mp} - t_{(m-1)p} = r_1 h_1 + r_2 h_2$ with $r_1 + r_2 = p$, $r_1, r_2, p \in \mathbb{N}$, and the sampling instants satisfy

$$t_{k+1} - t_k = \begin{cases} h_1, & (m-1)p \leq k < (m-1)p + r_1, \\ h_2, & (m-1)p + r_1 \leq k < mp. \end{cases} \tag{22}$$

By taking the same Lyapunov function $W(t)$ in Theorem 1, and following the similar steps, one may obtain that

$$W(t) \leq -aW(t) + b(t) \sup_{-h_2 \leq s \leq 0} W(t) \tag{23}$$

where $a = c - \rho$, and $b(t)$ switches between two values b_1 and b_2 due to the change of the sampling period, i.e., $b_1 = h_1 c(\rho_0 + c + c\sigma) + c\sigma$, and $b_2 = h_2 c(\rho_0 + c + c\sigma) + c\sigma$. Similarly, it yields that $\bar{b} = \bar{h}c(\rho_0 + c + c\sigma) + c\sigma$, $\bar{h} = (r_1 h_1 + r_2 h_2)/p$ standing for the average sampling period, and $\mu = \tilde{h}c(\rho_0 + c + c\sigma)$ with $\tilde{h} = \max_i |h_i - \bar{h}|$.

Consequently, another theorem can be derived by using Lemma 1.

Theorem 3. *For the master system (3) and slave system (4), when the sampling periods follows (22), and ensures*

$$a - \bar{b} > b_2(e^{\mu h_2} - 1), \tag{24}$$

the control scheme (6) together with (7) suffices to guarantee synchronization. Also, Zeno phenomenon is naturally excluded during the control process.

Remark 6. In most existing works, the data-sampling based event-triggered control demands that the sampling periods are less than certain threshold value, such as the schemes given in [4,5], and estimate the admissible sampling period to guarantee synchronization. In contrast, Theorem 3 shows that, despite the existence of larger sampling periods, synchronization is also possible if the sampling periods suffice to validate (24).

4 Numerical Simulations

This section provides some numerical examples to demonstrate the validity of the theoretical results. A modified Chua's circuit is used, which is described by [5]

$$\begin{cases} \dot{x}_1(t) = \alpha(x_2(t) - x_1(t) - p(x_1(t))), \\ \dot{x}_2(t) = x_1(t) - x_2(t) + x_3(t), \\ \dot{x}_3(t) = -\beta x_2(t), \end{cases} \tag{25}$$

where $p(x_1) = m_1 x_1 + \frac{m_0 - m_1}{2}(|x_1 + 1| - |x_1 - 1|)$. The system exhibits chaotic dynamics under the below parameters, $\alpha = 2.8$, $\beta = 3.1$, $m_0 = -1.34$ and $m_1 = -0.73$. It is easy to verify that Assumption 1 always holds for $\rho = 2.26$ and $\rho_0 = 8.63$. Following the master-slave framework, the slave system with control (6) can be readily obtained. Here, three different cases are investigated, respectively.

For the ease of illustration, $E(t) = \|y(t) - x(t)\|$ is defined to show the evolution of the synchronization error, and the fourth-order Runge-Kutta algorithm is used to solve the controlled system.

Case 1: According to Theorem 1, by taking the control parameters $c = 4$ and $\sigma = 0.05$, one may compute the largest admissible sampling period $h = 0.03$. In our simulation, the sampling periods are allowed to change, and set as $0.01 + 0.02 * rand$ with $rand$ being a random number taken from the interval $(0, 1)$, satisfying $t_{k+1} - t_k \leq h$. The evolution of the synchronization error and the resultant triggering instants for Case I are illustrated in Fig. 1. In specific, the lower panel displays the instants at which the triggers occur, and the inter-event time between two consecutive triggers, denoted by the heights of the stems. As is shown, synchronization error vanishes rapidly, and Zeno phenomenon is successfully ruled out due to the existence of data-sampling. In this simulation, the resultant average inter-event time can be computed, and its value is 0.0215.

Case 2: We consider a triggering condition with dynamic triggering parameters, where the parameter σ switches between two values and follows

$$\sigma(t) = \begin{cases} 0.03, \ t \in [mT, mT + \eta T), \\ 0.09, \ t \in [mT + \eta T, (m+1)T), \end{cases} \tag{26}$$

where $T = 1$ and $\eta = 0.7$. One may readily find that the time average of $\sigma(t)$, $\bar{\sigma} = 0.048$, and thus $\tilde{\sigma} = 0.042$, and the dynamic triggering parameter meets condition (21) with $h = 0.03$.

In our simulation, the sampling period follows the pattern in Case I, i.e., $0.01 + 0.02 * rand$. As a consequence, the synchronization error and the resultant triggering instants are illustrated in Fig. 2, and a remarkable observation is that the triggers appear less frequently when a larger parameter, σ_2, is in effect. The average value of the inter-event time is computed as 0.0231, which evidently helps to reduce the triggers and thus alleviates the burden of actuators than that in Case I.

Case 3: We consider the scenario that the sampling period is frequently larger than the bound found in Case 1. Assume $h_1 = 0.01$ and $h_2 = 0.05$, and the sampler samples with period h_1 for 3 times and then samples with period h_2 for 1 time repetitively. Thus, $r_1 = 3, r_2 = 1$ and $p = 4$. One may compute that the average sampling period is $\bar{h} = 0.02$, and $\tilde{h} = 0.03$, and thus $\bar{b} = 1.5685$,

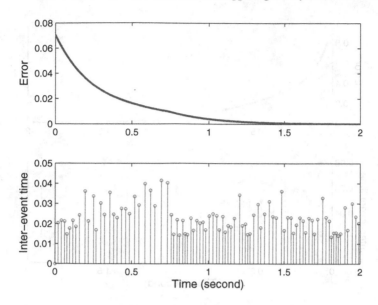

Fig. 1. Time history of the synchronization error and the resultant triggering instants in Case 1.

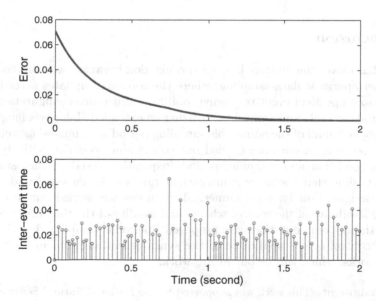

Fig. 2. Time history of the synchronization error and the resultant triggering instants in Case 2.

$\mu = 1.1975$, which fulfills the condition (24). The evolution of the synchronization error and the resultant triggering instants for Case 3 are illustrated in Fig. 3, and the average inter-event time is 0.0354, which is larger than that in the aforementioned two cases.

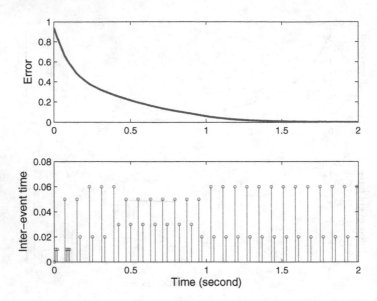

Fig. 3. Time history of the synchronization error and the resultant triggering instants in Case 3.

5   Conclusion

This work achieves the master-slave synchronization between two nonlinear systems under aperiodic data-sampling, where the controller updates according to certain state-dependent event-triggering condition. With such configurations, the synchronization problem is solved by utilizing an extended Halanay's inequality, and the upper bound of the admissible sampling period is estimated as well. Furthermore, some extensions are carried out to examine scenarios with dynamic triggering condition and sampling period frequently exceeding the estimated bound, revealing that the average inter-event time can be enlarged under some specified designs. Finally, some numerical examples are carried out to demonstrate the feasibility of the control scheme and validity of the theoretical results. It is worth to point out that the proposed aperiodic sampling based event-triggered control is also applicable to the synchronization problem for coupled large networks, which will be our future work.

Acknowledgment. This work was supported by the National Natural Science Foundation of China under Grant 62173163.

Conflict of Interest. The authors declare that they have no conflict of interest.

References

1. Dimarogonas, D.V., Frazzoli, E., Johansson, K.H.: Distributed event-triggered control for multi-agent systems. IEEE Trans. Autom. Control **57**(5), 1291–1297 (2011)
2. Fridman, E.: A refined input delay approach to sampled-data control. Automatica **46**(2), 421–427 (2010)
3. Gu, K., Chen, J., Kharitonov, V.L.: Stability of time-delay systems. Springer Science Business Media (2003). https://doi.org/10.1007/978-1-4612-0039-0
4. Guo, G., Ding, L., Han, Q.L.: A distributed event-triggered transmission strategy for sampled-data consensus of multi-agent systems. Automatica **50**(5), 1489–1496 (2014)
5. Han, Z., Tang, W.K.S., Jia, Q.: Event-Triggered Synchronization for Nonlinear Multi-Agent Systems with Sampled Data. Regular Papers, IEEE Transactions on Circuits and Systems I (2020)
6. Hu, W., Liu, L., Feng, G.: Cooperative output regulation of linear multi-agent systems by intermittent communication: a unified framework of time-and event-triggering strategies. IEEE Trans. Autom. Control **63**(2), 548–555 (2017)
7. Jia, Q., Han, Z., Tang, W.K.S.: Synchronization of multi-agent systems with time-varying control and delayed communications. IEEE Trans. Circuits Syst. I Regul. Pap. **66**(11), 4429–4438 (2019)
8. Jia, Q., Mwanandiye, E.S., Tang, W.K.S.: Master-slave synchronization of delayed neural networks with time-varying control. IEEE Trans. Neural Netw. Learn. Syst. **32**(5), 2292–2298 (2021)
9. Jia, Q., Tang, W.K.S.: Consensus of multi-agents with event-based nonlinear coupling over time-varying digraphs. IEEE Trans. Circuits Syst. II Express Briefs **65**(12), 1969–1973 (2018)
10. Jia, Q., Tang, W.K.S.: Event-based tracking consensus for multiagent systems with volatile control gain. In: IEEE Transactions on Cybernetics (2020)
11. Li, C., Xu, C., Sun, W., Xu, J., Kurths, J.: Outer synchronization of coupled discrete-time networks. Chaos: An Interdisc. J. Nonlinear Sci. **19**(1), 013106 (2009)
12. Li, C., Liao, X., Huang, T.: Exponential stabilization of chaotic systems with delay by periodically intermittent control. Chaos: An Interdisc. J. Nonlinear Sci. **17**(1), 013103 (2007)
13. Liu, B., Lu, W., Chen, T.: Stability analysis of some delay differential inequalities with small time delays and its applications. Neural Netw. **33**, 1–6 (2012)
14. Martens, E.A., Thutupalli, S., Fourrière, A., Hallatschek, O.: Chimera states in mechanical oscillator networks. Proc. Natl. Acad. Sci. **110**(26), 10563–10567 (2013)
15. Pecora, L.M., Carroll, T.L.: Synchronization in chaotic systems. Phys. Rev. Lett. **64**(8), 821 (1990)
16. Pisarchik, A.N., Feudel, U.: Control of multistability. Phys. Rep. **540**(4), 167–218 (2014)
17. Suykens, J., Curran, P., Chua, L.: Master-slave synchronization using dynamic output feedback. Int. J. Bifurcation Chaos **7**(03), 671–679 (1997)
18. Wen, G., Chen, M.Z., Yu, X.: Event-triggered master-slave synchronization with sampled-data communication. IEEE Trans. Circuits Syst. II Express Briefs **63**(3), 304–308 (2015)
19. Xue, C., Jiang, N., Lv, Y., Qiu, K.: Secure key distribution based on dynamic chaos synchronization of cascaded semiconductor laser systems. IEEE Trans. Commun. **65**(1), 312–319 (2016)

20. Jia, Q., Tang, W.K.S.: Event-based tracking consensus of multiagent systems with volatile control gain. IEEE Trans. Cybern. **52**(7), 6603–6614 (2022)
21. Yang, X.S., Huang, Y.: Complex dynamics in simple hopfield neural networks. Chaos: An Interdisc. J. Nonlinear Sci. **16**(3), 033114 (2006)
22. Yang, X., Liu, Y., Cao, J., Rutkowski, L.: Synchronization of coupled time-delay neural networks with mode-dependent average dwell time switching. IEEE Trans. Neural Netw. Learn. Syst. **31**(12), 5483–5496 (2020)
23. Yu, J., Hu, C., Jiang, H., Teng, Z.: Synchronization of nonlinear systems with delays via periodically nonlinear intermittent control. Commun. Nonlinear Sci. Numer. Simul. **17**(7), 2978–2989 (2012)

Two-Stream 3D MobileNetV3 for Pedestrians Intent Prediction Based on Monocular Camera

Yi Jiang[1], Weizhen Han[1], Luyao Ye[2], Yang Lu[1], and Bingyi Liu[1,3]([✉])

[1] Wuhan University of Technology, Wuhan, China
{jy0918,hanweizhen,ly_work,byliu}@whut.edu.cn
[2] City University of Hong Kong, Hong Kong, China
luyaoye2-c@my.cityu.edu.hk
[3] Wuhan University of Technology Chongqing Research Institute, Chongqing, China

Abstract. Recent developments in the field of autonomous vehicles have led to a renewed interest in combining deep learning and autonomous driving, aiming to support vision-related safety applications in Intelligent Transportation System (ITS). Pedestrians intent prediction is critical in improving the efficiency and safety of autonomous driving since complex traffic scenarios are ubiquitous, specifically facing pedestrian intrusion and other emergencies. Substantial studies have focused on using skeleton features as input to the neural network to extract pedestrians' features. However, most of them cannot avoid the loss of feature points and design a bloated neural network. In this paper, we propose a spatio-temporal MobileNetV3 based on Two-Stream to predict pedestrian intention. Based on the pedestrian images cropped by the bounding box, we use RGB images to generate optical flow to replace the pedestrian's skeleton features as the network's input. With the benefit of 3D depthwise separable convolution, our network's Multiply-Accumulate Operations (MACs) is one-fifth of the baseline model. Finally, we conduct experiments on real traffic datasets to verify our proposed method's effectiveness in different road environments.

Keywords: Pedestrians intent prediction · MobileNet · Two-stream network

1 Introduction

In a World Health Organization (WHO) report [1] on road safety in 2018, approximately 1.35 million people die in traffic accidents worldwide each year. More than half of these deaths are pedestrians, bicyclists, and motorcyclists. These vulnerable road users cannot protect themselves and rely on the driver's judgment in an emergency. However, with Level 5 autonomous driving, the Automated Driving System (ADS) will autonomously make decisions without the driver's intervention. Due to the rapid development of deep learning, ADS based on various neural

H. Zhang et al. (Eds.): NCAA 2022, CCIS 1638, pp. 247–259, 2022.
https://doi.org/10.1007/978-981-19-6135-9_19

networks can obtain environmental information through multiple sensors [2]. Fusing and processing this collected sensor data combined with V2X technology to expand the vehicle's sensing range can guide the vehicles to make the best behavioral decisions to avoid a collision [3,4]. For vulnerable road users, ADS combines object detection and tracking algorithms [5] to determine the location of pedestrians in front of the car based on monocular cameras. However, ADS does not effectively avoid traffic accidents by relying on pedestrian localization alone. It lacks a deep understanding of pedestrian behavior and intention, as human drivers do. Specifically, pedestrians often inform drivers of their crossing intentions through their eyes, hand gestures, and other behaviors. Such non-verbal cues are not easy for ADS to understand, and incorrect predictions of pedestrian intent can lead to traffic accidents. Therefore, it is critical to enable ADS to quickly and accurately understand and predict pedestrian behavior and intention.

At present, there are two main approaches for predicting pedestrian intention. The first approach is an end-to-end network based on consecutive frames of RGB images. Through the object detection and object tracking algorithm [6], the bounding box of each pedestrian in successive frames is obtained. Then an end-to-end convolutional network (ConvNet) is trained directly on pedestrian images to predict pedestrian crossing intention. In the real world, our human eyes can easily capture the dynamic changes of objects or scenes, and humans receive not separate image data but continuous video data. It is essential to make neural networks understand human thinking. Compared with the past continuous frame processing model based on 2D convolution, 3D convolution [7] can directly obtain the Spatio-temporal features between consecutive images. The ConvNets based on 3D convolution can drop the fully connected or Long Short-Term Memory (LSTM) [8] layers after the convolution layer to obtain the prediction results directly.

The second approach relies on pedestrian pose estimation algorithms to obtain the intrinsic connection between pedestrian skeleton features. Concretely, the human pose estimation algorithm [9] gets the pedestrian's keypoints, such as the head, arms, legs, etc. Specifically, the distance and angle between keypoints of the pedestrian, such as the distance between legs and the amplitude of arm swing, are used to sense the current pose of the pedestrian [10] and predict whether the pedestrian has the intention to cross. Compared with the first approach, the model based on pedestrian skeleton features needs to detect eighteen or twenty-five human keypoints using a human pose estimation algorithm on top of the object tracking. It is prone to miss detection and deviation in crowded road environments or poor lighting conditions. Besides, the first method using only RGB images does not have an advantage in capturing and understanding the subtle movements of pedestrians.

In this paper, we propose the Two-Stream 3D MobileNetV3 model and use the Joint Attention for Autonomous Driving (JAAD) [11] and Pedestrian Intention Estimation (PIE) [12] datasets for the experiments. Both of these datasets contain pedestrians' bounding box and crossing intention labels. Our main contributions are as follows: 1) To reduce the computational cost and parameters of 3D convolution, we extend 2D depthwise separable convolution [13] to 3D depthwise separable convolution. 2) We adopt an image enhancement algorithm to capture environmental information and ensure the invariance of pedestrian body proportions. Experiments show that this image enhancement algorithm improves the prediction by about 3% of the AP score over direct image rescaling. 3) Our experiment shows that the fusion of RGB and optical flow extracted pedestrian motion features can improve the accuracy of pedestrian crossing intention prediction.

The following section is a summary of related work in recent years. Section 3 is the proposed model algorithm. The specific steps and results of the experiments are in Sect. 4. Section 5 describes the conclusion and future work.

2   Related Work

In the past research on pedestrians, most of the work was about pedestrian trajectory prediction [14,15]. However, the datasets used in these works mostly took the form of a bird's-eye view. Pedestrian data is collected in fixed scenarios, mainly on the sidewalks along the roads. Compared with the pedestrian trajectory datasets, intent prediction for pedestrians uses video data collected by the onboard camera. Pedestrian images taken from a flat view have more diversified information, such as the pedestrian bounding box and pedestrian skeleton feature. In addition, predicting whether a pedestrian intends to cross the road will directly affect the decision of neighboring vehicles. For example, they can change lanes or slow down in advance to avoid traffic accidents.

[11] proposed the JAAD dataset containing pedestrian intention label. They simplified pedestrian intention prediction to a binary classification problem: Cross or Not Cross. Compared with the commonly used KITTI dataset [16], the JAAD dataset contains a large amount of pedestrian, pedestrian behavior label, road environment, and different climate data. They used the AlexNet to extract spatial features as the input of the linear SVM model to predict pedestrian intention. Similar to the combination of deep learning and traditional machine learning algorithms, [10] utilized the classic model VGG16 and human pose estimation algorithm [9] to extract skeleton features. They finally used the random forest to predict. [17] replaced the traditional CNN with GCN and transformed the scattered feature points into a graph based on the location information of the key parts of the pedestrian. [18] optimized based on DenseNet, and proposed ST-DenseNet that integrates spatio-temporal features. The advantages of densely connected blocks further improve the accuracy of pedestrian intention prediction. Other work reconstructed the scene frame of the entire picture [19] and then used the reconstructed scene frame to predict pedestrian intention.

None of the above studies have considered the advantages of optical flow for pedestrian behavior understanding. In the field of human behavior recognition, the Two-Stream network has been verified to have better recognition results than the RGB stream model alone. [20] used Spatial Steam ConvNet and Temporal Steam ConvNet to extract the features of single-frame RGB and multi-frame optical flow, and finally fused the results of the two ConvNets. After comparing standard video feature extraction networks, [21] proposed the Two-Steam

Inflated 3D ConvNets, which took advantage of the 3D convolution and merged the features of multi-frame RGB and multi-frame optical flow for pedestrian motion recognition. In this paper, we conduct experiments on the challenging JAAD and PIE datasets. Through our proposed Two-Stream 3D MobileNetV3 model, fusing the data feature of RGB and optical flow can improve the accuracy of pedestrian intention prediction. Meanwhile, to avoid the high computational cost and training parameters, we further simplify the MobileNetV3 network structure and replace the 2D depthwise separable convolution with 3D depthwise separable convolution. Our proposed network can be suitable for spatio-temporal feature extraction of continuous Two-Stream images.

3   System Model

The critical point of pedestrian intention prediction depends not only on the current pedestrian status but also on the influence of the surrounding environment on pedestrians. In this paper, we retained the proportional feature of the pedestrian and expanded the scope of cropped images. The cropped image contains information about the pedestrian and the surrounding environment. Through the ConvNet, the feature of the pedestrian and the environment can be extracted together. Meanwhile, we use 3D depthwise separable convolution instead of the recurrent neural network to process video data. We fuse RGB stream and optical flow to improve the accuracy of pedestrian intention prediction. Two-Stream features can better capture pedestrian motion patterns than a single RGB stream.

3.1   3D Depthwise Separable Convolution Vs. 3D Standard Convolution

In the past, a standard method of processing video data was to use 2D convolution to process images and recurrent neural network to extract sequence features. With the introduction of 3D convolution, the 3D CNN can directly process video data. An increasing amount of work [7,21] uses 3D convolution to replace the network structure of CNN and RNN. Nevertheless, the traditional 3D convolution has a high computation load when processing high-dimensional image features and is not suitable for onboard devices. According to the formula of convolution cost [22], the computational cost of 3D convolution is:

$$C_{in} \cdot K \cdot K \cdot K \cdot C_{out} \cdot W_{out} \cdot H_{out} \cdot T_{out} \tag{1}$$

where C_{in} is the channel size of the input, K is the size of the convolution kernel, C_{out} is the number of output channels, W_{out}, H_{out}, and T_{out} represent the output width, height, and time length. We use depthwise separable convolution instead of standard convolution. When the convolutionkernel size is

3×3, the computation of standard convolution is about 8–9 times that of 2D depthwise separable convolution [22]. We extend the 2D depthwise separable convolution into 3D. The depthwise separation convolution divides convolution into two parts: depthwise convolution and pointwise convolution. The convolution kernel of depthwise convolution is $K \times K \times K \times C_{in}$. Each convolution kernel only acts on its corresponding channel, and each channel of the input feature is independent. The computational cost is:

$$C_{in} \cdot K \cdot K \cdot K \cdot W_{out} \cdot H_{out} \cdot T_{out} \tag{2}$$

The depthwise convolution does not fuse the input features across channels. Therefore, the pointwise convolution will linearize the features across all output channels to compensate for the lost cross-channel features. The kernel size of pointwise convolution is $1 \times 1 \times 1 \times C_{out}$, and its primary function is to expand or reduce the number of input feature channels. The originally discretized channel features are fused through $1 \times 1 \times 1$ convolution to activate cross-channel information exchange. The pointwise convolution has a computational cost of:

$$C_{in} \cdot C_{out} \cdot W_{out} \cdot H_{out} \cdot T_{out} \tag{3}$$

According to Eqs. (1), (2) and (3), 3D depthwise separable convolution reduces the cost than ordinary 3D convolution:

$$\frac{C_{in} \cdot K \cdot K \cdot K \cdot W_{out} \cdot H_{out} \cdot T_{out} + C_{in} \cdot C_{out} \cdot W_{out} \cdot H_{out} \cdot T_{out}}{C_{in} \cdot K \cdot K \cdot K \cdot C_{out} \cdot W_{out} \cdot H_{out} \cdot T_{out}} = \frac{1}{C_{out}} + \frac{1}{K^3} \tag{4}$$

Usually, C_{out} is a large number (32, 64, 256, etc.), and the kernel size K is generally 3. Therefore, compared with ordinary 3D convolution, using 3D separable convolution can reduce the computational cost by at least twenty times. In the following sub-sections, we will discuss our proposed model (shown in Fig. 1) in more detail.

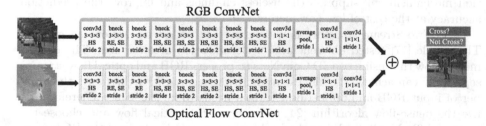

Fig. 1. Two-Stream 3D MobileNetV3 for pedestrian intention prediction.

3D Bneck Block

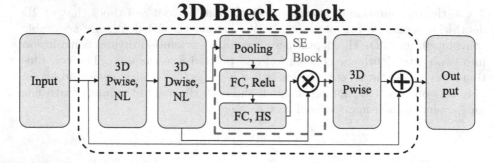

Fig. 2. 3D Bneck block. NL indicates the use of a nonlinear activation function. HS denotes h-swish activation function.

3.2   Two-Stream 3D MobileNetV3

The MobileNetV3 network, which processes 2D pictures, is replaced with 3D depthwise separable convolution to process video data. We delete the middle layer with the same input size in the original MobilenetV3 and replace the 555 kernel in the first half of the network with 333 to avoid the time dimension from being compressed too fast and losing sequence features. The 2D bneck block in MobileNetV3 is composed of inverted residual and Squeeze-and-Excite Block [23]. Our model uses 3D bneck blocks, as shown in Fig. 2. Unlike the traditional residual mode, deepwise convolution is an operation performed independently in different channels. After reducing the input channel dimension, the conventional residuals will reduce the number of channels acted by the deepwise convolution, which is unfavorable for extracting the spatial features of the image. The inverted residual structure preferentially expands the input channels, and more channels allow deepwise convolution to obtain richer spatial features. In the 3D bneck block, we expand and reduce the number of channels by two 3D pointwise convolutions. A Squeeze-and-Excite block assigns different weights to each channel before the dimensions of the channels are reduced. This simple attention mechanism will suppress the useless channels and improve the prediction accuracy at the cost of low parameters.

The Two-Stream structure is widely used in pedestrian action recognition. The article [17] tested different action classification architectures. The experiments show that using two 3D ConvNets to process RGB and optical flow streams separately can achieve the highest recognition accuracy. Furthermore, the number of input RGB and optical flow frames needs to be the same. Therefore, we use the dense-flow algorithm [24] to generate the optical flow and choose the improved 3D MobileNetV3 as the backbone network. We trained the RGB ConvNet and optical flow ConvNet separately and averaged their prediction results during the test.

4 Experiments and Results

The following section introduces the JAAD and PIE dataset used in the experiment and the pedestrian image enhancement algorithm. By analyzing the metric of AP score, parameters, and MACs, we verify the advantages of our proposed model on the task of pedestrian intent prediction.

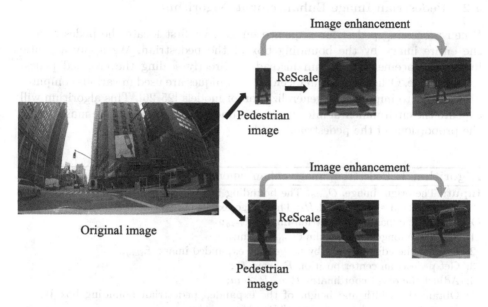

Fig. 3. Pedestrian image preprocessing comparison results.

4.1 Training and Testing Dataset

We selected the JAAD and PIE datasets for experiments. The JAAD dataset focuses on collecting pedestrian data while covering different road scenarios, weather conditions, and light intensity variations. It consists of 346 video data captured by high-resolution monocular cameras. The duration of each video is about 5 to 10 s, and the frame rate of the video is 30fps with 1920 W × 1080 H. We take the first 250 videos as the training dataset and the last 96 videos as the test dataset. We keep the pedestrian sequences whose bounding box width exceeds 60 pixels and are not occluded. We selected 16 consecutive frames of pedestrian images as the input to the network. After the above processing, the training dataset contains 20,657 pedestrian crossing samples and 9,919 pedestrian non-crossing samples. We randomly selected 9,919 pedestrian crossing samples among those 20,657 samples to address the problem of label imbalance. In the test dataset, the pedestrian crossing samples are 4618, which are also the same as the non-crossing samples. The PIE dataset consists of over 6 h video captured by a monocular camera inside the vehicle in HD format at 30fps. Unlike the JAAD dataset, the PIE dataset has complete information about the pedestrians in the continuous long-duration video. We processed the PIE dataset similarly, with 77,262 samples in the training dataset and 34,182 samples in the testing dataset. We trained our model using the Adam optimizer with a learning rate of 0.001 and batch size of 8 samples for 25 training epochs on NVIDIA RTX 3090 GPU.

4.2   Pedestrian Image Enhancement Algorithm

When processing pedestrian sequence images, we first locate the pedestrian in the entire image by the bounding box of the pedestrian. We apply a gentle image enhancement algorithm instead of directly scaling the cropped pedestrian's image. Other image enhancement techniques are used in various computer vision tasks to improve the generalization of models [25,26]. This algorithm will capture the environmental information around the pedestrian while maintaining the proportions of the pedestrian's body.

Algorithm 1. Pedestrian image enhancement

Input: The origin image, O_{img}; The bounding box of the pedestrian, P_b; The target image output size, W_t and H_t; The image expansion factors α;

Output: the expanded pedestrian image, P_{img};

1: Select the longest side of the O_{img}, get m;
2: Expand the edge of O_{img} by m, get the expanded image E_{img};
3: Get pedestrian center position P_c from P_b;
4: Adjust the offset coordinates $P_c \leftarrow P_c + m$;
5: Obtain the width and height of the expanded pedestrian bounding box $W_e \leftarrow expand(P_b, \alpha)$ and $H_e \leftarrow expand(P_b, \alpha)$;
6: Determine the crop size as S_c and the side with the larger ratio as M_s;
7: **if** $H_e/H_t > W_e/W_t$ **then**
8:     $S_c \leftarrow H_e$;
9:     $M_s \leftarrow H_t$;
10: **else**
11:     $S_c \leftarrow W_e$;
12:     $M_s \leftarrow W_t$;
13: **end if**
14: Ensure P_{img} contains P_b, $S_c \leftarrow min(S_c, P_c[0]/W_t * M_s * 2 - 1, P_c[1]/H_t * M_s * 2 - 1, (E_{img}[0] - P_c[0])/W_t * M_s * 2 - 1, (E_{img}[1] - P_c[1])/H_t * M_s * 2 - 1)$
15: Crop the expanded pedestrian img $P_{img} \leftarrow crop(E_{img}, S_c, M_s)$
16: **return** P_{img};

With the enhanced pedestrian image, the pedestrian remains in the center of the image. The outward extension crop strategy allows the image to include information about the current pedestrian environment. The ConvNets can learn the implied spatial feature relationship between pedestrians and the environment. Meanwhile, this algorithm can alleviate the situation that the offset of the bounding box leads to the inability to include the whole pedestrian.

The comparison results of direct scaled and enhanced pedestrian images are shown in Fig. 3.

4.3   Performance Analysis

In the experiments, to evaluate the performance of our proposed model, we choose three baseline models for comparison. The four models are trained and tested with the same training and testing dataset split from the JAAD and PIE datasets to conduct a fair comparison experiment. The pedestrian images are processed in the same way using the label data of the pedestrian bounding box provided in the datasets. We use the AP score as our evaluation metric because the AP score is widely used to evaluate model performance in pedestrian intent prediction [18,27]. The AP score is the area between the precision-recall curve and the coordinate axis, and is the weighted average of the precision under different recall.

Table 1. The performance of our proposed model is compared with other models in the literature under different conditions of preprocessing pedestrian image sequences.

Approch	Average precision			
	JAAD		PIE	
	Rescale	Keep proportions	Rescale	Keep proportions
ResNet50+LSTM [27]	82.53	89.73	92.52	95.76
I3D-RGB	83.86	89.93	94.06	95.71
I3D-OF	87.01	89.07	93.81	96.25
I3D [21]	88.88	92.99	94.98	96.88
ST-DenseNet [18]	87.22	91.26	93.94	96.25
Ours-RGB	83.31	91.21	93.23	95.92
Ours-OF	86.58	89.53	93.18	96.46
Ours	**89.18**	**94.09**	**95.22**	**97.37**

The first baseline model is a CNN+RNN network structure. It uses a feature extraction network of the classical ResNet50 [28], deletes the fully connected layer at the end of the ResNet50, merges it with the LSTM network, and finally outputs the prediction through the fully connected layer. As a many-to-one RNN module, the network extracts the features of multiple frames together as the input of the RNN network. Compared with the first baseline model, I3D relies only on 3D convolution to accomplish feature extraction and result prediction. The I3D network consists of two networks, I3D-RGB and I3D-OF. The I3D-RGB network extracts object and scene appearance features, and the other network extracts motion features. The I3D network applies the InceptionV1 module [29] to increase the depth and width of the network. The third baseline model is optimized based on the 2D DenseNet model. It is mainly composed of three Dense-Block-3D and two Transition-Layer-3D. This network outperforms other 3D ConvNets in terms of parameters due to using a large number of 1*1*1 convolutions.

In Table 1, we compare the effects of two different image processing methods on pedestrian intent prediction results. The first method is to scale the cropped pedestrian image to 224*224 directly. The other is to enhance the pedestrian image and then crop it according to the size of the pedestrian bounding box. Both approaches use consecutive 16 frames of pedestrian images as input for all models. The Two-Stream 3D MobileNetV3 network achieves the highest AP scores in ReScale and Keep Proportions experiments on experiments with two datasets. Among the three baseline models, I3D uses the same Two-Stream structure. Nonetheless, our model has a higher AP score than I3D, proving that the 3D bneck structure can extract more accurate high-dimensional pedestrian features. We recorded the experimental results of both networks using RGB images and optical flow images alone, and the experimental results demonstrated that the AP score of two-stream fusion is higher than that of the single-stream network. The ST-DenseNet model uses Dense-Block-3D, an optimized version of the block in ResNet50. The AP score of ST-DenseNet is significantly better than the ResNet50+LSTM model. Additionally, in Fig. 4, we show the visualization results of our algorithm for the pedestrian prediction task.

(a)   (b)   (c)

Fig. 4. Qualitative result of our proposed model. (a) A pedestrian is about to cross the road. (b) A pedestrian is walking straight at the vehicle (c) Two pedestrians are crossing the road, and another pedestrian is standing on the other side. Our proposed model can accurately predict the crossing intention of pedestrians in different scenarios.

To verify the influence of the pedestrian environment, we tested different image expansion factors α on the JAAD dataset and compared the whole scene direct scaling in Table 2. The experimental results show that the AP score is very low when scaling the whole scene directly, both in RGB and optical flow. Because there may be various types of pedestrians with different intents in the scene, this input method cannot predict the intent of each pedestrian in a targeted way. With different image expansion factors α, when α equals 0, only pedestrians are present in the whole image. And as α keeps expanding, the image will contain richer environmental information. In the experimental results comparing different α, the prediction model performs best when α is equal to

0.4, and too much or too little environmental information will affect the AP score of the model.

Table 2. Comparison experiments of model performance is under different image expansion factors α.

Approch	Average precision				
	$\alpha = 0$	$\alpha = 0.2$	$\alpha = 0.4$	$\alpha = 0.6$	scene
Ours-RGB	90.24	90.90	**91.21**	86.83	57.51
Ours-OF	87.43	88.97	**89.53**	87.52	71.51
Ours	92.01	92.65	**94.09**	90.80	69.29

4.4 Parameters and MACs Analysis

In Table 3, we list the parameter quantities and MACs of all models. It can be found that the ResNet50+LSTM and I3D models using ordinary 3D convolution have more than ten times the MACs and dozens of times more parameters than our proposed model. Although our model has more parameters than ST-DenseNet, in terms of computation, we are only 20% of the computation of ST-DenseNet. Such low parameters and low computational complexity make our model more lightweight.

Table 3. The number of parameters and MACs analysis of our proposed model in comparison to other baseline models.

Approch	Parameters	MACs
ResNet50+LSTM [27]	46.23M	39.98G
I3D [21]	25.56M	53.46G
ST-DenseNet [18]	**0.98M**	6.14G
Ours	1.04M	**1.24G**

5 Conclusion

In this paper, we introduce a Two-Stream 3D MobilenetV3 network for pedestrian intent prediction. We conduct experiments on the JAAD and PIE datasets containing various road scenes to demonstrate the advantages of our proposed network. The Two-Stream architecture can extract deeper pedestrian motion features than using only RGB images. Preprocessing pedestrian images with data enhancement instead of direct scaling can extract the implied spatial features between pedestrians and the environment and improve the AP score of the model by about 3%. Compared with the classic I3D network structure, our model is lightweight, and the AP score is still higher than it. In future work, we will focus on the interaction between pedestrians and Other road users to improve the interpretability of the model further.

Acknowledgments. This work was supported by Open Project of Chongqing Vehicle Test & Research Institute (No. 20AKC18), Natural Science Foundation of Chongqing of China (cstc2021jcyj-msxm4262), and Research Project of Wuhan University of Technology Chongqing Research Institute (ZL2021-05, ZD2021-04).

References

1. Organization, W.H., et al.: Global Status Report on Road Safety 2018: Summary. World Health Organization, Tech. rep. (2018)
2. Rasouli, A., Tsotsos, J.K.: Autonomous vehicles that interact with pedestrians: a survey of theory and practice. IEEE Trans. Intell. Trans. Syst. (TITS) **21**(3), 900–918 (2019)
3. Liu, B., Deng, D., Rao, W., Wang, E., Xiong, S., Jia, D., Wang, J., Qiao, C.: CPA-MAC: a collision prediction and avoidance MAC for safety message dissemination in MEC-assisted VANETs. IEEE Trans. Netw. Sci. Eng. **9**(2), 783–794 (2021)
4. Liu, B., et al.: A region-based collaborative management scheme for dynamic clustering in green VANET. In: IEEE Transactions on Green Communications and Networking (2022)
5. Luo, W., Xing, J., Milan, A., Zhang, X., Liu, W., Kim, T.K.: Multiple object tracking: a literature review. Artif. Intell. (AI) **293**, 103448 (2021)
6. Kapania, S., Saini, D., Goyal, S., Thakur, N., Jain, R., Nagrath, P.: Multi object tracking with uavs using deep sort and yolov3 retinanet detection framework. In: Proceedings of the 1st ACM Workshop on Autonomous and Intelligent Mobile Systems. pp. 1–6 (2020)
7. Tran, D., Bourdev, L., Fergus, R., Torresani, L., Paluri, M.: Learning spatiotemporal features with 3d convolutional networks. In: Proceedings of IEEE International Conference on Computer Vision (ICCV) (2015)
8. Hochreiter, S., Schmidhuber, J.: Long short-term memory. Neural Comput. **9**(8), 1735–1780 (1997)
9. Cao, Z., Hidalgo, G., Simon, T., Wei, S.E., Sheikh, Y.: Openpose: realtime multi-person 2d pose estimation using part affinity fields. IEEE Trans. Pattern Anal. Mach. Intelligence (TPAMI) **43**(1), 172–186 (2019)
10. Fang, Z., López, A.M.: Intention recognition of pedestrians and cyclists by 2D pose estimation. IEEE Trans. Intell. Trans. Syst. (TITS) **21**(11), 4773–4783 (2019)
11. Rasouli, A., Kotseruba, I., Tsotsos, J.K.: Are they going to cross? a benchmark dataset and baseline for pedestrian crosswalk behavior. In: Proceedings of the IEEE International Conference on Computer Vision Workshops (CVPRW), pp. 206–213 (2017)
12. Rasouli, A., Kotseruba, I., Kunic, T., Tsotsos, J.K.: Pie: A large-scale dataset and models for pedestrian intention estimation and trajectory prediction. In: Proceedings of the IEEE/CVF International Conference on Computer Vision (ICCV), pp. 6262–6271 (2019)
13. Howard, A., et al.: Searching for mobilenetv3. In: Proceedings of the IEEE/CVF International Conference on Computer Vision, pp. 1314–1324 (2019)
14. Alahi, A., Goel, K., Ramanathan, V., Robicquet, A., Fei-Fei, L., Savarese, S.: Social lstm: Human trajectory prediction in crowded spaces. In: Proceedings of the IEEE Conference on Computer Vision and Pattern Recognition (CVPR), pp. 961–971 (2016)

15. Gupta, A., Johnson, J., Fei-Fei, L., Savarese, S., Alahi, A.: Social gan: Socially acceptable trajectories with generative adversarial networks. In: Proceedings of the IEEE Conference on Computer Vision and Pattern Recognition (CVPR), pp. 2255–2264 (2018)
16. Geiger, A., Lenz, P., Stiller, C., Urtasun, R.: Vision meets robotics: The kitti dataset. Int. J. Robot. Res. (IJRR) 32(11), 1231–1237 (2013)
17. Cadena, P.R.G., Yang, M., Qian, Y., Wang, C.: Pedestrian graph: Pedestrian crossing prediction based on 2d pose estimation and graph convolutional networks. In: Proceedings of 2019 IEEE Intelligent Transportation Systems Conference (ITSC), pp. 2000–2005. IEEE (2019)
18. Saleh, K., Hossny, M., Nahavandi, S.: Real-time intent prediction of pedestrians for autonomous ground vehicles via spatio-temporal densenet. In: Proceedings of 2019 International Conference on Robotics and Automation (ICRA), pp. 9704–9710. IEEE (2019)
19. Gujjar, P., Vaughan, R.: Classifying pedestrian actions in advance using predicted video of urban driving scenes. In: Proceedings of 2019 International Conference on Robotics and Automation (ICRA), pp. 2097–2103. IEEE (2019)
20. Simonyan, K., Zisserman, A.: Two-stream convolutional networks for action recognition in videos. arXiv preprint arXiv:1406.2199 (2014)
21. Carreira, J., Zisserman, A.: Quo vadis, action recognition? a new model and the kinetics dataset. In: proceedings of the IEEE Conference on Computer Vision and Pattern Recognition (CVPR), pp. 6299–6308 (2017)
22. Howard, A.G., et al.: Mobilenets: Efficient convolutional neural networks for mobile vision applications. in arXiv preprint arXiv:1704.04861 (2017)
23. Hu, J., Shen, L., Sun, G.: Squeeze-and-excitation networks. In: Proceedings of the IEEE conference on computer vision and pattern recognition (CVPR), pp. 7132–7141 (2018)
24. Wang, S., Li, Z., Zhao, Y., Xiong, Y., Wang, L., Lin, D.: denseflow. https://github.com/open-mmlab/denseflow (2020)
25. Chen, Y., Wang, Z., Peng, Y., Zhang, Z., Yu, G., Sun, J.: Cascaded pyramid network for multi-person pose estimation. In: Proceedings of the IEEE conference on computer vision and pattern recognition (CVPR), pp. 7103–7112 (2018)
26. Redmon, J., Farhadi, A.: Yolov3: An incremental improvement. arXiv preprint arXiv:1804.02767 (2018)
27. Lorenzo, J., Parra, I., Wirth, F., Stiller, C., Llorca, D.F., Sotelo, M.A.: RNN-based pedestrian crossing prediction using activity and pose-related features. In: Proceedings of 2020 IEEE Intelligent Vehicles Symposium (IV), pp. 1801–1806. IEEE (2020)
28. He, K., Zhang, X., Ren, S., Sun, J.: Deep residual learning for image recognition. In: Proceedings of the IEEE conference on computer vision and pattern recognition (CVPR), pp. 770–778 (2016)
29. Szegedy, C., Liu, W., Jia, Y., Sermanet, P., Reed, S., Anguelov, D., Erhan, D., Vanhoucke, V., Rabinovich, A.: Going deeper with convolutions. In: Proceedings of the IEEE Conference on Computer Vision and Pattern Recognition (CVPR), pp. 1–9 (2015)

Design of Portrait System for Road Safety Based on a Dynamic Density Clustering Algorithm

Chenxing Li, Yongchuan Cui, and Chengyu Hu[✉]

School of Computer Science, China University of Geosciences, Wuhan, China
{chengxingli,cugcuiyc,huchengyu}@cug.edu.cn

Abstract. Traffic accidents are a safety issue that has received widespread attention. Practical mining of the causes of traffic accidents can lay the foundation for early warning of traffic accidents. The emerging user profiling technology can solve this problem very effectively, and the clustering algorithm is the key to realizing user profiling. In this paper, we have designed a portrait system for road safety based on dynamic density clustering algorithms and have developed an architecture and functional requirements for the system. To do this, we use an actual data sample representing road accidents in the United Kingdom (UK) between 2005 and 2014 for data mining and use the dynamic density clustering algorithm to cluster the dataset, discover the data flow trends of safety accidents from 2005 to 2014, discuss the road safety accident-prone scenarios, and combine the system business with realistic road safety management.

Keywords: System design · Data mining · Dynamic density clustering algorithm

1 Introduction

Traffic safety accidents have been causing much harm to individuals, families, and countries. According to the 2018 Global Status Report on Road Safety [1], published by WHO in December 2018, the number of deaths due to traffic accidents in 2016 was up to 1.35 million, and 50 million people were injured. The economic losses from these traffic accidents represent 1–3% of the gross domestic product of most countries, and the multifaceted losses from road traffic injuries will continue to increase in the future if adequate measures are not taken.

In order to reduce the harm caused by road traffic accidents, it is necessary to analyze the traffic accident data and identify the critical risk factors of traffic accidents. This paper hopes to complete the road safety portrait system design by extracting the essential features. The influencing factors in traffic accidents and mining the Spatiotemporal features related to traffic accidents better understand the development trend. Characteristics of road accidents can help develop appropriate methods and policies to improve safety and reduce road traffic injuries.

Nowadays, researchers often use data mining techniques to analyze the influencing factors in road safety accidents, among which decision trees and rules, non-linear regression and classification methods, and neural networks have been widespread, and

H. Zhang et al. (Eds.): NCAA 2022, CCIS 1638, pp. 260–272, 2022.
https://doi.org/10.1007/978-981-19-6135-9_20

researchers have achieved good results with these techniques, identifying the main triggers and associations between factors in road safety accidents. However, there has been a lack of real breakthroughs in this technology from theoretical to application. Therefore, we collected vehicle accident data from 2005 to 2014 in the UK in this paper. We extracted the main factors affecting road safety in this time series by analyzing these data and finally designed a road safety portrait system.

Our contributions are two: first, we have completed the design of the road safety portrait system, which provides a greater help for the extraction of road traffic safety accident features; second, the dynamic density clustering algorithm we designed overcomes the limitation of the traditional clustering algorithm in the process of analyzing safety accident data in the face of the existence of a continuous dynamic evolution process on the time series, and fully explores the evolution of data between time slices at the same time, the computation of the parts changed data and the local adjustment of the class cluster structure significantly reduces the computational effort of the clustering process.

The rest of the paper is organized as follows. In the second part of the paper, the architecture design of the safety incident mapping system is reviewed, and the algorithm design aspects of safety incident data mining are reviewed. In the fourth part, the dynamic density clustering algorithm results are presented and discussed; in the last part, an overview of the entire work is given, and future work is anticipated.

2 Literature Review

2.1 Algorithms for Security Incident Data Mining

In recent years, there has been a great deal of research effort in developing road safety theory and the factors affecting road safety. At first, researchers mostly used regression analysis techniques to model crash severity, such as logistic and ordered probit models [2–4]. However, these statistical models are more limited in dealing with complex, large-scale crash data because they rely on their model assumptions, which may lead to misestimation of model results due to unsupportable model assumptions and ultimately lead to unsatisfactory performance [5–7]. Therefore, we need to use data mining methods to extract information from large datasets of traffic accidents to achieve high accuracy in a relatively short period.

Recently, researchers have progressively started to use data mining methods to extract the features of traffic accidents. Zhonggui and Zhang [8] developed a three-step spatial data mining approach to identify road clusters directly. They conceptualize the spatial relationship between road and collision road and use clustering and outlier analysis to identify road clusters with high-frequency crashes. Li, Zhibin, et al. developed an SVM model to predict the injury severity associated with individual crashes [9]. They considered crash injury severity modeling a classification problem, classifying crashes into different categories based on severity. The SVM model first maps the input vector into a high-dimensional feature space. By choosing a non-linear mapping a priori, the SVM model constructs an optimal separation hyperplane in this high-dimensional space and divides the results into groups while maximizing the margins between linear decision boundaries. A learning model is generated from the training set with crash-related variables as input and the severity of crash damage as output.

In addition to these data mining methods being applied to traffic crash information extraction, clustering is also often used to analyze factors influencing driver injuries in traffic safety crashes. Li, Zhenning, et al. [10] used K-means clustering analysis and hierarchical Bayesian models to examine the severity of driver injuries in crashes associated with intersections. They used K-means cluster analysis to divide the entire dataset into homogeneous subsets and then used a hierarchical Bayesian random intercept model to study the K-means divided subtypes and the overall sample to examine cross-level interactions between crash level and vehicle/driver level variables. Chang, Fangrong, et al. [11] used latent class clustering for crash data. Unlike nonparametric clustering techniques, such as K-means, latent class clustering is a probability-based modeling technique. It requires distributional assumptions on the outcome variables, which assumes that the entire data is divided into proprietary latent classes by an unobserved or latent categorical variable. Each crash's category membership is inferred from its calculated probability of belonging to a cluster based on observed exogenous variables. Taamneh, et al. [12] used hierarchical clustering to group traffic crash data to organize their collected dataset into six different forms, each with different clusters. Subsequently, artificial neural network models were constructed for each generated form of accident clusters, and it was tested that the accuracy of the models using clustering was significantly higher than that of the unclustered models.

2.2 System Design of Safety Incident Profiling System

In order to prevent traffic safety accidents in advance and guide traffic decision-making and management optimization, many scholars have designed portrait systems for related safety risks. Deng, et al. [13] constructed a holographic portrait of bus safety operation characteristics based on multisource data such as Zhenjiang city bus vehicle warning. A BP neural network-based bus driver fatigue and driving risk prediction model was constructed from time, space, and driver factors. The constructed bus safety operation portrait system is divided into time distribution, spatial distribution, speed distribution, driver characteristics distribution in early warning data statistical analysis, and people, vehicle, road, the environment in risk cause analysis. After constructing the system, the frequent periods of abnormal vehicle warnings and the environmental conditions were obtained through simulation experiments. Tirtha et al. [14] proposed a model system to identify different traffic event duration profiles according to the event type. It is composed of a scaled multinomial logit model, and a grouped generalized ordered logit model system for the incident duration is adapted to the impact of observed and unobserved effects on event type and event duration. It can generate guidelines for event duration for essential variables such as event type, location, and time of day. In this way, the detour can be made in case of an accident through the identification of information signs.

3 Methodology

3.1 System for Security Incident Data Mining

Density-based spatial clustering of applications with noise (DBSCAN) is an algorithm that determines the clustering structure through the closeness of the sample distribution.

It examines the continuity between samples from the perspective of sample density and continuously expands the clusters based on connectable samples to obtain the final clustering results. For dense traffic accident data sets with high noise and unknown sample set distribution, we need to use the DBSCAN algorithm to cluster the traffic accident data sets, which can find arbitrarily shaped clusters and has high immunity to noise points. In the following paper, we first outline the key definitions of the DBSCAN algorithm.

First, in order to determine whether a group of points are similar enough to be considered a cluster, we often use Euclidean distance to measure the distance between two points i and j:

$$dist(i, j) = \sqrt{\left(x_{i1} - x_{j1}\right)^2 + \ldots + \left(x_{im} - x_{jm}\right)^2} \tag{1}$$

where $i = (x_{i1}, x_{i2}, \ldots, x_{in})$ and $j = (x_{j1}, x_{j2}, \ldots, x_{jn})$ are two n-dimension data objects. After obtaining the distance metric, it is specified that each object in the cluster with a given radius (Eps) of the neighborhood must contain at least a minimum number of objects (MinPts), which means that the cardinality of the neighborhood has to exceed some threshold:

$$N_{Eps} = \left\{ q \in \frac{D}{dist(p, q)} < Eps \right\} \tag{2}$$

Here, D is the database of objects. If the neighborhoods of a point P at least contains a minimal number of points, and then this point is called core point:

$$N_{Eps}(P) > MinPts \tag{3}$$

Here Eps and MinPts are user-specified parameters that represent the radius of the neighborhood and the minimum number of points in the ϵ-neighborhood of the core point, respectively. It is completed the definition of the DBSCAN algorithm.

In the face of dynamic data, such as traffic accident datasets of multiple years and the clusters, are dynamically changing in each iteration. This paper uses a dynamic clustering algorithm based on density clustering so that the correlation of data and the inheritance of class structure in time series of traffic accident datasets of adjacent years can be exploited, which is more conducive to discovering the attribute changes of data points and the evolution process of class cluster structure.

For the dynamic clustering algorithm, the dynamic clustering can be carried out after the primary density clustering algorithm obtains the initial clustering. The dynamic clustering process includes three parts: element disappearance, element addition, and element displacement, where the element displacement process includes two sub-processes of element disappearance and addition. Therefore, the algorithm can generally be divided into two parts: element disappearance (ER) and element increase (EA). The following are the steps of the algorithm:

Algorithm Dynamic Density Clustering Algorithm

Input: S^i :data set at time t_i
  addQueue: element remove queue t_i to t_{i+1}
  removeQueue:element remove queue from t_i to t_{i+1}
  ptses:attributes of t_i stage points
  corePoints: t_i stage core points and the points contained around them
  ϵ: density clustering radius parameter
  minPts: density clustering minimum data threshold parameters

Output: cluster classification $C = \{C_1, C_2, \ldots, C_k\}$
1       $ER(S^i, \epsilon, minPts, removeQueue)$
2       $EA(S^i, \epsilon, minPts, addQueue)$
3   for i in *corePoints*:
4   for j in *corePoints*:
5   intersection ← *corePoints*[i] ∩ *corePoints*[j]
6   if i ≠ j and intersection ≠ Ø:
7   if two clusters containing a common core point:
8   *corePoints*[i] ← *corePoints*[i] ∪ *corePoints*[j]
9   delete *corePoints*[j]
10   else /* delete duplicate boundary points */
11   remove elements in intersection from *corePoints*[j]
12   return *corePoints*/* $\{C_1, C_2, \ldots, C_k\}$ */

ER Process:
Input: $S^i, \epsilon, minPts, removeQueue$
1   n ← 0, num ← length of *removeQueue*
2   while *removeQueue* ≠ null:
3   A ← *removeQueue*.pop(), Array ← []
4   if n < num:
5   delete A from S^i
6   if number of points within ϵ of point A > *minPts*:
7   delete *corePoints*[A]
8   else if contains core points within ϵ from point A
9   delete A from all *corePoints*
10   add the points within the radius of the circle with point A as the centre ϵ to Array
11   for i in Array:
12   if the attributes of i have changed:
13   change the attributes of i
14   add i to *removeQueue*
15   n++

EA Process:
Input: $S^i, \epsilon, minPts, addQueue$
1   n ← 0, num ← length of *addQueue*
2   while *addQueue* ≠ null:
3   A ← *addQueue*.pop(), Array ← []
4   if n < num:
5   add A to S^i
6   if number of points within ϵ of point A > *minPts*:
7   add A to *corePoints*
8   else if contains core points within ϵ from point A
9   add A to one of the *corePoints*
10   add the points within the radius of the circle with point A as the centre ϵ to Array
11   for i in Array:
12   if the attributes of i have changed:
13   change the attributes of i
14   add i to *addQueue*
15   n++

3.2   System Design of Safety Incident Profiling System

The road safety portrait system models the data of past road safety accidents, de-picts road safety accidents in different data dimensions. By modeling and analyzing the main attributes that affect road safety accidents, the system abstracts the seman-tic labels that are easy to understand and forms a comprehensive information picture of road safety accidents.

1)   System Architecture Design

The road safety portrait system relies on the essential operating environment to pro-vide linear scalable computing storage resources. It uses a distributed large data plat-form based on Hadoop architecture to support the above traffic accident portrait layer, functional service layer, as shown in Fig. 1. The system mainly includes:

a)   Hardware Resource: The underlying hardware consists of computing nodes, storage nodes, and control nodes.
b)   Data Infrastructure Layer: Because of the enormous volume and variety of traffic accident data, the processing speed of traffic accident data exceeds the capacity of traditional data processing technology, the data infrastructure layer designed in this paper is supported by a Hadoop-based distributed large data platform. The unstruc-tured data is transformed into effective data for subsequent correlation analysis, clustering analysis, etc., by collecting, storing, processing, and analyzing IoT data, Internet data, and external data.
c)   Traffic Accident Portrait Layer: This layer portrays traffic safety accidents from different dimensions, abstracts easy-to-understand semantic labels, and forms a complete picture of traffic accident information by modeling and analyzing the static accident attributes Road-Type, Vehicle-Type, and other information in traffic safety accidents and dynamic accident attribute Vehicle-Manoeuvre. For the real-time labeling design of this layer, this paper introduces a dynamic density clustering algorithm to realize the self-updating of the multi-stage traffic accident data labels.
d)   Functional Services Layer: This layer is the available service of the road safety portrait system that is finally displayed to the public. By supporting the portrait tags in the lower layer, the system can provide safety accident warnings, risk prediction, feature visualization, and other functions to the public. It can provide users with a traffic safety warning service in real-time when the accident risk is high to reduce the probability of accidents.

2)   Business Architecture Design

The business architecture diagram of the road safety incident mapping system consists of fron-tend, system business, business process, and business capability, as shown in Fig. 2.

a) Channels: The front-end of the system will consist of web, mobile web, mobile app, and in-vehicle software. It provides a wide range of system services for pedestrians, drivers, and related transportation departments.

b) System business: The system business is divided into two parts: accident statistics and traffic alert. The accident statistics part contains data visualization, portrait tag generation, interactive analysis, and multi-dimensional reports; the traffic alert part contains traffic analysis, label similarity matches, road risk assessment, and road safety alert.

c) Process: The process of the system is first to obtain the current road information, then mine the risk features in the current information through cluster analysis, perform a risk assessment and provide corresponding planning and warning by the system.

d) Ability: The system capabilities include road risk level management, alarm monitoring, and risk control. The road risk level management capability can provide traffic management departments with better management strategies, which can reduce the accident rate. The alarm monitoring capability can provide law enforcement officers faster assistance to accident victims. Risk control can provide pedestrians, drivers, and passengers with risk warnings to control risks by alerting them in situations where traffic accidents are more likely to occur.

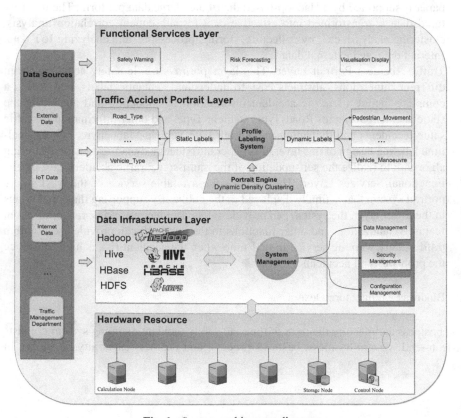

Fig. 1. System architecture diagram.

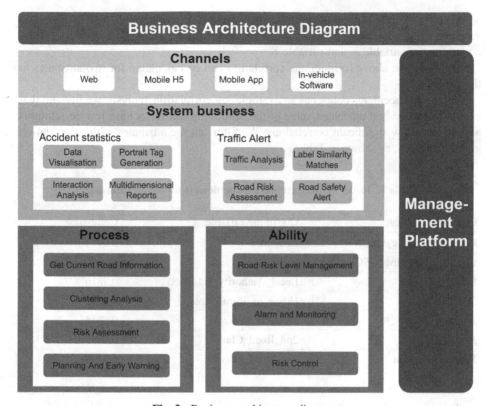

Fig. 2. Business architecture diagram.

4 Discussion

4.1 Data Description

UK police forces collect data on every vehicle collision in the UK on a form called Stats19. This study used a sample of actual data on traffic accidents in the UK from 2005 to 2014.

The sample contains three separate datasets, the accidents dataset, the casualties dataset, and the vehicles dataset. The accidents dataset contains 1780653 records with 32 attributes, the casualties dataset contains 2216720 records with 15 attributes, and the vehicles dataset contains 3004425 records with 22 attributes.

We need to select the key collision causes for these data while avoiding the selection of attributes with significant correlations. To do so, we need to perform correlation analysis on the variables in each of the three data.

For variables X and Y, the correlation coefficients of the two are Pearson correlation coefficients, according to the following equation:

$$\rho_{X,Y} = \frac{\text{Cov}(X,Y)}{\sqrt{\text{Var}(X)\text{Var}(Y)}} \qquad (4)$$

where $\rho_{X,Y}$ is the correlation coefficient of variable X and variable Y; Var(X) is the variance of the variable X; Var(Y) is the variance of the variable Y; Cov(X, Y) is the covariance of variable X and variable Y.

By using the above formula, the correlation can be calculated for two attributes in the data set.

Taking the accidents dataset as an example, by correlation analysis, we find out that except for individual attributes, most of the remaining attributes have low correlations and do not show significant correlations. We filter out the attributes with a correlation greater than 0.5, as shown in Table 1.

Table 1. Partial attribute correlation coefficients for the accident data.

One of the variables	One of the variables	Correlation
Location_Easting_OSGR	Longitude	0.9994
Location_Northing_OSGR	Latitude	1.0000
Police_Force	Local_Authority_(District)	0.9816
Speed_limit	Urban_or_Rural_Area	0.6847
Junction_Detail	Junction_Control	0.6815
Junction_Detail	2nd_Road_Class	0.6929
Junction_Control	2nd_Road_Class	0.9194

In this paper, these highly relevant attributes are processed, leaving only one of them in each group for subsequent selection. Similarly, the same operation is performed for the other two data sets.

After filtering out the attributes with high relevance by correlation analysis, this paper further filters the remaining attributes by combining the relevant studies on road safety accidents [11, 15]. Their study is to remove the redundant attributes in traffic accidents using hierarchical clustering and the latent class clustering, respectively, leaving the attributes with high information value. For the remaining attributes in the three data sets, this paper integrates these attributes into one data set to create a comprehensive record for each accident according to the common index of the three data sets. The composite record is shown in Table 2, and the final composite record consists of 6 attributes from the Accidents dataset, Casualties dataset, Vehicles dataset. These attributes are all discrete attribute values.

Table 2. Comprehensive record structure table.

Accidents data			Casualties data	Vehicles data	
Number_of_Vehicles	Number_of_Casualties	Road_Type	Casualty_Type	Vehicle_Manoeuvre	Vehicle_Type

4.2 Data Analysis

Due to the limitation of computing power, this paper makes a statistical sampling analysis on the combined data set and randomly selects 2000 samples every year from 2005 to 2014 for data analysis. After completing the sampling, we combined the dynamic density clustering algorithm designed in the previous paper to cluster the data while using Silhouette Coefficient [16] as the evaluation index of clustering results to assess the dispersion degree between clusters after clustering. In the clustering process, we choose the result with the largest Silhouette Coefficient to adjust the dynamic density clustering parameters Eps and MinPts. The Silhouette Coefficient for a certain sample point is calculated using the following equation:

$$s(i) = \frac{b(i) - a(i)}{max{a(i), b(i)}} \tag{5}$$

where i denotes the sample point i belonging to a cluster C_i, a_i denotes the average distance from the sample point i to all other points in the cluster it belongs to, and b_i denotes the average distance from the sample point i to all points in a cluster that does not contain it. After the calculation of the Silhouette Coefficient for a single sample point, the total Silhouette Coefficient of the clustering result is obtained by averaging all sample points.

Finally, t-SNE [17] algorithm was used for visual dimension reduction, and Fig. 3 was obtained.

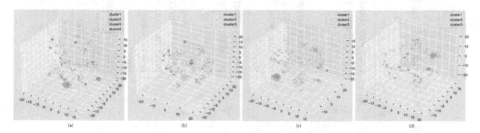

Fig. 3. Clustering results. (a) Clustering results for the 2005 traffic accident dataset. (b) Clustering results for the 2008 traffic accident dataset. (c) Clustering results for the 2011 traffic accident dataset. (d) Clustering results for the 2014 traffic accident dataset.

It can be seen that the sampled dataset eventually contains 3 to 4 clusters. After statistical analysis of all the samples in each cluster, the semantic descriptions of each attribute of each cluster are obtained, which is divided into a total of 5 types in Table 3. To facilitate the statistics of the changes in the traffic data for each year, the clustering results after labeling for each year from 2005 to 2014 are presented in Fig. 4 in the form of a Sankey diagram.

Table 3. Description of traffic accident attributes table.

Type	Number_of_Vehicles	Number_of_Casualties	Road_Type	Vehicle_Type	Vehicle_Manoeuvre	Casualty_Type
A	2	1	One-way streets predominant, a few two-way streets	Cars predominant, a few bicycles and motorbikes	Going ahead other predominant, a few rights turns	Car occupants predominant, a few cyclists
B	1	1	One-way streets predominant, a few two-way streets	Cars predominant, a few buses	Going ahead other predominant, a few others	Pedestrians predominant, a few car occupants
C	1	2	One-way streets predominant, a few two-way streets	Cars predominant, a few others	Going ahead other predominant, a few going ahead right or left-hand bend	Car occupants predominant, a few pedestrians
D	2	2	One-way streets predominant, a few two-way streets	Cars predominant, a few others	Going ahead other predominant, a few turn right	Car occupants predominant, a few others
E	3	1	One-way streets predominant, a few others	Cars predominant, a few bicycles	Going ahead other predominant, a few others	Car occupants predominant, a few others

From the diagram, we can see that the overall trend from 2005 to 2014 is towards more type A and less type C and D. In the last year of 2014, there even type E, that is, there are more vehicles involved in each accident. The accidents mainly involve cars and bicycles, and the injured are mainly car occupants and a few cyclists. It is probably due to the increasing number of vehicles as society develops. Therefore, for traffic accident prevention and control, we need to focus on the safety of cars and bicycles, especially car occupants and bicyclists are at higher risk of injury, and we need to pay special attention to the safety of traffic on one-way streets, where accidents are more likely to occur.

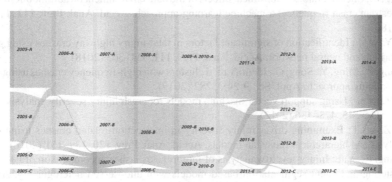

Fig. 4. Sankey diagram describing the changes in clustering results for each yearly intersection dataset.

5 Conclusion

In this paper, the traffic data from 2005 to 2014 in the UK were clustered and analyzed to dig out the data flow relationship of traffic accident types each year and find that driving cars and bicycles on one-way lines is an essential factor leading to traffic accidents. At the same time, this paper designs the road safety portrait system according to the results of data analysis. It also gives the system architecture and business architecture design, hoping to achieve the requirements of reducing traffic accident injury in a more reasonable way.

In the future, we hope to continue to improve the road safety portrait system to promote road traffic safety management in a more efficient and intelligent direction and further reduce the number of casualties due to traffic accidents.

Acknowledgment. This research was supported in part by the NSF of China (Grant No. 62073300, U1911205, 62076225). This paper has been subjected to Hubei Key Laboratory of Intelligent Geo-Information Processing, China University of Geosciences, Wuhan 430074, China.

References

1. World Health Organization. Global status report on road safety 2018: summary. No. WHO/NMH/NVI/18.20. World Health Organization (2018)

2. Mujalli, R.O., de Oña, J.: Injury severity models for motor vehicle accidents: a review. Proc. Inst. Civ. Eng. Transp. **166**(5), 255–270 (2013). Thomas Telford Ltd
3. Nasri, M., Aghabayk, K.: Assessing risk factors associated with urban transit bus involved accident severity: a case study of a Middle East country. Int. J. Crashworthiness **26**(4), 413–423 (2021)
4. Sze, N.-N., Wong, S.C.: Diagnostic analysis of the logistic model for pedestrian injury severity in traffic crashes. Accid. Anal. Prev. **39**(6), 1267–1278 (2007)
5. Yuan, Q., Chen, H.: Factor comparison of passenger-vehicle to vulnerable road user crashes in Beijing, China. Int. J. Crashworthiness **22**(3), 260–270 (2017)
6. Ding, C., Chen, P., Jiao, J.: Non-linear effects of the built environment on automobile-involved pedestrian crash frequency: a machine learning approach. Accid. Anal. Prev. **112**, 116–126 (2018)
7. Li, Y., et al.: Identification of significant factors in fatal-injury highway crashes using genetic algorithm and neural network. Accid. Anal. Prev. **111**, 354–363 (2018)
8. Zhang, Z., Ming, Y., Song, G.: Identify road clusters with high-frequency crashes using spatial data mining approach. Appl. Sci. **9**(24), 5282 (2019)
9. Li, Z., et al.: Using support vector machine models for crash injury severity analysis. Accid. Anal. Prev. **45**, 478–486 (2012)
10. Li, Z., et al.: Examining driver injury severity in intersection-related crashes using cluster analysis and hierarchical Bayesian models. Accid. Anal. Prev. **120**, 139–151 (2018)
11. Chang, F., et al.: Injury severity analysis of motorcycle crashes: a comparison of latent class clustering and latent segmentation based models with unobserved heterogeneity. Analytic Methods Accid. Res. **32**, 100188 (2021)
12. Taamneh, M., Taamneh, S., Alkheder, S.: Clustering-based classification of road traffic accidents using hierarchical clustering and artificial neural networks. Int. J. Inj. Contr. Saf. Promot. **24**(3), 388–395 (2017)
13. Deng, S., Yu, H., Lu, C.: Research on operation characteristics and safety risk forecast of bus driven by multisource forewarning data. J. Adv. Transp. **2020** (2020)
14. Tirtha, S.D., Yasmin, S., Eluru, N.: Modeling of incident type and incident duration using data from multiple years. Analytic Methods Accid. Res. **28**, 100132 (2020)
15. Maji, A., Velaga, N.R., Urie, Y.: Hierarchical clustering analysis framework of mutually exclusive crash causation parameters for regional road safety strategies. Int. J. Inj. Contr. Saf. Promot. **25**(3), 257–271 (2018)
16. Rousseeuw, P.J.: Silhouettes: a graphical aid to the interpretation and validation of cluster analysis. J. Comput. Appl. Math. **20**, 53–65 (1987)
17. Van der Maaten, L., Hinton, G.: Visualizing data using t-SNE. J. Mach. Learn. Res. 9(11) (2008)

An Ensemble Deep Learning Model Based on Transformers for Long Sequence Time-Series Forecasting

Jie Chu, Jingjing Cao[✉], and Yujia Chen

Wuhan University of Technology, Heping Street 1178, Wuhan 430063, China
bettycao@whut.edu.cn

Abstract. Accurate demand forecasts are commonly needed in real life to deploy future schedules of economic activities, such as merchandise sales and electricity consumption predictions. Transformer network has been demonstrated to have potential for time series forecasting in recent studies, however, practical tasks generally require long sequence forecasting outputs with limited length of inputs, which leads to high time complexity and large memory consumption. In this research, we propose a stacking ensemble model for increasing long sequence time-series forecasting accuracy which is based on three Transformer networks. The base learners include Autoformer, Informer and Reformer, which have different improvements on Transformer that enable our approach to improve forecast performance. Then, the results from base learners are applied to train the meta learner (MLP), and finally the trained meta-learner is applied to predict future demand. Experiments on two overt datasets reveal that our proposed ensemble Transformer model possesses great prediction accuracy while requires minor amount of computation. This approach may provide a fresh solution to the demand forecasting domain.

Keywords: Demand forecasting · Transformer · Self-attention · Ensemble learning

1 Introduction

The commodities trade and economic markets have been growing in recent years. Demand forecasting is an important component in many industries, such as online car-hailing [20], electricity supply system [7] and so on. Accurate demand forecasting can provide important help for decisions on promotion, pricing, replenishment and inventory planning of product or service. It can provide important information reference for enterprises to respond to market changes rapidly, which is one of the important tasks of enterprise management and the key to build an agile supply chain. Therefore, it is extremely relevant to improve the accuracy of demand forecasting.

© The Author(s), under exclusive license to Springer Nature Singapore Pte Ltd. 2022
H. Zhang et al. (Eds.): NCAA 2022, CCIS 1638, pp. 273–286, 2022.
https://doi.org/10.1007/978-981-19-6135-9_21

The purpose of demand forecasting is to acquire as much information as possible for predicting the amount at future points in time. Many scholars have presented forecasting models based on seasonality and periodicity of data throughout the last few decades. The traditional moving average model obtains highly fitted relational equations at the high computational cost and has poor reliability [10]. The key issue for demand forecasting is to boost the prediction capability, because historical data is often influenced by multiple factors that are regular, sudden or random. As a classical time-series forecasting model, ARIMA performs well on short-term univariate forecast with steady data, but has difficulty in adapting to long-term forecast and handling large variations data [15]. Models based on specific parameters and linear prediction may not fit the reality situations, so more sophisticated models are applied for demand forecasting, including Support Vector Machine (SVM) [11], Random Forests [3], Decision Trees [18]. Recurrent Neural Network (RNN) [13] and Long Short-term Memory (LSTM) [17] have also been devised and prominently improved to better estimate demand due to the rapid growth of neural networks in recent years. However, [25] proves that their performance and the inference speed decrease significantly with the increase of the prediction length.

In this work, we introduce Transformer networks to time series forecasting, which is a noteworthy neural network and the attention mechanism in Transformer enable to capture more related information. And we combine multiple Transformers with different attention mechanisms into an ensemble model to successfully complete various complex sequential forecast tasks. The following are the main contributions of our research:

- We propose a forecasting model based on Stacking ensemble learning that can fully integrate the advantages of each basic model to forecast long-term future demand.
- We apply multiple variants of Transformer as base learners that can be adapted to datasets with different characteristics.
- Experiments on two overt datasets illustrate the accuracy and dependability of our proposed model in demand forecasting.

The rest of this paper is assigned as follows: In Sect. 2, we cover the relevant study of demand forecasting and the evolution of Transformer. We introduce the overall framework of the Stacking ensemble model in Sect. 3. And then, we elaborate on every Transformer base learners, including Autoformer, Informer, Reformer in Sect. 4. Clarifying the experiments is in Sect. 5. And Sect. 6 concludes the work.

2   Related Work

2.1   Methods for Demand Forecasting

Based on the integration and analysis to a series of historical data, demand forecasting adopts specific methods to obtain demand volume for a future period.

Time series forecasting is thus the essence of demand forecasting. Depending on the model applied, the common methods of time series prediction can be divided into four types: mathematical statistics model, shallow machine learning model, deep learning model, and ensemble deep learning model.

Simple mathematical model calculates the mean or the most recent observation as the predicted value. GM(1,1), Multivariate polynomial regression, and ARIMA model are widely applied to forecast traffic flow [9], medical service [4] and supply chain [19]. These models have good interpretability, however, perform poor on highly stochastic and nonlinear data.

Representative shallow learning methods for time series prediction include KNN [12] and SVM [11]. This paper [11] provides a model for predicting electric load using an improved SVM on decomposed sub-series. These models can effectively capture the nonlinear law of demand. However, their feature learning ability is limited and the generalization ability is restricted to some extent.

The common approaches for deep learning models are RNN [23] and LSTM [17]. As the research progresses, some more novel networks are gradually proposed for time series forecasting [1]. Deep learning models perform outstanding on time-series data because of the deeper structure and emphasis on feature learning, but the prediction accuracy for non-smooth data with strong randomness still needs to be improved.

Ensemble learning incorporates multiple learners to accomplish the learning task. Bagging and Boosting are the two types of general ensemble learning methods. In [8], they designed an integrated prediction model with extreme gradient Boosting to make predictions at the group level and product level respectively. However, this ensemble strategy significantly increases the complexity. In [16], the author used a Deep Belief Network based on two Restricted Boltzmann Machine integrated with Bagging strategy to predict load demand. The simple weighted strategy makes the model depend on performance of base learners. Stacking ensemble strategy changes traditional weighted combination strategies in Bagging into a meta learner and remains the parallel base learners, thus can keep high accuracy and make the complexity of the integrated model not significantly higher.

2.2 Transformer for Forecasting

Transformer is first proposed by [21] for machine translation. It's also quite common in the field of natural language processing, where it's used for sentiment analysis [5]. [2] introduce Transformer to computer vision and improved image processing with concise networks. Transformer models have recently demonstrated great performance in capturing long-term dependency [25], the problem of high computational complexity is prominent in long sequences forecasting tasks. The improvements made by researchers on Transformer model mainly include fixed mode, learnable mode and low rank et al..

The earliest modification of self-attention is to limit view field through a local window and a fixed stride. [14] introduces a sparse matrix decomposition method that reduces the original time and memory growth with the square of the sequence length to linear growth. The learnable model classifies or clusters the input tokens to achieve a more optimal global view of the sequence while maintaining efficiency advantages. A hash-based similarity measurement is introduced in Reformer [6] to efficiently cluster the tokens into blocks. The low-rank approximation of the self-attention matrix can also be used to improve efficiency; the vital point is to assume a low-rank structure in the N × N matrix. Linformer [22] is a representative example for this type of approaches, as it uses a trainable kernel with a Gaussian distribution for low-rank approximation, decreasing memory cost by projecting the length dimension of keys and values to a lower dimensional representation.

3   Ensemble Forecasting Model

The time series forecasting task predicts the most possible future data as output by analyzing the characteristics of the input historical sequence. The feature of long sequence time-series forecasting is the longer length of output, which requires the capability of the model to effectively capture the long range dependent coupling relationship between the outputs and inputs. Therefore, we apply three Transformer-based models to make more accuracy prediction. As has been mentioned above, long term forecasting causes high computational complexity. To address these challenges, we design a Stacking ensemble forecasting model to integrate three base learners, and train the Multi-layer Perceptron as the meta learner. Parallel base learner structure makes the integrated model less complexity. The above three variants of Transformer are good at processing long sequence forecasting, the ensemble learning ideology to integrate them displays an excellent performance in accuracy. The ensemble forecasting framework can be divided into three steps:

Step 1: Data segmentation. Let original time series be $Y = \{y_1, \ldots, y_t,$ $\ldots, y_T\}$, where y_t denotes true demand volume at time t. According to temporal order, we divide the datasets into two parts, with the latest 10% of data collected as the holdout test set. While the 2/3 of the remaining 90% of data are used for training base models, and the later 1/3 are used for training meta model. When a sliding window with fixed size k slides through the time series data, a sub-sequence X_t can be obtained, and it will be feed into corresponding forecasting models.

Step 2: Individual forecast. Three Transformer variants with different attention modules predict the demand volume at time $t + k + 1$ as y_{t+k+1}^n, where $n = (1, 2, 3)$ denotes the nth individual model. We then assign weights to each base classifier using the averaging method.

Step 3: Ensemble forecast. The prediction results of base learners are fed to the meta model, and the MLP is trained on these predictions and the target variable. The trained dense layer predicts the demand volume at time $t + k + 1$ as Y_{t+k+1}'.

Figure 1 is the structure diagram of ensemble Transformers model. As the figure shows, this is an end-to-end network. Given a set of time series requirements data, the model can learn the characteristics of the data autonomously, and then make predictions.

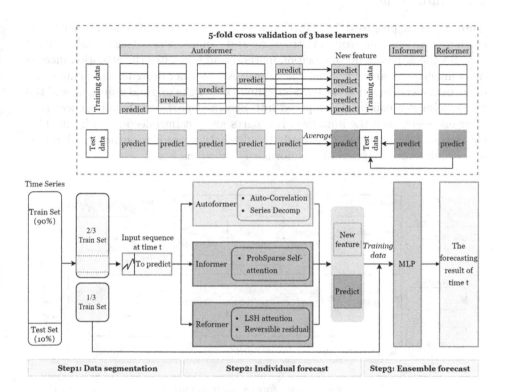

Fig. 1. Framework of ensemble transformer model

4   Transformer Based Learners

The main structure of the model consists of six connected encoders and decoders, where the attention mechanism plays a major role. It allows the network to compute correlations within data or features in parallel, and reduces the network's dependence on external information. The original Transformer has the problem of high storage complexity. L-length inputs/outputs cause L-quadratic computation and memory consumption. In this paper, we use three variants with simpler structures as the basic learners, which can effectively reduce the storage complexity. In this section, we present brief descriptions of individual base-learner algorithms.

The canonical self-attention mechanism in [21] is defined based on three tuple inputs—query, key and value matrices, which are derived from the input features themselves, and are vectors generated from the input features. For input vector whose dimension is d, the scaled dot-product is formulated by $\mathcal{A}(\mathbf{Q}, \mathbf{K}, \mathbf{V}) = Softmax\left(\mathbf{Q}\mathbf{K}^{\top}/\sqrt{d}\right)\mathbf{V}$, where $\mathbf{Q} \in \mathbb{R}^{L_Q \times d}, \mathbf{K} \in \mathbb{R}^{L_K \times d}, \mathbf{V} \in \mathbb{R}^{L_V \times d}$.

4.1 Autoformer

Autoformer is proposed in [24]. Compared with canonical Transformer, Autoformer remains the original structure and connections of the model, while changes the sequence decomposition method and renovates the attention mechanism in encoders and decoders. The overall Autoformer design is an architecture of encoder-decoder, with the encoder's input being the original sequence, and the whole sequence can be decomposed into Seasonal part and Trend-cyclical part after encoders. And the initialized two parts are put into decoder as input, plus the output from encoder, and finally the prediction result is output by Decoder. The architecture diagram of Autoformer is shown as Fig. 2.

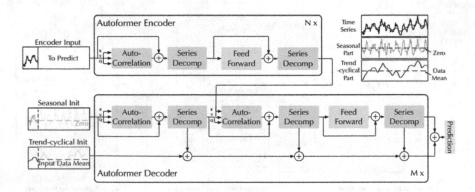

Fig. 2. Autoformer architecture

Autoformer decomposes a complex time series into sub-series with similarities. By adjusting the moving averages to smooth out cyclical fluctuations, then a sub-series reflecting the long-term trend is obtained, and the complete series is subtracted from this to obtain the seasonal sub-series. For input series $X \in \mathbb{R}^{L \times d}$ whose length is L, the decomposition process can be expressed by the following equations:

$$X_{\text{trend}} = \text{AvgPooling}(\text{Padding}(X)) \tag{1}$$
$$X_{\text{season}} = X - X_{\text{trend}} \tag{2}$$

where $X_{\text{trend}}, X_{\text{season}} \in \mathbb{R}^{L \times d}$ denote the sub-series decomposed from input series, and reflect long-term trend and seasonal features respectively.

In periodic sequences, the sub-sequences of each seasonal term are mostly similar. Auto-correlation mechanism is used to aggregate these shaped sub-sequences to try to replace the self-attention mechanism. Auto-correlation includes two steps: exploring the period-based dependencies and time delay aggregation. The fundamental concept is to detect as many useful periodicities in the time series as possible and then predict the most possible future series based on the similarity.

By limiting complex data interactions, storage complexity prominently reduces, and L-length inputs/outputs only cause L computation and memory consumption. It is worth noting that since Autoformer concerning more about recent data, it works better on regular and trending sequences.

4.2 Informer

Informer model is proposed by Haoyi Zhou et al. in 2021 [25]. The model is designed for long sequence time-series forecasting. In view of the high computational complexity, high memory consumption and slow output speed of the model, corresponding improvements are made. The architecture of Informer model is shown in Fig. 3.

The model can be separated into three portions based on their functions: encoder, decoder, and fully connected layer. The encoder on the left receives a large number of long sequence inputs (green sequence), and the ProbSparse self-attention in encoder is a newly proposed attention mechanism. Compared with the original self-attention mechanism, ProbSparse self-attention only computes the dot-product results of n queries and all keys:

$$\mathcal{A}(\mathbf{Q}, \mathbf{K}, \mathbf{V}) = Softmax\left(\frac{\overline{\mathbf{Q}}\mathbf{K}^{\top}}{\sqrt{d}}\right)\mathbf{V} \tag{3}$$

where $\overline{\mathbf{Q}}$ denotes the sparse matrix of query and it covers the top n dominating sparsity measurement queries and $n = c \cdot \ln L_Q$. Therefore, the time and memory complexity both reduce to $\mathcal{O}(L \log L)$.

In addition, the model added a self-attention distillation operation to make the Encoder into a pyramid-like structure to extract the dominant attention, thus significantly reducing the network size.

4.3 Reformer

Reformer model comes from [6] by Nikita Kitaev et al., which is designed for dealing with extremely long sequences. With locally sensitive hashing self-attention, Reformer achieves the time and memory complexity of $O(LlogL)$. To the best of our knowledge, Reformer performs well not only on text processing but also on long-sequence time-series forecasting.

We applied this model to be the base learner for demand forecasting in our paper. Specifically, Reformer utilizes locally sensitive hash attention and reversible residual layers to solve the memory consumption problem of long sequence prediction. The former is mainly adopted to reduce the complexity of self-attention computation, while the latter is for reducing the activation storage

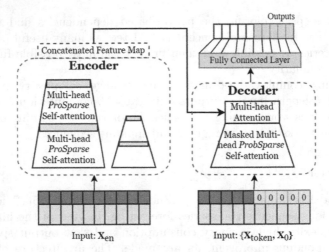

Fig. 3. Structure diagram of informer

of backpropagation in multi-layer models. As Fig. 4 shows, the dominant position in the attention matrix is sparse, however, self-attention ignores the sparsity leading to redundant computation. Firstly, the keys and values are sorted according to the hash bucket so that items with high similarity are aggregated. Then keep the balance of the number of keys and values in the same bucket and finally chunk the sorted attention matrix.

5   Experiments

5.1   Datasets

We conduct experiments on two publicly available datasets, ETTh1 and London Bike, which are frequently utilized in time series forecasting. ETTh1 is derived from the ETT (Electricity Transformer Temperature) dataset, which divides ETT into 1-hour intervals. The target of forecasting is oil temperature, and each data point has 6 attributes. London bike dataset has 17415 records, reflects changes of the number of shared bikes in London over time, which also includes features such as weather and wind speed. The overall data distribution images are shown in Fig. 5a and Fig. 5c, and some data fragments are shown in Fig. 5b and Fig. 5d. It can be observed that the overall data in ETTh1 have a clear downward trend, while data in London Bike remain generally flat, but show significant cyclicality.

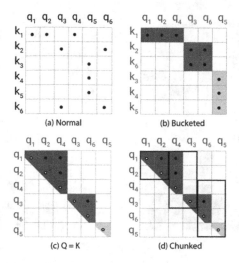

Fig. 4. Attention matrices

5.2 Indicator Metrics

To assess the recommended models' performance, we use two universally accepted metrics, one is the mean absolute error (MAE) and the other is mean square error (MSE). MAE refers to the mean value of the absolute between the predicted value of the model y_i the true value of the sample y. And MSE refers to the mean square of the between the predicted value of the model y_i and the ground true value of the sample y. The formulas are shown below:

$$\mathbf{MAE} = \frac{\sum |y - y_i|}{n} \tag{4}$$

$$\mathbf{MSE} = \frac{1}{n} \sum (y - y_i)^2 \tag{5}$$

5.3 Experiment Details

We train the above models in Ubuntu 18.04.6 LTS with Tensorflow 2.8.0 and CUDA 11.4. The hardware environment is Xeon(R) E5-2678 v3 of RAM 128 GB, and a single NVIDIA TITAN GPU. We use the Adam optimizer with the learning rate initially set to 0.0001 and decaying by a factor of 0.5 per epoch. With appropriate early stops, the total number of epochs is 4. In addition, the batch size of every comparison methods is 32.

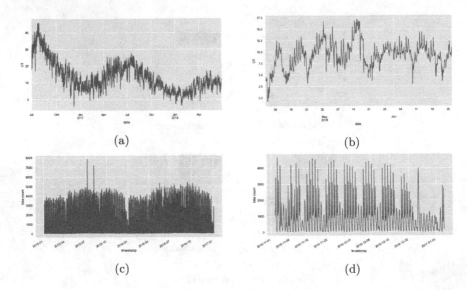

(a)   (b)

(c)   (d)

Fig. 5. Data distribution visualization of two used datasets: (a) for overall data of ETTh1, (b) for data fragments of ETTh1, (c) for overall data of london bike and (d) for data fragments of london bike

5.4 Experiment Results and Analysis

Based on the analysis in Section III, we summarize the time and storage complexity of four individual learners and the ensemble learner in Table 1. Since each base learner has a smaller memory cost compared to the standard Transformer, in terms of time consumption and memory occupation, we may conclude that our proposed ensemble approach works effectively.

Table 1. Comparison of time and memory consumption

Methods	Training		Testing
Metric	Time	Memory	Steps
Transformer	$O(L^2)$	$O(L^2)$	L
Autoformer	$O(LlogL)$	$O(LlogL)$	L
Reformer	$O(LlogL)$	$O(LlogL)$	L
Informer	$O(LlogL)$	$O(LlogL)$	1
Ensemble	$O(LlogL)$	$O(LlogL)$	L

The training data (90%) is also utilized to undertake a 5-fold forward chain cross-validation technique to assess the performance of each of the different models we used. To acquire the forecasting results, we designed four sets of tests for two datasets on Autoformer, Reformer, Informer and ensemble Transformer models respectively, and Table 2 presents the details.

Table 2. Performance measurement of models

Data	Output length	Indicators	Autoformer	Informer	Reformer	Transformer	Ensemble
ETTh1	96	MSE	0.9097	**0.6159**	0.7797	0.9133	0.6517
		MAE	0.7534	0.5549	**0.6422**	0.7651	0.6834
	192	MSE	0.6054	0.7262	0.9114	1.0618	**0.6013**
		MAE	**0.4861**	0.6187	0.7079	0.8291	0.5639
	336	MSE	0.7294	1.0417	1.0168	1.3725	**0.6952**
		MAE	0.5851	0.7881	0.7642	0.9685	**0.5620**
	720	MSE	0.8103	1.1571	1.2348	1.0756	**0.8046**
		MAE	**0.6051**	0.8468	0.8557	0.8182	0.6287
London bike	96	MSE	0.5506	0.6218	0.7814	0.6901	**0.4628**
		MAE	**0.5002**	0.5563	0.6217	0.6095	0.5043
	192	MSE	0.5705	0.7309	0.9473	0.8361	**0.5312**
		MAE	0.5125	0.6223	0.7089	0.6990	**0.4627**
	336	MSE	**0.6236**	1.2971	1.0018	1.0598	0.6721
		MAE	0.5300	0.9069	0.7271	0.7990	**0.4931**
	720	MSE	0.6624	1.0266	1.1252	1.1087	**0.6315**
		MAE	0.5560	0.7735	0.7957	0.8255	**0.5349**

For every dataset, we set the input length to 96 and check the error for four different output lengths: 96, 192, 336 and 720. As is evident from the results, all three base learners we employ on both datasets can show improved predictive capacity compared to vanilla Transformer algorithms, and the ensemble Transformers model performs superior forecasting on both MAE and MSE on both ETTh1 and London Bike dataset. More specifically, when given the fixed input length, ensemble Transformer achieves the lowest error for most prediction lengths. And in a few cases, although not the best, the difference between that and the lowest error is small. The advantage becomes more clear and the performance becomes more steady as the prediction sequence lengthens. We speculate that this is related to the structural properties of the model, where the longer the output sequence, the more demanding it is for the base learner, and the synthetically trained meta-learner performers better.

As shown in Fig. 6 and Fig. 7, when some base learners perform poorly on the data set, the ensemble model can take advantage of its advantages to reduce errors. Notably, we can see that this model also captures long periods of time in the future well, excelling in long series forecasting. However, each time there is a relatively large fluctuation, the forecast value has difficulty in reaching the peak of the fluctuation, which reveals the performance of the model needs to be further improved when demand is suddenly large.

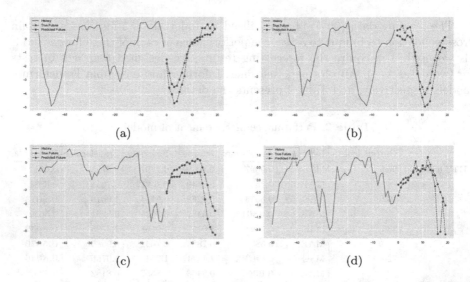

Fig. 6. Experiments of three base learners and ensemble learner on ETTh1 dataset: (a) for autoformer, (b) for informer, (c) for reformer and (d) for ensemble transformers

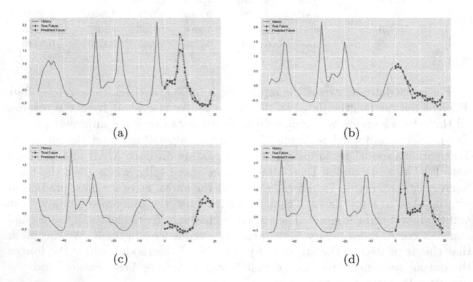

Fig. 7. Experiments of three base learners and ensemble learner on london bike dataset: (a) for autoformer, (b) for informer, (c) for reformer and (d) for ensemble transformers

6   Conclusion and Future Work

We investigate the long sequence time series problem in this study and design an ensemble model based on Transformer variants for demand forecasting. Specifically, we apply three models that are adept in handling long sequence forecasts as the base learners. The results of their integration are then used to train a new meta model, resulting in an ensemble prediction model with strong generalization capability. Our model has lower time and storage complexity. In addition, experiments on two public datasets demonstrate that the proposed model is effective.

We intend to make some attempts to improve the base models in future work, such as exploring the parts of the model that really work and adopting a more lightweight model, and testing performance with multiple datasets of different types and characteristics.

References

1. Bendaoud, N.M.M., Farah, N.: Using deep learning for short-term load forecasting. Neural Comput. Appl. **32**(18), 15029–15041 (2020). https://doi.org/10.1007/s00521-020-04856-0
2. Choi, M., Kim, H., Han, B., Xu, N., Lee, K.M.: Channel attention is all you need for video frame interpolation. In: AAAI, pp. 10663–10671. AAAI Press (2020)
3. Feng, Y., Wang, S.: A forecast for bicycle rental demand based on random forests and multiple linear regression. In: ICIS, pp. 101–105. IEEE Computer Society (2017)
4. Huang, Y., Xu, C., Ji, M., Xiang, W., He, D.: Medical service demand forecasting using a hybrid model based on ARIMA and self-adaptive filtering method. BMC Medical Informatics Decis. Mak. **20**(1), 237 (2020)
5. Jin, Y., Han, D.K., Ko, H.: Trseg: Transformer for semantic segmentation. Pattern Recognit. Lett. **148**, 29–35 (2021)
6. Kitaev, N., Kaiser, L., Levskaya, A.: Reformer: The efficient transformer. In: ICLR, OpenReview.net (2020)
7. Li, R., Chen, X., Balezentis, T., Streimikiene, D., Niu, Z.: Multi-step least squares support vector machine modeling approach for forecasting short-term electricity demand with application. Neural Comput. Appl. **33**(1), 301–320 (2020). https://doi.org/10.1007/s00521-020-04996-3
8. Lingelbach, K., Lingelbach, Y., Otte, S., Bui, M., Künzell, T., Peissner, M.: Demand forecasting using ensemble learning for effective scheduling of logistic orders. In: Ahram, T.Z., Karwowski, W., Kalra, J. (eds.) AHFE 2021. LNNS, vol. 271, pp. 313–321. Springer, Cham (2021). https://doi.org/10.1007/978-3-030-80624-8_39
9. Liu, B., Tang, X., Cheng, J., Shi, P.: Traffic flow combination forecasting method based on improved LSTM and ARIMA. Int. J. Embed. Syst. **12**(1), 22–30 (2020)
10. Liu, S., Ji, H., Wang, M.C.: Nonpooling convolutional neural network forecasting for seasonal time series with trends. IEEE Trans. Neural Networks Learn. Syst. **31**(8), 2879–2888 (2020)
11. Mallick, A., Singh, S.N., Mohapatra, A.: Data driven day-ahead electrical load forecasting through repeated wavelet transform assisted SVM model. Appl. Soft Comput. **111**, 107730 (2021)

12. Martínez, F., Frías, M.P., Pérez-Godoy, M.D., Rivera, A.J.: A methodology for applying k-nearest neighbor to time series forecasting. Artif. Intell. Rev. **52**(3), 2019–2037 (2019)
13. Masum, S., Liu, Y., Chiverton, J.: Multi-step time series forecasting of electric load using machine learning models. In: Rutkowski, L., Scherer, R., Korytkowski, M., Pedrycz, W., Tadeusiewicz, R., Zurada, J.M. (eds.) ICAISC 2018. LNCS (LNAI), vol. 10841, pp. 148–159. Springer, Cham (2018). https://doi.org/10.1007/978-3-319-91253-0_15
14. Parmar, N., et al.: Image transformer. In: ICML. Proceedings of Machine Learning Research, vol. 80, pp. 4052–4061. PMLR (2018)
15. Ponnoprat, D.: Short-term daily precipitation forecasting with seasonally-integrated autoencoder. Appl. Soft Comput. **102**, 107083 (2021)
16. Qiu, X., Ren, Y., Suganthan, P.N., Amaratunga, G.A.J.: Empirical mode decomposition based ensemble deep learning for load demand time series forecasting. Appl. Soft Comput. **54**, 246–255 (2017)
17. Siami-Namini, S., Tavakoli, N., Namin, A.S.: A comparison of ARIMA and LSTM in forecasting time series. In: ICMLA, pp. 1394–1401. IEEE (2018)
18. Sun, L., Xing, X., Zhou, Y., Hu, X.: Demand forecasting for petrol products in gas stations using clustering and decision tree. J. Adv. Comput. Intell. Intell. Informatics **22**(3), 387–393 (2018)
19. Svetunkov, I., Boylan, J.E.: State-space ARIMA for supply-chain forecasting. Int. J. Prod. Res. **58**(3), 818–827 (2020)
20. Teng, F., Teng, J., Qiao, L., Du, S., Li, T.: A multi-step forecasting model of online car-hailing demand. Inf. Sci. **587**, 572–586 (2022)
21. Vaswani, A., et al.: Attention is all you need. In: NIPS, pp. 5998–6008 (2017)
22. Verma, M.: Revisiting linformer with a modified self-attention with linear complexity. CoRR abs/2101.10277 (2021)
23. Wang, Z., He, L., Zhao, Y.: Forecasting the seasonal natural gas consumption in the US using a gray model with dummy variables. Appl. Soft Comput. **113**(Part), 108002 (2021)
24. Wu, H., Xu, J., Wang, J., Long, M.: Autoformer: decomposition transformers with auto-correlation for long-term series forecasting. In: NeurIPS, pp. 22419–22430 (2021)
25. Zhou, H., et al.: Informer: beyond efficient transformer for long sequence time-series forecasting. In: AAAI, pp. 11106–11115. AAAI Press (2021)

Multi-layer Integrated Extreme Learning Machine for Mechanical Fault Diagnosis of High-Voltage Circuit Breaker

Xiaofeng Li$^{(\boxtimes)}$, Tao Zhang, Wenyong Guo, and Sheng Wang

College of Power Engineering, Naval University of Engineering, Wuhan 430072, China
xiaofengli@whu.edu.cn

Abstract. Mechanical fault caused by abnormal joint clearance jeopardizes the safety of high-voltage circuit breakers (HVCB). Accurate fault diagnosis is of great significance to ensure the reliable service. Existing prediction approaches are generally based on individual classifiers, which may fail to achieve satisfactory performance, especially in multi-class fault diagnosis. In this paper, a multi-layer integrated extreme learning machine (IELM) is proposed. At first, clearance joint most likely to induce mechanical fault is found by finite element analysis. Then experiments under various working conditions are carried out, and the collected vibration is decomposed by variational mode decomposition. In feature extraction, the energy distribution in the different frequency domains is considered as the model input. In IELM, each ELM module in different layers is trained independently, and its optimal hidden layer number is adaptively searched. Then it is employed as a sub-classifier for the final diagnosis model. The fault diagnosis process of IELM is similar to a multi-stage filter since the data diagnosed by the previous layer is dynamically removed. It is demonstrated that the proposed method not only achieves the highest fault diagnosis accuracy than its counterparts but also maintains a certain recognition accuracy of unknown mechanical fault.

Keywords: High-voltage circuit breakers · Mechanical fault diagnosis · Extreme learning machine

1 Introduction

HVCB is the most important control and protection equipment in the grid system, whose service reliability greatly affects the safety of power delivery. Statistics from the International Council on Large Electric Systems (CIGRE) show that mechanical fault of HVCB has already become a key factor affecting the power delivery. More than 60% of power delivery accidents are caused by mechanical fault (e.g., joint clearance fault caused by wear and corrosion) [1–3]. Therefore, it is meaningful to investigate the intelligent fault diagnosis of HVCB.

For the fault diagnosis, many valuable works focus on fault feature extraction and intelligent diagnosis approach. For fault feature extraction, travel curve, operating coil current, and vibration signals are the most commonly state-related signal. Generally speaking, travel curve and operating coil current signals from experiments characteristics as low-frequency, and the extracted fault features mainly belong to time domain, such as the three-phase electrical contacts disconnection time, the end of acceleration, and other key time nodes. These signal features can ensure the diagnosis of insufficient control voltage and electromagnet core stuck. On the other hand, for health status detection of the internal mechanical structure of HVCB, it generally required features from high frequency domain [4–6]. The vibration signal consists of a series of complex high-frequency vibration waves generated by various components (e.g., electromagnet, motor, and operating spring). Its amplitude, frequency, energy distribution, and other signal features contain a large amount of structural state-related information [7]. Therefore, large percentages of the previous studies on the diagnosis of mechanical fault of HVCB were based on vibration signals. Thanks to these valuable works, some practical vibration signal processing methods, including empirical mode decomposition (EMD) [8, 9], local mean decomposition (LMD) [10], wavelet packet decomposition (WPD) [11, 12], singular value decomposition (SVD) [13] and variational mode decomposition (VMD) [14], etc. were widely used. Then different types of amplitude- and frequency-based vibration features, such as time-frequency entropy, permutation entropy, singular entropy, and energy entropy have been extracted as mechanical fault features. More specifically, Yang et al. [15] decomposed the vibration signal through EMD and then extracted the singular spectral entropy of different natural mode components as the mechanical fault in the diagnosis process. Chen et al. [16] decomposed the vibration signal of HVCB through VMD. After dynamic time warping analysis, effective natural modal components were selected, and then the high-dimensional permutation entropy was extracted as the eigenvalue of the vibration signal for fault diagnosis model building. In this study, VMD is used to decompose signals, since it is an adaptive signal decomposition method and has the advantage of being less prone to modal aliasing.

As for the intelligent diagnosis approach, it involves establishing a relationship between features and mechanical fault. With the development of artificial intelligence, machine learning-based intelligent classifier has been widely used for pattern recognition and classification [17]. Error feed-forward artificial neural networks likes BP and RBF were first applied to the diagnosis of mechanical fault such as base bolts looseness, mechanism stuck, and over-travel of HVCB [18]. With a limited training sample, support vector machine (SVM) was reported can still achieve satisfactory diagnosis accuracy [19]. Besides, after the kernel and penalty parameters optimization by annealing, genetic, and particle swarm algorithm, SVM can further improve its fault diagnosis performance than others. Therefore, Bi-class and multi-class types SVM were widely investigated for mechanical fault diagnosis of HVCB [20, 21]. Compared with feed-forward neural networks, the weights and offsets between the input and hidden layer of extreme learning machine (ELM) are set randomly, and its network parameters between the hidden and output layer are calculated by Moore-Penrose generalized inverse, which greatly reduces the computer burden [22]. Although machine learning methods have achieved great success in fault diagnosis, there still exist two potential issues. First, for multi-class fault diagnosis problems, the diagnosis accuracy is difficult to guarantee by individual classifiers. Second, it is difficult to obtain a complete range of fault samples due to the long-term immobility of HVCB during its service period. Previous studies of multi-class classification problems are mostly holding that all fault types are known, thus the diagnosis model is assumed to have been fully trained. However, recording all types of mechanical fault for the full model training is unrealistic. When an unknown fault arrives, it would be identified as a known state [23].

In this paper, a novel multi-layer integrated extreme learning machine called IELM is proposed for both known and unknown mechanical fault diagnosis of HVCB. VMD is first employed to decompose the vibration signal from the mechanical operation experiment on a real Zn12 HVCB. Afterward, the vibration energy distribution in different intrinsic mode function (IMF) components is extracted as the fault features for diagnosis model training and testifying. It is worth mentioning that the mode number directly affects the performance of VMD [24]. Although there are some reported techniques such as introducing cuckoo search algorithm into VMD to determine the mode number [25, 26]. It suffers from high computational complexity and is time-consuming. This paper adopts center frequency observation method for optimal mode number determination. Then handle the multi-class mechanical fault classification problems into multiple Bi-class classification issues. Multiple binary ELM modules are integrated into a multi-layer classifier, and the diagnosis task of different mechanical fault types is assigned to different ELM modules. The testing samples of various mechanical fault types are screened out one by one after the IELM. The comparative analysis demonstrates that the proposed approach has a certain improvement in mechanical fault diagnosis accuracy, and can also detect unknown fault.

2 Mechanical Fault Diagnosis Model Building

2.1 Extreme Learning Machine

ELM is developed for single hidden layer feed-forward neural networks, which contain three layers: input layer, hidden layer, and output layer, as shown in Fig. 1. During the training process of the ELM, the weight and bias parameters between the input and the hidden layer are randomly set and do not need to be adjusted. As for the weight parameters from the hidden layer to the output layer, they are determined according to Moore-Penrose (MP) generalized inverse analysis.

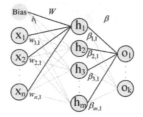

Fig. 1. The principle of ELM.

For better description, assume N training samples $(\mathbf{X}, \mathbf{t})$, where $X_i = [x_{i1}, x_{i2}...,x_{in}]^T \in R_n$, belongs to the n-dimensional input vector, and $t_i = [t_{i1}, t_{i2},..., t_{iq}]^T \in R_q$, is the q dimensional target vector. After n neurons input layer, for a neural network with m hidden neurons, the hidden layer output can be obtained as [22].

$$\sum_{i=1}^{m} \beta_i g(\mathbf{W_i} * \mathbf{X_j} + \mathbf{b_i}) = \mathbf{O_i}, \quad j = 1, 2, ..., n \tag{1}$$

where $\mathbf{W_i} = [w_{i,1}, w_{i,2},...w_{i,n}]^T$ and β represent the input-hidden weights for the network, and $\mathbf{b_i}$ is the unit bias of the i_{th} hidden neuron. $\mathbf{W_i} * \mathbf{X_j}$ stands for the inner product of $\mathbf{W_i}$ and $\mathbf{X_j}$. g is the activation function of the hidden layer, which is activated only once at the hidden layer in the whole learning process. The goal is to make the output approach zero error, which can be expressed as:

$$\sum_{j=1}^{N} \|\mathbf{O_j} - \mathbf{t_j}\| = 0 \tag{2}$$

After randomly specifying $\mathbf{W_i}$ and $\mathbf{b_i}$ in Eq. (1), there is β satisfies the following constraint:

$$\mathbf{H}\beta = \mathbf{T} \tag{3}$$

$$\mathbf{H} = \begin{bmatrix} g(W_1 * x_1 + b_1) & \cdots & g(W_m * x_1 + b_m) \\ \vdots & \cdots & \vdots \\ g(W_1 * x_n + b_1) & \cdots & g(W_m * x_n + b_m) \end{bmatrix}_{n \times m}, \beta = \begin{bmatrix} \beta_1^T \\ \vdots \\ \beta_m^T \end{bmatrix}_{m \times k}, \mathbf{T} = \begin{bmatrix} t_1^T \\ \vdots \\ t_n^T \end{bmatrix}_{n \times k} \tag{4}$$

where k is the number of output layers of the network, $\mathbf{H}$ is the output matrix of the hidden layer, and $\mathbf{T}$ is the target matrix of the training set. Since output matrix $\mathbf{H}$ is reversible, there is:

$$\beta = \mathbf{H}^+ \mathbf{T} \tag{5}$$

where $\mathbf{H}^+$ is the Moore-Penrose generalized inverse of the output matrix $\mathbf{H}$ for the hidden layer.

2.2 Multi-layer Integrated Extreme Learning Machine

The schematic diagram of the established IELM is shown in Fig. 2. Each layer of ELM module in the diagnostic model is trained separately, and the related structure parameters are adjusted based on its own diagnosis result. During the training process, the fault type that needs to be diagnosed is marked as the target condition, and all other training samples are marked as unknown conditions regardless of which state they really belong to. Each layer of ELM module is only responsible for the diagnosis of one specific kind of mechanical fault. Therefore, in the training process, based on the diagnosis result, the model parameters of each ELM module can be adjusted by optimization techniques like particle swarm and genetic algorithm which ensures the network's structure flexibility.

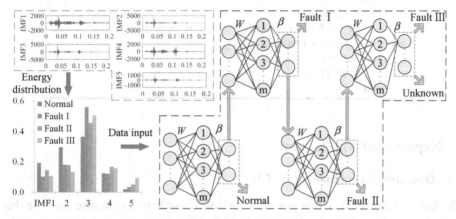

Fig. 2. The principle of IELM.

As Fig. 2 shows, when new data occurs, it is more likely to flow into the next layer of classifier until it is diagnosed as an unknown fault by the end ELM module. Besides, if enough new labeled samples are collected, it is only necessary to add a layer of ELM module to the head of the IELM model and perform separate training. There is no need to retrain other ELM modules, so the expansion of the model structure is also better. During the diagnosis process, all test samples will enter the first layer of ELM module. At this layer, the normal working condition data samples will be diagnosed as the target working condition, and the other samples will be classified as unknown type to enter the next ELM module. This process will continue until the last layer of ELM module completes the diagnosis task of the last mechanical fault type. The algorithm for IELM is shown in Algorithm 1.

Algorithm 1 for IELM
1:    **Input**: w, b, dataset, and the label
2:    **Output**: elm_model, predicted result
3   Data preprocessing: normalize the dataset, and divide it into training data and label set (x, X), as well as testing data and label set (y, Y)
4:    **For** i←1: layer of IELM
5:        **For** k =1: maximum hidden layer number
6:            model{i, k}=elm_train(w, b, k, x, X(i));
7:            f(k)=elm_predict(x, model(i, k));
8:            error(i, k)=sum(f(k)-X(i));
9:        **end**
10:       elm_model(i)= model(Find(minimum (error(i, k))));
11:   **end**
12:   **For** j← 1: layer of IELM
13:       sim_result (j)= elm_predict (y, Y, elm_model(j));
14:       T=find(sim_result (j)=target_type(j));
15:       y(:,T)=[];
16:       Y(:,T)=[];
17:   **end**

3   Experiment Application

3.1   Operation Experiment of HVCB

The fast operation of HVCB is generally controlled within tens of milliseconds. In that case, the structural dynamic characteristics are quite sensitive to the clearance joints. It has been proved that severe collisions between moving parts would be strengthened by small size joint clearance, thus deteriorating the operation quality of HVCB [27]. What's worse, under the influence of wear, corrosion, etc., the joint clearance size will increasingly deviate from the original design value, which leads to greater impact stress during mechanical operation and induces mechanical fault. For dynamic analysis, as shown in Fig. 3, this paper first establishes a finite element analysis model to analyze the dynamic response with clearance joints at different positions, as well as its influence on mechanical fault. For description, the center track of the shaft in a joint with the clearance size of 0.03mm is also given. It can be detected that, in the joint clearance space, the contact between the shaft and bushing is not evenly distributed along the circumference. A high-frequency alternation of contact and separation state occurs in the BC area, and the impact amplitude and alternating frequency are directly related to the joint clearance size. Therefore, the mechanical fault caused by clearance joint can be characterized by the energy statistics of the vibration signal in different frequency domains. Further dynamic sensitivity simulation analysis shows that clearance joints coded I, II, and III have greater effects on the health state of HVCB. Therefore, this paper will focus on the mechanical fault diagnosis caused by these three clearance joints.

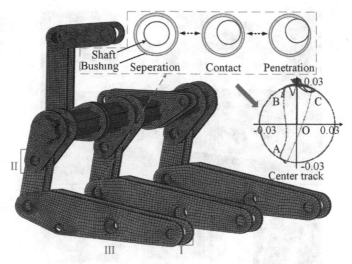

Fig. 3. FEA model and the center track of shaft in clearance joint.

A real HVCB is used to validate the effectiveness of the proposed method, and the related experimental setup is shown in Fig. 4. Three joint clearance fault conditions at different positions are constructed to conduct the mechanical operation tests and collect its vibration signal for mechanical fault diagnosis. During our experiment, the mechanical fault caused by the wear of clearance joint is simulated by changing the pin size. For that purpose, the abnormal joint clearance size in this paper is taken as 0.25 mm and 0.75 mm. As for the normal joint clearance size, it is controlled within 0.04 mm, which is negligible. Moreover, to avoid the mutual influence between two adjacent mechanical operations, the time interval of each opening and closing operation test of the HVCB is more than 3 min. The vibration signal of HVCB has a short time and a wide frequency domain which is characterized by a high degree of nonlinear and non-stationary. For accurate vibration signal measurement, two CCLD/IEPE shock accelerometers coming from Brüel&Kjær, type 8339, whose measurement range, sensitivity, and upper cut-off frequency are $\pm$ 10000 m/s^2, 0.25 mv/g, and 20 kHz, respectively, are adopted. Then they are screwed into the rack with mutually perpendicular axes. A set of Brüel&Kjær signal acquisition card types 3053-B-120 is employed for vibration collection, whose sampling frequency is set to 65536 Hz.

50 groups of vibration signals are collected per operation condition; hence, a total of 400 recordings of the three mechanical fault and normal operating conditions are collected. Then the obtained data is divided into training and testing dataset. It is important to note that, to guarantee an objective result, the testing data should not appear in the training dataset; otherwise, over-optimistic problems could emerge in the testing result.

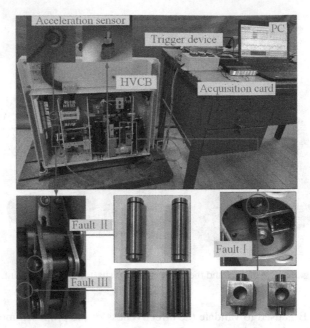

Fig. 4. Photograph of the experimental setup.

3.2 Feature Extraction by VMD

The vibration signal of the HVCB contains structural nonlinear health information, which is scattered in different frequency bands of the vibration signal. Therefore, VMD is used to decompose the vibration signal to obtain IMF. When using VMD, it should be noted that the number of decomposed modes (K) directly affects the algorithm performance and needs to be preset. However, there is no unified approach for the determination of the number of decomposed modes. In this study, the center frequency observation method is used to determine the optimal mode number. Specifically, the center frequency of each component obtained by VMD is arranged in order from small to large to determine the

Table 1. Center frequency of each IMF (Hz).

K	IMF1	IMF2	IMF3	IMF4	IMF5	IMF6	IMF7	IMF8	IMF9
3	6385	12168	18619						
4	6071	10127	12700	18697					
5	6071	10126	12695	18668	26572				
6	3625	64371	10181	12713	18675	26599			
7	3610	6402	10066	12306	14258	18804	26697		
8	3608	6397	10054	12256	14108	18267	20160	26894	
9	3555	6243	8992	10454	12400	14279	18308	20180	26905

optimal mode number. Table 1 shows the average center frequencies of normal vibration signal obtained by VMD while K varies from 3 to 9.

It can be observed that the highest center frequency of the obtained IMF is less than 20000 Hz while K equals 3 or 4, which is difficult to meet the effective signal eigenvalue extraction requirements. When K equals 5, the highest center frequency obtained by VMD rises to 26572 Hz, and the center frequency interval of each IMF component is relatively uniform to avoid modal aliasing problems. As the number of decomposition layers increases, no obvious increase occurs in the highest frequency. However, the center frequency difference between adjacent natural modal components is narrowed, and the phenomenon of modal aliasing will occur. Therefore, this paper takes K = 5 to perform VMD on the vibration signal to extract its eigenvalues. The energy distribution of the vibration signal in different IMF components can be calculated and normalized as follows:

$$\begin{cases} E_i = \int_{t_0}^{t_i} |A(t)|^2 dt \\ E = \sum_{1}^{10} E_i \end{cases} \tag{6}$$

$$P_i = \frac{E_i}{E} \tag{7}$$

where i is the serial number of IMFs. t_0 and t_i represent the start and end time of the collected vibration signal. A(t) and E_i denote the amplitude at different time points and signal energy of each IMF. E represents the total energy of the vibration signals of the two measuring points, and P_i stands for the energy proportion of i_{th} IMF in the total energy, that is, the energy distribution.

4 Result and Analysis

4.1 Mechanical Fault Identification

For comparative analysis, BP, SVM, ELM, and IELM are all used for the fault diagnosis. All of them are trained and tested by two dataset. In dataset A, 35 groups of samples are randomly selected to train the model, and the remaining 15 groups of data samples in the same joint clearance size level are employed for model testing. In dataset B, 50 groups of samples are considered as training set, then 35 groups of samples of different joint clearance size are randomly pick out as testing set. It is worth mentioning that data types 1 and 2 represent 0.25 mm and 0.75 mm clearance size respectively. Therefore, 1 → 2 means training the model by the input of 0.25 mm fault clearance size, then testing it by that of 0.75 mm fault clearance size and vice versa. Multiple trials are conducted to calculate the average diagnosis accuracy and the result comparison is listed in Fig. 5.

Figure 5(a) reveals that the average diagnostic accuracy of the four mechanical conditions by BP neural network is less than 90%. The diagnostic accuracy of the SVM model for the three mechanical faults is 91.7% and 93.3% in 1 → 1 and 2 → 2, and the diagnostic accuracy of the ELM model is 93.4% and 98.3 in 1 → 1 and 2 → 2, slightly inferior to that of IELM model with 95.0% and 98.3%. It can be detected from

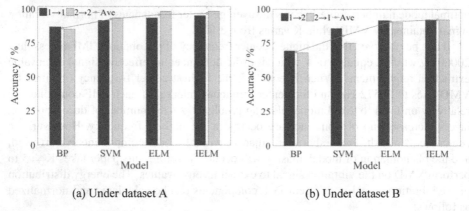

(a) Under dataset A   (b) Under dataset B

Fig. 5. Comparative diagnosis results of different model.

Fig. 5(b) that the diagnostic accuracy under different joint clearance sizes decreased significantly and the average accuracy of IELM is the highest, both at 92.1%. For the joint clearance fault diagnosis of HVCB, engineering applications often pay more attention to the diagnosis performance for different abnormal clearance sizes of the same joint position. In order to further verify the model's effectiveness, the specific performance of SVM, ELM, and IELM for each mechanical fault under $1 \rightarrow 2$ are also illustrated in Fig. 6 and Fig. 7.

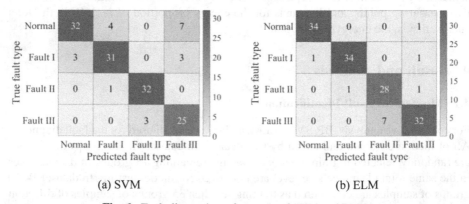

(a) SVM   (b) ELM

Fig. 6. Fault diagnosis performance of SVM and ELM.

As shown in Fig. 6, the model prediction distribution of each fault kind is listed in the confusion matrices, where the abscissa represents the true label of the data, and the ordinate represents the predicted label by the model. For SVM model, although its average diagnostic accuracy is the lowest, the diagnostic accuracy for joint clearance fault II is the highest. This phenomenon tells us that even the best-selected model may have poor performance under certain circumstances. Meanwhile, even the worst model may present relatively good performance for some kinds of mechanical fault types. Overall, it can be seen that the reported IELM can achieve the highest diagnostic accuracy.

According to Fig. 7, the four-layer IELM model can correctly diagnose 129 out of 140 test samples under $1 \rightarrow 2$. More specifically, for the first layer of ELM module, 1 group of Fault III is misdiagnosed as normal condition, and 2 groups of normal condition are misdiagnosed as unknown condition. Then the other 104 groups of test samples are labeled as unknown type and enter the next layer of ELM module. For the second layer of ELM module, 33 out of 35 groups of Fault I are correctly diagnosed, Fault II and Fault III are all diagnosed as unknown type and enter the next layer. It should be noted that, for the final layer of ELM module, 31 groups of Fault III have been correctly diagnosed, and 7 groups of operating conditions still have not received their final specific diagnosis result. It can be explained that the state-related feature of these 7 samples are unclear. This may be caused by several reasons such as signal processing error, environmental effects, and so on.

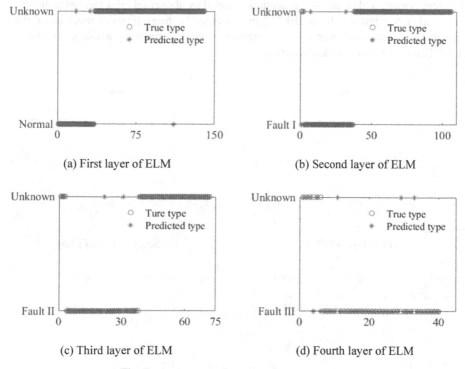

(a) First layer of ELM   (b) Second layer of ELM

(c) Third layer of ELM   (d) Fourth layer of ELM

Fig. 7. Fault diagnosis performance of IEML.

4.2 Detection of Unknown Fault

For the HVCB involved in this paper, the unknown mechanical fault diagnosis is unavoidable with the limited kinds of known samples in the training set, since it is impossible to construct a training set that contains all the fault types of HVCB's fault. Conventional individual classifiers are trained based on known type data, causing samples of unknown faults inevitably be misidentified as known states. By contrast, when unknown data appears in IELM, each ELM module is easier to diagnose it as unknown data and hand over the diagnostic task to the next layer. For testing the ability to diagnose unknown mechanical fault, a new mechanical condition is conducted, and other 40 times operation tests are carried out to collect the vibration signal for verification. Figure 8 describes the diagnosis performance of obtained IELM model under unknown condition. It can be detected that 29 out of the 40 groups of input samples are still diagnosed as other working conditions by the last ELM. Therefore, the overall accuracy for unknown data filtering is 72.5%. Compared with previous works, the recognition accuracy of unknown mechanical fault is lower. It should be noted that previous studies tend to investigate the diagnosis issue of totally different mechanical faults, like spring fatigue, looseness of base bolt, mechanical stuck, and so on. Thus, the feature data between different mechanical conditions is more differentiated. In this paper, although the studied mechanical faults are in different positions, they are still essentially caused by abnormal joint clearance and thus have higher consistency. If the IELM can be trained by more types of mechanical fault with greater feature differences, the generated decision boundary would own stronger capability to screen out unknown data.

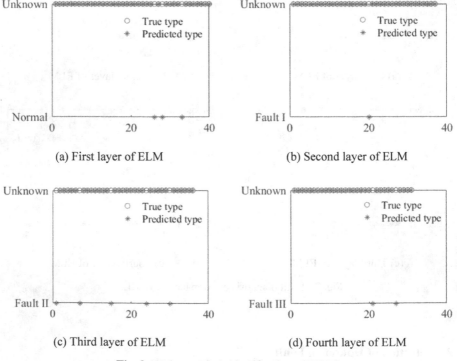

Fig. 8. Unknown fault identification of IELM.

Unlike existing individual classifiers, the reported IELM consists of multiple ELM modules. In IELM, the multi-class mechanical fault diagnosis issue is divided into simple Bi-class problem, which reduces the difficulty of classification and thus enhances the fault diagnosis performance. On the one hand, the reported IELM is more flexible to deal with multi-class mechanical fault diagnosis, as extra ELM modules can be directly added to the front of IELM and will not affect other modules. On the other hand, the proposed IELM own stronger robustness for multi-class classification issues. When one ELM module suffers from interruption of disturbance, the diagnosis result of other modules can still work fine, and the data not diagnosed will keep flowing into the next layer of classifier until be diagnosed as unknown data by the last ELM module. With the aid of human experience or other classifiers, further diagnosis of these unknown data can be performed, thereby reducing the misdiagnosis rate by traditional individual multi-class classifiers.

5   Conclusion

This study proposes a novel approach for mechanical fault diagnosis of HVCB, which can effectively extract precise features from vibration signals and improve fault identification performance. The conclusions can be drawn as follows:

(1) For eigenvalue extraction, through center frequency observation, the optimal mode number is determined as five, which can ensure efficient decomposition without obvious modal aliasing problems.

(2) The diagnostic comparison shows that the proposed IELM model has the highest diagnostic accuracy, with 95.0% and 98.3% at the same fault joint clearance size level and 92.1% at different fault joint clearance size level.

(3) The proposed IELM has a certain unknown data screening ability, whose experiment verification accuracy of unknown fault filtering is 72.5%. Besides, it is more robust and flexible to address multi-class mechanical fault diagnosis issues.

Overall, the IELM enhances the mechanical fault diagnosis of HVCB performance. However, the unknown mechanical fault identification accuracy is still not satisfactory. Therefore, some advanced optimizations will be studied in future work.

References

1. Janssen, A., Makareinis, D., Solver, C.: International surveys on circuit-breaker reliability data for substation and system studies. IEEE Trans. Power Delivery 29, 808–814 (2013)
2. Liu, Y., Zhang, G., Zhao, C., Qin, H., Yang, J.: Influence of mechanical faults on electrical resistance in high voltage circuit breaker. Int. J. Electr. Power Energy Syst. 129, 106827 (2021)
3. Razi-Kazemi, A., Niayesh, K.: Condition monitoring of high voltage circuit breakers: past to future. IEEE Trans. Power Delivery 36, 740–750 (2021)
4. Zhao, S., Wang, E., Hao, J.: Fault diagnosis method for energy storage mechanism of high voltage circuit breaker based on CNN characteristic matrix constructed by sound-vibration signal. J. Vibroengineering 21, 1665–1678 (2019)

5. Niu, W., Liang, G., Yuan, H., Li, B.: A fault diagnosis method of high voltage circuit breaker based on moving contact motion trajectory and ELM. Math. Probl. Eng. **2016**, 1–10 (2016)
6. Forootani, A., Afzalian, A., Ghohmshe, A.: Model-based fault analysis of a high-voltage circuit breaker operating mechanism. Turk. J. Electr. Eng. Comput. Sci. **25**, 2349–2362 (2017)
7. Liu, M., Wang, K., Sun, L., Zhen, J.: Applying empirical mode decomposition (EMD) and entropy to diagnose circuit breaker faults. Optik **126**, 2338–2342 (2015)
8. Jian, H., Hu, X., Fan, Y.: Support vector machine with genetic algorithm for machinery fault diagnosis of high voltage circuit breaker. Measurement **44**, 1018–1027 (2011)
9. Gao, W., Wai, R., Qiao, S., Guo, M.: Mechanical faults diagnosis of high-voltage circuit breaker via hybrid features and integrated extreme learning machine. IEEE Access **7**, 60091–60103 (2019)
10. Huang, N., Fang, L., Cai, G., Xu, D., Chen, H., Nie, Y.: Mechanical fault diagnosis of high voltage circuit breakers with unknown fault type using hybrid classifier based on LMD and time segmentation energy entropy. Entropy **18**, 322 (2016)
11. Ma, S., Chen, M., Wu, J., Wang, Y., Jia, B., Yuan, J.: Intelligent fault diagnosis of HVCB with feature space optimization-based random forest. Sensors **18**, 1221 (2018)
12. Li, X., Wu, S., Li, X., Yuan, H., Zhao, D.: Particle swarm optimization-support vector machine model for machinery fault diagnoses in high-voltage circuit breakers. Chin. J. Mech. Eng. **33**(1), 1–10 (2020). https://doi.org/10.1186/s10033-019-0428-5
13. Wang, D., Li, J., Memik, G.: User identification based on finger-vein patterns for consumer electronics devices. IEEE Trans. Consum. Electron. **56**, 799–804 (2010)
14. Dou, L., Wan, S., Zhan, C.: Application of multiscale entropy in mechanical fault diagnosis of high voltage circuit breaker. Entropy **20**, 325 (2018)
15. Yang, F., Sheng, G., Xu, Y., Qian, Y., Jiang, X.: Application of EEMD and high-order singular spectral entropy to feature extraction of partial discharge signals. IEEJ Trans. Electr. Electron. Eng. **13**, 1002–1010 (2018)
16. Chen, L., Wan, S.: Mechanical fault diagnosis of high-voltage circuit breakers using multi-segment permutation entropy and a density-weighted one-class extreme learning machine. Meas. Sci. Technol. **31**, 085107 (2020)
17. Xu, J., Zhang, B., Lin, S., Teng, Y.: Application of energy spectrum entropy vector method and RBF neural networks optimized by the particle swarm in high-voltage circuit breaker mechanical fault diagnosis. High Voltage Eng. **38**, 1299–1306 (2012)
18. Phyo, C., Zin, T., Tin, P.: Deep learning for recognizing human activities using motions of skeletal joints. IEEE Trans. Consum. Electron. **65**, 243–252 (2019)
19. Lin, L., Wang, B., Qi, J., Chen, L., Huang, N.: A novel mechanical fault feature selection and diagnosis approach for high-voltage circuit breakers using features extracted without signal processing. Sensors **19**, 288 (2019)
20. Yin, Z., Hou, J.: Recent advances on SVM based fault diagnosis and process monitoring in complicated industrial processes. Neurocomputing **174**, 643–650 (2016)
21. Huang, N., Chen, H., Cai, G., Fang, L., Wang, Y.: Mechanical fault diagnosis of high voltage circuit breakers based on variational mode decomposition and multi-layer classifier. Sensors **16**, 1887 (2016)
22. Huang, G., Zhu, Q., Siew, C.: Extreme learning machine: theory and applications. Neurocomputing **70**, 489–501 (2006)
23. Wan, S., Chen, L.: Fault diagnosis of high-voltage circuit breakers using mechanism action time and hybrid classifier. IEEE Access **99**, 146–157 (2019)
24. Dragomiretskiy, K., Zosso, D.: Variational mode decomposition. IEEE Trans. Signal Process. **62**, 531–544 (2014)
25. Lian, J., Liu, Z., Wang, H., Dong, X.: Adaptive variational mode decomposition method for signal processing based on mode characteristic. Mech. Syst. Signal Process. **107**, 53–77 (2018)

26. Ren, H., Liu, W., Shan, M., Wang, X.: A new wind turbine health condition monitoring method based on VMD-MPE and feature-based transfer learning. Measurement **148**, 106906 (2019)
27. Tian, Q., Flores, P., Lankarani, H.: A comprehensive survey of the analytical, numerical and experimental methodologies for dynamics of multibody mechanical systems with clearance or imperfect joints. Mech. Mach. Theory **122**, 1–57 (2018)

Wind Power Forecast Based on Multiple Echo States

Shaoxiong Zeng[1], Ruiqi Jiang[1], Zhou Wu[1(✉)], Xianming Ye[2],
and Zenghui Wang[3]

[1] School of Automation, Chongqing University, 174 Shazheng Street,
Chongqing 400044, Shapingba, China
{sxzeng,jiang_ruiqi,zhouwu}@cqu.edu.cn
[2] Department of Electrical, Electronic and Computer Engineering,
University of Pretoria, Pretoria 0002, South Africa
xianming.ye@up.ac.za
[3] School of Engineering, University of South Africa, Florida 1710, South Africa
wangz@unisa.ac.za

Abstract. Wind power is the most promising renewable energy for its
rich resources, low cost, and cleanliness. However, the intermittency of
wind power would put the safety of the power system at risk. An effective
method to solve this problem is making the power generation scheduling
through wind power forecast. Due to the volatility and complex temporal
dependence, it is a challenging task to predict wind power over multiple
time steps. In this paper, a novel method based on a chain echo state
network(CESN) is proposed to enhance mapping capability for multi-
step prediction. The multiple echo states of CESN is utilized to prevent
the error accumulation for multi-step prediction. Experimental results in
three cases demonstrate that the proposed method has promising performance on multi-step prediction and could prevent error accumulation
effectively.

Keywords: Echo state network · Multiple echo state · Multi-step
prediction · Wind power

1 Introduction

Greenhouse effect and environmental pollution are mainly caused by over consumption of exhaustible resources, such as oil, natural gas, and coal [1]. It is
necessary to develop alternative energy sources to alleviate the environmental
crisis. Renewable energy, such as wind, solar, and wave energy, are now considered viable alternatives to fossil energy. Wind power is one of the most promising
renewable energy technologies for its rich resources, low cost, and cleanliness [2].
In the year 2020, the global wind industry has grown up to 53% with more than
93GW wind facility installed [3], and the total amount of wind power exploration is continuously increasing in near future. As a result, wind power plays
an important role in the electricity supply.

© The Author(s), under exclusive license to Springer Nature Singapore Pte Ltd. 2022
H. Zhang et al. (Eds.): NCAA 2022, CCIS 1638, pp. 302–313, 2022.
https://doi.org/10.1007/978-981-19-6135-9_23

Due to the high stochasticity of wind, wind power is random, intermittent, and uncertain. These characteristics of wind power could cause an unsteady provision of electricity to power systems which puts the safety of the power system at risk. An effective method to solve this problem is making the power generation scheduling through wind power forecast [4]. When making the power generation schedules, more accurate information about the future could help entrepreneurs make a flexible commercial plan. Compared with the single-step forecast, the multi-step forecast could offer more information about the distant future for wind power planning.

There are two strategies for multi-step prediction including recursive strategy and Multi-Input Multi-Output (MIMO) strategy [5]. Single-step forecast models can only forecast one-step in one calculation. Thus, the recursive strategy is proposed for single-step models to make multi-step prediction by using the prediction value of the model as the input for next step prediction. Since the input set is composed of predicted values rather than actual observations, the result would suffer from accumulated errors. Given the input set, the MIMO model outputs a multi-step prediction in a single calculation to avoid accumulated errors of recursive strategy. This strategy has been used successfully for wind power multi-step forecast. Since the loss function of MIMO model is set according to multi-step prediction, the error accumulation of the recursive strategy does not occur. But it would cause an overall decrease in prediction accuracy. In recent years, various models have been proposed for wind power forecasts. Generally, models for wind power forecast are categorized into four types based on differences in modeling theory: physical models, statistical models, artificial intelligence-based models, and hybrid models [6].

The physical models, such as numerical weather prediction (NWP) [7] and weather researcher forecasting (WRF) [8], could use meteorological factors to realize multi-step prediction of wind speed. And a power curve model could be built using physical descriptions of particular wind farms. Then the multi-step wind power prediction is calculated by plugging wind speed prediction into the power curve model [9]. So the accuracy of wind power forecasting is constrained by the accuracy of NWP itself.

Compared with physical models, statistical models, including autoregressive (AR), autoregressive moving average (ARMA) [10], autoregressive integrated moving average (ARIMA) [11], and Gaussian regression model [12], are mostly used as single-step forecast model to achieve satisfactory accuracy. However, traditional statistical methods are only linear regression models and could not handle nonlinear problem effectively.

So to surpass the handicap many AI-based models have been widely used in wind power forecasting because of their nonlinear modeling abilities. There are some machine learning-based models, such as the support vector machine (SVM) [13], the least squares support vector machine (LSSTVM) [14], and the artificial neural networks (ANNs). The ANN of different structures have different models, such as the extreme learning machine (ELM) [15], the multi-layer perceptron (MLP) [16], the wavelet neural network (WNN) [17], the Elman network [18]. With the development of computer hardware, deep learning techniques have begun to have an impact on the field of wind power forecasts. Most commonly used techniques are auto-encoder [19], long-short-term memory (LSTM) [20, 21], recurrent neural network (RNN) [22], and the convolutional neural network (CNN) [23].

However, due to the intermittency of wind power, the aforementioned models may suffer from low prediction accuracy. So hybrid models are proposed to combine the advantages of different methods. Among the hybrid models, decomposition-based hybrid models are often used. Decomposing the time series into several sub-sequences could make the inputs of the forecast model reflect the wind power characteristics clearer [24]. Most commonly used decompose methods are Wavelet Decomposition (WD) [25,26], Wavelet Packet Decomposition (WPD) [27], Empirical Mode Decomposition (EMD) [28,29] and Variational Mode Decomposition (VMD) [15,30]. Unlike the EMD which is adaptive, VMD can determine the number of modal decomposition of the given sequence according to the actual situation, which is suitable for the frequency domain division of signals and the effective separation of each component.

So in this paper, we propose a new decomposition hybrid method based on a new echo state network (ESN) model called chain echo state network (CESN) to handle the aforementioned issues. First, wind power time series is pre-processed to handle the intermittency. Then, the processed time series are utilized to train the proposed CESN. The proposed CESN consists of multiple ESN modules connected to perform multi-step prediction. ESN is a simple and efficient recurrent neural network with three layers, including the input layer, hidden layer, and output layer. The hidden layer could map the temporal information of time series from low dimensional space to high dimensional state space. Thus, the mapping capabilities of ESN are stronger than other models. In addition, only the output weights need to be trained. With its simple training procedure and strong mapping capabilities, ESN has been successfully applied to nonlinear time series prediction. The key contributions of this paper can be categorized as follows.

1. A novel multi-step forecast model based on CESN is proposed to enhance the mapping capability with less time consumption. Compared with the original ESN, all neuron states generated by the previous step are utilized to prevent error accumulation.
2. Using three different wind power data sets, the prediction accuracy of the proposed model is evaluated by mean absolute error (MAE) and normalized root mean square error (nRMSE). The experimental results show that the proposed hybrid method is an effective method to handle the intermittency and prevent error accumulation for multi-step wind power forecast.

The rest of this paper is organized as follows. In Sect. 2, the concepts that constitute the building blocks of our proposed method are briefly reviewed. In Sect. 3, the proposed CESN is introduced. In Sect. 4, the performance metrics, detailed results from the designed scheme, and discussion are presented. At last the conclusion is provided in Sect. 5.

2   Preliminaries

2.1   Echo State Network

ESN is a simple and efficient approach for time series modeling due to the use of a distinctive recurrent network. It consists of an input layer with K input units, a reservoir layer with N internal units, and an output layer with L output units, as shown in Fig. 1. The reservoir is a dynamic system responsible for mapping the input sequence to fixed-size, high-dimensional state space. And the weights in the reservoir do not need to be trained because the weights in the reservoir are randomly initialized to a distribution that guarantees the echo state property. The only weight that needs to be trained is the weights of output.

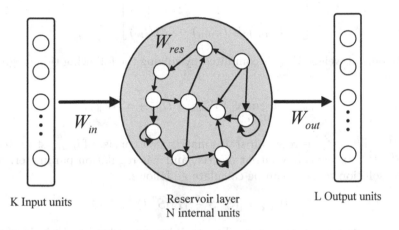

K Input units   Reservoir layer   L Output units
                       N internal units

Fig. 1. The framework of ESN

For a time series $e(1), \ldots, e(T) \in \mathbb{R}^K$, the training process of ESN can be briefly described as two steps: (1) to update the reservoir states based on the input, (2) to compute the output weight based on the reservoir. The state of the reservoir at each step t, which is denoted as $s(t)$, is updated according to

$$s(t) = (1 - \alpha)s(t - 1) + \alpha \left[\tanh\left(W_{in}e(t) + W_{res}s(t - 1)\right)\right], \qquad (1)$$

where $tanh()$ is the hyperbolic tangent function, $W_{in} \in \mathbb{R}^{N \times K}$ is the input-to-reservoir weight and $W_{res} \in \mathbb{R}^{N \times N}$ is the reservoir-to-reservoir weights, $\alpha \in [0, 1]$ is the leaky rate, which guarantee smooth change of reservoir status.

After the reservoir states are updated, it can be collected in a matrix S. The reservoir states matrix is denoted as:

$$S = \begin{bmatrix} s_1(1) & s_2(1) & \cdots & s_N(1) \\ s_1(2) & s_2(2) & \cdots & s_N(2) \\ \vdots & \vdots & \vdots & \vdots \\ s_1(l_{train}) & s_2(l_{train}) & \cdots & s_N(l_{train}) \end{bmatrix}_{l_{train} \times N}, \tag{2}$$

where l_{train} denote the length of training data set. The corresponding teacher signal vector matrix is collected as:

$$D = \begin{bmatrix} d_1(1) & d_2(1) & \cdots & d_L(1) \\ d_1(2) & d_2(2) & \cdots & d_L(2) \\ \vdots & \vdots & \vdots & \vdots \\ d_1(l_{train}) & d_2(l_{train}) & \cdots & d_L(l_{tr}) \end{bmatrix}_{l_{train} \times L}. \tag{3}$$

The output weight W_{out} is computed by solving the following ridge regression problem:

$$W_{out} = \arg\min_{W_{out}} \|SW_{out} - D\|_2^2 + \gamma \|W\|_2^2, \tag{4}$$

where $S \in \mathbb{R}^{l_{train} \times N}$ is reservoir states matrix that consist of l_{train} state vectors, $D \in \mathbb{R}^{l_{train} \times L}$ is expected output matrix, and γ is regulation parameter. The solution to Eq. 4 can be calculate as follows:

$$W_{out} = [(S^T S + \gamma I_d) S^T D]^T, \tag{5}$$

where I_d is the identity matrix. The training procedure of ESN is given in Algorithm 1.

Algorithm 1. ESN training procedure

Require: e: historical data; D: training output data
Ensure: W_{out}: output weight;
 1: Initialization: $\{W_{in}, W_{res}, \gamma, S(0)\}$
 2: **for** $t \leftarrow 1$ to l_{train}
 3:     $s(t) \leftarrow (1 - \alpha)s(t-1) + \alpha \left[\tanh\left(W_{in}e(t) + W_{res}s(t-1)\right)\right]$
 4: **end**
 5: $S \leftarrow \{s(1); s(2); \ldots; s(l_{train})\}$
 6: $W^{out} \leftarrow [(S^T S + \gamma I_d)^{-1} S^T D]^T$

---

3   Chain Echo State Network

When using ESN for multi-step forecast, the error is accumulated with temporal dependence changing over time. To improve the performance of ESN, We propose a new CESN to avoid error accumulation. Different from ESN, the CESN consists

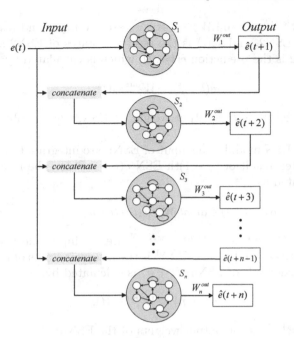

Fig. 2. The framework of CESN

of several ESN modules. The latter reservoir states are update by a input signal concatenated with last step prediction and original input. As shown in Fig. 2, ESN modules in the CESN are arranged in a chain. There are n ESN modules for n-step prediction, and each ESN module has K input units, N neuron in the reservoir layer, and 1 output unit. In the first ESN module, called ESN1, the reservoir states are updated by input $e(t)$ as:

$$s_1(t) = (1 - \alpha)s_1(t - 1) + \alpha \left[\tanh\left(W_1^{in}e(t) + W_1^{res}s(t - 1)\right)\right], \qquad (6)$$

where $W_1^{in} \in \mathbb{R}^{N \times K}$ and $W_1^{res} \in \mathbb{R}^{N \times N}$ are input and reservoir weights of ESN1 which are randomly initialized, and $s_1(t) \in \mathbb{R}^{N \times 1}$ denote reservoir states of ESN1 at time t. And the output weights are calculated by Eq. 5. So the forecast result is calculated by

$$\hat{e}(t + 1) = W_1^{out}s_1(t), \qquad (7)$$

where $W_1^{out} \in \mathbb{R}^{1 \times N}$ is output weights of ESN1.

For the second ESN module, which is called ESN2, the input is the integration of the current demand $e(t)$ and the output of ESN1 $\hat{e}(t + 1)$. So the reservoir states of ESN2 are updated as:

$$s_2(t) = (1 - \alpha)s_2(t - 1) + \alpha \left[\tanh\left(W_2^{in}(e(t); \hat{e}(t + 1)) + W_2^{res}s(t - 1)\right)\right] \qquad (8)$$

where $W_2^{in} \in \mathbb{R}^{N \times (K+1)}$ and $W_2^{res} \in \mathbb{R}^{N \times N}$ are the input and reservoir weights respectively of ESN2, and $s_2(t)$ is the reservoir states of ESN2 at time t. The output of ESN2 is the prediction $\hat{e}(t+2)$, which is calculated by:

$$\hat{e}(t+2) = W_2^{out} s_2(t), \tag{9}$$

where $W_2^{out} \in \mathbb{R}^{1 \times N}$ is the output weight of ESN2 that is obtained by ridge regression.

For the nth ESN module, the input of ESNn are integrated by the input $e(t)$ and the predicted result of $(n-1)$th ESN $\hat{e}(t+n-1)$. The reservoir states of ESNn are updated as:

$$s_n(t) = (1-\alpha)s_n(t-1) + \alpha tanh[W_n^{in}(e(t); \hat{e}(t+n-1)) + W_n^{res} s_n(t-1)], \tag{10}$$

where $W_n^{in} \in \mathbb{R}^{N \times (K+1)}$ and $W_n^{res} \in \mathbb{R}^{N \times N}$ are the input and reservoir weights of ESNn respectively, and $s_n(t) \in \mathbb{R}^{N \times 1}$ is the reservoir states of ESNn at time t. The predicted result of ESNn $\hat{e}(t+n)$ is calculated by:

$$\hat{e}(t+n) = W_n^{out} s_n(t), \tag{11}$$

where $W_n^{out} \in \mathbb{R}^{1 \times N}$ is the output weights of the ESNn.

The training procedure of CESN could be described in Algorithm 2, where $\hat{e}_l$ is the predicted output of lth ESN, $s_l(i)$ is the reservoir states of lth ESN modules at time i, and l_{train} is the length of the training set.

Algorithm 2. CESN training procedure

Require: e: historical data; O:expected output
Ensure: W^{out}: output weight;
 1: **for** $l \leftarrow 1$ to n
 2:     Initialization: $\{W_l^{in}, W_l^{res}, \alpha, \gamma, S_l(0)\}$
 3: **end**
 4: **for** $l \leftarrow 1$ to n
 5:     **for** $i \leftarrow 1$ to l_{train}
 6:         **if** $l = 1$
 7:             $s_l(i) = (1-\alpha)s_l(i-1) + \alpha \tanh(W_l^{in}e(i) + W_l^{res}s_l(i-1))$
 8:         **else**
 9:             $s_l(i) = (1-\alpha)s_l(i-1) + \alpha \tanh[W_l^{in}(e(i); \hat{e}_{l-1}(i)) + W_l^{res}s_l(i-1)]$
10:         **end**
11:     **end**
12:     $S_l \leftarrow \{s_l(1), s_l(2), \ldots, s_l(l_{tr})\}$
13:     $W_l^{out} \leftarrow [(S_l^T S_l + \gamma Id)^{-1} S_l^T O_l]^T$
14:     $\hat{e}_l \leftarrow (W_l^{out} S_l)$
15: **end**

---

4   Experimental Design and Discussion

4.1   Data Source and Performance Measures

In this section, the wind power dataset used to demonstrate the effectiveness of the proposed method was obtained from three different wind farms in Tanghe city, Henan Province, China. And the dataset contains wind power data at 10-minute intervals from October 2016 to September 2017.

In this study, two different indicators which are mean absolute error(MAE) and normalized root mean square error(nRMSE) are applied to evaluate wind power forecast error. These indicators can be formulated for step k as:

$$MAE_k = \frac{1}{T} \sum_{t=1}^{T} |e(t+k) - d(t+k)|, \tag{12}$$

$$nRMSE_k = \frac{\sqrt{\frac{1}{T} \sum_{t=1}^{T} (e(t+k) - d(t+k))^2}}{\sum_{t=1}^{T} e(t+k)/T} \times 100\% \tag{13}$$

where T is the length of testing horizon, $e(t+k)$ is the predicted value at step k, and $d(t+k)$ is the actual wind power at time $t+k$.

4.2   Comparison of CESN and Other Models

In this section, CESN is compared with ESN with recursive strategy, MIMO ANN, and MIMO LSTM to evaluate the accuracy and robustness of the proposed method. In this experiment, each training set with 4000 samples and testing set with 1000 samples are selected from three cases. The number of predicted steps is set to 10.

There are 10 ESN modules in CESN for 10-step prediction. And the parameters of each ESN are set to the same except the input size of the first module. The parameter in CESN are set as follows: the number of neurons in the reservoir is 90; the weights of the input layer are uniformly distributed in $[-0.1, 0.1]$; the spectral radius of reservoir weights is 0.8; the leaky decay rate is 0.8; the regulation parameter in ridge regression is 10^{-3}. The parameter of ESN with recursive strategy is the same as the first ESN module in CESN. The setting of ANN are as follows: the number of neurons in the hidden layer is 300; the output size is 10; the learning rate is 0.1 and the loss function is Mean Squared Error (MSE). The setting of LSTM are as follows: the number of neurons in the hidden layer is 50; the loss function is MAE and the learning rate is 0.01.

The average value of MAE and nRMSE of all steps for three different datasets are calculated and listed in Table 1. It can be seen that the CESN has higher accuracy with MAE in the range of 4.75–10.28 and nRMSE in the range of 6.03%-16.75%. The forecast error of CESN is much smaller than other models. The nRMSE of ANN varies in the range of 15.94%-30.69. The nRMSE of LSTM varies in the range of 15.24%-31.55%. So the average nRMSE of ANN and LSTM is similar. And the results of MAE and nRMSE of each step for three cases are shown in Figs. 3, 4, and 5. From the figures, it is obvious that the error of the proposed CESN increases slowly with the increase of steps. Comparing ESN with CESN, the prediction accuracy of ESN in the first step is about the same, but the error of the following multi-step prediction increases significantly from the second step. Apparently forecasting the future step based on the input that is prediction, the error would be accumulated during the recursive strategy. Although the error of ANN decreases or increase slowly in some steps, the error tends to increase and the overall prediction accuracy is low. The results illustrate that CESN has better performance than other models and could prevent the prediction error accumulation more effectively.

Table 1. Average statistical results with different forecast models.

Case	Model	MAE	nRMSE (%)	Training time (s)
I	CESN	**4.75**	**6.20**	**0.91**
	ESN	17.01	21.81	0.47
	ANN	13.23	15.94	128.19
	LSTM	11.56	17.10	1578.00
II	CESN	**4.40**	**6.03**	**0.90**
	ESN	35.29	36.77	0.47
	ANN	16.82	22.96	131.14
	LSTM	16.28	25.47	1601.04
III	CESN	**10.28**	**16.75**	**0.93**
	ESN	32.86	39.53	0.49
	ANN	22.73	30.69	134.99
	LSTM	16.28	31.55	1809.81

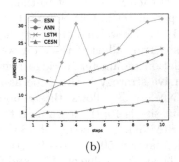

(a)   (b)

Fig. 3. Experimental results of each step in the term of (a) MAE and (b) nRMSE for case 1.

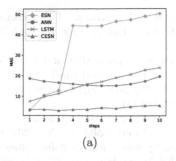

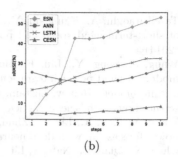

(a)   (b)

Fig. 4. Experimental results of each step in the term of (a) MAE and (b) nRMSE for case 2.

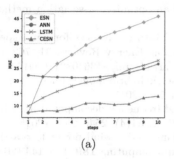

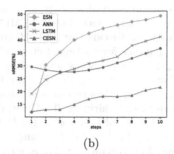

(a)   (b)

Fig. 5. Experimental results of each step in the term of (a) MAE and (b) nRMSE for case 3.

5   Conclusion

In this paper, a new wind power forecast method is proposed based on CESN. Relying on the powerful mapping capability, CESN has been successfully applied to the multi-step prediction. Experimental results show that the proposed CESN has the lowest forecast error. It demonstrates that CESN could have significant performance improvement and prevent the prediction error accumulation effectively.

Acknowledgements. This paper is supported by the National Key Research and Development Program of China (2021YFF0500903), and the National Natural Science Foundation of China (52178271).

References

1. Mikhaylov, A., Moiseev, N., Aleshin, K., Burkhardt, T.: Global climate change and greenhouse effect. Entrepreneurship Sustain. Issues **7**(4), 2897 (2020)
2. Dai, J., Yang, X., Wen, L.: Development of wind power industry in china: a comprehensive assessment. Renew. Sustain. Energy Rev. **97**, 156–164 (2018)
3. Council, G.W.E.: Gwec global wind report. Global Wind Energy Council (GWEC), Brussels, Belgium (2021)

4. Tascikaraoglu, A., Uzunoglu, M.: A review of combined approaches for prediction of short-term wind speed and power. Renew. Sustain. Energy Rev. **34**, 243–254 (2014)
5. Wang, J., Song, Y., Liu, F., Hou, R.: Analysis and application of forecasting models in wind power integration: A review of multi-step-ahead wind speed forecasting models. Renew. Sustain. Energy Rev. **60**, 960–981 (2016). https://www.sciencedirect.com/science/article/pii/S1364032116001441
6. Wang, Y., Zou, R., Liu, F., Zhang, L., Liu, Q.: A review of wind speed and wind power forecasting with deep neural networks. Appl. Energy **304**, 117766 (2021)
7. Chen, N., Qian, Z., Nabney, I.T., Meng, X.: Wind power forecasts using gaussian processes and numerical weather prediction. IEEE Trans. Power Syst. **29**(2), 656–665 (2014)
8. Zhao, J., Guo, Y., Xiao, X., Wang, J., Chi, D., Guo, Z.: Multi-step wind speed and power forecasts based on a wrf simulation and an optimized association method. Appl. Energy **197**, 183–202 (2017)
9. Jung, J., Broadwater, R.P.: Current status and future advances for wind speed and power forecasting. Renewable and Sustainable Energy Reviews 31, 762–777 (2014). https://www.sciencedirect.com/science/article/pii/S1364032116001441
10. Han, Q., Meng, F., Hu, T., Chu, F.: Non-parametric hybrid models for wind speed forecasting. Energy Convers. Manage. **148**, 554–568 (2017)
11. Yunus, K., Thiringer, T., Chen, P.: Arima-based frequency-decomposed modeling of wind speed time series. IEEE Trans. Power Syst. **31**(4), 2546–2556 (2015)
12. Yan, J., Li, K., Bai, E., Yang, Z., Foley, A.: Time series wind power forecasting based on variant gaussian process and tlbo. Neurocomputing **189**, 135–144 (2016)
13. Liu, A., Xue, Y., Hu, J., Liu, L.: Ultra-short-term wind power forecasting based on SVM optimized by GA. Power Syst. Protect. Contr. **43**(2), 90–95 (2015)
14. Zhang, Y., Wang, P., Ni, T., Cheng, P., Lei, S.: Wind power prediction based on LS-SVM model with error correction. Adv. Electr. Comput. Eng. **17**(1), 3–8 (2017)
15. Abdoos, A.A.: A new intelligent method based on combination of VMD and elm for short term wind power forecasting. Neurocomputing **203**, 111–120 (2016)
16. Yeh, W.C., Yeh, Y.M., Chang, P.C., Ke, Y.C., Chung, V.: Forecasting wind power in the mai liao wind farm based on the multi-layer perceptron artificial neural network model with improved simplified swarm optimization. Int. J. Electr. Power Energy Syst. **55**, 741–748 (2014)
17. Wang, J., Yang, W., Du, P., Niu, T.: A novel hybrid forecasting system of wind speed based on a newly developed multi-objective sine cosine algorithm. Energy Convers. Manage. **163**, 134–150 (2018)
18. Liu, H., Tian, H., Liang, X., Li, Y.: Wind speed forecasting approach using secondary decomposition algorithm and elman neural networks. Appl. Energy **157**, 183–194 (2015)
19. Mezaache, H., Bouzgou, H.: Auto-encoder with neural networks for wind speed forecasting. In: 2018 International Conference on Communications and Electrical Engineering (ICCEE), pp. 1–5. IEEE (2018)
20. Han, L., Jing, H., Zhang, R., Gao, Z.: Wind power forecast based on improved long short term memory network. Energy **189**, 116300 (2019)
21. Shahid, F., Zameer, A., Muneeb, M.: A novel genetic LSTM model for wind power forecast. Energy **223**, 120069 (2021)
22. Kisvari, A., Lin, Z., Liu, X.: Wind power forecasting-a data-driven method along with gated recurrent neural network. Renew. Energy **163**, 1895–1909 (2021)
23. Hong, Y.Y., Rioflorido, C.L.P.P.: A hybrid deep learning-based neural network for 24-h ahead wind power forecasting. Appl. Energy **250**, 530–539 (2019)

24. Qian, Z., Pei, Y., Zareipour, H., Chen, N.: A review and discussion of decomposition-based hybrid models for wind energy forecasting applications. Appl. Energy **235**, 939–953 (2019)
25. Catalão, J.P.S., Pousinho, H.M.I., Mendes, V.M.F.: Short-term wind power forecasting in Portugal by neural networks and wavelet transform. Renewable Energy **36**(4), 1245–1251 (2011)
26. Wang, D., Luo, H., Grunder, O., Lin, Y.: Multi-step ahead wind speed forecasting using an improved wavelet neural network combining variational mode decomposition and phase space reconstruction. Renew. Energy **113**, 1345–1358 (2017)
27. Meng, A., Ge, J., Yin, H., Chen, S.: Wind speed forecasting based on wavelet packet decomposition and artificial neural networks trained by crisscross optimization algorithm. Energy Convers. Manage. **114**, 75–88 (2016)
28. Liu, M.D., Ding, L., Bai, Y.L.: Application of hybrid model based on empirical mode decomposition, novel recurrent neural networks and the arima to wind speed prediction. Energy Convers. Manage. **233**, 113917 (2021)
29. Guo, Z., Zhao, W., Lu, H., Wang, J.: Multi-step forecasting for wind speed using a modified EMD-based artificial neural network model. Renew. Energy **37**(1), 241–249 (2012)
30. Zhang, Y., Liu, K., Qin, L., An, X.: Deterministic and probabilistic interval prediction for short-term wind power generation based on variational mode decomposition and machine learning methods. Energy Convers. Manage. **112**, 208–219 (2016)

UAV-Assisted Blind Area Pedestrian Detection via Terminal-Edge-Cloud Cooperation in VANETs

Qisen Zhang[1], Kai Liu[1], Luyao Jiang[1], Chengliang Zhong[1], Liang Feng[1],
Hao Zhang[2], and Ke Xiao[3(✉)]

[1] College of Computer Science, Chongqing University, Chongqing 400040, China
{qszhang,liukai0807,looyo,Zchengliang,liangf}@cqu.edu.cn
[2] College of Computer Science and Technology, Chongqing University of Posts
and Telecommunications, Chongqing 400065, China
zhanghao@cqupt.edu.cn
[3] College of Computer and Information Science, Chongqing Normal University,
Chongqing 401331, People's Republic of China
xiaoke@cqnu.edu.cn

Abstract. The collaboration of UAV (Unmanned Aerial Vehicle) and
VANETs (Vehicular Ad-hoc Networks) on enabling future intelligent
transportation systems has received great attention in both academia
and industries. In this paper, we make the first effort on presenting a
terminal-edge-cloud cooperation architecture for UAV assisted blind area
pedestrian detection in VANETs. On this basis, a task offloading algorithm is proposed, which offloads the computation task adaptively based
on heterogeneous computation, communication and storage capacities
of UAVs, vehicles, roadside infrastructures and the cloud server, aiming
at minimizing the total service delay by striking a best balance between
task computation delay and data transmission delay. Then, we implement
the system prototype and conduct hardware-in-the-loop experiments to
verify the effectiveness of the proposed algorithm. Finally, we test the
system performance in realistic VANETs environments and demonstrate
the feasibility of the proposed architecture and solution.

Keywords: Vehicle ad hoc network · Unmanned aerial vehicle · Task
offloading · Pedestrian detection · Terminal-edge-cloud cooperation

1 Introduction

Recent advances in wireless communication and intelligent networking technologies drive the development of vehicular ad hoc networks (VANETs), which forms
the basis of emerging intelligent transportation systems (ITSs) [1–4]. On the
other hand, with the prevalence of unmanned aerial vehicles (UAVs) based applications, such as environment inspection, intelligent surveillance, wireless communication, aerial photography, it is vision to form a brand new paradigm, in

which VANETs and UAVs are cooperated to realize innovative and powerful ITSs [5].

There have been a great number of studies on integrating UAV with VANETs for various purposes, including data forwarding [6–8], traffic monitoring [9–11], computation offloading [11–13] and trajectory optimization [14–16]. In data forwarding scenarios, UAVs are utilized as the relaying node, where the issues of data caching and trajectory planning are studied to maximize the network throughput. In traffic monitoring scenarios, UAVs make full use of mobility to monitor real-time conditions of the traffic network and report anomalies to the control center. In computation offloading scenarios, researches focus on reducing computational processing latency and saving UAV energy consumption by considering the cost of communication, storage and computing resource allocation between UAVs and VANETs. In trajectory optimization scenarios, the studies are dedicated to maximizing system throughput and reducing service delay when integrating UAVs with VANETs by planning the trajectory of UAVs.

Unlike previous studies, in this work, we consider the scenario where UAVs are cooperated with VANETs for pedestrian detection in blind areas of vehicles to avoid sudden encounter of pedestrians and potential collision via V2X (vehicle-to-everything) communication and terminal-edge-cloud cooperated task offloading. Although there have been great number of studies on terminal-edge-cloud cooperation and task offloading in VANETs [17–20], this work considers a unique application scenario with new characteristics. Specifically, a UAV with computation capability and communication interfaces is hovering in the air to monitor possible blind areas of driving vehicles, which is able to detect pedestrians based on a certain object detection model such as YOLO [21]. Meanwhile, the UAV can communicate with vehicles, roadside infrastructures and the cloud server via V2V (vehicle-to-vehicle), V2I (vehicle-to-infrastructure) and V2C (vehicle-to-cloud) communications, respectively. Accordingly, the UAV may offload the object detection task to the moving vehicles, the static edge node deployed at the roadside, as well as the cloud server by transmitting the monitored video frames to the corresponding node. Note that the UAVs, vehicles, edge servers and the cloud server have heterogeneous computation and communication capacities. On the other hand, the concerned application requires stringent time latency on pedestrian detection and warning to avoid potential accident. Therefore, it is expected to strike the best balance on task offloading so as to minimize the delay of pedestrian detection and warning.

With above motivation, this work first presents a terminal-edge-cloud cooperative architecture to support the UAV-assisted blind area pedestrian detection in VANETs. Then, an adaptive task offloading algorithm is proposed to best exploit the heterogeneous communication and computation capacities of UAVs, vehicles, edge nodes and the cloud server to minimize the system delay. Finally, we implement both the hardware-in-the-loop test platform and a realistic system prototype to validate the feasibility of the proposed solutions. The main contributions of this work are outlined as follows.

- We present a novel terminal-edge-cloud cooperative architecture for UAV-assisted blind area pedestrian detection in VANETs, which consists of terminal nodes (i.e., UAVs), mobile edge nodes (i.e., vehicles), static edge nodes (i.e., roadside computation units) and the cloud server. The UAV monitors the blind area and the pedestrian detection model such as YOLO can be deployed in terminal, mobile edge, static edge and cloud nodes with heterogeneous computation and communication capabilities. Depending on a certain task offloading algorithm, the UAV transmits video frames to the corresponding node for object detection via V2X communication, and alarm messages will be sent to the vehicle if potential pedestrian collision risks are detected.
- We propose an adaptive task offloading algorithm, which decides the node where the computation task is processed based on particular object detection models and real-time computation and communication resources available at the nodes. In particular, we model the task delay by analyzing the computation, communication and waiting delays for different offloading options. Then, two policies, namely delay driven and resource driven, are designed, which enable the adaptive task offloading.
- We conduct comprehensive performance evaluations via both hardware-in-the-loop testing and case study in realistic VANETs. Specifically, we implement two object detection models for pedestrian detection. Then, we set up a hardware-in-the-loop testbed, where Raspberry Pi, i5-4210H CPU RAM, Jetson Xavier NX and NVIDIA RTX3090 are adopted to emulate the computation of terminal, mobile edge node, static edge node and the cloud node, respectively. The Wi-Fi, DSRC (dedicated short range communication) and 4G communication interfaces are adopted to emulate V2X communications. The experimental results demonstrate the effectiveness of the proposed algorithm. Finally, we implement the system prototype in realistic VANETs environments, which further verifies the feasibility of the proposed solutions.

The remaining of this paper is organized as follows. Section 2 reviews the related work. Section 3 presents the system architecture. Section 4 proposes the algorithm. Section 5 implements system prototype and gives performance evaluation. Finally, Sect. 6 summarizes this work and discusses future research directions.

2   Related Work

In order to improve the connectivity of VANETs, a lot of researchers focused on multi-hop data relay using UAVs. Almasoud [6] considered a multi-Unmanned Aerial Vehicles (UAVs) system to convey collected data from isolated fields to the base station. In each field, a group of sensors or Internet of Things devices are distributed and send their data to one UAV. The UAVs collaborate in forwarding the collected data to the base station in order to maximize the minimum battery level for all UAVs by the end of the service time. Cheng et al. [7] proposed a method that leverages a group of UAVs or a single UAV as the relay node, which can collect information from the source node through UAVs, improve the network throughput, carry and forward information to the target node. Zhou et al. [8] formulated an aerial-ground cooperative vehicular networking architecture where multiple UAVs can form an air-to-ground network through air-to-air and air-to-ground communication. UAVs can be dispatched to areas of interest to collect information, and transmit auxiliary information to ground vehicles. Moreover, UAVs can also act as intermediate relays due to their flexible mobility when network partitions happen in the ground vehicular subnetwork.

Due to the distributed sensing and monitoring requirements in VANETs, a great number of studies focused on UAV-assisted information sensing and traffic monitoring. Motlagh et al. [9] leveraged UAV as the air traffic monitoring system to realize the full coverage of the network and reduce the congestion of ground infrastructures. Xu et al. [10] proposed a collision early warning (TCCW) algorithm based on track calibration and a customized communication protocol to realize the early traffic collision warning. Liu et al. [11] proposed a two-layer vehicular fog computing (VFC) architecture to discuss the synergistic effect of cloud node, static fog node and mobile fog node in VANETs, and further realized a prototype of traffic anomaly detection and warning system.

UAVs can be also integrated with VANETs in terms of task offloading to best exploit the heterogeneous communication and computation resources. In consideration of IMDs' limited battery capacity and UAV energy budget, Guo [12] studied the energy saving problem in UAV-enhanced edge by smartly making offloading decisions, allocating transmitted bits in both uplink and downlink. Liu et al. [11] built an offloading model with the goal of maximizing the completion rate of time-critical tasks. On this basis, an adaptive task offloading algorithm (ATOA) is proposed to adaptively categorizes all tasks into four types of pending lists, and then tasks in each list will be cooperatively offloaded to different nodes based on their features. Lyu et al. [13] proposed a new hybrid network architecture that leverages UAVs as an aerial mobile base station, which flies cyclically along the cell edge to offload data traffic for cell-edge mobile terminals.

There have been extensive studies on UAV trajectory planning for enhancing data transmission in VANETs. Fadlullah [15] proposed the optimized running trajectory algorithm, which enhances the system throughput of UAV network and reduces the data transmission delay which aiming at maximizing the overall network throughput, Wu et al. [16] formulated a joint caching and trajectory optimization (JCTO) problem to make decisions on content placement, content delivery, and UAV trajectory simultaneously. [14] focused on the former paradigm and studied a new UAV trajectory design problem subject to practical communication connectivity constraints with the GBSs, aiming to minimize the UAV's mission completion time by optimizing its trajectory, subject to a quality-of-connectivity constraint of the GBS-UAV link specified by a minimum receive signal-to-noise ratio target, which needs to be satisfied throughout its mission.

Distinguished from previous research work, this paper is devoted to presenting a terminal-edge-cloud cooperation framework for UAV-assisted pedestrian detection in blind area of vehicles, and proposing a tailored task offloading algorithm to strike the balance among UAVs, vehicles, roadside infrastructures and the cloud on task processing, so as to optimize system performance adaptively based on heterogeneous communication and computation resources.

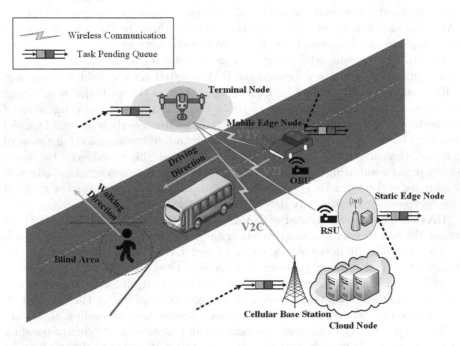

Fig. 1. Terminal-edge-cloud cooperative architecture for UAV-assisted blind area pedestrian detection in VANETs

3   System Architecture

As shown in Fig. 1, the presented terminal-edge-cloud cooperative architecture it is composed of four basic elements, including the UAV terminal, the vehicles acting as mobile edge node, the roadside units (RSU) acting as static edge node and the cloud server. Each element equips with certain computation and communication capacities, so that the UAV terminal is able to communicate with vehicles, RSUs and the cloud via V2X communication. With such an architecture, the general application scenario is described as follows. The UAV is deployed in the air, which can monitor blind areas of driving vehicles, such as the pedestrian shown in the circle. The monitored video frames can be processed based on a certain object detection model such as YOLO. The model can be deployed on UAV, vehicles, RSUs as well as the cloud for pedestrian detection. Based on a certain offloading algorithm, the UAV will transmit the video frames to the corresponding node for task computation via V2X communication. Finally, if potential pedestrian collision is detected, the warning message will be transmitted to the driving vehicle by the task computation node. The features of the four elements are elaborated below.

First, the terminal node (TN) is deployed in the air and monitors potential blind areas of driving vehicles. It is equipped with Wi-Fi, DSRC and 4G communication interfaces, which enables the V2V, V2I and V2C communications of the terminal. Meanwhile, the TN is equipped with certain computation capacity, which is typically the lowest among other nodes due to power and carry load limitation of UAVs. Second, the vehicles act as the mobile edge node (MEN), which equips with 4G and DSRC communication interfaces to enable V2V, V2I and V2C communications. The computation unit in MEN is typically higher than that in TNs due to more powerful energy supply of vehicles. Third, the static edge node (SEN) refers to the roadside infrastructure equipped with communication and computation units, such as RSUs. Typically, SEN can communicate with both TNs and MENs, and has greater computation capacities that both TNs and MENs than them. Finally, the cloud node (CN) refers to the cloud server connected through the cellular network such as 4G, which owns the strongest computation capacity. Nevertheless, due to the remote the cloud server and the limited cellular bandwidth, it usually suffers the highest data transmission delay.

With above-mentioned architecture, the service procedure and the characteristic of the system are described as follows. As shown in Fig. 1, the UAV is hovering on the air, monitoring the bind area of vehicle B. The pedestrian is about to cross the street, but cannot be observed by vehicle B due to the block of a parked bus A. In such a case, it is expected that the pedestrian can be detected automatically via the video frames of UAV and the warning messages would be sent to car B in time. The system flows are summarized below. 1) The UAV notifies the RSU that the blind area pedestrian service is on with the signaling message transmitted by V2I communication. 2) The RSU determines the node to which the task will be offloaded based on certain scheduling algorithm with the information of task requirement, as well as available computation resources and communication bandwidth of heterogeneous nodes. 3) The RSU notifies the UAV the scheduling result, and the UAV transmit its monitored video frames to the corresponding node via V2X communication. 4) The corresponding node executes the pedestrian detection task based on deployed models and further sends the warning message to the vehicle once a potential collision is detected.

Note that to maximize the utilization of heterogeneous computation and communication resources and minimize service delay in such a system architecture, it is imperative to adaptively offload the tasks to the TN, SEN, MEN and CN. For instance, if the computation task was processed locally by the TN, there is no extra overhead of data transmission. Nevertheless, there might be excessively long waiting time of task processing due to the limited computation capacity of UAVs. In contrast, if simply offloading all the tasks to the CN, the computation time could be minimized due to the powerful computation capacities of the cloud server, but the limited V2C communication bandwidth might become the bottleneck of the system, resulting in unexpected long delay of the system. In view of this, in the following of this work, we model the system delay by integratedly considering heterogeneous computation and communication capacities of the nodes and propose an adaptive task offloading algorithm to enhance system performance.

4   Proposed Algorithm

4.1   System Delay Model

Denote $S = \{s_1, s_2, ..., s_{|S|}\}$, $M = \{m_1, m_2, ..., m_{|M|}\}$, $U = \{u_1, u_2, ..., u_{|U|}\}$, $\{c\}$ as   the   set   of   static   edge   node, mobile edge node, terminal node and the cloud server, respectively. Denote $N = \{s_1, s_2, ..., s_{|S|}, m_1, m_2, ..., m_{|M|}, u_1, u_2, ..., u_{|U|}, c\}$ as the set of offloading nodes. The set of tasks sensed by a UAV is denoted by $W_j = \{w_{j1}, w_{j2}, ..., w_{j|W_j|}\}$. Each task can be represented by a 5-tuple $<\theta_{ji}, c_{ji}, \mu_{ji}, \vartheta_{ji}, \gamma_{ji}>$, which represent the input data size, the number of required CPU cycles, accuracy requirements, the generation time and the deadline, respectively. Hence, task w_{ji} must be completed between $[\vartheta_{ji}, \gamma_{ji}]$. Otherwise it is failed.

Each offloading node $n \in N$ is associated with five tuples $<C_n, B_n, \rho_n, S_n, R_n>$ which represent the radius of communication coverage, the total wireless bandwidth, the number of channels, the maximum storage capability, and the computation capability (i.e., the number of CPU cycles per unit time), respectively. We denote $con_{u,n,t} = 1$ means $u \in U$ can exchange messages with $n \in N$ within one-hop communication at time t. In addition, the distance between b and n is less than C_n. Otherwise $con_{b,n,t} = 0$. Note that $\{con_{u,u,t} = con_{u,c,t} = 1|0 < u < |U|, \forall t \in T\}$, which tasks in the system can be executed locally in the UAV, the UAV can always communicate with the cloud node.

Considering different neural network models can be applied for task computation with different time cost, we denote $H = \{h_1, h_2, ..., h_{|H|}\}$ to represent the weight of neural network models. Given task w_{ji} which is offloaded to node n using the neural network model h_k is utilized for inference, a binary variable x_{j,i,n,h_k} is defined as follows

$$x_{j,i,n,h_k} = \begin{cases} 1, \text{ task } w_{ji} \text{ is offloaded from } u_j \text{ to } n, \text{ using model } h_k \\ 0, \text{ otherwise} \end{cases} \quad (1)$$

with above definitions, we model the transmission delay, computation delay and waiting delay of the system, and then derive the task delay.

Transmission Delay: The task w_{ji} is transmitted to the corresponding offloading node at time t_{ji}, and its offloading period is $[t_{ji}, t_{ji} + T]$. A successful transmission of w_{ji} from u_j to n should be satisfied the following conditions: 1) u_j and m_j should keep connection with n during data transmission; 2) the input data θ_{ji} should not exceed current available storage resources of n; 3) the number of tasks offloaded to same offloading node should be less than the number of channels; 4) the total transmission delay should not exceed the task deadline. Denote $t_{b,j,i,n}^{trans}$ as the transmission delay from the node b to the offloading node n, which depends on the data size (θ_{ji}), the total bandwidth (B_n) and channels (ρ_n) of n, and is computed by $t_{b,j,i,n}^{\text{trans}} = \frac{\theta_{ji}}{B_n/\rho_n}$.

When task is executed in the cloud, given the bandwidth τ of the core network, the transmission delay is represented by $t_{b,j,i,c}^{trans} = \frac{\theta_{ji}}{\tau} + \frac{\theta_{ji}}{B_n/\rho_n}$.

When the offloading node returns the computation result to the vehicle, the total transmission delay t_{trans}^{total} can be computed by: $t_{\text{trans}}^{\text{total}} = t_{u,j,i,n}^{\text{trans}} + t_{n,j,i,m}^{\text{trans}}$.

Computing Delay: The computation capability of a node depends on the number of required CPU cycles, the number of CPU cycles per unit time and the neural network model selection, which is denoted by $t_{j,i,n}^{comp}$, formulated as follows $t_{j,i,n}^{\text{comp}} = \frac{c_{ji}}{R_n/h}$.

Waiting Delay: It represents the time that a task waits on the offloading node until it begins to be processed. Specifically, when a task is offloaded to n, its waiting time is the sum of the computation time of all the tasks in pending ahead which is: $t_{j,i,n}^{\text{wait}} = \frac{\sum_{l=1}^{|U|}\sum_{k=1}^{|W_l|} x_{l,k,n}^{before_{ji}} \cdot c_{l,k}}{R_n}$.

Which $x_{l,k,n}^{before_{ji}}$ represents the tasks processed before w_{ji} in the pending list. In general, cloud nodes have sufficient computing capacity and can support parallel computing, and thus we consider $t_{j,i,c}^{wait} = 0$.

Finally, the task delay for offloading w_{ji} to node n denoted by $t_{j,i,n}^{task}$, which is the sum of the total transmission delay, computing delay and waiting delay, which can be obtained by:

$$t_{j,i,n}^{\text{task}} = \sum_{n \in N} x_{j,i,n,h_k} \cdot \left(t_{\text{trans}}^{\text{total}} + t_{j,i,n}^{\text{comp}} + t_{j,i,n}^{\text{wait}} \right) \tag{2}$$

The objective is to minimize the average task delay by determining the offloading strategy:

$$\arg\min_{n,h_k} \frac{\sum_{j=1}^{|U|}\sum_{i=1}^{|W_j|}\sum_{n \in N} x_{j,i,n,h_k} \cdot t_{j,i,n}^{task}}{\sum_{j=1}^{|U|}|W_j|}$$

$$s.t.$$

$$C1 : x_{j,i,n,h_k} \in \{0,1\}$$

$$C2 : \sum_{n \in N} x_{j,i,n,h_k} = 1$$

$$C3 : \sum_{j=1}\sum_{i=1} \forall x_{j,i,n,h_k} \cdot \theta_{ji} \leq S_n, j \in |U|, i \in |W_j| \tag{3}$$

$$C4 : \sum_{j=1}\sum_{i=1} \forall x_{j,i,n,h_k} \leq \rho_n, j \in |U|, i \in |W_j|$$

$$C5 : \exists h \geq \mu_{ji}, h \in |H|$$

$$C6 : t_{j,i} + t_{j,i,n}^{\text{total}} \leq \gamma_{ji}$$

Algorithm 1: Delay driven policy update process

Input: Finite decision sets $\{\bar{x}_{j,i,n,h_k} | \forall n \in N, \forall h_k \in H\}$
Output: Final offloading scheme $\{x_{j,i,b,f} | b \in N, f \in H\}$

1 **for** $n \leftarrow 1$ **to** $|N|$ **do**
2   **for** $h_k \leftarrow 1$ **to** $|H|$ **do**
3        $t_{j,i,n}^{task} = \sum_{n \in N} x_{j,i,n,h_k} \cdot \left(t_{trans}^{total} + t_{j,i,n}^{comp} + t_{j,i,n}^{wait} \right);$
4   **if** $t_{cur} > t_{j,i,n}^{task}$ **then**
5           $x_{j,i,b,f} = x_{j,i,n,h_k};$
6           $t_{cur} = t_{j,i,n}^{task};$

7 **return** $\{x_{j,i,b,f} | b \in N, f \in H\}$

---

where C1 and C2 imply that each task must be offloaded and computed to one node; C3 represents the storage constraint; C4 represents that the number of tasks transmitted at the same time cannot exceed the channels of the offloading nodes; C5 means that the selected neural network model meets the accuracy requirement; C6 means that task should be completed before the deadline.

4.2   Adaptive Task Offloading Algorithm

In this part, we design an adaptive task offloading algorithm. The general idea is described as follows. First, the system estimates the task delay based on the current applied offloading policy. If it exceeded a predefined threshold, a greedy method is adopted to search a new offloading policy. There exist two policies, namely, a delay driven policy, aiming at minimizing overall task delay of the system, and a resource driven policy, aiming at maximizing resource utilization. Details are elaborated below.

Delay Driven Policy: Given a certain offloading policy, denote the task delay as t_{cur}. When a new task with data θ_{ji} and computing requirement c_{ji} arrives, we estimate the offloading delay t_{aver} based on the delay model derived above. When the difference between t_{cur} and t_{aver} exceeds a predefined threshold t_{diff}, the algorithm starts to search a new offloading policy. Specifically, it traverses each node n and each neural network model, and compute the total task delay $t_{j,i,n}^{task}$. Then, the policy with the minimum $t_{j,i,n}^{task}$ is chosen. The pseudo-code of such a policy is shown as follows.

Resource Driven Policy: Considering maximizing resource utilization, taking computing resource availability as an example. Specifically, given a certain offloading policy, denote the computing delay as t_{cur}^{comp}, when the R_n of the nodes in the system increases, the algorithm starts to search a new offloading policy. The specific update strategy is similar to the task driven delay strategy. The pseudo-code of such a policy is shown as follows.

Algorithm 2: Resource driven policy update process

Input: Finite decision sets $\{\bar{x}_{j,i,n,h_k} | \forall n \in N, \forall h_k \in H\}$
Output: Final offloading scheme $\{x_{j,i,b,f} | b \in N, f \in H\}$
1 **for** $n \leftarrow 1$ **to** $|N|$ **do**
2   **for** $h_k \leftarrow 1$ **to** $|H|$ **do**
3      $t_{j,i,n}^{comp} = \frac{c_{ji}}{R_n / h}$;
4   **if** $t_{cur}^{comp} > t_{j,i,n}^{comp}$ **then**
5        $x_{j,i,b,f} = x_{j,i,n,h_k}$;
6        $t_{cur}^{comp} = t_{j,i,n}^{comp}$;
7 **return** $\{x_{j,i,b,f} | b \in N, f \in H\}$

(a)   (b)   (c)

Fig. 2. Positive sample of data set

5 Performance Evaluation

5.1 Pedestrian Detection Model Implementation

First, we introduce the implementation of the pedestrian detection model, which
forms the basis of task offloading concerned in this work. In particular, YOLOv4
[21] and SSD [22] are trained for pedestrian detection, where YOLOv4 is a
relatively larger scale network compared with SSD, requiring higher computation overhead. In contrast, SSD is a one-stage detection algorithm, which syn-
chronously implements object detection and classification. The lightweight and

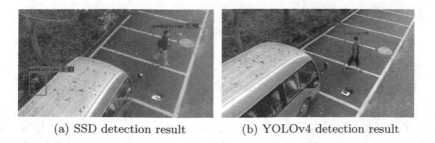

(a) SSD detection result   (b) YOLOv4 detection result

Fig. 3. Verification of the pedestrian detection models

fast inference characteristics of SSD make it more suitable for mobile device. Since UAV is adopted to monitor and detect pedestrian, we adopt the training data set from CUHK Square Data Set [23]. Where the videos are taken from the campus of Chinese University of Hong Kong, and the training data set with pedestrian information is shown in Fig. 2. 2000 frames are randomly selected from the video, where 90% are used for training and the rest 10% are used for model verification. For YOLOv4, setting batch size to 32, Average Recall expectation tends to 1, steps to 8000. For SSD, setting batch size to 24, and IOU threshold to 0.98, steps to 5000.

We have tested the two models in the campus of Chongqing University. The video frames are collected from the UAV, and the scenario is utilized to simulate potential blind areas of vehicles as shown in Fig. 3. The detection results of SSD and YOLOv4 are shown in Fig. 3(a) and Fig. 3, respectively. We have collected 200 testing frames, and the precision for YOLOv4 and SSD are 95% and 90%. Given the hardware settings of Intel(R) i5-4210H CPU, the average inference time of SSD and YOLOv4 are 150 ms and 400 ms, respectively.

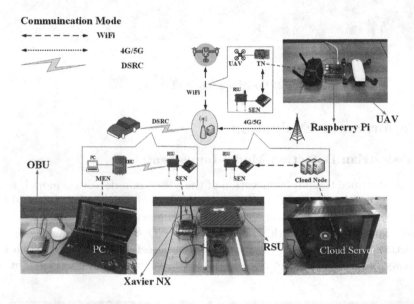

Fig. 4. Hardware-in-the-loop Testbed

5.2   Hardware-in-the-Loop Testbed

In order to demonstrate the effectiveness of terminal-edge-cloud cooperative architecture for UAV-assisted blind area pedestrian detection in VANETs, and the superiority of the proposed algorithm, we have implemented a hardware-in-the-loop testbed. As shown in Fig. 4, the terminal node is implemented by

DJI Spark UAV with Raspberry Pi 4B+. The static edge node consists of the computation unit of Jet- son Xavier NX and the communication unit of MK5 RSU. For the mobile edge node, the hardware settings for the computation unit include Intel(R) Core(TM) 2.90 GHz i5-4210H CPU and RAM 8 GB, while the MK5 OBU is adopted as the communication interface. Finally, the cloud node is equipped with NVIDIA RTX3090 with cellular network communication interface. In this way, the V2V and V2I communications are enabled by DSRC and the V2C communication is enabled by 4G cellular network.

In addition, the videos taken from realistic outdoor scenarios by UAV are adopted as the input. To evaluate the task delay, all the devices have been synchronized during the testing. For instance, when the task is offloaded to SEN, recording the time stamp t_1 when the UAV transmits the video frame, and the timestamp t_2 when the frame is received by SEN. Then, the waiting time t_{wait} and the execution time at the SEN t_{detect} are collected, followed by recording the timestamp t_3 when SEN starts to transmit the result to the vehicle, and the timestamp t_4 when the vehicle receives the frame. Accordingly, the transmission delay t_{trans} is $(t_2-t_1)+(t_4-t_3)$, and the computing delay t_{comp} is $(t_{detect}+t_{wait})$. Finally, the task delay can be obtained by $t_{comp} + t_{trans}$.

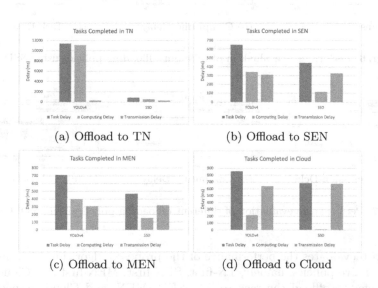

(a) Offload to TN   (b) Offload to SEN

(c) Offload to MEN   (d) Offload to Cloud

Fig. 5. Delay comparison by offloading SSD and YOLOv4 to different nodes

5.3 Experimental Results

Figure 5 compares the system performance by offloading the two models (i.e. SSD and YOLOv4) to different nodes. As noted, the computing delay dominates the system performance when offloading the two models to TN, namely, the computing delay contributes to most parts of the task delay. This is because UAV equips with the Raspberry PI, which has the lowest computation capacities compared with other nodes. On the other hand, note that the YOLOv4 takes more than 10000 ms for the inference, which apparently cannot meet the real-time requirement of the application. Therefore, it verifies the necessity of a adaptive task offloading in such a scenario. Meanwhile, the shortest transmission delay is achieved by offloading to TN for both the model, since the videos are sensed by UAV, and only the final detection results need to be transmitted to the vehicle. In contrast, both SSD and YOLOv4 achieve the shortest computing delay and the longest transmission delay when offloading to the cloud, due to the highest performance of the NVIDIA RTX3090, but the lowest bandwidth of cellular network. Finally, as observed, the shortest task delay for both SSD and YOLOv4 is achieved when they are offloaded to the SEN in this scenario.

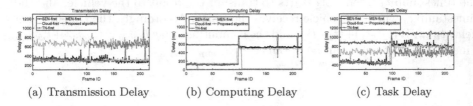

| (a) Transmission Delay | (b) Computing Delay | (c) Task Delay |

Fig. 6. Performance evaluation of different offloading policies using SSD

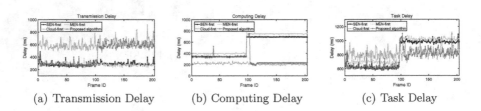

| (a) Transmission Delay | (b) Computing Delay | (c) Task Delay |

Fig. 7. Performance evaluation of different offloading policies using YOLOv4

Figure 6 evaluates the performance of the proposed algorithm by comparing with four naive policies, namely, TN-first, SEN-first, MEN-first and Cloud-first, which prefer to offload the task to TN, SEN, MEN, and Cloud, respectively. When the storage constraint is violated, the task will be offloaded to the cloud. The SSD is adopted in this set of experiments. As noted, for the proposed algorithm, although its transmission delay may go as high as cloud-first, and its computing delay may be consistent with TN-first for a very short period of time, it can quickly adaptive to the dynamic system workload, and achieve the minimum average task delay. Similar observations can be found in Fig. 7, where the YOLOv4 model is adopted for task offloading by different algorithm.

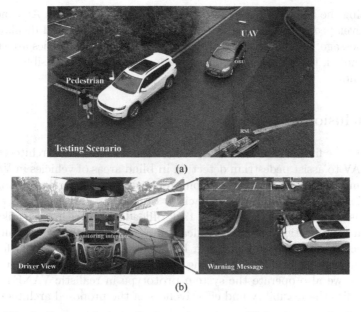

(a)

(b)

Fig. 8. Case study in realistic environments (Color figure online)

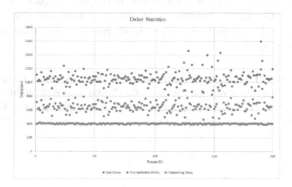

Fig. 9. Realistic environments delay

5.4 Case Study in Realistic Environments

To further evaluate the feasibility of the designed architecture and the proposed algorithm, we build the system prototype and carry out experiments in realistic VANETs environments, as shown in Fig. 8. The UAV is hovering in the air and monitoring the blind area of the red vehicle, whose view is block by the white vehicle. Accordingly, potential collision might happen for the pedestrian who is walking in front of the white vehicle and is about to cross the road. In this testing, the TN (i.e. UAV) communicates with SEN (i.e. RSU) via WIFI network and transmits sensed videos to the SEN. The YOLOv4 is deployed at SEN for object detection. The detected result will be sent to the MEN (i.e. vehicle) via

DSRC, using the RSU at the SEN and the OBU at the MEN. An Android-based APP is developed and deployed in the vehicle for receiving and displaying the warning message, as shown in Fig. 8(b). Finally, the delay statistics are shown in Fig. 9. As noted, the average task delay is about 1 s, which is feasible in realistic environments.

6   Conclusion and Future Work

In this work, we first present a terminal-edge-cloud cooperative architecture that enables UAV to assist pedestrian detection in blind areas of vehicles in VANETs. Then, we model the system delay by comprehensively considering heterogeneous computation and communication capacities of the TNs, MENs, SENs and CN. On this basis, we propose an adaptive task offloading algorithm to strike a best balance on computation and communication delay and enhance overall system performance. Finally, we implement a hardware-in-the-loop test platform and evaluate the system performance using two different object detection models. In addition, we also operate the system prototype in realistic VANETs, which further verifies the feasibility and effectiveness of the proposed architecture and algorithm.

In the future study, first, the pedestrian trajectory prediction technical will be investigated to give higher warning precision. In addition, the laser radar will be further integrated to enhance system performance on situational awareness and detection robustness.

Acknowledgement. This work was supported in part by the National Natural Science Foundation of China under Grant Nos. 62172064 and 61872049, the Chongqing Young-Talent Program (Project No. cstc2022ycjh-bgzxm0039) and the Venture & Innovation Support Program for Chongqing Overseas Returnees (Project No. cx2021063).

References

1. Liu, K., Xu, X., Chen, M., Liu, B., Wu, L., Lee, V.C.: A hierarchical architecture for the future internet of vehicles. IEEE Commun. Mag. **57**(7), 41–47 (2019)
2. Gurugopinath, S., Sofotasios, P.C., Al-Hammadi, Y., Muhaidat, S.: Cache-aided non-orthogonal multiple access for 5G-enabled vehicular networks. IEEE Trans. Veh. Technol. **68**(9), 8359–8371 (2019)
3. Liang, L., Ye, H., Yu, G., Li, G.Y.: Deep-learning-based wireless resource allocation with application to vehicular networks. Proc. IEEE **108**(2), 341–356 (2019)
4. Hadiwardoyo, S.A., Calafate, C.T., Cano, J.C., Ji, Y., Hernández-Orallo, E., Manzoni, P.: Evaluating UAV-to-car communications performance: from testbed to simulation experiments. In: 2019 16th IEEE Annual Consumer Communications & Networking Conference (CCNC), pp. 1–6. IEEE (2019)

5. Lakew, D.S., Sa'ad, U., Dao, N.N., Na, W., Cho, S.: Routing in flying ad hoc networks: a comprehensive survey. IEEE Commun. Surv. Tutor. **22**(2), 1071–1120 (2020)

6. Almasoud, A.M., Selim, M.Y., Alqasir, A., Shabnam, T., Masadeh, A., Kamal, A.E.: Energy efficient data forwarding in disconnected networks using cooperative UAVs. In: 2018 IEEE Global Communications Conference, pp. 1–6. IEEE (2018)

7. Cheng, C.M., Hsiao, P.H., Kung, H., Vlah, D.: Maximizing throughput of UAV-relaying networks with the load-carry-and-deliver paradigm. In: 2007 IEEE Wireless Communications and Networking Conference, pp. 4417–4424. IEEE (2007)

8. Zhou, Y., Cheng, N., Lu, N., Shen, X.S.: Multi-UAV-aided networks: aerial-ground cooperative vehicular networking architecture. IEEE Veh. Technol. Mag. **10**(4), 36–44 (2015)

9. Motlagh, N.H., Bagaa, M., Taleb, T.: UAV-based IoT platform: a crowd surveillance use case. IEEE Commun. Mag. **55**(2), 128–134 (2017)

10. Xu, X., Liu, K., Xiao, K., Feng, L., Wu, Z., Guo, S.: Vehicular fog computing enabled real-time collision warning via trajectory calibration. Mob. Netw. Appl. **25**(6), 2482–2494 (2020). https://doi.org/10.1007/s11036-020-01591-7

11. Liu, C., Liu, K., Guo, S., Xie, R., Lee, V.C., Son, S.H.: Adaptive offloading for time-critical tasks in heterogeneous internet of vehicles. IEEE Internet Things J. **7**(9), 7999–8011 (2020)

12. Guo, H., Liu, J.: UAV-enhanced intelligent offloading for Internet of Things at the edge. IEEE Trans. Ind. Inf. **16**(4), 2737–2746 (2019)

13. Lyu, J., Zeng, Y., Zhang, R.: UAV-aided offloading for cellular hotspot. IEEE Trans. Wireless Commun. **17**(6), 3988–4001 (2018)

14. Zhang, S., Zeng, Y., Zhang, R.: Cellular-enabled UAV communication: a connectivity-constrained trajectory optimization perspective. IEEE Trans. Commun. **67**(3), 2580–2604 (2018)

15. Fadlullah, Z.M., Takaishi, D., Nishiyama, H., Kato, N., Miura, R.: A dynamic trajectory control algorithm for improving the communication throughput and delay in UAV-aided networks. IEEE Netw. **30**(1), 100–105 (2016)

16. Wu, H., Lyu, F., Zhou, C., Chen, J., Wang, L., Shen, X.: Optimal UAV caching and trajectory in aerial-assisted vehicular networks: a learning-based approach. IEEE J. Sel. Areas Commun. **38**(12), 2783–2797 (2020)

17. Liu, K., Xiao, K., Dai, P., Lee, V.C., Guo, S., Cao, J.: Fog computing empowered data dissemination in software defined heterogeneous VANETs. IEEE Trans. Mob. Comput. **20**(11), 3181–3193 (2020)

18. Liu, C., Liu, K., Ren, H., Xu, X., Xie, R., Cao, J.: RtDS: real-time distributed strategy for multi-period task offloading in vehicular edge computing environment. Neural Comput. Appl. 1–15 (2021). https://doi.org/10.1007/s00521-021-05766-5

19. Wang, J., Liu, K., Li, B., Liu, T., Li, R., Han, Z.: Delay-sensitive multi-period computation offloading with reliability guarantees in fog networks. IEEE Trans. Mob. Comput. **19**(9), 2062–2075 (2019)

20. Zhou, Y., Liu, K., Xu, X., Liu, C., Feng, L., Chen, C.: Multi-period distributed delay-sensitive tasks offloading in a two-layer vehicular fog computing architecture. In: Zhang, H., Zhang, Z., Wu, Z., Hao, T. (eds.) NCAA 2020. CCIS, vol. 1265, pp. 461–474. Springer, Singapore (2020). https://doi.org/10.1007/978-981-15-7670-6_38

21. Bochkovskiy, A., Wang, C.Y., Liao, H.Y.M.: YOLOv4: optimal speed and accuracy of object detection. arXiv preprint arXiv:2004.10934 (2020)

22. Liu, W., et al.: SSD: single shot multibox detector. In: Leibe, B., Matas, J., Sebe, N., Welling, M. (eds.) ECCV 2016. LNCS, vol. 9905, pp. 21–37. Springer, Cham (2016). https://doi.org/10.1007/978-3-319-46448-0_2
23. Wang, M., Li, W., Wang, X.: Transferring a generic pedestrian detector towards specific scenes. In: 2012 IEEE Conference on Computer Vision and Pattern Recognition, pp. 3274–3281. IEEE (2012)

A Survey of Optimal Design of Antenna (Array) by Evolutionary Computing Methods

Xin Zhang[1] and Zhou Wu[2(✉)]

[1] Tianjin Key Laboratory of Wireless Mobile Communications and Power Transmission, Tianjin Normal University, Tianjin 300387, China
ecemark@tjnu.edu.cn
[2] College of Automation, Chongqing University, Chongqing 400044, China
zhouwu@cqu.edu.cn

Abstract. To satisfy the increasing requirements of big data transmission, mutual interference and wide band, the optimal design of antenna and antenna array has drawn great interests and attentions of researchers. This paper summarizes and classifies state-of-the-art optimal antenna designs by using evolutionary computing (EC) methods. Antenna designs are classified based on three aspects. First, based on array type, they are divided into single antenna design, linear array design and planar array design. Second, based on the number of optimization objectives, they are divided into single objective, bi-objective, and three objective methods. Third, approaches are divided based on real world scenarios. Furthermore, a benchmark is built for the designs of linear antenna array and planar antenna array. Such benchmark can be used as baseline to assist researchers to verify or create more powerful EC methods. Simulation results show that the benchmark is scalable and reliable for testing the performance of EC methods.

Keyword: Antenna design · Survey · Benchmark · Evolutionary computing

1 Introduction

Antenna and antenna array play a very important role in wireless data transmission [1, 2]. Many real world scenarios need antennas to provide high quality of service (QoS). To meet the increasing demand of big data communication, antenna design becomes a very important research topic. Combining multiple antennas to form an array antenna can better meet the needs of practical applications. The design and analysis of antenna systems to meet specific needs often boils down to optimization problems, and then find appropriate optimization algorithms to solve these problems.

EC refers to a set of methods or techniques to resolve practical application problems that traditional methods become useless for some reasons [3]. EC methods are mainly comprised of evolutionary algorithm (EA) and swarm intelligence (SI) methods [4]. To better understand the recent advances of EC for antenna designs, this paper attempts to perform a comprehensive survey.

H. Zhang et al. (Eds.): NCAA 2022, CCIS 1638, pp. 331–344, 2022.
https://doi.org/10.1007/978-981-19-6135-9_25

EC methods for antenna designs, in this paper, are summarized and classified from three aspects. First, recent researches are divided based on antenna array types. Second, they are divided based on the number of optimization objectives. Third, they are divided based on application scenarios. Furthermore, this paper presents two benchmarks for antenna array designs. The purpose of benchmarks is to test the effectiveness of EC methods.

Section 2 presents the theory and examples of antenna designs. Section 3 presents the survey of EC methods for antenna designs. Section 4 presents the benchmark, and its simulation results. Conclusion is made in Sect. 5.

2 Antenna Array Design Model

Consider a linear antenna array with M elements. All elements are spread with equal distance. An example is given in Fig. 1 (a).

Denote $x_i(t)$ as the output of element i. The output signal of antenna array is expressed as:

$$y = \mathbf{w}^T \mathbf{x} = [w_1, \cdots, w_M][x_1, \cdots, x_M]^T \tag{1}$$

where $\mathbf{w}$ is the weight vector of $\mathbf{x}$. To f power, the problem is modeled as [5]:

$$\min J = E[y^2] = E[\mathbf{w}^T \mathbf{x}\mathbf{x}^T \mathbf{w}] = \mathbf{w}^T \mathbf{R} \mathbf{w}$$
$$s.t. \qquad A(0) = 1 \tag{2}$$
$$A(\theta_k) = g_k, k = 1, 2, \cdots, M - 1$$

where g_k is the peak size of sidelobe k, and $A(\theta_k)$ is:

$$A(\theta_k) = \sum_{k=0}^{M-1} w_k \exp\left(-\frac{j2\pi dk}{\lambda}\right) \sin \theta \tag{3}$$

Also, we have:

$$A(\theta_k) = \mathbf{w}^T \mathbf{a}(\theta_k) = \mathbf{a}(\theta_k)^T \mathbf{w} \tag{4}$$

Denote $\mathbf{G} = [g_0, g_1, \cdots, g_{M-1}]^T$. Hence, the constraints of model (2) becomes $\mathbf{G} = \mathbf{A}\mathbf{W}$, where

$$\mathbf{A} = \begin{bmatrix} \mathbf{a}^T(\theta_1) \\ \mathbf{a}^T(\theta_2) \\ \vdots \\ \mathbf{a}^T(\theta_M) \end{bmatrix}. \tag{5}$$

Consider a linear antenna array with M elements. All elements are spread with unequal distance. An example is given in Fig. 1 (b). They can be laid out anywhere within the length of aperture L. That is nonuniform antenna array. By restricting sidelobe level, the antenna array is expressed as [6]:

$$\min \quad \sum_{i=1}^{M} I_i e^{j\varphi_i} e^{jkwd_i} p_i$$
$$s.t. \quad 0 < d_c \le (d_i - d_k), 1 \le k < i \le M$$
$$w = \cos\theta - \cos\theta_0 \qquad (6)$$
$$\sum_{i=1}^{M-1} d_i = L$$

Minimization model (6) can also be solved by Lagrange multiplier method. For large scale antenna array, model (6) becomes large scale problem, which would be difficult to be solved by traditional methods. They can be solved by CI methods.

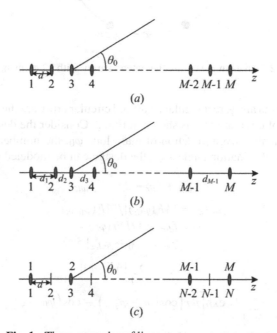

Fig. 1. Three examples of linear antenna array system.

Consider a linear antenna array with M elements. All elements are spread with unequal distance, where the element positions are thinned to grids. An example is given in Fig. 1 (c). Suppose N grid locations are available within the length of aperture L. The design is to place M elements on N locations. For such case, the problem is expressed as [7]:

$$\min \quad -\left| \sum_{i=1}^{N} \delta_i I_i e^{j\varphi_i} e^{jkwx_i} p_i \right| / |MP|$$
$$s.t. \quad x_i \in \{0, 1\}, 1 \le i \le N \qquad (7)$$
$$w = \cos\theta - \cos\theta_0$$

where MP denotes peak level of mainlobe. The objective of model (7) is to minimize sidelobe level and maximize main sidelobe level. The decision variables are integer values; hence the problem can be solved by integer programming methods or CI methods.

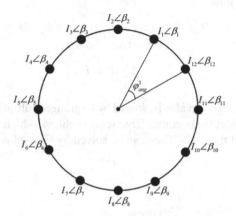

Fig. 2. An example of circular antenna array with 12 elements.

For planar antenna array, rectangular array and circular array are the most commonly type. An example of circular array is shown in Fig. 2. Consider the design of a circular array with M elements. Given direction of main lobe φ_0, the number of zero control directions num, and destination angle φ_{des}, the design can be modeled as:

$$\begin{aligned}
\min_{x \in \Omega} \quad & f_{sll} + f_{dir} + f_{des} + f_{null} \\
s.t. \quad & f_{sll} = |AF(\varphi_{sll})|/|AF(\varphi_{\max})| \\
& f_{dir} = 1/Dir(\varphi_0) \\
& f_{des} = |\varphi_0 - \varphi_{des}| \\
& f_{null} = \sum_{k=1}^{num} |AF(\varphi_k)|
\end{aligned} \tag{8}$$

$$AF(\varphi) = \sum_{m=1}^{M} I_m \exp\left[jkr\left(\cos\left(\varphi - \varphi_{ang}^m\right) - \cos\left(\varphi_0 - \varphi_{ang}^m\right) \right) + \beta_m \right]$$

Three examples of linear antenna array and a circular array example are introduced above as well as the associated mathematical model. Their optimization models can also be changed by adding other objectives or constraints. Moreover, decision variables can also be increased or decreased based on users' requirements. No matter what expressions are used, such models can be solved by EC methods.

3 Evolutionary Computing Methods for Antenna Array Design

EC methods proposed before the twentieth century include genetic algorithm (GA), evolutionary strategy (ES), particle swarm optimization (PSO), estimation of distribution algorithm (EDA), differential evolution (DE), ant colony optimization (ACO), etc. [3]. EC methods proposed in the past two decades include harmony search(HS), artificial bee colony (ABC), biogeography-based optimization (BBO), neighborhood field optimization (NFO), Teaching-learning-based optimization (TLBO), cat swarm optimization (CSO), firefly algorithm (FA), bacterial foraging optimization (BFO), bat algorithm (BAT), flower pollination algorithm (FPA), firefly algorithm (FA), bacterial foraging optimization (BFO), Moth flame optimization (MFO), most valuable play algorithm (MVP), brain storm optimization (BSO), cuckoo search (CS), symbiotic organisms search (SOS), etc. [4]. There are heuristic methods that utilize random factors but not recognized as EC methods like pattern search, simulated annealing, light beam search (LBS), normal boundary intersection (NBI), etc.

Two or more EC methods can be merged to form a new method. Such method is called memetic algorithm. Moreover, MA can also be the hybrid of EC methods with non-EC methods. MA is able to take advantages of several EC or non-EC methods.

The paradigms of EC methods have been extended to solve single objective or multiobjective optimization problems. EC methods for multiobjective optimization problems include non-dominated sorting genetic algorithm II (NSGA-II), multiobjective evolutionary algorithm based on decomposition (MOEA/D), multiobjective multi-verse optimizer (MOMVO), etc.

The state-of-the-art researches of evolutionary computing methods are summarized in this section for antenna array design. The related works are found through search engine like Google scholar, IEEEXplore, SpringerLink, ScienceDirect, Taylor & Francis Online, etc. Note that it is impossible to collect all related works, as papers continue to emerge. The related works summarized in this section are all published in 2017, 2018 and 2019. It is necessary to declare that the authors have no bias when collecting related works such that an objective summary is made in this section.

3.1 Classification Based on Array Type

Antennas are classified to single antenna, multiple antennas, linear antenna array, and planar antenna array. Single antenna means a transmitting antenna or a receiving antenna. Multiple antennas means multiple transmitting antennas, multiple receiving antennas, or multiple both transmitting and receiving antennas. Linear antenna array means an array of antennas arranging in a line shape as shown in Fig. 1. Planar antenna array means an array of antennas arranging in a planar shape.

The state-of-the-art researches are summarized in Table 1. There are 21 papers studying single antenna. As shown in Table 1, there are 3 papers studying multiple antennas, there are 16 papers studying linear antenna array, and there are 16 papers studying planar antenna array.

From the above summary, it can be seen that single antenna draws greatest attention and has more researches than other categories. Linear array and planar array has similar research amount, while multiple antennas has the least researches. Some researches study more than one category. For example, single antenna, linear array and planar array are optimized in [51], as the authors give an introduction and tutorial of MOO algorithms.

Table 1. Summary of researches based on array type. N/A means not available.

Array type	Number of elements	Evolutionary computing methods
Single antenna	1 [6, 8, 11–13, 24, 39, 41, 47, 48, 50, 52, 53, 55, 57], 2 [42], 2 [45], 2 [46, 54], 3 [51]	hybrid ABC and DE [8], GA [9], adaptive BFO [10], GA [11], GA and BAT [12], PSO [13], adaptive variable differential ABC [6], NSGA-II and surrogate-based MOEA [39], DE [41], continuous and discrete BSO [42], parallel surrogate model-assisted hybrid DE [45], NSGA-II [46], MOEA/D [47], binary PSO [48], MOEA/D genetic operator [50], seven MOO algorithms [51], hybrid ICA with ACO [52], adaptive PSO [53], PSO [54], GA [55], hybrid BFO with PSO [57]
Multiple antennas	N/A [14], 2 [15], 4 [43]	centralized GA, PSO, ACO, SA, HS, and TLBO [14], MOEA/D [15], MOEA/D-DE and MOEA/D-GO [43],
Linear antenna array	25 and 101 [7], 10 [16], 12 [17], 12, 16, 20 [18], 10 [19], 12–16 [20], 4 [21], 10, 28, 32 [22], 12, 20, 22, 26, 28, 32 [23], 20 [24], 25 [25], 10 to 100 with interval 10 [26], 16 [40], 6 [51], 10, 11 and 24 [56], 10, 30 and 50 [59]	Adaptive memetic PSO [7], LBS based NBI [16], PSO [17], CSO [18], ecosystem PSO [19], BAT [20], DE [21], FPA [22, 23], PSO with velocity mutation [24], hybrid DE with convex programming [25], ABC with similarity induced search [26], various DE algorithms [40], seven MOO algorithms [51], CS [56], contraction adaptive PSO [59]

(*continued*)

Table 1. (*continued*)

Array type	Number of elements	Evolutionary computing methods
Planar antenna array	4 to 12 with interval 2 [27], 4, 6, 8 [28], 10 [6], 12, 18, 24 [29], 57 [30], 8 [31], 22 to 30 with interval 2 [32, 33], 18, 24 [34], 36 to 60 with interval 6 [35], 28 [36], 16 [37], 316 [38], 8 and 10 [44], reflectarray [49], 18, 30 and 105 [58]	MFO [27], MVP [28], adaptive variable differential ABC [6], efficient BBO [29], comprehensive learning PSO [30], asynchronous PSO [31], CSO [32], FA [33], wavelet mutation based novel PSO [34], craziness based PSO [35], GA [36, 37], MOMVO [38], DE and PSO [44], MQC10 BBO [49], SOS [58]

The second column of Table 1 gives the number of elements tested in associated papers. It can be seen that 7 papers test more than 50 elements [7, 26, 30, 35, 38, 58, 59], while 47 papers test antenna array with elements <50. This means that large scale antenna array or massive MIMO antennas still deserves studying in the future.

3.2 Classification Based on the Number of Optimization Objectives

In general, optimization problems can be classified to single objective problems and multiobjective problems. Multiobjective problems mean two or more objectives to be optimized. In this paper, an accurate classification is made based on the number of objectives. The state-of-the-art researches are classified based on the number of optimization objectives. They are single objective, bi-objective, and three objectives. As researches on four or more objectives are not detected, they are not listed in Table 2.

Table 2. Summary of researches based on the number of optimization objectives.

Number of objectives	Evolutionary computing methods
Single objective	[6–11, 13, 14, 17–30, 32–36, 40–42, 44, 45, 48, 49, 52–59]
Bi-objectives	LBS based NBI [16], asynchronous PSO [31], kriging-based MOEA [124], MOMVO [38], NSGA-II [46], MOEA/D [47], seven MOO algorithms [51]
Three objectives	MOEA/D [15], NSGA-II and surrogate-based MOEA [39], MOEA/D-DE and MOEA/D genetic operator [43, 50]

As shown in Table 2, the ratio of single objective, bi-objective and three objectives is 41:7:4. It can be seen that most researches focus on single objective optimization problem. Only a few researches concern three objective optimization problems. To our best knowledge, no researches concerning objectives ≥ 4 is found. This means that multiobjective antenna optimization design still deserves studying in the future. Moreover, more optimizing objectives need to be considered or detected in the future.

3.3 Classification Based on Real World Scenario

Antenna is generally used for wireless communications, which includes a good many scenarios. Real world scenarios include worldwide interoperability for microwave access

Table 3. Summary of researches based on real world scenarios. N/A means not available.

Real world scenario	Working frequency	Evolutionary computing methods
WiMAX	2.5 GHz and 3.5 GHz [8], 3.5 GHz [47]	hybrid ABC and DE [8], MOEA/D [47]
WLAN	5.2 GHz [8], 5 GHz [36], 2.4 GHz, 5.2 GHz and 5.8 GHz [47]	hybrid ABC and DE [8], GA [36], MOEA/D [47],
Satellite communication and earth observation mission	N/A [9], 7.3 GHz [10]	GA [9], adaptive BFO [10],
Aircraft	8 MHz, 18 MHz, 28 MHz	MOEA/D [15]
Wireless communication	4.2–11.6 GHz [17], 2 GHz [18], 5.88 GHz [21], 1.9 GHz and 2.4 GHz [42], 700 MHz [53], 24 GHz and 60 GHz [54], 3.49 GHz and 4.73 GHz [57]	[17, 18, 21, 42, 53, 54, 57]
Digital video broadcasting-terrestrial	470 MHz–790 MHz	PSO with velocity mutation [24]
Indoor UWB communication	3.1 GHz–10.6 GHz	Asynchronous PSO [31]
MIMO communication	2.4 GHz and 5 GHz [36], 2.6 GHz–3.9 GHz [43], 2.4 GHz and 5 GHz [44], 30 GHz [49], 28 GHz [52],	GA [36], MOEA/D-DE and MOEA/D-GO [43], DE and PSO [44], MQC10 BBO [49], hybrid ICA with ACO [52],
Microwave wireless power transmission	12.5 GHz–16 GHz	MOMVO [38]
Compact UWB communication	3.1 GHz–10.6 GHz [39], 2 GHz–12 GHz [48]	NSGA-II and surrogate-based MOEA [39], binary PSO [48],
Radio frequency identification	915 MHz [50]	MOEA/D genetic operator [50]

(WiMAX), wireless local area network (WLAN). WiMAX requires antenna array having working frequency at 2.5 GHz or 3.5 GHz. WLAN requires frequency at 2.4 GHz, 5 GHz, 5.2 GHz or 5.8 GHz. Table 3 summarizes the scenarios of state-of-the-art researches. References are not listed in Table 3, given working frequency was not reported in the papers. In [50], MOEA/D with genetic operators is used to optimize the design of radio frequency identification (RFID) antennas.

It can be seen from Table 3 that optimized antenna array can be applied to many scenarios. The working frequency ranges from 8 MHz to 60 GHz. The fifth generation (5G) wireless communication, MIMO and UWB draw greatest attention of researchers. There are still researches studying WiMAX [8, 47], WLAN [8, 36, 47], and RFID [50]. Currently, 5G wireless communication, massive MIMO, UWB communication, and simultaneous wireless information and power transfer (SWIPT) are very hot topics. Thus, it is expected that such researches will continue in the future.

4 Benchmark and Simulation

Different research studies have different element setting; even they focus on solving the same antenna array type. In this case, the studies could hardly provide useful information for users to choose EC methods. Thus, a benchmark is reported in this section to test EC methods. An advantage of such benchmark is the scalability. That is any element number can be set to simulate and compare EC methods.

4.1 Linear and Planar Antenna Array Design Benchmark

Both examples are implemented in Matlab. The source code of the benchmark can be obtained upon request. GA, DE, PSO and NFO are implemented as baseline to test the source code. In the simulation, the number of elements for linear array design is 10. The number of function evaluations is set to 10000 for each method. The population size is set to 50. Hence, each method evolves 200 generations before termination. Each method is executed 31 times to gain a statistical meaningful result. Note that algorithmic parameters may affect the performance of EC methods [60]. In the simulation, the target is to show the benchmark property instead of comparing EC methods. Algorithmic parameters of GA, DE, PSO and NFO are set based on existing researches [6, 7, 61].

Table 4. Statistics of the results found by the GA, DE, PSO, and NFO methods for linear array. N/A means not available.

Method	Minimum	Median	Maximum	Mean	Std	C (U test)
GA	−6.456	−5.978	−5.695	−6.015	1.079	99.95%
DE	−6.325	−6.084	−5.832	−6.090	1.083	99.95%
PSO	−6.851	−6.274	−5.713	−6.284	1.152	99%
NFO	−6.802	−6.417	−5.777	−6.393	1.155	N/A

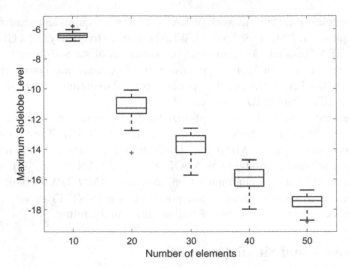

Fig. 3. Box plot of the results attained by the NFO methods for linear array with different elements.

The minimum, median, maximum, mean and standard deviation (std) values of the results are given in Table 4. Note that array design is a minimization problem in the simulation. It can be seen from Table 4 that NFO attains the best performance in terms of median and mean values of the results. PSO is the best in terms of minimum value of the results. DE is the best in terms of maximum value of the result. GA is the best in terms of std value of the results.

To show the significance of the results, Wilcoxon rank-sum test (U test) is taken to compare the results of two methods. As NFO attains the best median results, U test is performed by NFO against other methods. The confidence interval C is set to 99.95% or 99% (if 99.95% fails). It can be seen from Table 4 that NFO is significant better than PSO with 99% confidence. NFO is significant better than GA and DE with 99.95% confidence.

4.2 Scalability of Benchmark

For linear array, the number of elements is set to 10, 20, 30, 40, and 50. The NFO method is used in the optimal design. For linear array, the result is shown in Fig. 3. The figure shows the box plot of the results over 31 executions.

It can be seen from Fig. 3 that the benchmark works well with different elements. The NFO method is able to find more and more better MSLL values along with the increase of elements. The std values of NFO also become larger when the number of elements increases. The above results show that the benchmark is scalable with the number of elements. This is useful for users to verify the performance of EC methods.

5 Conclusion

This paper summarizes the state-of-the-art researches on EC methods for antenna designs. Moreover, the related works are classified and discussed from three aspects.

Furthermore, this paper presents linear array and planar array benchmark for antenna array designs. The purpose is to test the effectiveness of EC methods and to assist users choose powerful EC methods.

In the future, there are interesting topics and perspectives to be further studied:

(1) Antenna design under a whole system is an important step towards practical applications. As both transmitting and receiving antennas are used in practical applications, optimal design is highly preferred by modeling both transmitting and receiving antennas.

(2) The studies of many multiobjective optimization evolutionary algorithms are very hot in EC fields. There is seldom research on antenna design considering objectives ≥ 4. Researches optimizing many objectives of antenna design have good perspective.

(3) Microwave antennas are able to realize wireless power transmission (WPT), which is very useful for battery based devices like wireless sensors. Optimal antenna design plus efficient WPT is also highly preferred.

(4) How to reduce the number of evaluations of EC methods under the condition of good effectiveness deserves further studying.

Acknowledgments. This paper was supported in part by the Supported by the National Key Research and Development Program of China (2022YFE0198900, 2021YFF0500903), and the National Natural Science Foundation of China (52178271, 61901301), and in part by the Tianjin Higher Education Creative Team Funds Program.

References

1. Balanis, C.A.: Antenna Theory Analysis and Design, 3rd edn. Wiley, Hoboken (2005)
2. Milligan, T.A.: Modern Antenna Design, 2nd edn. Willey, Hoboken (2005)
3. Baeck, T., Fogel, D.B., Michalewicz, Z.: Handbook of Evolutionary Computation. CRC Press, Boca Raton (1997)
4. Olariu, S., Zomaya, A.Y.: Handbook of Bioinspired Algorithms and Applications. Chapman and Hall/CRC Press, New York (2005)
5. Shah, I.U., Durrani, T.S.: Constraint adaptive natural gradient algorithm (CANA) for adaptive array processing. In: 17th European Signal Processing Conference (EUSIPCO 2009), Glasgow, Scotland, pp 884–888 (2009)
6. Zhang, X., Zhang, X., Wang, L.: Antenna design by an adaptive variable differential artificial bee colony algorithm. IEEE Trans. Magn. **54**(3), Article No. 7201704 (2018).
7. Zhang, X., Zhang, X.: Thinning of antenna array via adaptive memetic particle swarm optimization. EURASIP J. Wirel. Commun. Netw. **2017**, Article No. 183 (2017)
8. Ustun, D., Akdagli, A.: Design of band–notched UWB antenna using a hybrid optimization based on ABC and DE algorithms. AEU Int. J. Electron. Commun. **87**, 10–21 (2018)
9. Liu, R., Guo, H., Liu, R., Wang, H., Tang, D., Deng, Z.: Structural design and optimization of large cable–rib tension deployable antenna structure with dynamic constraint. Acta Astronaut. **151**, 160–172 (2018)

10. Gupta, N., Saxena, J., Bhatia, K.S.: Design optimization of CPW-fed microstrip patch antenna using constrained ABFO algorithm. Soft Comput. **22**(24), 8301–8315 (2017). https://doi.org/10.1007/s00500-017-2775-4
11. Kaur, G., Rattan, M., Jain, C.: Design and optimization of PSI slotted fractal antenna using ANN and GA for multiband applications. Wirel. Pers. Commun. **97**, 4573–4585 (2017)
12. Kaur, G., Rattan, M., Jain, C.: Optimization of Swastika slotted fractal antenna using genetic algorithm and bat algorithm for S-band utilities. Wirel. Pers. Commun. **97**(1), 95–107 (2017). https://doi.org/10.1007/s11277-017-4495-6
13. Manoochehri, O., Salari, M.A., Darvazehban, A.: A short broadband monopole antenna. AEU Int. J. Electron. Commun. **83**, 240–244 (2018)
14. FallahHoseini, M., Rafeh, R.: Proposing a centralized algorithm to minimize message broadcasting energy in wireless sensor networks using directional antennas. Appl. Soft Comput. **64**, 272–281 (2018)
15. Wang, C., Chen, Y., Yang, S.: Application of characteristic mode theory in hf band aircraft-integrated multiantenna system designs. IEEE Trans. Antennas Propag. **67**(1), 513–521 (2019)
16. An, S., Yang, S., Ren, Z.: Incorporating light beam search in a vector normal boundary intersection method for linear antenna array optimization. IEEE Trans. Magn. **53**(6), Article No. 7001304 (2017)
17. Shandilya, M., Pawar, S.S., Chaurasia, V.: Design and optimization of a non-cross feed printed log periodic dipole array antenna using particle swarm optimization. AEU Int. J. Electron. Commun. **93**, 172–181 (2018)
18. Ram, G., Mandal, D., Ghoshal, S.P., Kar, R.: Null placement in time-modulated linear antenna arrays of dipole element. IETE J. Res. **63**(3), 403–412 (2017)
19. Liu, J., Ma, D., Ma, T., Zhang, W.: Ecosystem particle swarm optimization. Soft. Comput. **21**(7), 1667–1691 (2016). https://doi.org/10.1007/s00500-016-2111-4
20. Das, A., Mandal, D., Ghoshal, S.P., Kar, R.: An efficient side lobe reduction technique considering mutual coupling effect in linear array antenna using BAT algorithm. Swarm Evol. Comput. **35**, 26–40 (2017)
21. Chatterjee, A., Mondal, T., Patanvariya, D.G., Jagannath, R.P.K.: Design of a wide bandwidth and high efficiency circularly polarized microstrip patch antenna using coaxial feeding technique. AEU Int. J. Electron. Commun. **83**, 549–557 (2018)
22. Chakravarthy, V.V.S.S.S., Chowdary, P.S.R., Panda, G., Anguera, J., Andújar, A., Majhi, B.: On the linear antenna array synthesis techniques for sum and difference patterns using flower pollination algorithm. Arab. J. Sci. Eng. **43**(8), 3965–3977 (2017). https://doi.org/10.1007/s13369-017-2750-5
23. Singh, U., Salgotra, R.: Synthesis of linear antenna array using flower pollination algorithm. Neural Comput. Appl. **29**(2), 435–445 (2016). https://doi.org/10.1007/s00521-016-2457-7
24. Zaharis, Z.D., Gravas, I.P., Lazaridis, P.I., Glover, I.A., Antonopoulos, C.S., Xenos, T.D.: Optimal LTE-protected LPDA design for DVB-T reception using particle swarm optimization with velocity mutation. IEEE Trans. Antenna Propag. **66**(8), 3926–3935 (2018)
25. Cui, C.-Y., Jiao, Y.-C., Zhang, L.: Synthesis of some low sidelobe linear arrays using hybrid differential evolution algorithm integrated with convex programming. IEEE Antennas Wirel. Propag. Lett. **16**, 2444–2448 (2017)
26. Wang, L., Zhang, X., Zhang, X.: Antenna array design by artificial bee colony algorithm with similarity induced search method. IEEE Trans. Magn. **55**(6), Article No. 7201904 (2019)
27. Das, A., Mandal, D., Ghoshal, S.P., Kar, R.: Concentric circular antenna array synthesis for side lobe suppression using moth flame optimization. AEU Int. J. Electron. Commun. **86**, 177–184 (2018)
28. Bouchekara, H.R.E.-H., Orlandi, A., Al-Qdah, M., de Paulis, F.: Most valuable player algorithm for circular antenna arrays optimization to maximum sidelobe levels reduction. IEEE Trans. Electromagn. Compat. **60**(6), 1655–1661 (2018)

29. Sun, G., Liu, Y., Liang, S., Wang, A., Zhang, Y.: Beam pattern design of circular antenna array via efficient biogeography-based optimization. AEU Int. J. Electron. Commun. **79**, 275–285 (2017)
30. Ismaiel, A.M., Elsaidy, E., Albagory, Y., Atallah, H.A., Abdel-Rahman, A.B., Sallam, T.: Performance improvement of high altitude platform using concentric circular antenna array based on particle swarm optimization. AEU Int. J. Electron. Commun. **91**, 85–90 (2018)
31. Banerjee, S., Mandal, D.: Array pattern optimization for steerable circular isotropic antenna array using cat swarm optimization algorithm. Wirel. Pers. Commun. **99**(3), 1169–1194 (2018). https://doi.org/10.1007/s11277-017-5171-6
32. Chien, W., Chiu, C.-C., Cheng, Y.-T., Liao, S.-H., Yen, H.-S.: Multi-objective optimization for UWB antenna array by APSO algorithm. Telecommun. Syst. **64**(4), 649–660 (2016). https://doi.org/10.1007/s11235-016-0197-8
33. Banerjee, S., Mandal, D.: Array pattern optimization for a steerable circular isotropic antenna array using the firefly algorithm. J. Comput. Electron. **16**(3), 952–976 (2017). https://doi.org/10.1007/s10825-017-1049-9
34. Bera, R., Mandal, D., Kar, R., Ghoshal, S.P.: Non-uniform single-ring antenna array design using wavelet mutation based novel particle swarm optimization technique. Comput. Electr. Eng. **61**, 151–172 (2017)
35. Sahoo, P.K., Mandal, D.: Swarm intelligence-based optimal design to minimize the sidelobe level of concentric circular antenna array with element failures. IETE J. Res. **64**(4), 559–568 (2017)
36. Soltani, S., Lotfi, P., Murch, R.D.: Design and optimization of multiport pixel antennas. IEEE Trans. Antenna Propag. **66**(4), 2049–2054 (2018)
37. Liu, Q., Qi, S.-S., Yin, Q., Wu, W.: Frequency-scanning dual-beam parallel-plate waveguide continuous transverse stub antenna array with sidelobe suppression. IEEE Antenna Wirel. Propag. Lett. **17**(7), 1228–1232 (2018)
38. Li, X., Luk, K.M., Duan, B.: Multiobjective optimal antenna synthesis for microwave wireless power transmission. IEEE Trans. Antenna Propag. **67**(4), 2739–2744 (2019)
39. Koziel, S., Bekasiewicz, A.: Comprehensive comparison of compact UWB antenna performance by means of multiobjective optimization. IEEE Trans. Antenna Propag. **65**(7), 3427–3436 (2017)
40. Deb, A., Roy, J.S., Gupta, B.: A differential evolution performance comparison: comparing how various differential evolution algorithms perform in designing microstrip antennas and arrays. IEEE Antennas Propag. Mag. **60**(1), 51–61 (2018)
41. Hu, C., Zeng, S., Jiang, Y., Sun, J., Sun, Y., Gao, S.: A robust technique without additional computational cost in evolutionary antenna optimization. IEEE Trans. Antenna Propag. **67**(4), 2252–2259 (2019)
42. Aldhafeeri, A., Rahmat-Samii, Y.: Brain storm optimization for electromagnetic applications: continuous and discrete. IEEE Trans. Antenna Propag. **67**(4), 2710–2722 (2019)
43. Lu, D., Wang, L., Yang, E., Wang, G.: Design of high-isolation wideband dual-polarized compact MIMO antennas with multiobjective optimization. IEEE Trans. Antenna Propag. **66**(3), 1522–1527 (2018)
44. Medina, Z., Reyna, A., Panduro, M.A., Elizarraras, O.: Dual-Band performance evaluation of time-modulated circular geometry array with microstrip-fed slot antennas. IEEE Access **7**, 28625–28634 (2019)
45. Liu, B., Akinsolu, M.O., Ali, N., Abd-Alhameed, R.: Efficient global optimisation of microwave antennas based on a parallel surrogate model-assisted evolutionary algorithm. IET Microwaves Antennas Propag. **13**(2), 149–155 (2019)
46. Easum, J.A., Nagar, J., Werner, P.L., Werner, D.H.: Efficient multiobjective antenna optimization with tolerance analysis through the use of surrogate models. IEEE Trans. Antenna Propag. **66**(12), 6706–6715 (2018)

47. Dong, J., Qin, W., Wang, M.: Fast multi-objective optimization of multi-parameter antenna structures based on improved BPNN surrogate model. IEEE Access **7**, 77692–77701 (2019)
48. Palanisamy, T.S.C.A., Murugappan, M.: Joint optimisation of ground, feed shapes with material distributive topology of patch in UWB antennas using improved binary particle swarm optimisation. IET Microwaves Antennas Propag. **12**(12), 1967–1972 (2018)
49. Pirinoli, P., Beccaria, M., Massaccesi, A.: MQC10-BBO optimization applied to multi-beam antenna design. In: 13th European Conference on Antennas and Propagation, Krakow, Poland, pp. 1–3. IEEE (2019)
50. Ding, D., Xia, J., Yang, L., Ding, X.: Multiobjective optimization design for electrically large coverage. IEEE Antennas Propag. Mag. **60**(1), 27–37 (2018)
51. Nagar, J., Werner, D.H.: Multiobjective optimization for electromagnetics and optics: an introduction and tutorial based on real-world applications. IEEE Antennas Propag. Mag. **60**(6), 58–71 (2018)
52. Jian, R., Chen, Y., Chen, T.: Multi-Parameters unified-optimization for millimeter wave microstrip antenna based on ICACO. IEEE Access **7**, 53012–53017 (2018)
53. Tang, M.-C., Chen, X., Li, M., Ziolkowski, R.W.: Particle swarm optimized, 3-D-printed, wideband, compact hemispherical antenna. IEEE Antennas Wirel. Propag. Lett. **17**(11), 2031–2035 (2018)
54. Zhang, B., Rahmat-Samii, Y.: Robust optimization with worst case sensitivity analysis applied to array synthesis and antenna designs. IEEE Trans. Antenna Propagat. **66**(1), 160–171 (2018)
55. Smith, J.S., Baginski, M.E.: Thin-wire antenna design using a novel branching scheme and genetic algorithm optimization. IEEE Trans. Antenna Propag. **67**(5), 2934–2941 (2019)
56. Khodier, M.: Comprehensive study of linear antenna array optimisation using the cuckoo search algorithm. IET Microwaves Antennas Propag. **13**(9), 1325–1333 (2019)
57. Gupta, N., Saxena, J., Bhatia, K.S.: Optimized metamaterial-loaded fractal antenna using modified hybrid BF-PSO algorithm. Neural Comput. Appl. **32**(11), 7153–7169 (2019). https://doi.org/10.1007/s00521-019-04202-z
58. Dib, N.: Design of planar concentric circular antenna arrays with reduced side lobe level using symbiotic organisms search. Neural Comput. Appl. **30**(12), 3859–3868 (2017). https://doi.org/10.1007/s00521-017-2971-2
59. Zhang, X., Lu, D., Zhang, X., Wang, Y.: Antenna array design by a contraction adaptive particle swarm optimization algorithm. EURASIP J. Wirel. Commun. Netw. **2019**, Article No. 57 (2019)
60. Phan, H.D., Ellis, K., Barca, J.C., Dorin, A.: A survey of dynamic parameter setting methods for nature-inspired swarm intelligence algorithms. Neural Comput. Appl. **32**(2), 567–588 (2019). https://doi.org/10.1007/s00521-019-04229-2
61. Wu, Z., Chow, T.W.S.: Neighborhood field for cooperative optimization. Soft Comput. **17**(5), 819–834 (2013)

A Span-Based Joint Model for Measurable Quantitative Information Extraction

Di Mo[1], Bangrui Huang[1], Haitao Wang[2], Xinyu Cao[2], Heng Weng[3], and Tianyong Hao[4(✉)]

[1] School of Artificial Intelligence, South China Normal University, Guangzhou, China
{modi,2021023276}@m.scnu.edu.cn
[2] China National Institute of Standardization, Beijing, China
{wanght,caoxy}@cnis.ac.cn
[3] State Key Laboratory of Dampness, Syndrome of Chinese Medicine, The Second Affiliated Hospital of Guangzhou University of Chinese Medicine, Guangzhou, China
wengh@gzucm.edu.cn
[4] School of Computer Science, South China Normal University, Guangzhou, China
haoty@m.scnu.edu.cn

Abstract. As a set of essential information related to the quantity of objects, measurable quantitative information widely exists in various texts. This paper proposes a new span-based joint model with lexicon enhanced BERT for measurable quantitative information extraction, which incorporates both measurable quantitative information recognition and association within one model. A standard dataset containing 3,106 Chinese text sentences and a clinical dataset containing 1,359 Chinese electronic medical records are used for evaluation. Experiment results show that our joint model achieves an F1-score of 97.44% on the clinical dataset and outperforms baseline models by an improvement of 3.2% for recognition and 5.1% for association in F1-score on the standard dataset.

Keywords: Measurable quantitative information · Joint model · Span · Information extraction

1 Introduction

Measurable quantitative information (MQI) widely exists in various texts, such as clinical trials, medical articles, industrial production standards and electronic medical records [1]. As the significant information in medical field, MQI is helpful for clinical decision support, disease risk prediction and disease monitoring [2–4]. The demands for accurate measurable quantitative information from unstructured texts are increasing in industry and academia, which require the reliable extraction [1]. Specifically, MQI consists of four basic elements, including entity, numeral, unit and comparison operation [5]. According to ISO, the combination of numeral and corresponding unit is represented as quantity [6]. Thus, the measurable quantitative information extraction is usually divided into two sub-tasks, measurable quantitative information recognition and measurable quantitative information association. Among them, the first task is to extract entities, numbers, units in the text, and the second task is to identify the association between each entity and its corresponding quantity.

Although the extraction of MQI has gained great attention from industry and Natural Language Processing (NLP) researchers, it still faces the following challenges. Firstly, since the lack of clear word boundary, it is difficult to extract accurate and complete entities of MQI [7]. Secondly, due to the multiple expressions of professional term, it causes difficulty in extracting MQI which may contain nested entity. Existing work splits the measurable quantitative extraction into two interdependent sub-tasks and handle these two tasks in a pipeline manner, which makes the task easy to deal with and each component can be more flexible [5]. However, it neglects the correlation between the two sub-tasks and each task is an independent model [8]. Furthermore, the performance of relation classification is affected by the results of entity recognition which also lead to erroneous delivery [9].

This paper is aimed to propose a span-based joint model for measurable quantitative information extraction. Different from the traditional pipelined methods, the joint learning framework is to extract MQI and associate them together using a single model. Unlike previous work based on sequence labels [5], the method of span-based is introduced to enhance the prediction of boundary information and solve the nested of entity recognition [10].

The major contributions of this paper lie in the following aspects. 1) A new span-based method is proposed to jointly implement measurable quantitative information recognition and association. The joint extraction model reduces the error transmission, and the experiment shows that our joint extraction model has better performance in MQI recognition than the pipelined extraction model based on sequence labeling. 2) By combining lexical information with BERT, word features and contextual features are enhanced to improve the extraction of measurable quantitative information. 3) Our model achieved the highest performance for recognition and association compared with other baseline methods.

2   Related Work

Compared with the rapid development of other element extraction, the research of measurable quantitative information extraction is developing steadily. Previous extraction methods are mainly divided into the following three categories: (1) base rule and pattern, (2) traditional machine learning methods, and (3) deep learning methods.

In the early stage, there were few studies on measurable quantitative information extraction and most of them were rule-based methods. Valx uses semantic knowledge and domain knowledge to extract and construct numerical laboratory comparison statements [10]. The F1-measure value of type 2 diabetes test is 97.80%. The combination of pattern learning and rule-based methods to identify and extract measurable quantitative information, which achieves an F1-measure of 92.90% [11]. Heuristic rule and pattern learning are integrated to extract and normalize temporal expression with an F1-score

91.10% on the English clinical requests, and an F1-score 93.60% on the Chinese discharge summaries, respectively [12]. Rule-based approaches typically have higher precision but lower recall rates, requiring experts to invest a lot of time manually designing rules.

Based on feature engineering, machine learning is applied to extract measurable information to balance precision and recall. Structural Support Vector Machines (SSVMs) combine the strengths of CRF [26] and Support Vector Machines (SVMs) to identify treatment of discharge summaries and detection entities from medical problems [13]. Gruss et al. [14] apply Naive Bayes classifiers to extract and categorize numerical expressions. Random Forest is used for entity-quantity association [5]. However, feature engineering is time-consuming and laborious.

Due to the advantages in automatic feature extraction, deep learning methods have attracted extensive attention. Li et al. [15] combine medical dictionaries and part-of-speech into Bi-LSTM-CRF to improve clinical named entity recognition, which achieves 85.40% F1-score. Liu et al. [5] present an extended Bi-LSTM-CRF model with external features integrated by domain knowledge information and position characteristics. The methods yield to be an F1-measure of 94.27% for measurable quantitative information recognition. However, these methods of feature enhancement do not take into account the fusion of lexicon information and character information.

Zhang et al. [16] first consider integrating word information into character-based sequence encoders for explicit modeling of word features and prove the effectiveness of lattice structure for named entity recognition. Since then, some studies have proposed lexicon information fusion methods based on CNN and GNN respectively, but they have not solved the problem of long-distance dependence [20, 21].

Recently, the emergence of pre-training models (PTMs), such as Bidirectional Encoder Representation (BERT) [22] has brought NLP into a new era [17]. Zhang et al. [18] use word embedding pretrained by BERT as input feature to Bi-LSTM-CRF for named entity recognition and achieves F1-scores of 93.53%. Liu et al. [19] combine the advantages of lexicon information and pre-training model (BERT) for Chinese sequence labeling tasks. However, these studies do not consider the measurable quantitative information.

3   The Joint Model for MQI Extraction

We propose a new span-based model with lexicon-enhanced BERT to jointly carry out measurable quantitative information recognition and association. The lexicon information and BERT is combined to enhance the extraction of measurable quantitative information. Hybrid representation is used to further identify the link types of measurable quantitative information. The overall framework of the model is show in Fig. 1. The framework consists of three components: pre-processing, measurable quantitative information joint extraction, and comparison relation recognition.

The pre-processing step includes: eliminating blanks, line breaks, full-corner and half-corner conversion in sentences, and removing garbled characters. Garbled characters are mostly Chinese low-frequency characters and odd characters, such as ' 乂 ', ' 屮', ' 丿', ' 厂', ' ■' and so on. The garbled characters generated by Optical Character

Recognition (OCR) recognition errors frequently destroy the original information and are not conducive to subsequent model training, leading to the degradation of model performance. We use a dictionary matching method to screen the garbled characters, and build a garbled dictionary containing 77 garbled characters by manually screening original document corpus. The other steps are described as follows in details.

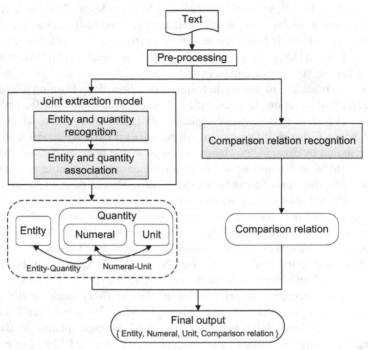

Fig. 1. The framework architecture of our joint model for measurable quantitative information extraction.

3.1 Comparison Relation Recognition

Comparison relation is an important part of measurable quantitative information. A comparison relation usually exists in a sentence in the form of a specific phrase or a comparative symbol near the numeral, such as '不超过' (*no more than*), '<' (*less than*) and so on. Therefore, we use a rule-based approach with a predefined dictionary to extract comparison relations [5].

As the scope of corpus expands, the contents of comparative relation words also vary dramatically. Therefore, in order to fully recognize comparison relation, we expand the predefined dictionary using a mask-based approach. Inspired by BERT's Mask strategy [22] in the pre-training stage, we also Mask comparative words in sentences, and then use BERT to predict similar words. For example, we mask '不超过' (*no more than*) and get a mask word '不[MASK]超过' (*no [MASK] more than*). Then, by using BERT to predict [MASK], candidate words are generated, such as '不应超过' (*shall not exceed*), '不可超过' (*not more than*), '不能超过' (*cannot exceed*), '不会超过' (*will not exceed*) and so on. For the generated similar words, we use the previous 10 words as a reference, and select the appropriate words to add to the dictionary, filter out the inconsistent words. For example, the '不会超过' listed above will be filtered, which does not match the '不超过' comparison relation.

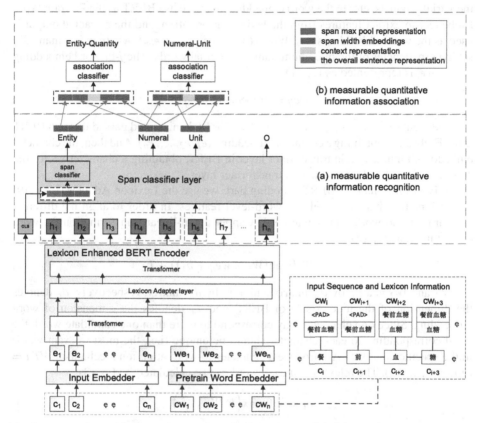

Fig. 2. The main architecture for our span-based joint model for measurable quantitative information extraction.

3.2   Measurable Quantitative Information Recognition

As shown in part (a) of Fig. 2, measurable quantitative information recognition consists of three steps: Word Embedding, Lexicon Enhanced BERT encoding and MQI Span classifying. The Lexicon Enhanced BERT Encoder is the core of our model, Which is constructed based on BERT [22] and Lexicon Adapter [19], with 12 layers of transformer and 2 layers of lexicon adapter. We use the BERT-Base-Chinese checkpoint from huggingface[1] to initialize BERT.

Giving an input sequence $S_{char} = (c_1, c_2, ..., c_n)$, the lexicon information is a words sequence $S_{word} = (cw_1, cw_2, ..., cw_n)$ of all the potential words inside the sequence by matching the S_{char} with a predefine lexicon D. D is a multi-field dictionary expanded from a medical dictionary [5], with terms including drugs, clinical indicators, chemicals, industrial products, etc. In the word embedding task, we use BERT as the Pretrain Word Embedder to extract features from the lexicon information, and the extracted output is used as the word embeddings we_i. $We_i = (we_{i1}, we_{i2}, ..., we_{im})$, where m is a manually set hyperparameter of the maximum num of potential words. The j-th word embedding we_{ij} in we_i is represented as Eq. (1).

$$we_{ij} = \text{embedder}^w (cw_{ij}) \tag{1}$$

The input sequence S_{char} is tokenized by BERT tokenizer and passed through BERT Input Embedder, obtaining a character embedding $(e_1, e_2, ..., e_n)$. And then, the character embedding input a certain transformer layer in BERT, obtaining a character vector (hc 1, hc 2, ..., hc n). c denotes the c-th transformer layer.

In Lexicon Enhanced BERT encoding part, we use the Lexicon Adapter (LA) [19] to combine the character-level and word-level features. In order to align the different of character and word representation, LA apply a non-linear transformation for word embeddings, as Eq. (2).

$$v_{ij} = W_2 (\tanh(W_1 \times we_{ij} + b_1)) + b_2 \tag{2}$$

W_1 and W_2 is a d_c by d_w matrix and a d_c by d_w matrix, respectively. d_c denotes the dimension of the hidden size of BERT, and d_w denotes the dimension of word embedding. Since one character may correspond to more than one candidate word, the word corresponding to each character should be unique (i.e., the most relevant word) in the specific context. LA introduce a character-to-word attention mechanism. $VT\, i = (v_{i1}, v_{i2}, ..., v_{im})$. The relevance of each word can be calculated by Eq. (3).

$$a_i = \text{softmax}\left(h_i^c W_{attn} V_i^T\right) \tag{3}$$

W_{attn} is the weight matrix of bilinear attention. Then the weight sum of all words can be calculated by Eq. (4) and combine the weighted lexicon information with character information by Eq. (5).

$$z_i = \sum_{j=1}^{m} a_{ij} v_{ij} \tag{4}$$

$$\tilde{h}_l = h_i^c + z_i \tag{5}$$

After lexicon enhanced BERT encoding, span classifying task is to recognize the span of entity, numeral, unit and O (means non-MQI) in sequence. We use a lightweight

---

[1] https://huggingface.co/bert-base-chinese.

reasoning based on BERT embedding [24] for span classifying, which detects entity, numeral and unit among all spans (or subsequences). For example, the Chinese sequence '体温高达39摄氏度' (*Body temperature up to 39 °C*) maps to the spans ('体'), ('体', '温'), ('摄', '氏'), ('摄', '氏', '度'), etc.. As shown in Fig. 2, the span representation input to the span classifier consists of three parts: BERT encoding of span, span-width embedding, and the representation of the overall sentence.

$$x^s = \text{maxpool}(h_i, h_{i+1}, \ldots, h_{i+k}) \oplus w_{k+1} \oplus c \tag{6}$$

As shown in Eq. (6), the span encodings are combined using a max-pool fusion, $\text{maxpool}(h_i, h_{i+1}, \ldots, h_{i+k})$. The span-width embedding w_{k+1} from a dedicated embedding matrix [23]. The representation of the token '[CLS]' represent the overall sentence c. Where $\oplus$ denotes concatenation. Finally, the span representation is fed into a softmax classifier, as shown in Eq. (7).

$$\widehat{y^s} = \text{softmax}(W^s \cdot x^s + b^s) \tag{7}$$

3.3 Measurable Quantitative Information Association

As shown in part (b) of Fig. 2, we pair off spans with measurable information as candidate association pairs, and then used an association classifier to identify the relations between them. And the span that doesn't contain measurable quantitative information, we filter it out. We define the link type between an Entity and a correlated numeral as 'Entity-Quantity' and the link type between a numeral and a corresponding unit as 'Numeral-Unit'. In this association task, we use hybrid representation of candidate association pairs to further identify link types. The hybrid representation consists of three parts: the encoding of spans (e^r), span-width embeddings (e^w), and the representation of the context between two spans of each candidate pair ($c(s_1, s_2)$) [24], which can be calculated by Eq. (8), Eq. (9) and Eq. (10), respectively.

$$e^r = \text{maxpool}(h_i, h_{i+1}, \ldots, h_{i+k}) \oplus \text{maxpool}(h_j, h_{j+1}, \ldots, h_{j+l}) \tag{8}$$

$$e^w = w_{k+1} \oplus w_{l+1} \tag{9}$$

$$c(s_1, s_2) = \text{maxpool}(h_{i+k+1}, h_{i+k+2}, \ldots, h_{j-1}) \tag{10}$$

$$x^r = e^r \oplus c(s_1, s_2) \oplus e^w \tag{11}$$

Finally, the hybrid representation of candidate pair (x^r) pass through a single-layer association classifier:

$$\widehat{y^r} = \sigma(W^r \cdot x^r + b^r) \tag{12}$$

σ denotes a sigmoid of size 2. Any link type of candidate pair with a score $\geq \theta$ (a confidence threshold) is considered activated. If none is activated, the two spans of the candidate pair are considered unrelated.

4   Evaluation and Results

4.1   Datasets

The performance of our model is evaluated on the Clinical Quantitative Information (CQI) and the Standard Quantitative Information (SQI) datasets. The statistics of the two datasets is shown in Table 1. In particular, '#Sent' denotes the count of sentences, '#Ent' represents the total number of Entities, and '#Rel' represents the count of associations.

Table 1. The statistics of the datasets.

Dataset	Type	Train	Dev	Test	Total
CQI	#Sent	4,735	290	2,813	7,838
	#Ent	11,373	609	6,391	18,373
	#Rel	7,552	404	4,245	12,201
SQI	#Sent	1,919	119	1,068	3,106
	#Ent	1,462	90	992	2,544
	#Rel	1,013	62	686	1,761

Clinical Quantitative Information (CQI). This is an electronic medical record dataset from a burn injury department of a three-A hospital in mainland China. The dataset is also used as an evaluation dataset of previous work [5]. The original dataset is tagged by using a BIO scheme. Compared with BIO tagging scheme, we add a span-based scheme which more focus on the boundary of component information. Table 2 lists an example of annotation with both the BIO and the span-based tagging schemes as a comparison.

Table 2. The annotation of using BIO and span-based tagging schemes on an example text 体温 36 °C' (*The body temperature is* 36 °C)'.

Character	体	温	3	6	°C	。
BIO Label	B-Entity	I-Entity	B-Num	I-Num	B-Unit	O
Index	0	1	2	3	4	5

	Span Type	Start Index	End Index
Span Label1	Entity	0	1
Span Label2	Num	2	3
Span Label3	Unit	4	4

Standard Quantitative Information (SQI). This is a measurable information dataset constructed from 30 full Chinese national standards text provided by The China National Institute of Standardization[2]. This dataset is tagged by using a span-based scheme and randomly shuffled into three subgroups as training dataset (62%), developing dataset (3%) and testing dataset (35%).

4.2 Evaluation Metrics

The performance of models is evaluated using widely applied metrics including Precision, Recall and F1-score. Precision is the proportion of true positives in all predicted positive samples, defined as Eq. (13). Recall is the proportion of true positives in all positive samples, defined as Eq. (14). TP (true positive) is the number of positive samples predicted to be positive, and FN (false negative) is the number of positive samples predicted to be negative. FP (false positive) is the number of negative samples predicted to be positive. The F1-score is the harmonic average of the precision and recall, defined as Eq. (8).

$$Precision = \frac{TP}{TP + FP} \tag{13}$$

$$Recall = \frac{TP}{TP + FN} \tag{14}$$

$$F1\text{-}score = \frac{2 \times Precision \times Recall}{Precision + Recall} \tag{15}$$

4.3 Results

To validate the proposed joint measurable quantitative information extraction model, five experiments are conducted. The first experiment is to verify the effectiveness of our lexicon- enhanced model in joint extraction by comparing to two baseline methods, including SPN4RE [25] and SpERT [24]. Table 3 presents the experiment results, in which 'P@REC', 'R@REC', 'F1@REC' represent precision, recall and F1-score at recognition respectively, while 'P@ASS', 'R@ASS', 'F1@ASS' represent precision, recall and F1-score at association respectively.

Our model outperforms other baseline methods on all evaluation metrics for both recognition and association tasks, achieving an F1-score 97.44% on recognition of CQI, an F1-score 96.79% on association of CQI. Compared with the other two span-based baseline methods, our model integrates lexicon information and has better performance in recognition, which achieves 85.72% F1-score. With the improvement of entity quality, the association performance is further improved and achieves 83.17% F1-score.

---

[2] https://www.cnis.ac.cn/pcindex/

Table 3. Evaluation results comparing to the baseline methods on the test dataset.

Dataset	Methods	P@REC (%)	R@REC (%)	F1@REC (%)	P@ASS (%)	R@ASS (%)	F1@ASS (%)
CQI	SPN4RE	93.58	93.80	93.69	93.58	93.80	93.69
	SpERT	96.48	97.75	97.11	96.14	96.80	96.47
	Our model	**96.71**	**98.17**	**97.44**	**96.60**	**96.98**	**96.79**
SQI	SPN4RE	66.90	67.49	67.19	66.90	67.49	67.19
	SpERT	83.03	81.85	82.44	82.13	74.34	78.04
	Our model	**84.95**	**86.64**	**85.72**	**84.63**	**82.02**	**83.17**

To test the effectiveness of the proposed span-based joint model for measurable quantitative information recognition, we compared six baseline methods based on sequence labeling. The two methods adding external features are based on medical external features, while the SQI dataset is a multi-field dataset, so experimental results for these two methods on SQI are unavailable. As shown in Table 4, 'P@CQI', 'R@CQI', 'F1@CQI' represent precision, recall and F1-score at recognition on CQI dataset respectively. As well as, 'P@SQI', 'R@SQI', 'F1@SQI' represent precision, recall and F1-score at recognition on SQI dataset respectively. Whether on the CQI dataset or SQI dataset, our span-based model outperforms the best baseline model by 2% on all metrics.

Table 4. The comparison of our model with the sequence labeling method.

Methods	P@CQI (%)	R@CQI (%)	F1@CQI (%)	P@SQI (%)	R@SQI (%)	F1@SQI (%)
Bi-LSTM-CRF	93.67	93.79	93.73	54.58	62.70	58.36
CRF + external features	93.54	94.61	94.08	–	–	–
Bi-LSTM-CRF + external features	93.81	94.74	94.27	–	–	–
Lattice-LSTM	96.05	94.53	95.28	70.41	74.60	72.44
LGN	94.41	95.47	95.08	67.77	75.06	71.23
LEBERT	94.81	96.99	95.89	81.74	84.42	83.06
Our model	**96.71**	**98.17**	**97.44**	**84.95**	**86.64**	**85.72**

The third experiment is to evaluate the robustness and stability of the proposed span-based joint model with different sizes of training data. We randomly sampled the two datasets at the proportions of 20%, 40%, 60% and 80% respectively, and evaluate the performance of recognition and association on the testing dataset. The recognition performance of our model is shown by the solid line in Fig. 3, and its association performance is shown by the dotted line in Fig. 3. Our model presents a stable performance in extracting Entity, Numeral, and Unit from the CQI dataset. As the CQI dataset size increases from 20% to 80%, the F1-measure of Unit increases from 96.92% to 98.02%. As well as, the F1-measure of Entity increased by 3% from 94.77%. The F1-measure of Numeral stabilizes around 97%. As the size of the dataset increases, the performance of our model in extracting Entity, Numeral and Unit from the SQI dataset steadily improves. Different from CQI, SQI contains more complex measurable quantitative information, and the training data of SQI is also less. When 20% of random sampling is taken, the training set of SQI contains only 384 sentences. When the random sampling reaches 40%, the performance of the model is stable and improves with the increase of training data.

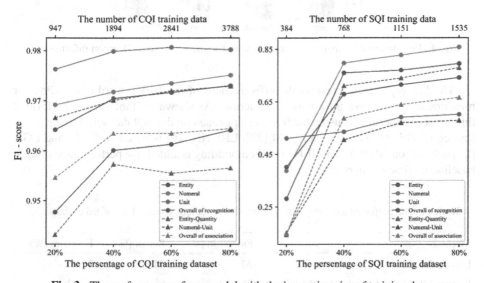

Fig. 3. The performance of our model with the increasing size of training data.

The dotted line in Fig. 3 presents the stable association performance of our model. On the CQI dataset, the association performance of the model is stable. When the size of the training set is only 20%, The F1-measure of 'CQI-Overall' still achieves 95.46%. Moreover, training with 80% dataset, the F1-measure of 'CQI-Overall' is 96.44%. Due to the small amount of SQI dataset, the association performance of our model is stable after training with more than 40% of the training set.

The fourth experiment verifies the validity of the fusion dictionary information by comparing the differences in MQI recognition performance of the models on SQI dataset. The experimental results are shown in Fig. 4, where 'without lexicon information' represents the recognition performance of the model without lexicon information. The comparison shows that the model with lexicon information improves the recognition of Entity the most, with the F1 score increasing from 74.37% to 83.01%. Meanwhile, the fusion of lexicon information also helps the recognition of numeral and unit.

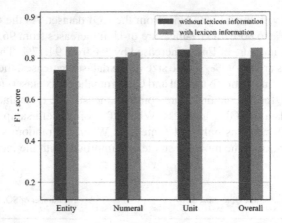

Fig. 4. Performance comparison of models with the incorporation of lexicon information.

The last experiment presents the effectiveness of the BERT word embedding for measurable quantitative information extraction. As shown in Table 5, we performed three sets of experiments using the baseline approach on the SQI dataset. Among them, Lattice-LSTM [16] is based on LSTM [27], LRCNN [20] is based on CNN, and LGN [21] is based on GNN. When BERT word embedding is added, the performance of each baseline method is improved.

Table 5. The performance comparison of baseline methods using BERT word embedding.

Methods	Precision (%)	Recall (%)	F1-score (%)
Lattice-LSTM	70.41	74.60	72.44
Lattice-LSTM + BERT word embedding	85.26	80.11	82.61
LRCNN	66.06	75.05	70.27
LRCNN + BERT word embedding	74.74	78.68	76.66
LGN	67.77	75.06	71.23
LGN + BERT word embedding	81.36	72.31	76.57

5   Conclusions

This paper proposed a span-based joint model for measurable quantitative information extraction with lexicon enhanced BERT. Based on fusing lexicon information and character information, our model uses BERT embedding and span-based inference methods to jointly extract MQI, fusing word features and contextual features while reducing erroneous delivery. Experiments were conducted to evaluate the effectiveness of the proposed joint model through a comparison with a list of baseline methods. The results showed that our span-based joint model obtained the best performance in all the evaluation metrics, demonstrating our span-based extraction model is more efficient than the previous model based sequence labeling for measurable quantitative information extraction. Our method of fusing contextual information and generating sentence embeddings is too simple, and BERT has not been pre-trained on relevant domain datasets. How to generate word embeddings and sentence embeddings that are effective for measurable quantitative information extraction is a problem that we need to solve in the future.

Acknowledgements. The work is supported by grants from National Natural Science Foundation of China (No. 61871141), Natural Science Foundation of Guangdong Province (2021A1515011339), Presidential Foundation of China National Institute of Standardization (522022Y-9406), and Collaborative Innovation Team of Guangzhou University of Traditional Chinese Medicine (No. 2021XK08).

References

1. Hao, T., We, Y., Qiang, J., Wang, H., Lee, K.: The representation and extraction of quantitative information. In: Proceedings of the 13th Joint ISO-ACL Workshop on Interoperable Semantic Annotation (ISA-13) (2017)
2. Maguire, A., Johnson, M.E., Denning, D.W., Ferreira, G.L.C., Cassidy, A.: Identifying rare diseases using electronic medical records: the example of allergic bronchopulmonary aspergillosis. Pharmacoepidemiol. Drug Saf. **26**(7), 785–791 (2017)
3. Frost, D.W., Vembu, S., Wang, J., Tu, K., Morris, Q., Abrams, H.B.: Using the electronic medical record to identify patients at high risk for frequent emergency department visits and high system costs. Am. J. Med. **130**(5), 601-e17 (2017)
4. Lossio-Ventura, J.A., et al.: Towards an obesity-cancer knowledge base: Biomedical entity identification and relation detection. In: 2016 IEEE International Conference on Bioinformatics and Biomedicine (BIBM), pp. 1081–1088 (2016)
5. Liu, S., Nie, W., Gao, D., Yang, H., Yan, J., Hao, T.: Clinical quantitative information recognition and entity-quantity association from Chinese electronic medical records. Int. J. Mach. Learn. Cybern. **12**(1), 117–130 (2020). https://doi.org/10.1007/s13042-020-01160-0
6. Hao, T., Wang, H.: Semantic annotation framework (SemAF)—Part 11: Measurable Quantitative Information (MQI). ISO/DIS 24617-11, International Organization for Standardization (2021)
7. Wong, K.F., Li, W.J., Xu, R.F., Zhang, Z.S.: Introduction to Chinese natural language processing. Synt. Lect. Hum. Lang. Technol. **2**(1), 1–148 (2009)
8. Zheng, S., Wang, F., Bao, H., Hao, Y., Zhou, P., Xu, B.: Joint extraction of entities and relations based on a novel tagging scheme. arXiv preprint arXiv:1706.05075 (2017)
9. Li, Q., Ji, H.: Incremental joint extraction of entity mentions and relations. In: Proceedings of the 52rd Annual Meeting of the Association for Computational Linguistics, pp. 402–412 (2014)
10. Hao, T., Liu, H., Weng, C.: Valx: a system for extracting and structuring numeric lab test comparison statements from text. Methods Inf. Med. **55**(03), 266–275 (2016)
11. Liu, S., Pan, X., Chen, B., Gao, D., Hao, T.: An automated approach for clinical quantitative information extraction from Chinese electronic medical records. In: Siuly, S., Lee, I., Huang, Z., Zhou, R., Wang, H., Xiang, W. (eds.) HIS 2018. LNCS, vol. 11148, pp. 98–109. Springer, Cham (2018). https://doi.org/10.1007/978-3-030-01078-2_9

12. Hao, T., Pan, X., Gu, Z., Qu, Y., Weng, H.: A pattern learning-based method for temporal expression extraction and normalization from multi-lingual heterogeneous clinical texts. BMC Med. Inform. Decis. Mak. **18**(1), 15–25 (2018)

13. Tang, B., Cao, H., Wu, Y., Jiang, M., Xu, H.: Clinical entity recognition using structural support vector machines with rich features. In: Proceedings of the ACM Sixth International Workshop on Data and Text Mining in Biomedical Informatics, pp. 13–20 (2012)

14. Gruss, R., Abrahams, A.S., Fan, W., Wang, G.A.: By the numbers: the magic of numerical intelligence in text analytic systems. Decis. Support Syst. **113**, 86–98 (2018)

15. Li, L., Zhao, J., Hou, L., Zhai, Y., Shi, J., Cui, F.: An attention-based deep learning model for clinical named entity recognition of Chinese electronic medical records. BMC Med. Inform. Decis. Mak. **19**(5), 1–11 (2019)

16. Zhang, Y., Yang, J.: Chinese NER using lattice LSTM. In: Proceedings of the 56th Annual Meeting of the Association for Computational Linguistics (Volume 1: Long Papers), pp. 1554–1564 (2018)

17. Qiu, X., Sun, T., Xu, Y., Shao, Y., Dai, N., Huang, X.: Pre-trained models for natural language processing: a survey. Science China Technol. Sci. **63**(10), 1872–1897 (2020). https://doi.org/10.1007/s11431-020-1647-3

18. Zhang, X., et al.: Extracting comprehensive clinical information for breast cancer using deep learning methods. Int. J. Med. Inform. **132**, 103985 (2019)

19. Liu, W., Fu, X., Zhang, Y., Xiao, W.: Lexicon enhanced Chinese sequence labeling using BERT adapter. In: Proceedings of the 59th Annual Meeting of the Association for Computational Linguistics and the 11th International Joint Conference on Natural Language Processing (Volume 1: Long Papers), pp. 5847–5858 (2021)

20. Gui, T., Ma, R., Zhang, Q., Zhao, L., Jiang, Y.G., Huang, X.: CNN-based Chinese NER with lexicon rethinking. In: IJCAI, pp. 4982–4988 (2019)

21. Gui, T., et al.: A lexicon-based graph neural network for Chinese NER. In: Proceedings of the 2019 Conference on Empirical Methods in Natural Language Processing and the 9th International Joint Conference on Natural Language Processing (EMNLP-IJCNLP), pp. 1040–1050 (2019)

22. Devlin, J., Chang, M.-W., Lee, K., Toutanova, K.: BERT: pre-training of deep bidirectional transformers for language understanding. In: Proceedings of the 2019 Conference of the North American Chapter of the Association for Computational Linguistics: Human Language Technologies, Volume 1 (Long and Short Papers), pp. 4171–4186 (2019)

23. Lee, K., He, L., Lewis, M., Zettlemoyer, L.: End-to-end neural coreference resolution. In Proceedings of the 2017 Conference on Empirical Methods in Natural Language Processing, pp. 188–197 (2017)

24. Eberts, M., Ulges, A.: Span-based joint entity and relation extraction with transformer pre-training. In ECAI **2020**, 2006–2013 (2020)

25. Sui, D., Chen, Y., Liu, K., Zhao, J., Zeng, X., Liu, S.: Joint entity and relation extraction with set prediction networks. In: AAAI (2021)

26. Lafferty, J., McCallum, A., Pereira, F.C.: Conditional random fields: probabilistic models for segmenting and labeling sequence data. In: Proceedings of ICML, vol. 3(2), pp. 282–289 (2001).

27. Chen, X., Qiu, X., Zhu, C., Liu, P., Huang, X.J.: Long short-term memory neural networks for Chinese word segmentation. In: Proceedings of the 2015 Conference on Empirical Methods in Natural Language Processing, pp. 1197–1206 (2015)

Bayesian Optimization Based Seq2Seq Network Models for Real Estate Price Prediction in Hong Kong

Yonglin Liu, Zeqiong Wu, Choujun Zhan, and Hu Min[✉]

School of Electrical and Computer Engineering, Nanfang College Guangzhou, Guangzhou, Guangdong, China
zchoujun2-c@my.cityu.edu.hk, minghu032@gmail.com

Abstract. Accurate house price prediction plays an important role in the real estate market. This study develops Seq2Seq models based on Bayesian Optimization using Hong Kong Island estate price dataset from January 1, 1996 to May 13, 2021. In addition, the currency information is also collected and employed in the Seq2Seq models. In comparison, deep learning algorithm including CNN, GRU, LSTM and Broad Learning System (BLS) are used in this study. To assess the performance of the models, Root mean square error (RMSE), Mean absolute error (MAE), R-Square (R^2) and Median Absolute Deviation (MAD) of the models are determined. Experiments demonstrate that Seq2Seq models are more suitable for estate prices prediction problems.

Keywords: House price forecasting · Seq2Seq · Attention · Bayesian optimization · Hong Kong Island

1 Introduction

A growing body of research considers the importance of the real estate market in economic growth. The rapid development of the real estate market is highly related to the increasing growth of the national economy [7,30,33]. Researchers have developed different techniques for forecasting the estate price. For instance, it has been developed to study the relationship between house characteristics and house price [5,6,14,25,36]. Harding et al. firstly adopted the ordinary least square (OLS) model and Hedonic model to investigate the bargaining effects of vacant and occupied houses by using house sales samples from 1993 to 1999 in the US. The results indicated that controlling the bargaining power is important in Hedonic price models. In addition, the bargaining power affects the negotiated price [16]. Hedonic price models take into account the heterogeneous character of real estate by Maurer et al. This study covers 84,686 transactions from January

H. Zhang et al. (Eds.): NCAA 2022, CCIS 1638, pp. 359–371, 2022.
https://doi.org/10.1007/978-981-19-6135-9_27

1990 to December 1999 in the French real estate market [23]. However, Hedonic price models are limited in their ability to capture nonlinear relationships because they are based on linear analysis. Therefore, researcher try to find other methods/techniques, which require fewer data.

In recent years, artificial intelligence algorithms have significantly developed. Researchers found that the artificial neural network (ANN) [19,22], AdaBoost [24] and SVM [15] can better capture the house characteristics and show better performance in predicting house price. Selim confirmed that ANN can better predict Turkish house prices. Experimental results show that the fuzzy neural network prediction model has a strong function approximation ability and can predict house prices based on the quality of data [26]. The C4.5, RIPPER, Naive Bayesian, and AdaBoost in machine learning are used by Park and Bae as a research method to predict whether the closing price of a house will be higher or lower than the listing price [24]. Their research showed that machine learning algorithms could improve the predictability of house prices models.

Additionally, the task of developing house prices prediction models with deep learning methods have also been explored by researchers. Chen et al. used housing price data from January 2004 to October 2016 to predict the average house prices of Beijing, Shanghai, Guangzhou, and Shenzhen in November and December 2016 [9]. They compared the LSTM network and the Autoregressive Integrated Moving Average (ARIMA) in that study, whose results show that the LSTM model has good prediction performance for time series. In another study, ARIMA, LSTM and HYBRID (ARIMA, LSTM) model have been used to predict housing sales, which results show that HYBRID model has the best performance among these three models [29]. Jozefowicz et al. examined the GRU and LSTM models and discovered that the GRU model can obtain equivalent outcomes to the LSTM model [18]. Furthermore, CNN performs well in time series prediction [13]. Nowadays, existing researches have demonstrated that using encoder-decoder architectures improves real-time prediction performance [11,12,17]. Du et al. proposed sequence-to-sequence (Seq2Seq) model for forecasting air quality multivariate time series by using the LSTM based encoder-decoder architecture [11]. The results revealed that Seq2Seq model can accurately estimate multivariate time series. Jin et al. developed time-series data modeling an encoder-decoder architecture with a gated recurrent units (GRU) with high robustness, which includes the temporal attention layer to capture the key features of input data [17]. Nevertheless, their encoder-decoder methods have a problem in that they only focus on one recurrent unit. Hence, in order to compare which recurrent unit is superior, we develop the Seq2Seq framework, which used encoder-decoder architectures with GRU and LSTM recurrent units, respectively.

However, hyperparameters have a significant impact on the performance of the Seq2Seq architecture. In general, optimization hyperparameters techniques (such as the genetic algorithm, Bayesian optimization algorithm, and differential evolution algorithm) are the most widely utilized conventional algorithms, and are commonly used in combination with other algorithms [35]. Ahn et al. proposed an extended Ridge Regression model, which combines ridge regression with a genetic algorithm (GA-Ridge). In order to verify the performance of the GA-Ridge model, an experimental design was conducted on the Korea real estate market [1]. In another study, the hybrid GA and support vector machine (GA-SVM) model, which utilizes GA to select optimal hyper-parameters of the SVM, was developed by Gu et al. for residential price prediction [15]. Among them, the GA is used to optimize the parameters of the SVM. Xia et al. employed Bayesian hyperparameters to optimize the hyperparameters of XGBoost for credit scoring, and the findings demonstrated that it outperforms random search, grid search and manual search [31]. Xiong et al. suggested a stacked generation model that consists of various regression models for predicting house price, each of which uses bayesian optimization to find the appropriate hyperparameters [32]. Li and Tian utilized a multiobjective differential evolution (MODE) method by integrating the bagging concept into the MODE evolution process [21]. Therefore, developing an effective technique for determining the optimal model hyperparameters is a critical step in resolving the problem, which is why we consider the bayesian optimization algorithm in this work.

In conclusion, accurate house price prediction is important for studying the real estate market. This work focuses on forecasting real estate market prices and considering currency information in Hong Kong Island. A hybrid method combining Seq2Seq-GRU/LSTM with Bayesian Optimization algorithm is proposed. In this method, the hyperparameters are optimized by the Bayesian Optimization algorithm. In addition, four other neural network models, including CNN, GRU, LSTM and broad learning system models, are adopted as the baseline model for comparison. The following are the major contributions of this study:

- We develop the Seq2Seq model based on GRU and LSTM recurrent units to predict house prices, and incorporate an attention mechanism to capture the nonlinear relationship between features and house prices.
- Utilising Bayesian optimization algorithm to select the hyperparameters of the Seq2Seq-GRU/LSTM and enhances the model prediction performance.

The remainder of this paper is organized as follows. Section 2 introduces the dataset. The details of the proposed Bayesian Optimization based Seq2Seq models are presented in Sect. 3. In Sect. 4, the experimental results are described, while Sect. 5 presents conclusion of the entire study and indicates the future work.

2   Dataset

This study predicts the Hong Kong Island estate price based on two classes of features: housing price information and currency information. These two kinds of datasets are derived from the information platform of the Hong Kong real estate market (https://www.28hse.com/) and the Hong Kong Monetary Authority (https://www.hkma.gov.hk/). The housing information dataset contains seventeen basic characteristics of housing unit, including the transaction date, the district, the estate, the change of price, the gross area, the usable area, the unit price of usable area, the unit price of gross area, the total transaction for the unit, the occupancy date, the school district, the annual total transaction, the estate contract form of agreement, the estate contract form of assignment, the estate contract form of speculation, and the longitude and latitude of the estate.

The currency information includes a combination of Hong Kong money supply indicators. There are ten indicators, including Legal tender notes and coins in hands of public, Demand deposits with licensed banks, money supply M1, Savings deposits with licensed banks, Time deposits with licensed banks, NCDs issued by licensed banks and held by public, money supply M2, Deposits with RLBs and DTCs, NCDs issued by RLBs and DTCs and held by public, and money supply M3.

With the growth of the money supply, the demand in the real estate market also grows, which affects the growth of real estate prices [34]. Figure 1 shows the Pearson correlation plot [3] that was created to determine and present the correlation between the selected variables and the house price. It can seen that the target variable (unit price of gross area) is significantly correlated with the change of price, transaction date, the unit price of usable area, M1, M2, M3, Time deposits with licensed banks, Demand deposits with licensed banks, Savings deposits with licensed banks, Legal tender notes and coins in hands of public and NCDs issued by licensed banks and held by public.

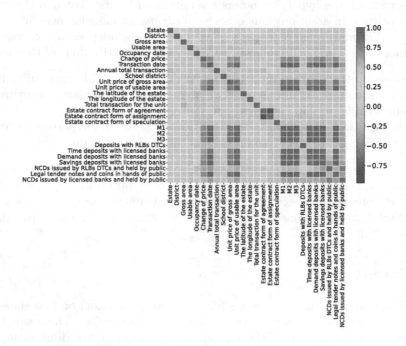

Fig. 1. The heatmap of Pearson correlation coefficient

3   Methodologies

Here, hybrid methods combining Seq2Seq-GRU/LSTM with Bayesian Optimization algorithm is proposed. The framework of the proposed methods is shown in Fig. 2. In the flowing part, we give a detailed description of the proposed methods.

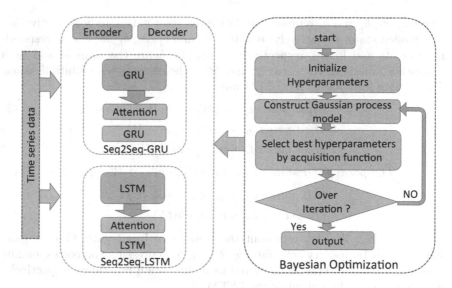

Fig. 2. Seq2Seq-GRU/LSTM with Bayesian optimization.

3.1   Seq2Seq

Sequence-to-sequence (Seq2Seq) was first proposed in 2014 [10, 28], which is utilized to improve the performance of machine translation systems. The Seq2Seq model consists of three components, including encoder, intermediate vector, and decoder. The main idea of adopting Seq2Seq to solve the problem is to use a deep neural network model, which can be LSTM or other recurrent units. In this study, GRU and LSTM are used as recurrent units. Mapping a sequence with a cyclic unit as input to a sequence with another cyclic unit as output. This process consists of two links: Encoder input and Decoder output. The encoder is responsible for encoding the sequence into a fixed length vector, which is the intermediate vector. The intermediate vector is used as the input of the decoder, while the variable-length vector is output.

Given the time series $x \in R^{1 \times T}$, the update formula of GRU and LSTM unit is as follows.

- The update formula for the GRU unit:

$$z_t = \sigma(W_z \cdot h_{t-1} + x_t \cdot b_z)$$
$$r_t = \sigma(W_r \cdot h_{t-1} + x_t \cdot b_r)$$
$$\hat{h}_t = tanh(W[r_t \bigotimes h_{t-1}; x_t] + b_{\hat{h}}) \qquad (1)$$
$$h_t = (1 - z_t) \bigotimes h_{t-1} + z_t \bigotimes \hat{h}_t$$

where z_t, r_t, $\hat{h}_t$ and h_t represent update gate, reset gate, candidate activation and hidden state, respectively; W_z, W_r, W and b_z, b_r, $b_{\hat{h}}$ are the corresponding weights and biases, respectively; σ and $tanh$ are the sigmoid and tanh activation functions, respectively; and $\bigotimes$ is the element-wise multiplication.
- The update formula for the LSTM unit:

$$Input\ gate : i_t = \sigma(W_i \cdot x_t + W_i \cdot h_{t-1} + b_i) \qquad (2)$$

$$Output\ gate : o_t = \sigma(W_o \cdot h_{t-1} + W_o \cdot x_t + b_o) \qquad (3)$$

$$Forget\ gate : f_t = \sigma(W_f \cdot x_t + W_f \cdot h_{t-1} + b_f) \qquad (4)$$

$$Temporary\ cell\ state : \widetilde{C_t} = tanh(W_c \cdot h_{t-1} + W_c \cdot x_t + b_c) \qquad (5)$$

$$Cell\ state : C_t = f_t \cdot C_{t-1} + i_t \cdot \widetilde{C_t} \qquad (6)$$

$$Output : o_{lstm} = o_t \cdot tanh(C_t) \qquad (7)$$

where W_i, W_o, W_f, W_c represent the weights of Input gate, Output gate, Forget gate, Temporary cell state; b_i, b_o, b_f, b_c represents the corresponding bias; σ and $tanh$ are the sigmoid and tanh activation functions, respectively; o_{lstm} represents the output of the LSTM.

3.2   The Attention Mechanism

Since encoding the information of the entire input sequence into a fixed-length vector is a bottleneck to improve the performance of the encoder-decoder structure, Bahdanau et al. propose a solution mechanism called attention mechanism [2]. Let the output target of Seq2Seq be y_t, then, the conditional probability of hidden layer of GRU and LSTM is $p(y_t|y_1, y_2, .., y_{t-1}, x) = g(y_{t-1}, s_t, c_t)$, where s_t is the hidden state of GRU and LSTM at time t. s_t can be derived by $s_t = f(s_{t-1}, y_{t-1}, c_t)$. Each conditional probability depends on a different context vector c_t. Here, c_t depends on the sequence annotation $a_1, a_2, ..., a_T$, which is the input sequence of the encoder. Each sequence and annotation a_t contains information about the entire input sequence. Focus on the part surrounding the ith word of the input sequence. c_t can be derived by:

$$c_t = \sum_{k=1}^{T} a_{tk} a_k \qquad (8)$$

$$a_{tk} = \frac{exp(e_{tk})}{\sum_{i=1}^{T} exp(e_t i)} \qquad (9)$$

$$e_{tk} = f(hide_{t-1}, a_k) \qquad (10)$$

where a_{tk} determines the importance of a_k to the context vector c_t when generating the output sequence at t time. $hide_{t-1}$ is the hidden state of the decoder GRU and LSTM at time $t - 1$, $f()$ is a function of the relationship scoreused to compute both h_{t-1} and a_k.

3.3 Bayesian Optimization (BO)

The Bayesian optimization is an optimization algorithm, which can help automatically adjust the hyperparameters of the model [27]. The main idea is to use a prior function to build a model to approximate the posterior distribution of an unknown function and use the acquisition function to select the best hyperparameters. The ultimate goal is to reduce the number of training while improving the effect of the model. Bayesian optimization assumes that the hyperparameters of the model have a functional relationship with the objective function. In this study, the Bayesian Optimization algorithm is used to select hyperparameters in Seq2Seq-GRU/LSTM models. Here, prior function is a Gaussian process, acquisition function is Expected Improvement (EI), and objective function is the evaluation index R^2 of the model. Therefore, our Bayesian optimization goal is to maximize the objective function $f(x)$:

$$f(x) = \arg\max_{x \in X} R^2(x)$$
$$R^2 = 1 - \frac{\sum(y^i - \hat{y}^i(x))^2}{\sum(y^i - \overline{y})^2} \tag{11}$$

where x is one hyperparameters of the model, and X is the hyperparameter set. y^i is the true house price of the i-th sample, $\hat{y}^i$ is the predicted house price of the i-th sample, and $\overline{y}$ is the average house price.

4 Experimental Results

Table 1. The predictive results of the proposed Seq2Seq-GRU/LSTM methods and the other methods.

Model	RMSE	MAE	R^2	MAD
CNN	1386.8306	893.9697	0.895	588.1768
GRU	1316.5019	827.0515	0.9054	505.0254
LSTM	1308.9468	816.6931	0.9065	507.3019
BLS	1324.6894	789.5121	0.9046	472.0937
Seq2Seq-GRU	**1237.7833**	**746.4658**	**0.9165**	**447.0731**
Seq2Seq-LSTM	1260.1745	772.29	0.9133	457.9627

Here, we forecast the real estate price in Hong Kong Island. The dataset covers 25 years from January 1996 to May 2021. We develop Seq2Seq-GRU and Seq2Seq-LSTM for developing housing price prediction models, and use CNN [20], GRU [18], LSTM [4] and BLS [8,37] as baseline models for comparison. In addition, we add BO algorithm to adjust the models parameters. In order to evaluate the predictive performance of the model, the following evaluation indicators are employed: RMSE, MAE, R^2 and MAD.

$$RMSE = \sqrt{\frac{1}{m}\sum_{i=1}^{m}(y^i - \widehat{y}^i)^2}$$

$$MAE = \frac{1}{m}\sum_{i=1}^{m}|y^i - \widehat{y}^i|$$

$$R^2 = 1 - \frac{\sum(y^i - \widehat{y}^i)^2}{\sum(y^i - \bar{y})^2}$$

$$MAD = \frac{1}{m}\sum_{i=1}^{m}median_m(y^i - \widehat{y}^i)$$

(12)

where m is the number of samples; y^i is the real house price of the i-th sample; $\widehat{y}^i$ is the predicted house price of the i-th sample; $\bar{y}$ is the average house price; k is the number of sample features; and $median_m()$ represents the median value of m samples.

The hyperparameters of the model were optimised using the Bayesian optimization algorithm. The hyperparameters set of the Bayesian optimization algorithm are depicted in Table 2.

Then, we show the results of the proposed BO-based Seq2Seq-GRU/LSTM method and the other four neural network models, including CNN, GRU, LSTM and BLS. The detailed results are given in Table 1. The prediction result of the Seq2Seq-GRU, Seq2Seq-LSTM, CNN, GRU, LSTM and BLS are shown in Fig. 3. Obviously, the proposed Seq2Seq-GRU model shows the best performance among all of the six models according to the four evaluation indicators, while CNN shows the worst performance. The RMSE, MAE, R^2 and MAD of the Seq2Seq-GRU model is 1237.7833, 746.4658, 0.9165 and 447.0731, respectively. Additionally, GRU-based Seq2Seq model outperform LSTM-based Seq2Seq model in terms of predictive performance.

To verify whether the Seq2Seq-GRU/LSTM models have the ability to capture nonlinear relationship from real estate market data, we compare the Seq2Seq-GRU/LSTM with CNN, GRU, LSTM and BLS. As shown in Fig. 3, obviously, the Seq2Seq-GRU/LSTM models have better prediction results than other baseline models, indicating that the importanceof modeling the attention and recurrent units, generally have better prediction performance. Moreover, the Seq2Seq-GRU/LSTM models perform well than those based on a single model (GRU, LSTM). This is mainly due to the GRU and LSTM models are more difficult to handle complex time series data than Seq2Seq-GRU and the Seq2Seq-LSTM model.

In detail, the RMSE, the MAE and the MAD of the Seq2Seq-GRU model are reduced by approximately 5.98%, 9.74% and 11.5% compared with the GRU model, and the R^2 are approximately 1.21% higher than that of GRU. The RMSE, the MAE and the MAD of the Seq2Seq-LSTM model are approximately 3.73%, 5.44% and 9.73% lower than that of the LSTM model, while the R^2 of Seq2Seq-LSTM model is improved by 0.75%.

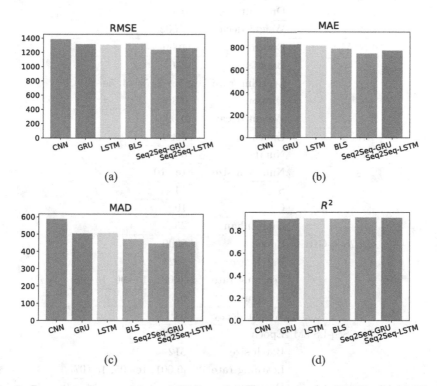

Fig. 3. Deep learning models and BLS model score result: (a) RMSE; (b) MAE; (c) MAPE; (d) R^2.

Table 2. Bayesian optimization hyperparameter results

Model	Hyperparameter	Values
CNN	Epoch	4000
	Batch size	256
	Learning rate	[5e−2, 5e−4, 5e−6]
	Dropout	0.3
	Weight decay	0.0001
GRU	Epoch	8000
	Batch size	128
	Learning rate	[1e−3, 1e−5, 1e−7]
	Dropout	0.4
	Weight decay	0.002
LSTM	Epoch	4000
	Batch size	256
	Learning rate	[5e−2, 5e−4, 5e−6]
	Dropout	0.3
	Weight decay	0.0001
BLS	Epoch	4000
	NumFea	0.4
	NumWin size	1e−10
	S	14
	C	10
	NumEnhan	25
Seq2Seq-GRU	Epoch	8000
	Batch size	1024
	Learning rate	[0.001, 1e−06, 1e−08]
	Dropout	0.5
	Weight decay	0.01
Seq2Seq-LSTM	Epoch	20000
	Batch size	512
	Learning rate	[0.001, 1e−05, 1e−07]
	Dropout	0.5
	Weight decay	0.01

5   Conclusion

This research proposed a BO-based Seq2Seq-GRU/LSTM models for estate price forecasting in Hong Kong Island. On the one hand, we use the BO algorithm to find the optimal hyperparameters of Seq2Seq models. On the other hand, the Seq2Seq models, including Seq2Seq-GRU and Seq2Seq-LSTM, are introduced to capture the nonlinear relationship of time series. Eventually, the Seq2Seq-GRU model shows the best predictive performance compared with the CNN model, the GRU model, the LSTM model and the BLS model. In summary, Seq2Seq models with BO algorithm perform well for estate price forecasting. Moreover, Our empirical framework should not be difficult to implement and has the potential to be adapted for forecasting housing prices in other regions or nations. But this study was limited to only Hong Kong Island of property, so the Seq2Seq-GRU model need to be retrained using datasets from other regions or countries. Further research should be carried out to establish other state-of-art deep learning algorithms such as Transformer and GNN models to predict housing prices.

Acknowledgment. This work was supported by Natural Science Foundation of Guangdong Province, China (2020A1515010761).

References

1. Ahn, J.J., Byun, H.W., Oh, K.J., Kim, T.Y.: Using ridge regression with genetic algorithm to enhance real estate appraisal forecasting. Expert Syst. Appl. **39**(9), 8369–8379 (2012)
2. Bahdanau, D., Cho, K., Bengio, Y.: Neural machine translation by jointly learning to align and translate. arXiv preprint arXiv:1409.0473 (2014)
3. Benesty, J., Chen, J., Huang, Y., Cohen, I.: Pearson correlation coefficient. In: Benesty, J., Chen, J., Huang, Y., Cohen, I. (eds.) Noise Reduction in Speech Processing. STSP, vol. 2, pp. 1–4. Springer, Heidelberg (2009). https://doi.org/10. 1007/978-3-642-00296-0_5
4. Bengio, Y., Simard, P., Frasconi, P.: Learning long-term dependencies with gradient descent is difficult. IEEE Trans. Neural Netw. **5**(2), 157–166 (1994)
5. Bin, O.: A prediction comparison of housing sales prices by parametric versus semi-parametric regressions. J. Hous. Econ. **13**(1), 68–84 (2004)
6. Bourassa, S.C., Haurin, D.R., Haurin, J.L., Hoesli, M., Sun, J.: House price changes and idiosyncratic risk: the impact of property characteristics. Real Estate Econ. **37**(2), 259–278 (2009)
7. Case, B., Goetzmann, W.N., Rouwenhorst, K.G.: Global real estate markets-cycles and fundamentals. Technical report, National Bureau of Economic Research (2000)
8. Chen, C.P., Liu, Z.: Broad learning system: an effective and efficient incremental learning system without the need for deep architecture. IEEE Trans. Neural Netw. Learn. Syst. **29**(1), 10–24 (2017)
9. Chen, X., Wei, L., Xu, J.: House price prediction using LSTM. arXiv preprint arXiv:1709.08432 (2017)
10. Cho, K., et al.: Learning phrase representations using RNN encoder-decoder for statistical machine translation. arXiv preprint arXiv:1406.1078 (2014)
11. Du, S., Li, T., Horng, S.J.: Time series forecasting using sequence-to-sequence deep learning framework. In: 2018 9th International Symposium on Parallel Architectures, Algorithms and Programming (PAAP), pp. 171–176. IEEE (2018)
12. Du, S., Li, T., Yang, Y., Horng, S.J.: Multivariate time series forecasting via attention-based encoder-decoder framework. Neurocomputing **388**, 269–279 (2020)

13. Gehring, J., Auli, M., Grangier, D., Yarats, D., Dauphin, Y.N.: Convolutional sequence to sequence learning. In: International Conference on Machine Learning, pp. 1243–1252. PMLR (2017)

14. Goodman, A.C.: Hedonic prices, price indices and housing markets. J. Urban Econ. **5**(4), 471–484 (1978)

15. Gu, J., Zhu, M., Jiang, L.: Housing price forecasting based on genetic algorithm and support vector machine. Expert Syst. Appl. **38**(4), 3383–3386 (2011)

16. Harding, J.P., Knight, J.R., Sirmans, C.: Estimating bargaining effects in hedonic models: evidence from the housing market. Real Estate Econ. **31**(4), 601–622 (2003)

17. Jin, X.B., et al.: Deep-learning forecasting method for electric power load via attention-based encoder-decoder with Bayesian optimization. Energies **14**(6), 1596 (2021)

18. Jozefowicz, R., Zaremba, W., Sutskever, I.: An empirical exploration of recurrent network architectures. In: International Conference on Machine Learning, pp. 2342–2350. PMLR (2015)

19. Lam, K.C., Yu, C., Lam, K.: An artificial neural network and entropy model for residential property price forecasting in Hong Kong. J. Prop. Res. **25**(4), 321–342 (2008)

20. Lawrence, S., Giles, C.L., Tsoi, A.C., Back, A.D.: Face recognition: a convolutional neural-network approach. IEEE Trans. Neural Netw. **8**(1), 98–113 (1997)

21. Li, K., Tian, H.: A bagging based multiobjective differential evolution with multiple subpopulations. IEEE Access **9**, 105902–105913 (2021)

22. Limsombunchai, V.: House price prediction: hedonic price model vs. artificial neural network. In: New Zealand Agricultural and Resource Economics Society Conference, pp. 25–26 (2004)

23. Maurer, R., Pitzer, M., Sebastian, S.: Hedonic price indices for the Paris housing market. Allg. Stat. Arch. **88**(3), 303–326 (2004). https://doi.org/10.1007/s101820400173

24. Park, B., Bae, J.K.: Using machine learning algorithms for housing price prediction: the case of Fairfax County, Virginia housing data. Expert Syst. Appl. **42**(6), 2928–2934 (2015)

25. Rosen, S.: Hedonic prices and implicit markets: product differentiation in pure competition. J. Polit. Econ. **82**(1), 34–55 (1974)

26. Selim, H.: Determinants of house prices in Turkey: hedonic regression versus artificial neural network. Expert Syst. Appl. **36**(2), 2843–2852 (2009)

27. Snoek, J., Larochelle, H., Adams, R.P.: Practical Bayesian optimization of machine learning algorithms. arXiv preprint arXiv:1206.2944 (2012)

28. Sutskever, I., Vinyals, O., Le, Q.V.: Sequence to sequence learning with neural networks. In: Advances in Neural Information Processing Systems, pp. 3104–3112 (2014)

29. Temur, A.S., Akgün, M., Temur, G.: Predicting housing sales in Turkey using ARIMA, LSTM and hybrid models (2019)

30. Wilson, I.D., Paris, S.D., Ware, J.A., Jenkins, D.H.: Residential property price time series forecasting with neural networks. In: Macintosh, A., Moulton, M., Preece, A. (eds.) Applications and Innovations in Intelligent Systems IX, pp. 17–28. Springer, London (2002). https://doi.org/10.1007/978-1-4471-0149-9_2

31. Xia, Y., Liu, C., Li, Y., Liu, N.: A boosted decision tree approach using Bayesian hyper-parameter optimization for credit scoring. Expert Syst. Appl. **78**, 225–241 (2017)

32. Xiong, S., Sun, Q., Zhou, A.: Improve the house price prediction accuracy with a stacked generalization ensemble model. In: Hsu, C.-H., Kallel, S., Lan, K.-C., Zheng, Z. (eds.) IOV 2019. LNCS, vol. 11894, pp. 382–389. Springer, Cham (2020). https://doi.org/10.1007/978-3-030-38651-1_32

33. Xu, T.: The relationship between interest rates, income, GDP growth and house prices. Res. Econ. Manag. 2(1), 30–37 (2017)

34. Xu, X.E., Chen, T.: The effect of monetary policy on real estate price growth in China. Pac. Basin Finance J. 20(1), 62–77 (2012)

35. Zhan, C., Tse, C.K., Fu, Y., Lai, Z., Zhang, H.: Modeling and prediction of the 2019 coronavirus disease spreading in China incorporating human migration data. PLoS ONE 15(10), e0241171 (2020)

36. Zhan, C., Wu, Z., Liu, Y., Xie, Z., Chen, W.: Housing prices prediction with deep learning: an application for the real estate market in Taiwan. In: 2020 IEEE 18th International Conference on Industrial Informatics (INDIN), vol. 1, pp. 719–724. IEEE (2020)

37. Zhan, C., Zheng, Y., Zhang, H., Wen, Q.: Random-forest-bagging broad learning system with applications for COVID-19 pandemic. IEEE Internet Things J. 8(21), 15906–15918 (2021)

Ensemble Learning for Crowdfunding Dynamics: JingDong Crowdfunding Projects

Hu Min[1], Kaihan Wu[1(✉)], Minghao Tan[1], Junyan Lin[1], Yufan Zheng[2], and Choujun Zhan[1(✉)]

[1] Nanfang College · Guangzhou, Guangzhou, Guangdong 510970, China
wkaihan168@gmail.com, zchoujun2-c@my.cityu.edu.hk
[2] Huangpu Institute of Materials, Guangzhou, Guangdong 510970, China

Abstract. As an emerging internet financing model with high efficiency, low cost, diversified returns, and a small investment, crowdfunding is sought after by entrepreneurs and investors. However, many crowdfunding projects are faced with the risk of low success rates and failure to reach the financing target within the specified period. Therefore, the prediction of crowdfunding project financing results and multi-model comparison are important ways to improve the project success rates and reduce market risk. First, we collected project data of JingDong (JD) crowdfunding platform for preprocessing and analyzed the characteristics of successful projects. Then, we use ensemble learning and traditional machine learning models to predict the daily amount of crowdfunding with grid search to obtain the optimal hyperparameters of each model. Several evaluation metrics are then employed to assess the performance of the model. The experimental results demonstrate that the Extra Tree Regression (ETR) ensemble model achieves the best prediction performance, with a coefficient of determination(R^2) of 90.1%, when forecasting the daily crowdfunding fundraising. Furthermore, the ensemble learning model showed significant advantages in other evaluation indicators, indicating its potential in forecasting the financing amount of crowdfunding projects.

Keywords: Crowdfunding financing prediction · Correlation coefficient · Extra Tree Regression · Grid search

1 Introduction

The earliest crowdfunding in the world can be traced back to the 18th century, while modern web-based crowdfunding began in the 21st century. The biggest difference between modern crowdfunding and previous crowdfunding is the emergence of online platforms, where people raise money for their ideas and projects, which has become a business model. In 2014, around $16.2 billion was invested on the global crowdfunding market. In 2015, that investment doubled to $34.4

H. Zhang et al. (Eds.): NCAA 2022, CCIS 1638, pp. 372–386, 2022.
https://doi.org/10.1007/978-981-19-6135-9_28

billion. By 2016 it was over $50 billion. In the coming years, this financing model will grow steadily and become the main method of project financing [21]. However, a number of crowdfunding projects can not reach their expected financing target. In other words, the crowdfunding success rate is not as ideal as everybody thinks [7]. Crowdfunding project originators, platforms, and investors are eager to know the financing results of crowdfunding projects before they are launched [22]. Therefore, exploring the key factors affecting the success of crowdfunding and establishing a crowdfunding project prediction model is an effective way to improve the success rate of crowdfunding project financing. Predicting the amount of crowdfunding before the launch of a project can help entrepreneurs formulate a strategy for launching the project and help investors select high-quality projects.

Most crowdfunding platforms adopt the e-commerce model, providing a platform for cash-strapped, creative entrepreneurs to present their products and ideas to the public on the open web. The public can then invest in their favorite projects or activities on crowdfunding platforms [25]. Due to the different uses of crowdfunding projects, the way of return also varies, which can be roughly divided into four categories: reward-based, donation-based, lending-based and equity-based [2,25]. In the case of reward crowdfunding, the funding model is "all or nothing". When the project reaches the target amount of funding within a specified period of time, the project originator will receive funds to implement their idea. If the target money is not reached, the funds raised are returned to the investors [3].

In addition, the rapid development of crowdfunding mode is accompanied by some problems that need to be solved. Online financing results are influenced by many factors. According to empirical research, the number of investors, the investment amounts, and the pledge targets have the most impact on crowdfunding performance. The finance objective, project type, and investment quota all have varying degrees of influence [23]. The participation intentions of investors have a substantial effect on the project's success. The communication between entrepreneurs and online users and the release of online advertisements can promote the willingness of investors to participate in the project and promote the success of project financing [12]. In the case of research projects themselves, the main influencing factors include a project description, image, and video. That is, the originator can use appropriate project descriptions (text, images, and videos) to improve the financing success rate [13]. Crowdfunding projects mainly reflect project quality information through text description, so the language style of text description has an important impact on the quality of the project [16].

Currently, many scholars have conducted different researches on the business model of crowdfunding. Kamath et al. constructed multiple supervised learning models (including neural network, random forest, Bayesian algorithm, and decision tree) to predict the success rate of crowdfunding activities and found that neural networks performed better [9]. Wang et al. used the method of multi-model comparison to predict the success of Kickstarter projects and established five machine learning models, including decision tree and random forest, and a deep learning model, MLP. MLP model of deep learning obtained the best prediction effect with an accuracy of 92.3%, followed by the decision tree model with an accuracy of 90.8% [22]. Li et al. used BP neural network to evaluate the reasonable range of crowdfunding objectives and estimated that the financing target that incentive crowdfunding projects in China can accommodate is about RMB 1 million. This helps project founders to set reasonable financing goals [14]. The prediction model proposed by Peng et al. is based on the extreme gradient lifting algorithm, compared with the other four algorithms (Classification, Regression Tree, K-NearestNeighbor, Linear Regression, and Artificial Neural Network), using grid search optimization parameters. Results show that the machine learning algorithm has potential in predicting the fundraising amount of medical crowdfunding projects [17]. Most researchers use classification algorithms to predict the success of the crowdfunding industry. However, few people have studied regression algorithms to predict the financing amount of crowdfunding.

For originators and investors, research on the enhancement of the success rate of crowdfunding project financing and its affecting factors is of major importance. For originators, if enterprises know in advance which the appropriate project release time and duration can significantly improve the crowdfunding success rate, they would value these aspects to improve financing performance and reduce opportunity costs. By forecasting the amount of financing, investors would avoid high-risk, failure-prone projects and use limited funds to support projects with higher success rates, thereby increasing the possibility of being rewarded for their investment. In this study, based on our constructed Jing-Dong time-series dataset, we use Spearman correlation analysis to select the data input features of the prediction model and put the filtered features into Decision Tree(DT) [4], K-Nearest Neighbors(KNN) [11], Extra Tree Regression(ETR) [6], Gradient Boosting Decision Tree(GBDT) [5], Random Forest (RF) [1], Support Vector Regression(SVR) [20], Adaboost(ADA) [24], Catboost(CAT) [18], Light-GBM(LGB) [10] nine machine models trained. Among them, the best hyperparameters of the model are obtained through grid search. The effectiveness of the models is evaluated based on the Mean Absolute Error(MAE), Root Mean Squared Error(RMSE), Coefficient of Determination(R^2), Adjusted R-Square(R^2_{adj}) and Mean Absolute Deviation(MAD). The main contributions of this work as follows:

- Develop a JD crowdfunding platform research dataset that includes 8,139 projects from May 2018 to August 2020. This Dataset is scheduled to be disclosed in the future.
- Exploring the influence of project launch time and crowdfunding duration on the success rate of crowdfunding.
- Use classical Machine Learning, and Ensemble Learning to predict the daily financing amount of crowdfunding and compared the predicted results.

2 Data Description

The data used in the experiment came from the JD crowdfunding platform, one of the most influential and most active crowdfunding platforms in China. We collected data from JD Finance and established a crowdfunding project dataset containing 8139 projects spanning from May 2018 to August 2020. We have six original features, including project id, project name, cumulative crowdfunding amount, cumulative crowdfunding percent, crowdfunding start date, crowdfunding raise time. Since text features cannot be used as input features of the forecasting model, we removed the project id and project name features. Based on these original features, we can infer the subsequent increased features:

– Daily crowdfunding amount:

$$\Delta M(t) = M(t) - M(t-1). \tag{1}$$

– Crowdfunding end date:

$$x_{ED} = x_{SD} + x_{RT} \tag{2}$$

All the input features are summarized in Table 1, for each day t, there are 6 features in the research dataset. Note that these input features can be classified into two categories:

$$\begin{aligned} x_\theta &= \{x_{SD}, x_{ED}, x_{RT}\}, \\ x(t) &= \{M(t), P(t), \Delta M(t)\} \end{aligned} \tag{3}$$

where $x_\theta \in \mathbb{R}^{3 \times 1}$ represents constant features, which is time independent, while $x(t) \in \mathbb{R}^{3 \times 1}$ stands for time-varying features.

Table 1. Input features for machine learning models.

	Original features
$M(t)$	crowdfunding amount at time t
$P(t)$	crowdfunding percent at time t
x_{SD}	crowdfunding start date
x_{RT}	crowdfunding raise time
	Augmented features
x_{ED}	crowdfunding end date
$\Delta M(t)$	daily crowdfunding amount

Here, we provide m day forecasts for n consecutive days based on Machine learning and Eensemble learning models [26,27]. The prediction model is:

$$\hat{y}(t+n) = f(x(t-m), x(t-m+1), \cdots, x(t-1), x_\theta), \tag{4}$$

where $\hat{y}(t+n)$ represents the predicted value, while $f(\cdot)$ represents the Machine learning and Eensemble learning models. Next, the prediction problem can be formulated as:

$$\min_{f(\cdot)} \sum \|y(t+n) - \hat{y}(t+n)\|_2^2$$
$$s.t. \quad \hat{y}(t+n) = f\left(x(t-m), x(t-m+1),\right.$$
$$\cdots, x(t-1), x_\theta\right), \tag{5}$$

where function $y(t)$ is the actual value.

3   Data Exploratory Analysis

Due to the excessive data of crowdfunding projects, it is inevitable that there will be data errors and missing in the process of acquiring data. As Fig. 1(a) shows, the financing percent of some crowdfunding projects reached several hundred times in April 2019. The crowdfunding projects corresponding to these abnormal data belong to welfare goods projects on the JD crowdfunding platform, which has the problem of too low target amount and too high financing percent. In order to avoid the impact of these crowdfunding projects on the overall data analysis, they are considered anomalous and excluded. The final data set of crowdfunding research was 6998 items.

Crowdfunding is a rewarding project. When the funding percentage is greater than 1, the crowdfunding is successful and the project originator receives the funding proceeds; otherwise, it fails. According to this characteristic, we conducted a statistical analysis of all projects in 2018–2020. As shown in Fig. 1(b), 58.4% of the projects are in a successful state and 41.6% of the projects are in a failed state. This shows that in the domestic crowdfunding market, successful projects account for the majority, but there is still a risk of failure for nearly half of the projects.

The project originator also needs to consider the best time for the project to go online. We analyzed the relationship between the average financing amount of crowdfunding every month and the number of projects and combined the monthly crowdfunding success rate to select the best crowdfunding month. Figure 2(a) shows the relationship between the monthly financing amount of crowdfunding projects and the number of projects. January has the highest average crowdfunding funding amount with more than 160,000 and 400 projects, followed by March with an average funding amount of about 120,000 and more than 200 projects. This indicates that the founders and investors of crowdfunding were active in these two months. However, in June, August, October, and December, the number of crowdfunding projects reached more than 400 with a funding amount of only 40,000–50,000. There are many crowdfunding projects, but the financing amount is small, indicating that the investors have a low willingness to invest in projects, which generally reflects the dull crowdfunding industry in these months. Especially, the average financing amount of crowdfunding was the lowest in November, only about 30000.

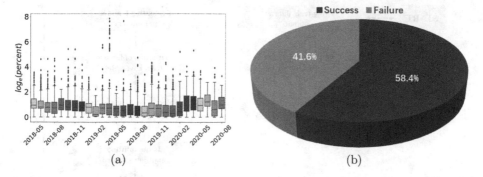

Fig. 1. (a) Percent box diagram of the project in 2018–2020. The x-axis represents the month. The y-axis is the percent of financing after the natural logarithm. (b) State distribution chart of all JD crowdfunding projects.

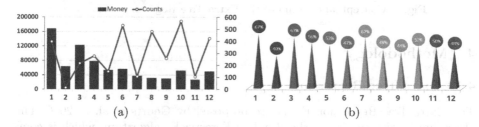

Fig. 2. (a) The average amount of crowdfunding in different months and the number of crowdfunding projects. The x-axis is months of 2019, The blue column represents the average amount of financing per month, and the red line represents the number of projects per month. (b) The success rate of crowdfunding projects in different months. The x-axis is months of 2019. Each cone represents the success rate for each month. (Color figure online)

To better understand which months are the most suitable for crowdfunding, we have calculated the success rate of crowdfunding projects for each month. Figure 2(b) shows that the highest success rate of crowdfunding projects is in January, with 66%, and the lowest is in February, with a success rate of only 39%. Overall, the project success rate declined month by month compared to January and stabilized at around 47% after June. A comprehensive analysis of the above shows that January and March are the best months for crowdfunding. While choosing when to start a project, the project originator can pick between these two months.

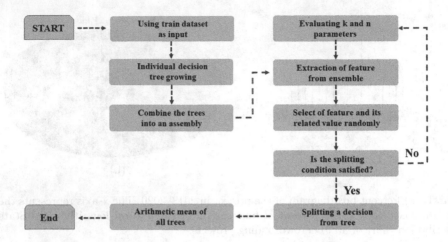

Fig. 3. Conceptual diagram of the Extra Tree Regression algorithm

4   Methodology

4.1   Extra Tree Regression

The Extra Tree Regression(ETR) are proposed by Geurts et al. in 2006. The algorithm is very similar to the Random Forest(RF) algorithm, which is composed of many decision trees [6]. The Extra Tree Regression(ETR) framework is a decision tree-based ensemble learning method. It performs regression by feature bagging and fitting an ensemble of decision trees. Figure 3 shows a conceptual diagram of the ETR algorithm.

For regression, the Random Forest (RF) uses two steps, bootstrapping and bagging. In the bootstrapping step, a set of decision trees is made up of trees that grow independently from each random sample using the training dataset. After achieving the ensemble, the bagging process is divided into two phases, the decision tree nodes are divided, and several random training data subsets are selected in the initial bagging process. By selecting the optimal subset and its corresponding values, the decision-making process is completed [19].

Breiman considered the RF to be a series of decision trees, in which $D(x, \vartheta_n)$ indicates the n th predicting tree, where ϑ_n indicates an uniform independent distribution vector selected before the growth of the tree. All the predicting trees are combined and averaged in an ensemble of trees of $D(x)$ constructed using Eq. 6 [1]

$$D(x, \vartheta_1, ..., \vartheta_n) = \frac{1}{N} \sum_{n=1}^{N} D(x, \vartheta_n) \tag{6}$$

The ETR algorithm is an extension of the RF, with two key differences. Firstly, ETR uses the entire training data set to train the decision tree in the set, rather than using a subset of the training data set for bootstrap sampling. Secondly, in the segmentation stage, the optimal feature is not selected from the specified features, but randomly selected, and its corresponding value for node segmentation. In ETR algorithm, the splitting process is controlled by two parameters, which are k(number of features randomly picked at each node) and n_{min}(minimum sample size required for node splitting) respectively. These two parameters improve the accuracy of the ETR and reduce the overfitting of the model [8,15].

4.2 Spearman's Correlation Coefficient

The correlation coefficient is a statistical measure of the strength of the relationship between the relative two variables. Spearman's correlation coefficient is defined as the pearson correlation coefficient between rank variables. The difference is that it is a method of evaluating the monotonic relationship between two continuous or ordinal variables. In a monotonic relationship, variables tend to change at the same time, but not necessarily at a constant rate. Its value range is between -1 and $+1$, -1 means complete negative correlation, $+1$ means complete positive correlation, and 0 means no linear correlation. Spearman correlation coefficient of X and Y is defined as:

$$\rho(X,y) = \frac{\sum_{i=1}^{n}\left(X_i - \bar{X}\right)\left(y_i - \bar{y}\right)}{\sqrt{\sum_{i=1}^{n}\left(X_i - \bar{X}\right)^2}\sqrt{\sum_{i=1}^{n}\left(y_i - \bar{y}\right)^2}} \tag{7}$$

where X_i is the feature value of sample i, y_i is the predicted value of sample i, $\bar{X}$ is the mean value of the feature value, $\bar{y}$ is the mean of the predicted value.

5   Experiment and Results

5.1   Correlation Analysis

This experiment uses the time-series dataset of crowdfunding projects based on the JD crowdfunding platform with 6 features. However, not all features are suitable to be used to predict the amount of financing. Therefore, using correlation analysis to select relevant features from the original features for data-driven modeling. Figure 4 is the absolute value distribution of spearman's correlation coefficient between predictor variables and feature variables. As shown on the kernel density curve, there is a trough near 0.4. Based on observations, we set 0.4 as the threshold for feature selection. Here, the Spearman correlation coefficient of three features, including crowdfunding amount, crowdfunding percent, and daily crowdfunding amount, is greater than 0.4 and are adopted as the model input. The final dataset period is from May 2018 to August 2020, and 2020 is set as the split point between the training and test sets. Thus, 94,278 samples of data are in the training set and 28,747 samples of data are in the test set, a ratio of 7:3. In the field of machine learning, data standardization is needed to address comparability between data indicators in order to eliminate the effects

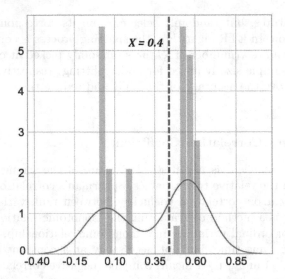

Fig. 4. The absolute value distribution of Spearman's correlation coefficient between predictor variables and feature variables. The x-axis is the correlation coefficient after taking the absolute value. The y-axis is the frequency. The red dashed line represents the selected threshold. (Color figure online)

of different dimensions and units between feature data. Z-score normalization is used to standardize the selected features:

$$x_i^* = \frac{x_i - \bar{x}}{s} \tag{8}$$

where $\bar{x} = \frac{1}{n} \sum_{i=1}^{n} x_i$, $s = \sqrt{\frac{1}{n-1} \sum_{i=1}^{n} (x_i - \bar{x})^2}$.

5.2   Model Adjustment

Before training the machine learning model, how to find the optimal hyperparameters is very critical. In this work, we use a grid search method to search the optimal hyperparameters, which finds the optimal model hyperparameters in the provided array of hyperparameters. In this way, the model can have more data for training and testing by simply dividing the dataset into a training set and a test set. The hyperparameters to be adjusted for each model are as follows:

- DT have criterion(C), min_samples_split($MinS_s$), min_samples_leaf($MinS_l$) and splitter(S).
- KNN have weights(w), n_neighbors(nn), leaf_size(ls) and P.
- ETR have criterion(C), min_samples_split($MinS_s$), min_samples_leaf($MinS_l$) and n_estimators(ne).
- GBDT have learning_rate(lr), min_samples_split($MinS_s$), loss, n_estimators(ne) and sub-sample($SubS$).

- RF have criterion(C), min_samples_split($MinS_s$), min_samples_leaf($MinS_l$) and n_estimators(ne).
- SVR have kernel(K).
- ADA have learning_rate(lr), loss and n_estimators(ne).
- CAT have learning_rate(lr), iterations(i) and l2_leaf_reg(l2).
- LGB have learning_ rate(lr), boosting_type(bt), n_estimators(ne), colsample_bytree(cb) and lambda_l2(Lam_{l2}).

Through grid search, the optimal hyperparameters of each model are shown in Table 2.

Table 2. Model optimal hyperparameters

Model	Hyperparameters				
DT	C = MSE	$MinS_s = 3$	$MinS_l = 2$	S = best	–
KNN	w = distance	nn = 9	ls = 25	P = 2	–
ETR	C = MSE	$MinS_s = 6$	$MinS_l = 1$	ne = 90	–
GBDT	lr = 0.05	$MinS_s = 1$	loss = 1	ne = 120	$SubS = 2$
RF	C = MSE	$MinS_s = 2$	$MinS_l = 2$	ne = 120	–
SVR	K = linear	–	–	–	–
ADA	lr = 0.1	loss = square	ne = 40	–	–
CAT	lr = 0.1	i = 1300	l2 = 1	–	–
LGB	lr = 0.2	bt = gbdt	ne = 2000	cb = 0.8	$Lam_{l2} = 1$

5.3   Predict Results

The hyperparameters obtained by grid search are used to set the parameters of the machine learning model. In our work, we used machine learning, including DT, KNN, SVR and Ensemble learning, including ETR, GBDT, RF, ADA, CAT, LGB. The above 9 models perform data modeling predictions on daily crowdfunding amounts. We extracted 6998 crowdfunding projects sample that has been released for more than 3 years (2018–2020) from the JD crowdfunding time-series database. These data take 2020 as the cut-off point. The data before 2020 is used as the training set, and the data after 2020 is used as the test set, with a ratio of 7:3.

To evaluate the performance of the prediction model, regression evaluation matrics are used: Root Mean Squared Error(RMSE), Mean Absolute Error(MAE), Coefficient of Determination(R^2), Median Absolute Error(MAD) and Adjusted R^2. Each evaluation indicators is calculated as follows:

$$RMSE = \sqrt{\frac{1}{n} \sum_{i=1}^{n} (y_i - \hat{y}_i)^2} \tag{9}$$

$$MAE = \frac{1}{n} \sum_{i=1}^{n} |y_i - \hat{y}_i| \tag{10}$$

$$SS_{res} = \sum_{i=1}^{n} (y_i - \hat{y}_i)^2 ,$$

$$SS_{tot} = \sum_{i=1}^{n} (y_i - \bar{y})^2 , \tag{11}$$

$$R^2 = 1 - \frac{SS_{res}}{SS_{tot}} = 1 - \frac{\sum_{i=1}^{n} (y_i - \hat{y}_i)^2}{\sum_{i=1}^{n} (y_i - \bar{y})^2}$$

$$MAD = \mathrm{median} \left(|y_1 - \hat{y}_1|, \cdots, |y_n - \hat{y}_n| \right) \tag{12}$$

$$R_{adj}^2 = 1 - \left[\frac{\left(1 - R^2\right)(n-1)}{n - k - 1} \right]. \tag{13}$$

Suppose the test set has n samples, k is a feature of the model input, y_i is the sample representing the actual value, $median_n(.)$ represents the median of n samples, SS_{res} is the residual sum of squares, and SS_{tot} is the total sum of squares.

Table 3. Results of evaluation indicators for each model

Model	$RMSE$	MAE	R^2	MAD	R_{adj}^2
DT	2110.815	415.656	0.666	26.532	0.665
KNN	2937.794	520.717	0.353	83.087	0.352
ETR	**1148.722**	**178.384**	**0.901**	**22.151**	**0.901**
GBDT	3138.616	1144.355	0.261	882.464	0.261
RF	2023.724	360.317	0.693	38.380	0.692
SVR	3446.197	583.573	0.109	90.242	0.109
ADA	4168.541	2595.455	−0.302	2663.574	−0.303
CAT	2504.401	516.888	0.529	122.160	0.529
LGB	2152.493	634.298	0.652	197.182	0.652

Table 3 shows the results of evaluation indicators for each. Among them, the R^2 and RMSE of ETR reached 0.901 and 1148.722, which show the best predictive performance. In addition, to validate the effect of feature selection, we evaluated the performance when all features were used as input against when selected features were used as input. The results indicate that the predictive performance of the model decreases when all inputs are used. The R^2 of the ETR model lowers to 0.859, while the RMSE rises to 1368.487. This further demonstrates that selecting features based on correlation analysis can improve the model's prediction performance to some extent. In this experiment, RMSE and R^2 were mainly used as evaluation indicators. In order to better compare the prediction performance among models, the optimal evaluation results of each model were extracted, as shown in Fig. 5.

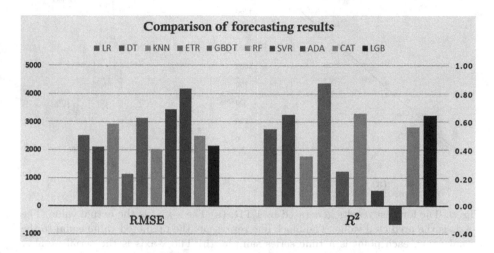

Fig. 5. Comparison of $RMSE$ and R^2 results for different models.

We can see Fig. 5 strikes a comparison of RMSE and R^2 values for every model. Lower the RMSE value higher the performance of the prediction model. It can be observed that the ETR's RMSE values are significantly lower when compared to other models' approaches. It is a clear indicator that ETR outperforms other models, such as the RF model and DT model. The closer the R^2 value is to 1, the better the model fit is. It can be observed that ETR has a high R^2 value with an accuracy close to 90.1%, indicating that the ETR model has high accuracy. According to Table 3, it can be observed that three models such as DT, RF, and LGB have a strong predictive performance and do not differ significantly from the ETR model. In contrast to the simple KNN and SVR models, the predicting accuracy of the ETR ensemble model is greatly enhanced, and the error is relatively low. In addition, the MAD indicator of the ETR model is 22,151, which demonstrates its great robustness. The ensemble-based algorithm predicts the fundraising performance of reward-based crowdfunding projects the best overall.

5.4   Experimental Results by ETR

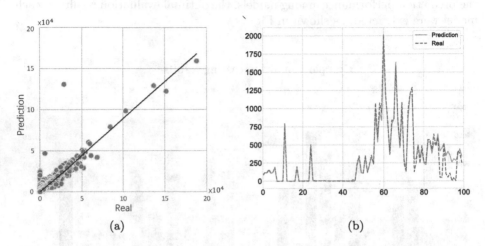

(a)   (b)

Fig. 6. The forecast results generated by ETR. (a) The x-axis is the actual value. The y-axis is the predicted value. The black line represents the predicted value equal to the actual value, each point is a time series sample. (b) The x-axis is the serial number of crowdfunding samples. The y-axis is the daily financing amount. The blue dotted line represents the actual value. The orange solid line represents the predicted value of ETR. (Color figure online)

Experimental results show that ETR achieves the best value among all the comparison algorithms. All the evaluation indicators of ETR are all the best (shown in Table 3). Figure 6 shows the predicted value and actual value of all the cases. We plot a scatter plot of the ETR predictions Fig. 6(a). The deviation of the prediction can be known by the position of the observation point. We found that the model predictions are all near the black line. This result is still ideal. We use a partial result of the ETR model prediction to plot a prediction Fig. 6(b). The orange line is the predicted value, and the blue line is the actual value. The figure demonstrates that the forecast results correspond well with the actual findings and that the ETR model has produced accurate predictions. The results show that the predicted value can accurately obtain the amount of a short-term crowdfunding project.

6   Conclusions and Future Work

Online crowdfunding has gradually become accepted by modern human society and has become one of the main ways of raising funds for businesses. However, online crowdfunding has a great deal of uncertainty, which can lead to investors not being able to accurately assess the risks of a crowdfunding project. Investors need an effective and timely tool to help them hedge their investment risks.

In this study, we constructed a crowdfunding research dataset containing 6,998 projects from May 2018 to August 2020 based on the JingDong crowdfunding platform. We then conducted an exploratory analysis of the release time of crowdfunding projects, crowdfunding duration, and success rates, and concluded that crowdfunding projects launched in January and March were more likely to be successful and raise more funds. Ultimately, we propose to use machine learning and ensemble learn models to drive the daily crowdfunding amount time series prediction model based on the crowdfunding research dataset. The experimental results demonstrate that ensemble learning outperforms classical machine learning in terms of prediction accuracy, and the Extra Tree Regression performs best in the evaluation of this prediction task, with a good fit and robustness, with a coefficient of determination of 0.901 and a mean square error of 1148.722. The prediction models constructed can effectively and accurately provide information on the future trend of each crowdfunding project, thus helping investors adjust their strategies to avoid risks in a timely manner.

Considering that this is an exploratory study, even though several algorithms have been compared and considered, there are still some limitations that must be acknowledged. In this study, only one crowdfunding platform was considered for the study, while in the future we will include data from more different types of crowdfunding platforms to validate the modeling of ensemble learning in the crowdfunding domain. And, we also consider the use of unsupervised techniques to explore the potential relationships between crowdfunding projects.

Acknowledgement. This work was supported by Natural Science Foundation of Guangdong Province, China (2020A1515010761).

References

1. Breiman, L.: Random forests. Mach. Learn. **45**(1), 5–32 (2001)
2. Chung, J., Lee, K.: A long-term study of a crowdfunding platform: Predicting project success and fundraising amount. In: Proceedings of the 26th ACM Conference on Hypertext and Social Media, pp. 211–220 (2015)
3. Cumming, D.J., Leboeuf, G., Schwienbacher, A.: Crowdfunding models: Keep-it-all vs. all-or-nothing. Financial Management **49**(2), 331–360 (2020)
4. Freund, Y., Mason, L.: The alternating decision tree learning algorithm. In: ICML, vol. 99, pp. 124–133. Citeseer (1999)
5. Friedman, J.H.: Greedy function approximation: a gradient boosting machine. Annals of statistics, pp. 1189–1232 (2001)
6. Geurts, P., Ernst, D., Wehenkel, L.: Extremely Randomized trees. Mach. Learn. **63**(1), 3–42 (2006)
7. Guo, Y., Zhou, X., Zhan, C., Zeng, Y., Zhong, L.: Prediction and analysis of success on crowdfunding projects. In: Proceedings of the 2020 4th International Conference on Electronic Information Technology and Computer Engineering, pp. 785–789 (2020)
8. John, V., Liu, Z., Guo, C., Mita, S., Kidono, K.: Real-time lane estimation using deep features and extra trees regression. In: Bräunl, T., McCane, B., Rivera, M., Yu, X. (eds.) PSIVT 2015. LNCS, vol. 9431, pp. 721–733. Springer, Cham (2016). https://doi.org/10.1007/978-3-319-29451-3_57

9. Kamath, R., Kamat, R.: Supervised learning model for kickstarter campaigns with r mining. Int. J. Inf. Technol., Model. Comput. (IJITMC) **4**, (2016)

10. Ke, G., et al.: Lightgbm: A highly ecient gradient boosting decision tree. In: Advances in Neural Information Processing Systems, vol. 30 (2017) Lightgbm: A highly ecient gradient boosting decision tree. In: Advances in Neural Information Processing Systems, vol. 30 (2017)

11. Kramer, O.: K-nearest neighbors. In: Dimensionality reduction with unsupervised nearest neighbors, pp. 13–23. Springer (2013). https://doi.org/10.1007/978-3-642-38652-7

12. Kraus, S., Richter, C., Brem, A., Cheng, C.F., Chang, M.L.: Strategies for reward-based crowdfunding campaigns. J. Innov. Knowl. **1**(1), 13–23 (2016)

13. Kromidha, E., Robson, P.: Social identity and signalling success factors in online crowdfunding. Entrepreneurship Regional Develop. **28**(9–10), 605–629 (2016)

14. Li, H., Chen, X., Zhang, Y., Hai, M.: Prediction of financing goal of crowdfunding projects. Procedia Comput. Sci. **139**, 108–113 (2018)

15. Mishra, G., Sehgal, D., Valadi, J.K.: Quantitative structure activity relationship study of the anti-hepatitis peptides employing random forests and extra-trees regressors. Bioinformation **13**(3), 60 (2017)

16. Parhankangas, A., Renko, M.: Linguistic style and crowdfunding success among social and commercial entrepreneurs. J. Bus. Ventur. **32**(2), 215–236 (2017)

17. Peng, N., Zhou, X., Niu, B., Feng, Y.: Predicting fundraising performance in medical crowdfunding campaigns using machine learning. Electronics **10**(2), 143 (2021)

18. Prokhorenkova, L., Gusev, G., Vorobev, A., Dorogush, A.V., Gulin, A.: Catboost: unbiased boosting with categorical features. In: Advances in neural information processing systems, vol. 31 (2018)

19. Sharafati, A., Asadollah, S.B.H.S., Hosseinzadeh, M.: The potential of new ensemble machine learning models for effluent quality parameters prediction and related uncertainty. Process Saf. Environ. Prot. **140**, 68–78 (2020)

20. Smola, A.J., Schölkopf, B.: A tutorial on support vector regression. Stat. Comput. **14**(3), 199–222 (2004)

21. Testa, S., Nielsen, K.R., Bogers, M., Cincotti, S.: The role of crowdfunding in moving towards a sustainable society. Technol. Forecast. Soc. Chang. **141**, 66–73 (2019)

22. Wang, W., Zheng, H., Wu, Y.J.: Prediction of fundraising outcomes for crowdfunding projects based on deep learning: a multimodel comparative study. Soft. Comput. **24**(11), 8323–8341 (2020). https://doi.org/10.1007/s00500-020-04822-x

23. Yao, H., Zhang, Y.: Research on influence factors of crowdfunding. Int. Business Manage. **9**(2), 27–31 (2014)

24. Ying, C., Qi-Guang, M., Jia-Chen, L., Lin, G.: Advance and prospects of adaboost algorithm. Acta Autom. Sinica **39**(6), 745–758 (2013)

25. Yu, P.F., Huang, F.M., Yang, C., Liu, Y.H., Li, Z.Y., Tsai, C.H.: Prediction of crowdfunding project success with deep learning. In: 2018 IEEE 15th international conference on e-business engineering (ICEBE), pp. 1–8. IEEE (2018)

26. Zhan, C., Tse, C.K., Fu, Y., Lai, Z., Zhang, H.: Modeling and prediction of the 2019 coronavirus disease spreading in china incorporating human migration data. PLoS ONE **15**(10), e0241171 (2020)

27. Zhan, C., Zheng, Y., Zhang, H., Wen, Q.: Random-forest-bagging broad learning system with applications for covid-19 pandemic. IEEE Internet Things J. **8**(21), 15906–15918 (2021)

Cage Mass Center Capture for Whirl Analysis Using an Improved MultiResUNet from the Multimodal Biomedical Image Segmentation

Zhaohui Yang[1](✉) (iD), Xiaoliang Niu[1] (iD), Tianhua Xiong[1] (iD), and Ningning Zhou[2]

[1] School of Aeronautics, Northwestern Polytechnical University, Xi'an 710072,
People's Republic of China
zhaohui@nwpu.edu.cn, {niuxl,xiongth}@mail.nwpu.edu.cn
[2] Beijing Key Laboratory of Long-Life Technology of Precise Rotation and Transmission
Mechanisms, Beijing Institute of Control Engineering, Beijing 100094,
People's Republic of China

Abstract. The severe whirling of cage is the main failure mode of space bearing. However, it is difficult to capture whirl motion without changing the cage structure. This paper aims to capture whirl motion of cage precisely with high-speed technology and Semantic Segmentation Algorithms. An improved MultiResUNet for the multimodal biomedical image segmentation is trained with 20 high-speed cage rotational pictures, 5 pictures are validation set, then 1000 cage mask images during rotation are obtained. The cage mass center for 1000 mask images·is calculated with Center of Mass (CoM) operation for whirling analysis. To verify the effectiveness of our model, the results of whirl orbit and Y displacement are compared by MultiResUNet and Improved MultiResUNet. Additionally, our model is also compared with TEMA Motion which is a commercial tracking software. The results show that our model can correctly predict the whirl trend that the whirl frequency is consistent with cage rotation frequency and whirl radius is a fixed value, and is high-precision trajectory capture that max deviation is 0.0118 mm from real cage mass center to prediction.

Keywords: Space bearing · Cage · Whirl motion · Improved MultiResUNet

1 Introduction

The space bearing is the core component of the space actuator, and its life and reliability directly affect the performance of the space actuator. The lubrication mechanism of this bearing is different from that of ground bearing, which mainly relies on the lubricating oil pressed into the porous material cage to overflow and lubricate. This lubrication mechanism determines that the failure of the space bearing is often caused by the wear of the cage [1], so the stability study of cage is necessary for space bearing. Regarding the stability of the cage, the capture of cage mass center motion (whirl motion) trajectory

H. Zhang et al. (Eds.): NCAA 2022, CCIS 1638, pp. 387–400, 2022.
https://doi.org/10.1007/978-981-19-6135-9_29

and the study of trajectory characteristics are the most straightfoward and most widely used methods.

Kingsbury et al. [2, 3] first proposed the cage whirl motions to explain the cage movement during a stable operation or squeal, and the method of the acquirement of the whirl motion is to establish a dynamic model and then perform a differential solution. Niu et al. [4] used Gupta P.K.'s model to study the whirl motion of ball bearing cage under solid lubrication conditions, he used the same technique as Kingsbury, but added to the model the contact force of the cage with the surrounding elements. The relationships of whirl orbit and radius with contact force are detailed discussed by them. Crawford R. Meeks [5] established an angular contact bearing dynamics model that describes cage-ring and cage-ball contact as inelastic contact that simulates the six degrees of freedom of the cage. Meanwhile, the different geometric parameters of the cage, the coefficient of frictional traction and the effect of speed on collision and friction are studied. These methods all have obvious defects. Although they can analyze the stability of the cage from the trend, they cannot provide a certain standard to judge the performance of the bearing because they simplify the force of the cage and trajectory of whirl motion is unauthentic.

Test technology is the only way to obtain the real whirl motion trajectory. Choe et al. [6, 7] designed a test rig to investigate the dynamic behavior of ball bearing in a cryogenic environment. He used two fiber optic displacement sensors at a 90° interval that is used to measure the cage whirling amplitude, so the whirl motion trajectory can be acquired. However, the cage geometry has been changed for ease of measurement. Chen [8] used the two mutually perpendicular laser displacement sensors to obtain the whirl motion of cage. Due to the particularity of this sensor, only large bearings can be measured. Arya [9] et al. used high precision proximity probes to measure the whirl orbit of cage. Two probes spaced 90° apart in the circumferential direction were used to measure the radial whirl motion and three probes spaced 120° apart were used to measure the axial whirl motion. However, to reliable and accurate measurement, the custom standoffs were designed to be mounted on the surface of cage. References [10–12] both used the displacement sensor to measure the whirl orbit, but the cage geometry structure was changed by them. To sum up, the cost of accurate measurement of the whirl orbit is sacrificing the original geometry of the cage. Fortunately, with the maturity of high-speed photography technology, more scholars have studied the cage stability from the perspective of image. Abele et al. [13–15] used a high-speed camera perpendicular to the test rig to capture images of test bearing. The basic method of capture of whirl orbit is the edge detection algorithm, to ensure the accuracy of detection, a complex pre-processing is required. Yang et al. [16] developed a high-precision bearing test rig to evaluate the stability of cage. They used high-speed photography technology and target marking technology to shoot photo and then used the TEMA motion to process the photo to get the whirl orbit. Although the complex pre-processing is solved, but the target marking lead to poor repeatability of the test.

The proposal of FCN [17] has made great progress in semantic segmentation. Most of the current successful network structures [18, 19] are the rethinking of fully convolutional neural network. However, they are often unfriendly to small datasets. Even the application of Transformer [20, 21], which have been quite hot recently, have been shown to be more efficient for large datasets. Fortunately, a lightweight network U-net [22] is very friendly to small datasets.

Inspired by these semantic segmentation algorithms, this paper develops a whirl orbit capture method based on MultiResUNet [23] which is an improved version of one proposed in [16]. This method masks the part between the inner of the cage and the outer of the mandrel, and then the network is trained using the labeled data to obtain the weight parameters of high cross-combination ratio. Finally, the cage mass center is calculated using the Center-of-Mass (CoM) operation. The remainder of this paper is mainly organized as follows.

2 Test Rig and Image Processing

2.1 Test Rig

Figure 1 shows a high-precision bearing test rig, is used to measure the stability of the cage. The test rig includes an axial loading device, radial loading device, precision spindle, light source equipment, high-speed camera and test bearing. For the detailed operation of the test rig, please refer to [16].

Fig. 1. High-precision bearing test rig

2.2 Image Processing

Reference [16] uses the TEMA motion to process the images captured by the high-speed camera. Figure 2 shows the image that needs to be processed, the eight tracking spots are marked in the surface of cage, and these spots are round with black. It is mention that these spots are on the same circumference, usually three points can determine a circle, and we marked up to 8 points in study [16], which improves the accuracy of cage mass center capture. More details can be found in in reference [16]. After each image is processed, the Cartesian coordinates of the corresponding number of target spots are output. As shown in Fig. 3, we use the ordinary least squares to fit the cage mass center, and its quality is evaluated by its deviation from the true cage mass center. The broken line is the circle fitted by the ordinary least squares, its centroid is the dot in the middle of Fig. 3, and the blue square is the tracking spot. To sum up, each image has a centroid and the centroid is the cage mass center of bearing. Therefore, we obtain 5000 images in 0.5 s that the sample frequency is 10000 Hz, Fig. 4 shows the whirl orbit (The motion trajectory of cage mass center).

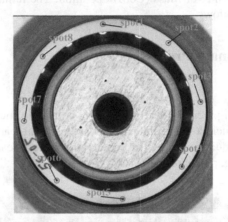

Fig. 2. Target mark on cage surface

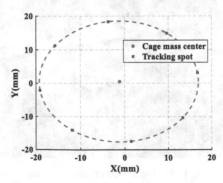

Fig. 3. The cage mass center calculation using ordinary least squares

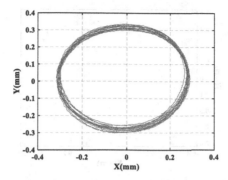

Fig. 4. Whirl orbit

With the development of semantic segmentation algorithms, many problems in industry have been solved. However, the primary task of capturing centroids using semantic segmentation algorithms is to annotate accurate labels. Figure 5 shows the labels used in this paper, binary images. The part between the inner ring of the cage and the outer ring of the precision spindle is 1, and the rest of the picture is 0. The spatial relationship between cage and spindle can further help us distinguish between whirl motion and cage rotation motion. The cage rotation frequency is a constant value proportional to the spindle rotation frequency, but the whirl frequency is a variable value, which is related to the stable state of the cage. Figure 6 shows the orbit of centroid of cage inner ring using 30 labeled images.

Fig. 5. Annotation of label

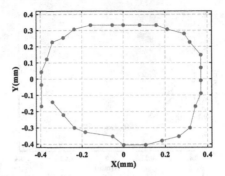

Fig. 6. Whirl orbit using 30 labeled images

3 Improved Unet and Hyperparameter Tunning

3.1 Proposed Network

In this paper, we used the Improved Unet approach to replace the TEMA Motion tracking to obtain the cage mass center. MultiResUnet is a variant of Unet, Unet was originally designed to solve segmentation tasks at the cell level. Figure 7 shows the overall structure of Unet, which is an encoder-decoder structure. However, the skip connection of the Unet network ignores higher-level semantic information, and MultiResUnet adds a residual network to the skip connection (B of Fig. 7) to solve this problem, as shown in Fig. 8. Additionally, instead of A in Fig. 7, MultiResUnet used the MultiRes block, the block is the combination of inception network and residual network. As shown in Fig. 9, the architecture includes 3 3×3 convolutional blocks whose outputs are concatenated together, and a 1×1 convolutional layers are added to comprehend some additional spatial information. It is worth mentioning that the number of 3 3 * 3 convolution blocks is an arithmetic sequence (1:2:3).

To make the network more suitable for our study, we adjust network layers accordingly and the loss function is the binary cross entropy. We found the double descent of validation accuracy after using the MultiResUNet. The reason is some parameters are import for the network, so this phenomenon occurs as epochs increase. The phenomenon is more like overfitting, but the difference is that it always tends to stabilize at the end. Even so, the phenomenon occurs so often, which affects the convergence of the results. In Fig. 10, we added a dropout to each residual block of Res path (C), and our research found that dropout block reduced the frequency of the multiple descent, the results can be explained through the dropout abandoned the sensitive parameters for the network. Figure 11 shows the distinction between original ResBlock (ORB) and improved ResBlock (IRB), the validation accuracy with IRB converges faster than with ORB.

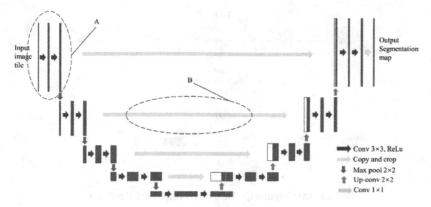

Fig. 7. Unet architecture

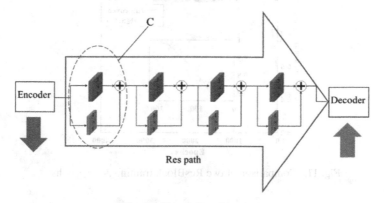

Fig. 8. Res path architecture

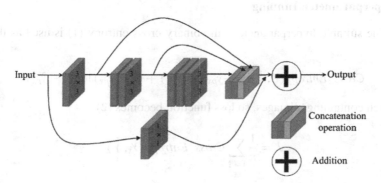

Fig. 9. MultiRes block

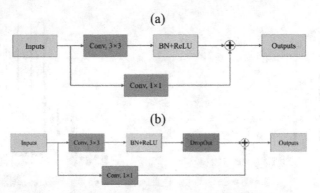

Fig. 10. (a) Original ResBlock, (b) Improved ResBlock

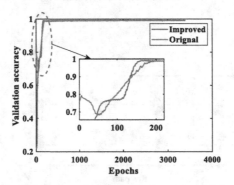

Fig. 11. Comparison of two ResBlock training 3400 epochs

3.2 Hyperparameter Tunning

To find the suitable hyperparameters, the binary cross entropy (1) is used as the loss function.

$$Cross\_Entropy(Y, \hat{Y}) = -y_{px} \log(\hat{y}_{px}) + (1 - y_{px}) \log(1 - \hat{y}_{px}) \tag{1}$$

For a batch containing n images, so loss function becomes (2).

$$J = \frac{1}{n} \sum_{i=1}^{n} Cross\_Entropy(Y_i, \widehat{Y}_i) \tag{2}$$

where y_{px} is the real pixel value, $\widehat{y}_{px}$ is the predicted value

We study the parameter of the initial learning rate by five different values, 0.5, 0.05, 0.005, 0.0005, 0.00005. The learning rate function for epochs is (3). The convergence results of validation accuracy are shown in Fig. 12 and the validation accuracy is expressed in (4). It can be seen that the green, purple and yellow polylines fluctuate severely in iterations 100 to 300, the accuracy suddenly drops sharply and then rises. The overall trend of the orange and blue polylines is smooth, and they first experience a small convergence in iteration 0 to 100. However, all polylines converge around 0.99, and the initial learning rate is 0.005 has the biggest mean and smallest variance. Therefore, the initial learning rate of 0.005 is defined in our network training process.

$$\begin{cases} lr = lr_{initial}, \ if \ (epochs \le 200) \\ lr = learning\_rate \times 0.1^{(int((epochs-100)/100))}, \ if \ (epochs > 200) \end{cases} \tag{3}$$

$$Val\_accuracy = (TP + TN)/(TP + FN + FP + FN) \tag{4}$$

where, TP is the number of true positives, FP is the number of false positives, FN is the number of false negatives, TN is the number of true negatives

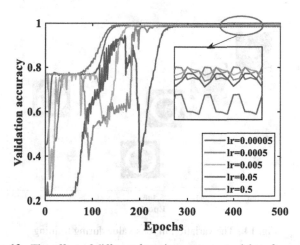

Fig. 12. The effect of different learning rates on model performance

In addition, we also studied the batch size after fixing a learning rate of 0.005, the batch size is set to 2, 4, 6 and 8. As shown in Fig. 13, with the epochs of iterations increases, the accuracy of all polylines gradually increases and stabilizes in a small interval, except the orange polylines. The blue, yellow and purple polylines both are suitable for our network, but the blue purple is smoother than others. Therefore, the batch size equals to 2 is defined in our network.

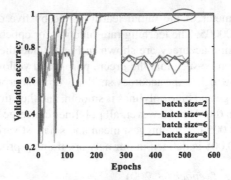

Fig. 13. The effect of different batch size on model performance

Compared with our processing data, the use of binary cross entropy, a learning rate of 0.005 and a batch size of 2 already make the model achieve good performance, so this paper will not make too many comparisons with other loss functions. Figure 14 shows the variation in loss during training and the corresponding processing results.

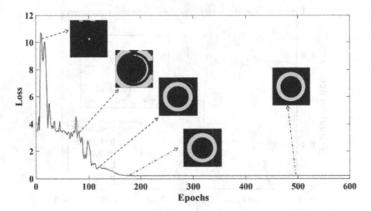

Fig. 14. The variation in loss value during training

4   Model Validation and Comparison

4.1   Model Validation

To verify the validity of the proposed model, we prepared the following experiments. The test bearing is 7004, the sample frequency of high-speed camera is 10000 Hz, the spindle rotation speed is 5000 rpm, the radial force is 0 N and the axial force is 90 N. Since labeling is a difficult task, we randomly selected 20 images from the 1000 images captured by the camera as the training set, 5 as the validation set, and 5 as the test set. It is worth mentioning that the data of the training set should be rotated exactly one circle, which improves the generalization ability of the network.

After training the data using the method proposed in this paper, we obtained the ring mask image, then filled the ring inside, and finally used the CoM operation (5) to obtain the centroid.

$$x_m = \frac{\sum\limits_{i=1}^{n} w_i x_i}{\sum\limits_{i=1}^{n} w_i}, y_m = \frac{\sum\limits_{i=1}^{n} w_i y_i}{\sum\limits_{i=1}^{n} w_i} \tag{5}$$

where w_i is the pixel of a position, x_i is x-coordinate of a location, y_i is y-coordinate of a location.

Figure 15 shows the distinction between Improved Unet and MultiResUNet. It is seen that our model both the periodicity of the whirl motion and the whirl orbit test has achieved better results than MultiResUNet. The whirl orbit capturing using MultiResUNet is chaotic, which is contrary to our previous research 16. The whirl motion orbit generated by the axial force of 90 N and 5000 rpm is relatively stable, so the reason can be attributed to the relatively poor generalization ability of the model.

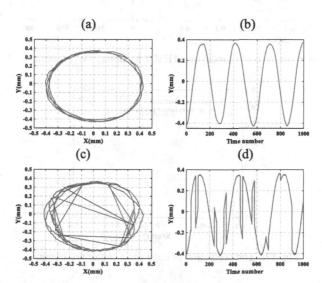

Fig. 15. Whirl orbit and Y displacement of cage. (a) Whirl orbit (Improved Unet), (b) Y displacement of cage (Improved Unet), (c) Whirl orbit (MultiResUNet), (d) Y displacement of cage (MultiResUNet).

4.2   Comparison with TEMA Motion

Figure 16 shows the difference between TEMA Motion and our model. The whirl orbit using TEMA Motion capture is eccentric and the whirl radius of different revolutions is bigger than using our model. The reason can be explained as follows. Firstly, it is difficult to ensure that the eight target spots are on the same circle. Then, with the rotational speed increases, the target spots are more difficult to capture, and the erroneous capture of each spot can have a large impact on the results. However, rotational trend is accurate.

Figure 17 shows the prediction accuracy of the 5 test set images, and it is expressed by (6) and (7). It can be seen that the left position and the right position are bigger than others, the deviations are 0.0118 mm and 0.0113 mm respectively. And it is less than the radial runout of spindle (20 μm), so we can think that the error generated by our method has little effect on the evaluation of cage stability.

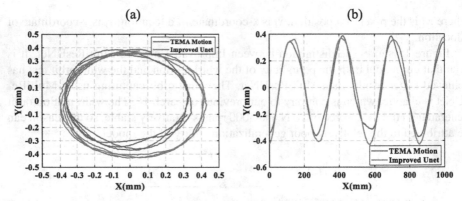

Fig. 16. Comparison between our model and TEMA Motion (a) Whirl orbit, (b) Y displacement

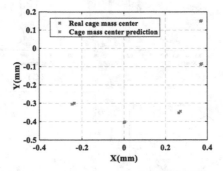

Fig. 17. The prediction of cage mass center prediction

$$deviation = \sqrt{(x_{prediction} - x_{test})^2 + (y_{prediction} - y_{test})^2} \qquad (6)$$

$$prediction\_metrics = MAX\,(deviation) \qquad (7)$$

5   Conclusion

To improve the accuracy of cage centroid capture, this paper proposed a method that accurately captures cage mass center trajectories driven by deep learning. We take inspiration from the field of medical segmentation due to its small sample size and high accuracy. The Improved MultiResUNet is used to train our sample data, we not only predict the trend of whirl motion but also ensure the high prediction accuracy.

Compared with TEMA Motion, our model gets rid of the need for target points, so that there is no eccentricity error and repeatability error as long as your network is trained well enough.

In the future, we will further explore the appropriate network and label processing methods to ensure higher accuracy and provide help for the life prediction of high-precision bearings.

References

1. Ning, F.: Space Bearing Reliability. Chemical Industry Press, Beijing (2020)
2. Kingsbury, E., Walker, R.: Motions of an unstable retainer in an instrument ball bearing. Trans. ASME **116**(2), 202–208 (1994). https://doi.org/10.1115/1.2927197
3. Kingsbury, E.P.: Torque variations in instrument ball bearings. ASLE Trans. **8**, 435–441 (1965). https://doi.org/10.1080/05698196508972113
4. Niu, L., Cao, H., He, Z., Li, Y.: An investigation on the occurrence of stable cage whirl motions in ball bearings based on dynamic simulations. Tribol. Int. **103**, 12–24 (2016). https://doi.org/10.1016/j.triboint.2016.06.026
5. Meeks, C.R.: The dynamics of ball separators in ball bearings—Part II: results of optimization study. ASLE Trans. **28**, 288–295 (1985). https://doi.org/10.1080/05698198508981623
6. Choe, B., Lee, J., Jeon, D., Lee, Y.: Experimental study on dynamic behavior of ball bearing cage in cryogenic environments, Part I: effects of cage guidance and pocket clearances. Mech. Syst. Signal Process. **115**, 545–569 (2019). https://doi.org/10.1016/j.ymssp.2018.06.018
7. Choe, B., Kwak, W., Jeon, D., Lee, Y.: Experimental study on dynamic behavior of ball bearing cage in cryogenic environments, Part II: effects of cage mass imbalance. Mech. Syst. Signal Process. **116**, 25–39 (2019). https://doi.org/10.1016/j.ymssp.2018.06.034
8. Chen, S., Chen, X., Li, Q., Jiaming, G.: Experimental study on cage dynamic characteristics of angular contact ball bearing in acceleration and deceleration process. Tribol. Trans. **64**, 45–52 (2020). https://doi.org/10.1080/10402004.2020.1790706
9. Arya, U., Sadeghi, F., Conley, B., Russell, T., Peterson, W., Meinel, A.: Experimental investigation of cage dynamics and ball-cage contact forces in an angular contact ball bearing. Proc. Inst. Mech. Eng. Part J J. Eng. Tribol., 1–13 (2022). https://doi.org/10.1177/13506501221077768
10. Gupta, P.K., Dill, J.F., Bandow, H.E.: Dynamics of rolling element bearings experimental validation of the DREB and RAPIDREB computer programs. J. Tribol. **107**, 132–137 (1985). https://doi.org/10.1115/1.3260989
11. T. Sakaguchi, K. Harada: Dynamic analysis of cage stress in tapered roller bearings using component-mode synthesis method. ASME J. Tribol. **131**, 011102:1–011102:9 (2009). https://doi.org/10.1115/1.3002326
12. Babbick, T., Sauer, B.: Analytische and experimentelle Verifikationsmethoden anhand eines Walzlagerkafig-Simulationsmodells. In: 5th Colloquium, Dresden, Germany, pp. 119–131 (2017)
13. Abele, E., Holland, L., Nehrbass, A.: Image acquisition and image processing algorithms for movement analysis of bearing cages. J. Tribol. **138**(2), 021105:1–021105:7 (2016). https://doi.org/10.1115/1.4031792
14. Abele, E., Holland, L., Hönig, P.: Image acquisition and image processing algorithm for movement analysis of bearings' rolling element. J. Tribol. **140**(1), 011103:1–011103:9 (2018). https://doi.org/10.1115/1.4037066

15. Abele, E., Holland, L.: Image-based movement analysis of bearing cages of cylindrical hybrid roller bearings. J. Tribol. **139**(6), 061101:1–061101:8 (2017). https://doi.org/10.1115/1.403 6320

16. Yang, Z., Li, C., Zhou, N., Zhang, J.: Research on the cage stability of high-precision ball bearing with image acquirement and error separation. Measurement **186**, 110149:1–110149:11 (2021). https://doi.org/10.1016/j.measurement.2021.110149

17. Long, J., Shelhamer, E., Darrell, T.: Fully convolutional networks for semantic segmentation. In: IEEE Conference on Computer Vision and Pattern Recognition (CVPR), pp. 3431–3440. IEEE (2017). https://doi.org/10.1109/CVPR.2015.7298965

18. Badrinarayanan, V., Kendall, A., Cipolla, R.: SegNet: a deep convolutional encoder-decoder architecture for image segmentation. IEEE Trans. Pattern Anal. Mach. Intell. **39**(12), 2481–2495 (2017). https://doi.org/10.1109/TPAMI.2016.2644615

19. Chen, L.-C., Zhu, Y., Papandreou, G., Schroff, F., Adam, H.: Encoder-decoder with atrous separable convolution for semantic image segmentation. In: Ferrari, V., Hebert, M., Sminchisescu, C., Weiss, Y. (eds.) ECCV 2018. LNCS, vol. 11211, pp. 833–851. Springer, Cham (2018). https://doi.org/10.1007/978-3-030-01234-2_49

20. Zheng, S., et al.: Rethinking semantic segmentation from a sequence-to-sequence perspective with transformers. In: Conference on Computer Vision and Pattern Recognition (CVPR), pp. 6877–6886. IEEE (2021). https://doi.org/10.1109/CVPR46437.2021.00681

21. Ji, Y., et al.: Multi-compound transformer for accurate biomedical image segmentation. In: de Bruijne, M., et al. (eds.) MICCAI 2021. LNCS, vol. 12901, pp. 326–336. Springer, Cham (2021). https://doi.org/10.1007/978-3-030-87193-2_31

22. Ronneberger, O., Fischer, P., Brox, T.: U-Net: convolutional networks for biomedical image segmentation. In: Navab, N., Hornegger, J., Wells, W.M., Frangi, A.F. (eds.) MICCAI 2015. LNCS, vol. 9351, pp. 234–241. Springer, Cham (2015). https://doi.org/10.1007/978-3-319-24574-4_28

23. Ibtehaz, N., Sohel Rahman, M.: MultiResUNet: rethinking the U-Net architecture for multimodal biomedical image segmentation. Neural Netw. **121**, 74–87 (2020). https://doi.org/10.1016/j.neunet.2019.08.025

A Scene Perception Method Based on MobileNetV3 for Bionic Robotic Fish

Ming Wang[✉], Xiaobin Du, Zheng Chang, and Kunlun Wang

Shandong Key Laboratory of Intelligent Buildings Technology, School of Information and Electrical Engineering, Shandong Jianzhu University, Jinan 250101, China
xclwm@sdjzu.edu.cn

Abstract. When facing practical underwater applications, how to improve environmental perception and autonomous decision-making ability in unstructured and complex dynamic underwater environment is a challenge for bionic robotic fish. In order to improve the environmental perception ability of bionic robot fish, a scene perception method of bionic robot fish based on MobileNetV3 is proposed in this paper. Firstly, the basic framework of bionic robotic fish scene perception based on MobileNetV3 network is designed. This method uses the transfer learning strategy to train the model, and optimizes the model according to the super parameters. Secondly, the scene perception and tracking control software platform of bionic robotic fish is developed and experiments are carried out. Thirdly, the robot fish scene perception algorithm based on SIFT-SVM is constructed and compared with the scene perception method based on MobileNetV3. The experimental results show that the scene perception method based on MobileNetV3 is feasible and effective, and can be applied to the environment perception of bionic robotic fish.

Keywords: Bionic robotic fish · Scene perception · Deep learning

1 Introduction

The development of underwater robots has had a major impact on marine exploration [1]. Underwater robots can perform tasks such as exploration, reconnaissance, target search and so on. Underwater robots are being used more and more widely. As one of the underwater robots, robotic fish has its own advantages. It has the advantages of small size, high mobility and low noise. In particular, it does not use propeller propulsion and has great concealment. On the one hand, the study of bionic robotic fish can explore new underwater vehicles, on the other hand, it can imitate fish swimming, so as to explore the swimming mechanism and drag reduction mechanism of fish. However, in order to promote the application of bionic robotic fish, the problem of how to perceive the environment must be effectively solved. The research on underwater environment perception and recognition of bionic robotic fish has attracted extensive attention of researchers. Vision, radar, ultrasound and other technologies have been used to perceive the environment of bionic robotic fish. This paper explores the application of deep learning technology to the environmental perception of bionic robotic fish.

© The Author(s), under exclusive license to Springer Nature Singapore Pte Ltd. 2022
H. Zhang et al. (Eds.): NCAA 2022, CCIS 1638, pp. 401–412, 2022.
https://doi.org/10.1007/978-981-19-6135-9_30

A deep learning method based on MobileNetV2 and transfer learning was proposed by Liu [2], which is installed in an underwater robot equipped with an embedded device system. They improved the real-time and accuracy of marine animal image classification. Yadira et al. processed a new dataset with dolphin, dolphin_pod, open_sea, and seabirds by using convolutional neural networks [3]. Their method also adopted transfer learning using the following models InceptionV3 and MobileNetV2. A deep neural network architecture based on the MobileNet was designed by Zhu to detect targets in 3D environment [4]. Deep learning has been more and more widely used. In recent years, convolutional neural network has also been used in more fields, such as remote sensing image processing [5–7], traffic scene segmentation [8, 9], facial feature detection [10–12], facial expression recognition [13, 14], license plate recognition [15], industrial scene detection [16], and so on. For the bionic robot fish, the best way to carry out scene recognition is to install radar or camera and other environmental sensing equipment. The goal of this paper is to classify and recognize the environment by using deep learning method for the sampled images. First of all, this paper proposes an underwater scene perception method of bionic robot fish based on MobileNetV3 network. Then, a scene perception algorithm based on SIFT-SVM is proposed. Finally, experimental tests are carried out. The experimental results showed the underwater scene perception method based on MobileNetV3 has better effect, and the accuracy rate reaches 90%.

The remain of the paper is arranged as follows. The deep learning network structure of MobileNetV3 is designed in Sect. 2. To evaluate the MobileNetV3 method, a SIFT-SVM method is adopted in the Sect. 3. Section 4 provide experiments and discussion with the MobileNetV3 method and the SIFT-SVM method for scene recognition. Finally, we provide conclusion remark and on-going work in the Sect. 5.

2 MobileNetV3 Network Structure

When the bionic robotic fish is applied underwater, it is very important to correctly perceive the surrounding environment. This is related to whether the robotic fish can successfully perform the scheduled task. If the environment cannot be perceived correctly, it will affect the obstacle avoidance, path planning and target tracking of bionic robotic fish. In this paper, the data sets of five underwater scenes are established: fish schools, giant fish, marine vegetation, coral reefs and shallow water islands and reefs. On this basis, combined with bionic robotic fish, an underwater scene perception system based on MobileNetV3 network is designed.

Robotic fish are equipped with visual sensors or auxiliary visual detection equipment. The vision module collects the image data of the surrounding environment. The image data will be input to the deep learning reasoning module deployed with MobileNetV3 network. The reasoning module performs scene perception and outputs classification results to provide basis for robot fish decision-making. The sensing results are then transmitted to the robot fish microcontroller, which makes independent decisions and sends motion instructions to the actuator. The actuator such as steering gear will act after receiving the control command. The communication module is responsible for the communication between the robot fish and the host computer, and uploads the underwater image and the motion state of the bionic robot fish to the management computer. The management computer monitors and displays various data. The control command sent by it is transmitted to the bionic robot fish through the communication module.

The amount of computation of convolution operation depends on the number of channels of the input characteristic graph. A large number of channels will bring a huge amount of computation. However, the performance of embedded microcontroller of bionic robotic fish is limited. In order to solve the above problems, MobileNetV3 network uses pointwise convolution and depthwise convolution to replace the traditional convolution.

Compared with ordinary convolution, depthwise splits the convolution kernel into a single channel. Under the condition that the depth of the input characteristic map does not change, the convolution operation is carried out for each channel respectively. In this way, the feature map output with the same depth as the input feature map can be obtained. Then the pointwise convolution is used to increase or reduce the dimension. Pointwise method utilization 1×1 convolution kernel. The number of channels is equal to the output dimension after depthwise convolution. Although the pointwise method has a large number of channels, its size is 1×1. The convolution kernel of 1×1 is conducive to reducing the amount of computation. The convolution network structure is shown in Fig. 1. The traditional convolution network structure is shown in Fig. 1(a). The network structure of depthwise and pointwise is shown in Fig. 1(b).

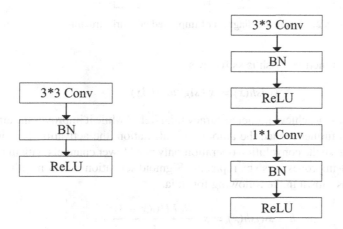

(a) traditional convolution   (b) convolution with depthwise and pointwise

Fig. 1. Schematic diagram of two convolution structures

A remarkable feature of MobileNetV3 is that it adopts neural network architecture search technology (NAS), which is similar to ManasNet. Firstly, it uses NAS technology with limited resources to search the network in the whole network and each module, which is called blockwise search. Then the NetAdapt hierarchical search algorithm is used to adjust each layer, especially the number of convolution cores of each layer. Because its search results show that the front or back layers in the network structure need a higher amount of computation. Therefore, some improvements are made to optimize the network speed under the condition of ensuring the network accuracy. In the last layers of the original network structure, 1×1 convolution maps feature to high-dimensional space, which not only obtains more feature information, but also increases the amount of calculation. Therefore, set 1×1 after the convolution is adjusted to the average pool layer, 1×1. The characteristic diagram required for convolution calculation is composed of 7×7 becomes 1×1. At the same time, the second half of the last bottleneck does not need to be used to reduce channels, so the amount of computation is reduced. The improved structure is shown in Fig. 2.

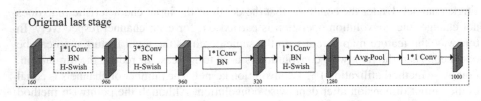

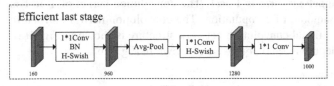

Fig. 2. Schematic diagram of improved network structure

The activation function swish is as follows:

$$swish(x) = x \cdot sigmoid(\beta x) \tag{1}$$

Swish function can achieve higher accuracy than ReLU while it improves the amount of calculation. To further reduce the amount of calculation, the activation function h-swish is used to make the convolution operation only need fewer channels with the same accuracy. H-swish improves swish by replacing Sigmoid activation function with variant ReLU function, as shown in the following formula.

$$h - swish(x) = s \cdot \frac{ReLU6(x+3)}{6} \tag{2}$$

MobileNetV3 has squeeze and excitation block (SE), which is used to construct the internal dependencies between channels of the feature map. Moreover, it adjusts the characteristics accordingly and improves the network accuracy.

MobileNetV3 includes two versions: large version and small version, which are suitable for devices with different computing performance. The network structure of large version and small version adopts a convolution layer to extract features in the initial part. The difference between the two lies in the number of bottlenecks and the specific settings of internal structure. In this paper, we apply MobileNetV3 large version to the scene perception of bionic robotic fish. Its network structure is shown in Table 1.

For the MobileNetV3 model, transfer learning strategy is used to train the model network. It carries out transfer learning on ImageNet dataset. The whole process includes

Table 1. Network structure of MobileNetV3 large version.

Input	Operator	#exp	#out	SE	Stride
$224 \times 224 \times 3$	Conv2d 3×3	–	16	–	2
$112 \times 112 \times 16$	G-bneck	16	16	–	1
$112 \times 112 \times 16$	G-bneck	48	24	–	2
$56 \times 56 \times 24$	G-bneck	72	40	–	1
$56 \times 56 \times 24$	G-bneck	72	40	1	2
$28 \times 28 \times 40$	G-bneck	120	80	1	1
$28 \times 28 \times 40$	G-bneck	240	80	–	2
$14 \times 14 \times 80$	G-bneck	200	80	–	1
$14 \times 14 \times 80$	G-bneck	184	80	–	1
$14 \times 14 \times 80$	G-bneck	184	112	–	1
$14 \times 14 \times 80$	G-bneck	480	112	1	1
$14 \times 14 \times 112$	G-bneck	672	160	1	1
$14 \times 14 \times 112$	G-bneck	672	160	1	2
$7 \times 7 \times 160$	G-bneck	960	160	–	1
$7 \times 7 \times 160$	G-bneck	960	160	1	1
$7 \times 7 \times 160$	G-bneck	960	160	–	1
$7 \times 7 \times 160$	G-bneck	960	160	1	1
$7 \times 7 \times 160$	Conv2d 1×1	–	960	–	1
$7 \times 7 \times 960$	AvgPool 7×7	–	–	–	–
$1 \times 1 \times 960$	Conv2d 1×1	–	1280	–	1
$1 \times 1 \times 1280$	FC	–	1000	–	–

two stages of training. In the first stage, only the last layer of the model is trained to obtain the first stage model; The second stage unfreezes all network layers and trains all network layers to make the neural network perform well on the data set. The MobileNetV3 based on two-stage training can make full use of ImageNet network model parameters and stronger generalization on self-built data sets.

3 Scene Perception Model Based on SIFT-SVM

To evaluate the MobileNetV3 model, a scene perception model based on SIFT-SVM for the comparative experiment in the next section. During model training, the feature matrix of all sample images of different scenes in the training set is extracted by SIFT method. Then the word set of BoW model is clustered and saved with unsupervised K-means method. The word set description of each scene obtained by clustering is used as the input data of SVM. Finally, the scene perception model is obtained by SVM classifier. SIFT is a local feature description algorithm in the field of image processing. The algorithm has considerable robustness to image scale translation, rotation, transformation scaling, brightness change and affine transformation. It is suitable for fast and accurate matching of underwater scene images. SIFT feature extraction is mainly divided into four basic steps:

(i) **Scale space extremum detection**, Search the scene image position at all scales, and recognize the possible interest points with scale rotation invariance based on Gaussian differential function.
(ii) **Key point positioning**, an accurate fitting model is used to determine the corresponding position scale of all candidate positions.
(iii) **Key point size and direction match**, based on the local gradient direction of the scene image, each key point is assigned one or more directions. The subsequent operations mainly transform the position, scale and direction of the key points.
(iv) **Key point description**, take a key point as the center and take the number of windows as $N \times N$. Then divide the window into $M \times M$ sub windows ($M < N$). The directions of all key points in different scene images depend on the directions of N/M seed points. Count the direction histograms contained in all sub windows to form SIFT feature description vector to obtain scene images with different key point descriptions.

SVM is a supervised training and learning model, which is widely used in pattern recognition. A special global hyperplane is used to classify the data set. The principle of minimizing structural risk is used to improve the training effect. Compared with other classifiers, SVM model has outstanding advantages in dealing with small samples, high latitude and nonlinear pattern recognition.

The SVM binary classification model can be mathematically expressed as the follow.

$$\begin{cases} \min_{w,b} \frac{1}{2}ww^T + C \sum_{i=1}^{l} \zeta_I \\ y_i(w^T x_i + b) \geq 1, \zeta_i \geq 0, i = 1, 2, \cdots, l \end{cases} \tag{3}$$

where (x_i, y_i) is sample space, $x_i \in R^n$, $y \in \{\pm 1\}$, ζ_i is the classification loss of the ith sample, C is the penalty function.

The kernel function maps the low dimensional input data to the high-dimensional space, so that the data samples can be divided in the high-dimensional feature space. The selection of kernel function will directly affect the final classification results. We adopted four types of kernel functions as follows.

Linear kernel function:

$$K(x, x_i) = x \cdot x_i \tag{4}$$

Polynomial kernel function:

$$K(x, x_i) = [(x \cdot x_i) + 1]_q \tag{5}$$

RBF kernel function:

$$K(x, x_i) = \exp(-\frac{||x - x_i||^2}{\sigma^2}) \tag{6}$$

Sigmoid kernel function:

$$K(x, x_i) = \tanh(v(x \cdot x_i) + c) \tag{7}$$

4 Experiments and Discussion

4.1 Experimental Environment

The experimental environment of this section is based on Tensorflow v2.4, Python v3.7, with a computer on the operating system is windows10 64 bits with Intel (R) core (TM) i5-9500F, 3.00 GHz. The default parameter settings of the MobileNetV3 model and the SIFT-SVM model are shown in Table 2. In the Table 2, epoch represents the number of training rounds, batch size represents the amount of images used in a batch size training, learning_rate represents the weight update rate, and input represents the input size of the image.

Table 2. Parameters of the MobileNetV3 model and the SIFT-SVM model.

Network type	Parameters	Values
MobileNetV3	Epoch	10
	Batchsize	16
	Learning_rate	0.0005
	Input	224 × 224
SIFT-SVM	Linear	Grid-based methods
	RBF	
	Poly	
	Sigmoid	

For the SIFT-SVM method, the original image data is first converted into gray image during training. Then a SVM classifier based on one of the four kernel functions is constructed. Taking the loss of mean square deviation as the parameter optimization index, the grid cross search algorithm is used to optimize the parameters in the process of classification model training.

In this section, the experiments are carried out by adjusting batch size, learning_ rate and other parameters for training the MobileNetV3 model. Adjusting the kernel function of SVM is used to train the SIFT-SVM model. The other parameters of the two models adopt the default parameters. On this basis, the effects of different models and different parameters on scene perception classification are compared. The accuracy and confusion matrix are used as evaluation indexes.

4.2   Experiment 1: Adjusting Batch Size for the MobileNetV3 Model

The experiment is to evaluate the image processing performance of the MobileNetV3 model by adjusting the parameter batchsize. Set the batchsize to 8, 16 and 64 respectively to verify the impact of different batch size inputs on underwater scene perception and classification, and conduct a comparative experiment on the testing data set. The experimental results are shown in Table 3 and Fig. 3. In Fig. 3, co_reef means coral reef, hu_fish stands for giant fish, ma_veta stands for marine vegetation, sh_reef stands for shallow water islands and reefs, sh_fish stands for a school of fish. The comparative experimental results show that the final training results of different batch sizes are similar. The best performance is achieved when the batchsize is 64.

Table 3. Classification accuracy of different batchsize.

Network type	Batchsize	Accuracy
MobileNetV3	8	0.9721
	16	0.9744
	64	0.9766

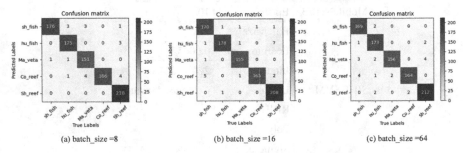

(a) batch_size =8   (b) batch_size =16   (c) batch_size =64

Fig. 3. Confusion matrix with different batchsize of the MobileNetV3 model.

4.3   Experiment 2: Adjusting Learning_rate for the MobileNetV3 Model

This experiment mainly evaluates the impact of different learning rates on underwater scene classification. In the experiment, the learning rates were 0.01, 0.001 and 0.0005 respectively. Experiments are carried out on the test set for the models with different learning rates. The experimental results are shown in Table 4 and Fig. 4. In Fig. 4, co_reef means coral reef, hu_fish stands for giant fish, ma_veta stands for marine vegetation, sh_reef stands for shallow water islands and reefs, sh_fish stands for a school of fish. The experimental results show that the underwater scene classification performance is the best when the learning rate is 0.0005. While the learning rate is 0.01, the classification performance is the worst.

Table 4. Classification accuracy of different learning_rate.

Network type	Learning_rate	Accuracy
MobileNetV3	0.0005	0.9744
	0.001	0.9666
	0.01	0.6974

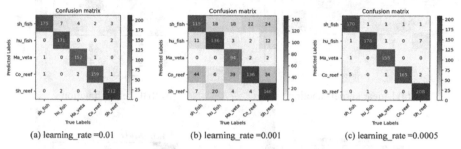

(a) learning_rate =0.01   (b) learning_rate =0.001   (c) learning_rate =0.0005

Fig. 4. Confusion matrix with different learning rates of the MobileNetV3 model.

4.4 Experiment 3: The SIFT-SVM

The experiments in this section are used to evaluate the scene classification performance of SVM classifier with different kernel functions. The accuracy of SIFT-SVM classifier with four different kernel functions is shown in Table 5. Normalize the classification results to obtain the confusion matrix as shown in Fig. 5. The grid on the diagonal corresponds to the accuracy of each category.

Table 5. Classification accuracy of different kernel functions.

Network type	Kernel function	Accuracy
SIFT-SVM	Linear	0.706
	BBF	0.712
	Poly	0.702
	Sigmoid	0.652

In Fig. 5, co_reef means coral reef, hu_fish stands for giant fish, ma_veta stands for marine vegetation, sh_reef stands for shallow water islands and reefs, sh_fish stands for a school of fish. It showed that the classification accuracy of classifiers with different kernel functions is about 80% on coral reefs and marine plants, about 70% on giant fish and shallow islands and reefs, and about 50% on fish schools. According to Table 5, the SVM classifier using RBF kernel function obtains relatively high accuracy. The average classification accuracy of different kernel functions is close. The overall average accuracy is about 70%.

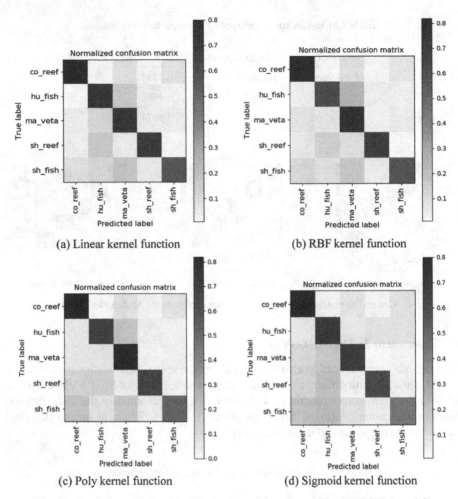

(a) Linear kernel function   (b) RBF kernel function

(c) Poly kernel function   (d) Sigmoid kernel function

Fig. 5. Confusion matrix with different kernel functions of the SIFT-SVM model.

Compared with three sets of scene classification experiments, the scene perception classification accuracy of bionic robotic fish based on MobileNetV3 model is higher. The accuracy of scene classification is more than 90%. It is superior to the underwater scene perception model based on the SIFT-SVM model. This verifies the feasibility of applying the MobileNetV3 model to bionic robot fish scene perception.

5   Conclusion

Scene perception is the key issue for the autonomous swimming of bionic robotic fish. This paper proposes an underwater scene perception method of bionic robot fish based on MobileNetV3 network. It adopts the transfer learning strategy for model training and optimizes the model according to the super parameters. In order to evaluate the performance of scene classification, this method is compared with the scene perception algorithm based on the SIFT-SVM. On this basis, experimental tests are carried out. The experimental results showed that the scene perception classification model based on MobileNetV3 has better performance.

The future work should be to transplant MobileNetV3 scene recognition software to the embedded system. It is configured into the bionic robot fish controller. In this way, the bionic robot fish perceives the surrounding environment automatically through machine vision. The ultimate goal is to realize autonomous swimming of bionic robotic fish.

Acknowledgement. This work is supported by the National Natural Science Foundation of China under Grant Nos. 62073196 and U1806204.

References

1. Zhou, Z., Liu, J., Yu, J.: A survey of underwater multi-robot systems. IEEE/CAA J. Autom. Sinica **9**(1), 1–18 (2021)
2. Liu, X., Jia, Z., Hou, X., Fu, M., Ma, L., Sun, Q.: Real-time marine animal images classification by embedded system based on MobileNet and transfer learning. In: OCEANS 2019, Marseille, France, pp. 1–5. IEEE (2019)
3. Quiñonez, Y., Lizarraga, C., Peraza, J., Zatarain, O.: Image recognition in UAV videos using convolutional neural networks. IET Softw. **14**(2), 176–181 (2020)
4. Zhu, J., Zhu, J., Wan, X., Wu, C., Xu, C.: Object detection and localization in 3D environment by fusing raw fisheye image and attitude data. J. Vis. Commun. Image Represent. **59**, 128–139 (2019)
5. Li, P., Che, C.: SeMo-YOLO: a multiscale object detection network in satellite remote sensing images. In: 2021 International Joint Conference on Neural Networks (IJCNN), Shenzhen, China, pp. 1–8. IEEE (2021)
6. Yu, D., Xu, Q., Guo, H., Zhao, C., Lin, Y., Li, D.: An efficient and lightweight convolutional neural network for remote sensing image scene classification. Sensors **20**(7), 1999 (2020)
7. Wang, J., He, X., Faming, S., Lu, G., Jiang, Q., Hu, R.: Multi-Size object detection in large scene remote sensing images under dual attention mechanism. IEEE Access **10**, 8021–8035 (2022)
8. Shuai, Y., Zhiyu, C., Shangdong, L., Mengxue, W., Feng, T., Yimu, J.: YoLite+: a lightweight multi-object detection approach in traffic scenarios. Procedia Comput. Sci. **199**, 346–353 (2022)
9. Wu, X., Qiu, T., Wang, Y.: Multi-object detection and segmentation for traffic scene based on improved Mask R-CNN. Chin. J. Sci. Instrum. **42**(07), 242–249 (2021). (in Chinese)
10. Lin, B., Mu, Y., Fu, Z., Li, C., Duan, X.: A network for detecting facial features during the COVID-19 epidemic. In: Zhang, D. (ed.) CCEAI 2021: 5th International Conference on Control Engineering and Artificial Intelligence (CCEAI), pp. 141–146. Association for Computing Machinery, New York (2021)
11. AL-Marghilani, A.A.: Target detection algorithm in crime recognition using artificial intelligence. CMC Comput. Mater. Continua **71**(1), 809–824 (2022)
12. Wan, Y., Zhang, M.X., Zhang, Y.A., Yao, L.: Research on unconstrained face recognition based on deep learning. In: 2020 International Conference on Big Data & Artificial Intelligence & Software Engineering (ICBASE), Bangkok, Thailand, pp. 219–227. IEEE (2020)
13. Nan, Y., Ju, J., Hua, Q., Zhang, H., Wang, B.: A-MobileNet: an approach of facial expression recognition. Alex. Eng. J. **61**(6), 4435–4444 (2022)

14. Su, C., Wang, G.: Design and application of learner emotion recognition for classroom. J. Phys. Conf. Ser. **1651**(1), 1–9 (2020)
15. Awalgaonkar, N., Bartakke, P., Chaugule, R.: Automatic license plate recognition system using SSD. In: 2021 International Symposium of Asian Control Association on Intelligent Robotics and Industrial Automation (IRIA), Goa, India, pp. 394–399. IEEE (2021)
16. Kong, Y., Han, S., Li, X., Lin, Z., Zhao, Q.: Object detection method for industrial scene based on MobileNet. In: 2020 12th International Conference on Intelligent Human-Machine Systems and Cybernetics (IHMSC), Hangzhou, China, pp. 79–82. IEEE (2020)

Data Enhancement Method Based on Generative Adversarial Network for Small Transmission Line Detection

Wenkong Wang[1], Weijie Huang[1](✉), Hongnan Zhao[2], Menghua Zhang[1], Jia Qiao[1], and Yong Zhang[1]

[1] School of Electrical Engineering, University of Jinan, Jinan 250000, China
{cse_huangwj,cse_qiaoj,cse_zhangy}@ujn.edu.cn,
zhangmenghua@mail.sdu.edu.cn
[2] Jinan Hospital, Jinan 250000, China

Abstract. Unmanned aerial vehicle (UAV) patrol inspection is an important means of transmission line detection. However, during UAV patrol, due to the UAV or the bad weather, there may be droplets on the camera lens, which will especially reduce the effect of small target detection. In this paper, a data enhancement method based on improved generative adversarial network (GAN) is proposed for automatically removing droplets from UAV images of transmission lines. In order to take more account of the context information of small targets when removing droplets, this method integrates the attention mechanism in the design of generator. Experiment shows that compared with the benchmark method, the proposed method can make the image after removing droplets more similar to the original image, and increase the structural similarity (SSIM) by 8.69%. The practical value of this method is further proved in the process of object detection of droplets removed images.

Keywords: Generative adversarial network · Attention mechanism · Transmission lines

1 Introduction

UAV inspection is an important way of transmission line defect detection. The intelligent detection of transmission line defects can be realized by photographing the image by UAV and then detecting the image by object detection [1] algorithm. A wide range of occlusion to the UAV image may be caused because of the existence of droplets. In addition, when the UAV takes photos in the process of line patrol, the camera should aim at the transmission line equipment. The existence of droplets will affect the focusing of the camera, and make the image background virtual, which make the loss of image details is serious.

H. Zhang et al. (Eds.): NCAA 2022, CCIS 1638, pp. 413–426, 2022.
https://doi.org/10.1007/978-981-19-6135-9_31

Therefore, the existence of droplets will lead to uneven image quality, affect the extraction and utilization of image information, and reduce the accuracy and reliability of object detection. In order to improve the overall quality of these degraded images and ensure the performance of detection algorithm, it is very important to automatically remove these droplets in UAV images caused by the bad weather.

GAN [2] is a deep learning model, which was first proposed in 2014. At present, it has a variety of development directions. The neural network learning method of GAN is different from the general neural network. It makes the two neural networks of generator and discriminator form an antagonistic relationship for antagonistic learning, and finally makes them reach Nash balance in the training process to realize parameter optimization. In this relationship, the generator is responsible for generating the image, and the discriminator is responsible for judging whether the image generated by the generator is a ground truth image. In order to pass the judgment of the discriminator, the generator will generate an image similar to the ground truth image as much as possible.

GAN can generate new data similar to the distribution of train dataset. For example, GAN can be used for automatic generation of animation style avatars, train dataset generation of other deep learning tasks, etc. In addition, it can also be used for super-resolution reconstruction, neural style and other tasks. In the gradual development, DCGAN [3] replaced the full connection layer in GAN with convolution layer, and made important changes to the structure of GAN. DCGAN is composed of convolution layer, batch normalization (BN) layer and leakyReLU activation function. Compared with the full connection layer structure, DCGAN has higher stability than GAN. Conditional generative adversarial network (CGAN) is a kind of GAN, which uses the additional information to constrain the model to ensure the similarity between the generated image and the ground truth image. Image translation is the task of acquiring images from one domain and converting them to make them have the style of images from another domain. This kind of network can successfully learn the image mapping of day and night, zebra and horse, four seasons transformation and so on. Currently, the effective conditional generation countermeasure networks mainly include pix2pix [4], Cycle GAN [5], etc.

In this paper, the data enhancement of UAV images is regarded as the task of image translation, and a CGAN-based data enhancement method is proposed for the removal of droplets from UAV images. The main contributions of this paper are as follows:

1) This paper creates a 1042 pairs of UAV paired image dataset of transmission line scene. Each pair of images includes an original image and an image with droplets. The dataset is collected in the real transmission line environment, with rich scenes and strong robustness. The defects in the scene include bird's nest and suspended foreign objects.
2) In order to make full use of the context information of targets, especially small targets, in the images, a generator integrating attention mechanism is

proposed in this paper to improve the effect of removing droplets. Experiments show that the similarity between the image generated by the improved generator and the original image is 8.69% higher than that of the benchmark algorithm in SSIM.

3) In this paper, the images before and after removing droplets are used to evaluate the effect of object detection, which proves the effectiveness of removing droplets to optimize the intelligent detection algorithm.

This paper is organized as follows. In Sect. 2, traditional and deep learning methods for removing droplets are reviewed. The proposed data enhancement method are detailed in Sect. 3. And the experimental details are detailed in Sect. 4. Finally, the conclusion is given in Sect. 5.

2   Previous Work

In the field of image processing, droplets removal of single image is an extremely complex technology. For a long time, there are few traditional methods to carry out relevant research. Wang et al. [6] used a guided filter to distinguish the low frequency part of the edges. Then, the observed image and the low frequency part are used as the input of the designed filter to obtain rough results. Then, the final refined result is achieved through the minimization operation between the rough result and the observed image. In [7], a scattering inpainting algorithm was proposed, which uses the scatter line to continuously scan the nearby pixels to fill the damaged region, but it needs to manually set the scanning region, and is easy to appear black spots due to the influence of light. At present, the effect of traditional methods on image processing with dense droplets is not ideal, and the background image covered by droplets cannot be repaired accurately.

The continuous innovation of deep learning and related technologies has greatly promoted the development of computer vision. It is widely used in the fields of rain and fog removal, and has great application potential in the field of droplet removal. In recent years, Xiang et al. [8] proposed a deep neural network architecture, namely feature supervised generation adversarial network (FS-GAN), which is used to remove rain from a single image. Shen et al. [9] proposed a multi branch attention generation adversarial network (called MBA-RainGAN) to completely eliminate the mixed rain. Raj et al. [10] used a condition to generate a countermeasure network, and on this basis, made targeted adjustments to the network architecture, which can well reduce the occlusion of haze in the image. Wang et al. [11] Summarized the current video and single image rain removal methods based on deep learning, and more elaborate branches of each method are presented.

A data enhancement method based on CGAN and attention mechanism is proposed in this paper, and is applied to the removal of droplets from a single image. The droplet removal image obtained by this method is closer to the ground truth image, and can improve the effect of object detection.

3   Method

The proposed network architecture mainly includes generator (G) and discriminator (D), as shown in Fig. 1. The proposed network architecture needs to input droplets image and ground truth image to complete the training process. The droplets image will be used as the input of the generator to obtain the generated image. In order to prevent the generator from learning a fixed mapping pattern, a small amount of random noise is added to the input to enhance its robustness. And then the generated image and the droplets image will be concated based on the channel dimension and put into the discriminator for judgment. Next, the discriminator gives the probability that the input contains the ground truth image. When the input does not contain a ground truth image, the discriminator expects to judge the probability as 0. However, the higher the similarity between the generated image and the ground truth image, the greater the probability gived by discriminator will be. In addition, the ground truth image and droplets image are also concated based on the channel dimension. Similarly, they are input to the discriminator for training, and the discriminator is expected to judge the probability as 1.

The generator generates an image y after removing droplets by inputting an image x with droplets and random noise z. Therefore, the task of generator can be expressed as Formula 1:

$$G : \{x, z\} \rightarrow y \tag{1}$$

where the random noise z is used to prevent the deterministic mapping between the input image and the generated image, which makes the mapping more robust.

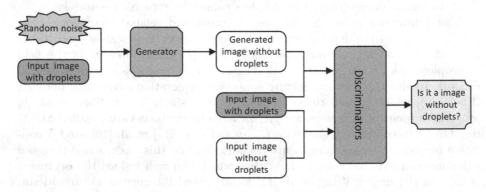

Fig. 1. The structure of proposed GAN. It is mainly composed of generator and discriminator.

The purpose of the generator is to generate an image after removing droplets that the discriminator can judge it as a ground truth image as much as possible after training. The purpose of the discriminator is to judge the image generated by the generator as non ground truth image as much as possible after training. This design makes the generator and discriminator form an relationship of adversarial optimization in training process.

3.1 Generator

In this paper, a improved unet256 [12] embedded with attention mechanism is designed as the generator, and the structure is shown in Fig. 2. The proposed generator is mainly composed of convolution (Conv) layer, batch normalization (BN) layer, transposed convolution (TransConv) layer, LeakyReLU and ReLU activation function, and squeeze-and-excitation (SE) module [13]. In the feature extraction network, the input image performs 8 convolution operations, and a total of 256 times of down sampling is performed on the image. In the feature fusion network, 8 transpose convolution layers perform 256 times up sampling on the input tensor. At the same time, the channel connection is established in the feature map with the same size in the process of up sampling and down sampling. By combining the semantic information in high-level feature maps and the fine-grained surface information in low-level feature maps, the generator well meets the needs of generating images for these two aspects of information.

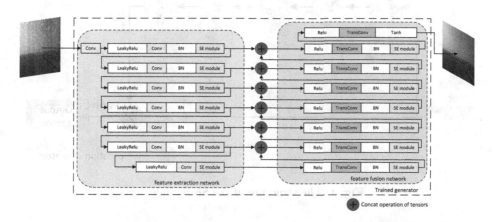

Fig. 2. The structure of the proposed generator.

Considering that the generation of images should not only be constrained at the pixel level, the proposed generator should take more full account of context information in order to achieve image translation closer to the ground truth image semantics. Therefore, SE attention modules are added on the basis of Unet256 to solve this problem.

In neural networks, attention mechanism is a technology that imitates cognitive attention. By adjusting the weight of each part of the tensor, the attention mechanism enhances some important parts of the input tensor and reduces others. Learning which part of the tensor is more important than others depends on the context information and is trained by gradient descent.

SE module is a convolution block structure composed of convolution layer, average pooling layer, relu and sigmoid activation function. Among them, the role of global average pooling is to aggregate features in the spatial dimension. Compared with the global max pooling, it can more comprehensively consider the information of all elements in the tensor. The structure of SE module is shown in Fig. 3. Assuming that the size of the input image is $W \times H \times C$, a channel attention map of $1 \times 1 \times C$ size is obtained through global average pooling and full connection layer, and then multiplied by the corresponding position of the original input image through skip connection, so as to give different attention weights to each channel. It aims to give the network the ability to perform dynamic channel feature recalibration and make the image features generated by the network more representative. SE module can significantly improve the classification and object detection network. ResNet [14] embedded in SE module once won the championship of Imagenet classification competition. It can improve the channel interdependence almost without calculating the cost.

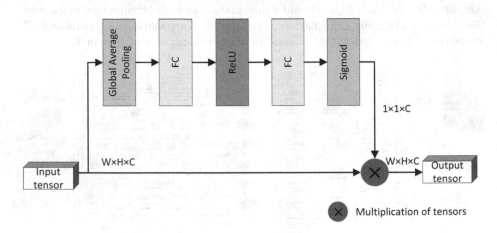

Fig. 3. The structure of the SE module.

3.2 Discriminator

In this paper, the patch generative adversarial network (Patch GAN) [15] with convolution structure is used as the discriminator. The discriminator attempts to divide the image into multiple patches for separate evaluation, and finally averages the scores. The advantage of this approach is that it can better take into account the influence of different parts of the image. The structure of Patch GAN is simple. By setting the number of layers of convolution layer manually, the input is mapped into a matrix with size $N \times N$ through multi-layer multiple convolution layers. The value in the matrix is the probability that each patch belongs to ground truth.

3.3   Loss Function

The loss function of the proposed algorithm mainly includes the loss function of GAN and L1 loss function.

The loss function of the generated countermeasure network can be expressed as Eq. 2:

$$\mathcal{L}_{cGAN}(G, D) = \mathbb{E}_{x,y}[\log D(x, y)] + \mathbb{E}_{x,z}[\log(1 - D(x, G(x, z)))] \qquad (2)$$

The loss function of GAN reflects the antagonism of generator and discriminator in the training process. The generator expects $D(x, G(x, z))$ to be as large as possible, that is, $\log(1 - D(x, G(x, z)))$ to be as small as possible, so that the value of the loss function is as small as possible. The discriminator expects $\log D(x, y)$ and $\log(1 - D(x, G(x, z)))$ to be as large as possible, so that the value of the loss function is as large as possible. An antagonistic mutual optimization relationship forms between the generator and discriminator because of this part of loss function. In addition, to realize image translation, the task of the generator is not only to deceive the discriminator to generate a certain type of image, but also to require the generated drops removal image to be close to the ground truth image. In order to ensure the similarity between the generated image and the target image, $L1$ loss function is added to constrain the image generated by the generator, as shown in Eq. 3:

$$\mathcal{L}_{L1}(G) = \mathbb{E}_{x,y,z}\left[\|y - G(x, z)\|_1\right] \qquad (3)$$

To sum up, the total loss function of the proposed algorithm can be expressed as Eq. 4:

$$G^* = \arg\min_G \max_D \mathcal{L}_{cGAN}(G, D) + \lambda \mathcal{L}_{L1}(G) \qquad (4)$$

where λ is the weight coefficient of the $L1$ loss that can be set manually, which is set to 100 in this paper.

4   Experiment

4.1   Dataset

The experimental data adopts the real transmission line defect dataset collected by UAV, including 1042 pairs of paired images with and without droplets. Some of example paired images are shown in (a), (b) and (c) in Fig. 4. The dataset are divided into 834 pairs of train dataset and 208 pairs of test dataset according to the ratio of 8:2. The defects in the transmission line images include bird's nest and suspended foreign objects, and mostly of them are small targets.

(a)   (b)   (c)

Fig. 4. (a), (b) and (c) are some of paired images in dataset. Each pair of images includes an image (the one above) with droplets and an original image (the one below) without droplet.

4.2 Training Process

The proposed model is trained on an NVIDIA geforce RTX 2060. The training parameters are set as follows. Adam gradient descent algorithm [16] is used for training. The batch size is set to 1, and a total of 200 epochs are trained. The initial learning rate is set to 0.0002, and the linear learning rate decay strategy is adopted in the 100th iteration. The flow chart of the training process is shown in Fig. 5. The loss function is updated alternately between the gradient descent step of discriminator and the gradient descent step of generator. When optimizing the generator, the back propagation gradient of the discriminator does not participate in the calculation, and vice versa.

The $L1$ loss in the loss function can reflect the similarity between the droplets removal image and the ground truth image. Figure 6 shows the change of $L1$ loss during training, in which the loss decreases and the similarity gradually increases.

4.3 Evaluation Method

SSIM is used to compare the similarity between two images (For example, x and y) based on brightness, contrast and structure. The flow chart of calculating SSIM of two images is shown in Fig. 7.

Taking image x as an example, take the average gray value μ_x is the estimation of brightness measurement, and the formula is shown in Eq. 5:

$$\mu_x = \frac{1}{N} \sum_{1}^{N} x_i \tag{5}$$

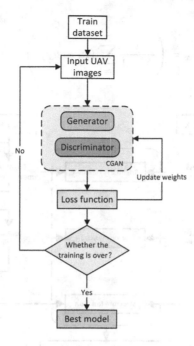

Fig. 5. Training flowchart of the proposed network.

Then, the average gray value μ_x is subtracted from the original image to obtain $x - \mu_x$. The standard deviation σ_x can then be used as the measured value of contrast, as is shown in Eq. 6:

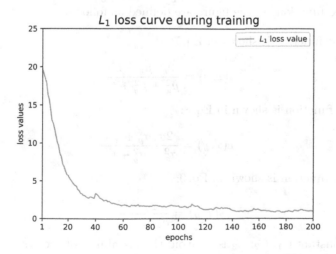

Fig. 6. Changes of $L1$ Loss during Training.

$$\sigma_x = \left(\frac{1}{N-1} \sum_{i=1}^{N} (x_i - \mu_x)^2 \right)^{\frac{1}{2}} \tag{6}$$

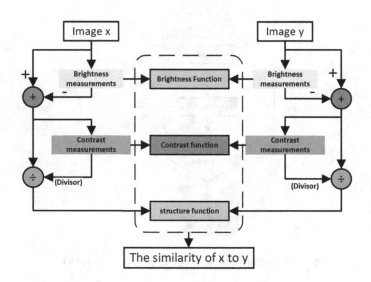

Fig. 7. The flow chart of calculating SSIM of x and y.

Next, the signal is divided by its own standard deviation, and the structure function is defined as the function of $\frac{(x-\mu_x)}{\sigma_x}$ and $\frac{(y-\mu_y)}{\sigma_y}$.

The three functions in the figure are defined as follows:

1) Brightness function is shown in Eq. 7:

$$l(x,y) = \frac{2\mu_x\mu_y + C_1}{\mu_x^2 + \mu_y^2 + C_1} \tag{7}$$

2) Contrast function is shown in Eq. 8:

$$c(x,y) = \frac{2\sigma_x\sigma_y + C_2}{\sigma_x^2 + \sigma_y^2 + C_2} \tag{8}$$

3) Structure function is shown in Eq. 9:

$$s(x,y) = \frac{\sigma_{xy} + C_3}{\sigma_x\sigma_y + C_3} \tag{9}$$

where, constant C_1, C_2, C_3 is to avoid the instability of the system when the numerator or denominator approaches 0.

The final calculation formula of SSIM is shown in Eq. 10:

$$SSIM(x,y) = [l(x,y)]^{\alpha}[c(x,y)]^{\beta}[s(x,y)]^{\gamma} \tag{10}$$

when α, β, γ set to 1 and take $C_3 = \frac{C_2}{2}$, the simplified form of calculating structural similarity is obtained, as shown in Eq. 11:

$$SSIM(x,y) = \frac{(2\mu_x\mu_y + C_1)(2\sigma_x\sigma_y + C_2)}{(\mu_x^2 + \mu_y^2 + C_1)(\sigma_x^2 + \sigma_y^2 + C_2)} \tag{11}$$

4.4 Results

Example of Data Enhancement. The effect of UAV defect image enhancement using the proposed network is shown in Fig. 8. It can be seen that after the processing of the proposed network, most of the droplets can be eliminated, which is worthy of application in subsequent tasks such as object detection.

Comparison of SSIM Value. The image enhancement effect of pix2pix algorithm is compared with that of the proposed method. Firstly, the images with droplets in the test dataset are input into the two networks respectively, and the SSIM value is calculated between the generated image and the original image without droplets in test dataset. Finally, the average value of all images in test dataset is taken as the comparison index. The experimental results are shown in Table 1. It can be seen that the images with droplets removed generated by the improved generator is closer to the original image without droplets than the benchmark algorithm, and has higher application potential.

Table 1. SSIM value comparison.

Algorithm	Average SSIM value in test dataset (%)
Pix2pix	79.63
The proposed method	88.32

Effect for Object Detection. The enhanced image can be used for subsequent applications, such as object detection tasks. Object detection is a typical computer vision task. Its purpose is to use the box to select the desired object. Faster R-CNN [17] is a typical two-stage object detection network.

Take the original images without droplets as the train dataset, and train the train dataset on Faster R-CNN network and use different test dataset to detect bird's nest and suspended foreign objects. The test results of using different test datasets to detect bird nests and suspended solids are shown in Table 2. It can be seen that when the network trained by original image without droplets is used to

Fig. 8. The first line is the images with droplets, the second line is the deleted droplets images generated from the network, and the third line is the original images without droplets taken for the UAV.

detect images with droplets, the average accuracy will be greatly reduced as seen in table shrinkage. The image processed by GAN can be close to the detection effect of original image, which has high application value.

Table 2. Effect for object detection.

Dataset	AP% (nest)	AP% (foreign objects)	mAP%
Original images without droplets	84.2	65.9	75.1
Images with droplets	63.5	42.7	53.1
Droplets removal images	80.7	61.8	71.3

5   Conclusion

A GAN integrated with attention mechanism is proposed in this paper and used for data enhancement of UAV images of small targets with transmission line defects. After embedding the SE module, the generator can adaptively learn the importance of tensor in each channel, and more fully consider the context information that the generated image should follow. Experiments show that the proposed method has an average similarity improvement of 8.69% on SSIM compared with the benchmark algorithm. In the further application of object detection, the UAV images with droplets removed is closer to the original image without droplets.

Acknowledgement. This work was supported in part by the Key R&D Project of Shandong Province under Grant No. 2022CXGC010503, the Youth Foundation of Shandong Province under Grant No. ZR202102230323, the National Natural Science Foundation for Young Scientists of China under Grant No. 61903155, and the Doctoral Scientific Fund Project under Grant No. xbs1910.

References

1. Zaidi, S.S.A., Ansari, M.S., Aslam, A., Kanwal, N., Asghar, M., Lee, B.: A survey of modern deep learning based object detection models. Digit. Sig. Process. **126**, 103514 (2022)
2. Aggarwal, A., Mittal, M., Battineni, G.: Generative adversarial network: an overview of theory and applications. Int. J. Inf. Manag. Data Insights **1**(1), 100004 (2021)
3. Radford, A., Metz, L., Chintala, S.: Unsupervised representation learning with deep convolutional generative adversarial networks. arXiv preprint arXiv:1511.06434 (2015)
4. Isola, P., Zhu, J.Y., Zhou, T., Efros, A.A.: Image-to-image translation with conditional adversarial networks. In: Proceedings of the IEEE Conference on Computer Vision and Pattern Recognition, pp. 1125–1134 (2017)
5. Zhu, J.Y., Park, T., Isola, P., Efros, A.A.: Unpaired image-to-image translation using cycle-consistent adversarial networks. In: Proceedings of the IEEE International Conference on Computer Vision, pp. 2223–2232 (2017)
6. Ding, X., Chen, L., Zheng, X., Huang, Y., Zeng, D.: Single image rain and snow removal via guided L0 smoothing filter. Multimed. Tools Appl. **75**(5), 2697–2712 (2016). https://doi.org/10.1007/s11042-015-2657-7
7. Hao, J., Jiang, M., Huang, Y., Bai, Y., Wang, Y.: Scatter inpainting algorithm for rain or snow removal in a single image. In: 2018 IEEE 4th International Conference on Computer and Communications (ICCC), pp. 1700–1704. IEEE (2018)
8. Xiang, P., Wang, L., Wu, F., Cheng, J., Zhou, M.: Single-image de-raining with feature-supervised generative adversarial network. IEEE Sig. Process. Lett. **26**(5), 650–654 (2019)
9. Shen, Y., et al.: MBA-RainGAN: multi-branch attention generative adversarial network for mixture of rain removal from single images. arXiv preprint arXiv:2005.10582 (2020)
10. Raj, N.B., Venketeswaran, N.: Single image Haze removal using a generative adversarial network. In: 2020 International Conference on Wireless Communications Signal Processing and Networking (WiSPNET), pp. 37–42. IEEE (2020)
11. Wang, H., Wu, Y., Li, M., Zhao, Q., Meng, D.: A survey on rain removal from video and single image. arXiv preprint arXiv:1909.08326 (2019)
12. Ronneberger, O., Fischer, P., Brox, T.: U-Net: convolutional networks for biomedical image segmentation. In: Navab, N., Hornegger, J., Wells, W.M., Frangi, A.F. (eds.) MICCAI 2015. LNCS, vol. 9351, pp. 234–241. Springer, Cham (2015). https://doi.org/10.1007/978-3-319-24574-4_28
13. Hu, J., Shen, L., Sun, G.: Squeeze-and-excitation networks. In: Proceedings of the IEEE Conference on Computer Vision and Pattern Recognition, pp. 7132–7141 (2018)

14. He, K., Zhang, X., Ren, S., Sun, J.: Deep residual learning for image recognition. In: Proceedings of the IEEE Conference on Computer Vision and Pattern Recognition, pp. 770–778 (2016)
15. Gui, J., Sun, Z., Wen, Y., Tao, D., Ye, J.: A review on generative adversarial networks: algorithms, theory, and applications. IEEE Trans. Knowl. Data Eng. **1**, 1 (2021)
16. Henry, M.: Review on gradient descent algorithms in deep learning approaches. J. Innov. Dev. Pharm. Tech. Sci. (JIDPTS) **4**, 91–95 (2021)
17. Ren, S., He, K., Girshick, R., Sun, J.: Faster R-CNN: towards real-time object detection with region proposal networks. In: Advances in Neural Information Processing Systems, vol. 28 (2015)

Improved Faster R-CNN Algorithm for Transmission Line Small Target Detection

Wenkong Wang[1], Ping Meng[2], Weijie Huang[1(✉)], Menghua Zhang[1], Jia Qiao[1], and Yong Zhang[1]

[1] School of Electrical Engineering, University of Jinan, Jinan 250000, China
{cse_huangwj,cse_qiaoj,cse_zhangy}@ujn.edu.cn,
zhangmenghua@mail.sdu.edu.cn
[2] AVIC Research Institute For SPECIAL Structures of Aeronautical Composites,
Jinan 250000, China

Abstract. Bird's nests, as well as suspended foreign objects such as plastic and rags, are serious potential safety hazards on transmission lines. Because the Bird's nests and suspended foreign objects in the unmanned aerial vehicle (UAV) images often belong to small targets with less pixels, more noise and easy to be disturbed, the detection of these objects puts forward higher requirements for the detection algorithm. In this paper, an deep learning-based algorithm are designed for the detection of these two kinds of small targets. By adding the attention mechanism module to the backbone network in this algorithm, the importance of each part of the feature map extracted from UAV image are refined in two different dimensions, and sufficient context learning are carried out to improve the detection of small targets. Further more, a post-processing algorithm based on Soft-NMS are designed to prevent small targets from being filtered and further improve the detection of small targets. Compared with the benchmark algorithm Faster R-CNN, the proposed algorithm achieves 4.7% improvement in average precision (AP).

Keywords: Transmission line detection · Attention mechanism · Soft-NMS

1 Introduction

In order to guarantee the safety and quality of transmission, transmission line inspection has become an essential work. Among the potential safety hazards of transmission lines, especially the bird's nests and all kinds of plastic or rag suspended foreign objects are easy to cause power accidents [1]. Due to the high

H. Zhang et al. (Eds.): NCAA 2022, CCIS 1638, pp. 427–441, 2022.
https://doi.org/10.1007/978-981-19-6135-9_32

transmission line and rugged terrain, manual inspection requires a lot of time and energy on the one hand, and there are inevitable safety risks on the other hand. UAV provides a new tool for transmission line inspection. However, the UAV pilot must take photos of each transmission line location from multiple angles, and then manually view the photos in turn. Such a workflow brings a huge workload. So it is meaningful to develop an intelligent defect detection algorithm to automatically detect the transmission defect images.

Object detection algorithm brings a new direction for UAV automatic inspection. Object detection task requires accurate positioning of target objects and accurate identification of target types [2]. Traditional object detection relies on manual design algorithm based on image processing. These methods have narrow applicability and great limitations. New impetuses on the development of object detection and deep learning are brought by convolutional neural network. Current deep learning-based object detection algorithms are mainly divided into one-stage and two-stage. The detection speed of the one-stage algorithms is faster, but the detection accuracy is not as high as the two-stage algorithms. Two-stage algorithms have higher accuracy and can also fit the real-time requirement. Considering the security requirements of patrol inspection, a novel two-stage algorithm is designed to detect the defects of transmission lines in this paper.

In the research of object detection algorithm, small target detection [3] has always been a research hotspot. In the COCO dataset [4], small targets are defined as areas less than 32×32 pixel object to be detected. Small target detection is always a problem worthy of study in the research of object detection because it contains fewer pixels, more noise and is easy to be disturbed by the background. The bird's nest on overhead lines, as well as the suspended foreign objects such as plastic bags and rags, are mostly small targets.

For the detection of the above two kinds of small targets in the transmission line, an attention model and Soft-NMS-based detection algorithm of bird's nest and suspended foreign objects in the transmission line is proposed in this paper. Experiments show that the designed method can effectively improve the detection effect of bird's nest and suspended foreign objects on transmission lines. The main contributions of this paper are listed below:

1) A transmission line scene dataset collected by UAV is created in this paper, which includes real bird's nest and suspended foreign objects such as rags and plastics. The dataset is collected from real substations and transmission lines, so it has high application value.
2) An attention model-based backbone network and Resnet is designed. The context information of the input feature map can be learned more effective using the proposed backbone network, to fully infer the location and category of small targets through other pixels, which is helpful to detect small targets such as bird's nest and suspended foreign objects.
3) A post-processing algorithm based on Soft-NMS are designed in this paper. The algorithm is used to avoid the missed detection because of the overlap of detection boxes with same category, and further improve the detect of small targets.

The rest of this paper is organized as follows. Small target detection algorithm and research on transmission line detection are reviewed in Sect. 2. In Sect. 3, the structure of the designed object detection algorithm is detailed. And the experimental process and results are described in Sect. 4. Finally, the conclusion is given in Sect. 5.

2 Previous Work

Improving the accuracy of small target detection is still a urgent problem to be solved in the field of target detection. Some studies have proved that combining different feature layers can improve the effect of small target detection. In [5], a light-weight architecture was proposed to build featurized image pyramid in an object detection method and the multi-scale features affected the predicted results by using an attention module. In [6], a feature refinement module were designed by Yang et al. to improve detection performance by getting more accurate features. The learning of context information was also very useful for detecting small targets. The attention-based method attempts to refine the importance of different parts in the feature map. The squeeze and excitation network (SENet) [7] can learn the weight of each channel in the input feature map and use this weight to determine the importance of the channel. The convolution block attention module (CBAM) [8] used the channel attention part and spatial attention part to calibrate the feature map to learn better feature representation. The gather-excite (GE) module [9] gathered features at the global level and used the gathered information to enhance important components of local features.

At present, some studies have tried to verify the effectiveness of deep learning in transmission line inspection. Liu et al. [10] enhanced partial features and weaken the expression of irrelevant features by aggregating the information of the feature map on the channel and spatial dimensions, to increase the average accuracy of yolov5-s in transmission line defect detection. Zheng et al. [11] expanded the dataset by adjusting the size of convolution kernel of Faster R-CNN and rotating the images, which greatly increases the accuracy of transmission line detection. Fan et al. [12] used random forest (RF) for image segmentation to realize insulator object recognition. Then the convolution neural network (CNN) method is used to classify the normal state and defect state of insulators. Finally, the defect location and identification is realized by Faster R-CNN. The above research improves the accuracy of transmission line detection from all categories. However, there is a lack of research on small target defect detection in transmission line. In this paper, the dataset of transmission line with most small targets such as bird's nest and suspended foreign objects is constructed, and the algorithm research is carried out for the above small target defects. An attention model and Soft-NMS-based transmission line small target detection algorithm is proposed to detect small targets such as bird's nest and suspended foreign objects.

3 Method

The diagram of overall designed detection network is shown in Fig. 1, which is composed of three main parts. The first part is the feature extraction network. Extracting feature map of UAV images after scaling and normalization is the main function of this part. The second part is the region proposal network (RPN), which mainly creating anchors and providing foreground region proposals to the feature map extracted from the first part. The third part is the detection network, which mainly receiving the feature map obtained in the first part and detect the bird's nests, as well as suspended foreign objects in the UAV image. It will output the location, label and confidence of these two types of defects.

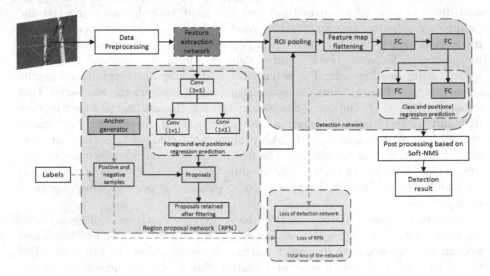

Fig. 1. The overall architecture of the designed object detection algorithm. It is mainly composed of three parts: feature extraction network, RPN and detection network.

3.1 Feature Extraction Network

The designed feature extraction network mainly includes backbone network (improved resnet50 [13]) and feature pyramid structure [14], as shown in Fig. 2.

Backbone. Small targets such as bird's nest and suspended foreign objects have few pixels, so it is difficult to extract effective features. In this paper, the backbone is designed to make the feature extraction network consider more small target context information. By integrating the attention mechanism, the backbone can learn what features should be paid more attention to better consider the context information.

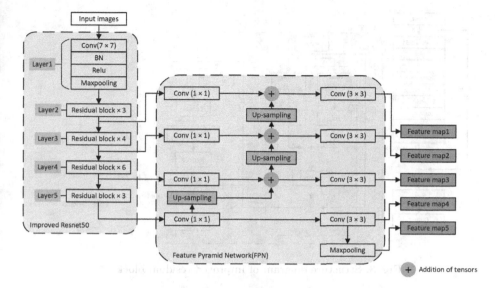

Fig. 2. Feature extraction network mainly includes the improved ResNet50 and feature pyramid structure.

The backbone is designed as an improved ResNet50 structure embedded with attention module. The backbone network mainly includes five parts: Layer1, Layer2, Layer3, Layer4 and Layer5. Among them, part Layer1 consists of the following structural sequence: a convolution layer, a batch normalization (BN) layer [15], a ReLU activation function [16] and a max pooling layer. Layer2, Layer3, Layer4 and Layer5 are composed of improved residual blocks integrated into attention mechanism with different parameters and the same structure. The number of residual blocks used are 3, 4, 6 and 3 respectively. The preprocessed UAV image will extract features through five parts in sequence: Layer1, Layer2, Layer3, Layer4 and Layer5.

The residual block integrated into the attention mechanism is the infrastructure of the designed backbone network, as shown in Fig. 3. Each residual block is mainly composed of convolution layer, BN layer, ReLU activation function and CBAM module. In addition, each residual block will retain the input feature information and fuse the input feature information with the output at the end to prevent gradient disappearance, gradient explosion and degradation when the network depth is too deep. Each residual block has been extracted mainly through the convolution layer with 3×3 kernel size. Therefore, this paper chooses to add convolution block attention module (CBAM) after this convolution layer to learn the context information around the small target, so as to better locate and classify the small targets.

CBAM is an effective attention module composed of convolution layer, full connection layer and activation function. CBAM is shown in Fig. 3. Assuming that the size of the input tensor is $C \times H \times W$, its channel attention and spatial

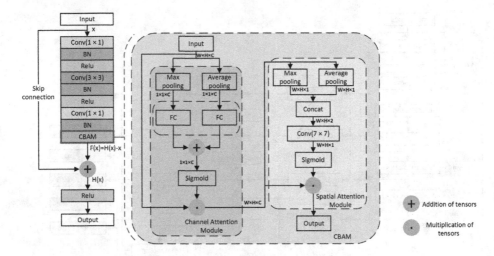

Fig. 3. Structure diagram of improved residual block.

attention maps will be calculated successively by CBAM and multiplied with the input feature map respectively to realize the parameter optimization considering the context information, as shown in the Eq. 1.

$$\mathbf{F'} = \mathbf{F} * \mathbf{M_c}(\mathbf{F}) * \mathbf{M_s}\left(\mathbf{F} * \mathbf{M_c}(\mathbf{F})\right) \tag{1}$$

where F is the input tensor. $\mathbf{F'}$ is the output tensor of learning context information. $\mathbf{M}_c()$ and $\mathbf{M}_s()$ are the acquisition processes of channel and spatial attention respectively. The two processes can be expressed as Eq. 2 and Eq. 3 respectively:

$$\mathbf{M_c}(\mathbf{F}) = \sigma(MLP(AvgPool(\mathbf{F})) + MLP(MaxPool(\mathbf{F}))) \tag{2}$$

where $AvgPool()$ and $MaxPool()$ are global average pooling operation and global max pooling operation respectively.

$$\mathbf{M_s}(\mathbf{F}) = \sigma\left(f^{7\times7}([AvgPool(\mathbf{F}); MaxPool(\mathbf{F})])\right) \tag{3}$$

where $f^{7\times7}$ is the convolution layer with 7×7 kernel size.

In the process of obtaining the channel attention map, the global max pooling operation are firstly performed in the spatial dimension to feature map F and outputs two $C \times 1 \times 1$ Tensor. Then, they are respectively sent to two connection (FC) layers and the output are added in pairs. Finally, the attention map and the feature map F are multiplied to obtain the feature map after channel attention correction. The process is shown in Eq. 2.

In the acquisition part of the channel attention map, the input tensor first performs the max pooling operation on the channel dimension to obtain two tensors with size $1 \times H \times W$, and then the two tensors are concated on the channel dimension to obtain tensor of $2 \times H \times W$. Next, a convolution operation with 7×7 kernel size is performed to the tensor, and the channel dimension of the tensor is reduced to 1. Then the spatial attention map is acquired after sigmoid function. The process is shown in Eq. 3. Finally, the attention map and the input tensor are multiplied to obtain the feature map considering both channel attention and spatial attention.

Feature Pyramid Structure. The shallow network can better understand the information contained in pixels, while deep networks can extract more accurate semantic information. Therefore, deep networks are more conducive to accurately detect larger targets. Faster R-CNN [17] only predicts the target on the final feature map. However, there are targets of different sizes in the image, which does not necessarily adapt to the depth of the feature map. In addition, as the network depth increases, it is also possible to lose the information of small targets in the down sampling process. The design idea of the feature pyramid used in this paper is to integrate the features extracted from the backbone network at different depths, and predict the targets at different depths at the same time, which can effectively increase the detection accuracy of small targets.

As shown in Fig. 2, FPN takes the feature map output by the Layer2, Layer3, Layer4 and Layer5 structures of the backbone as the input. Starting from the output of Layer5 layer with the most abundant semantic information, up sampling is carried out, and then added to the output of C4 layer. The fused feature map is further fused with the output of layer Layer3 in the same operation until it is fused with the output of Layer2. Each fused feature map further performs a convolution operation to generate feature maps (feature map1,2, 3 and 4) for predicting the targets. The deepest feature map also generates a higher-level feature map5 that is only used to generate anchor boxes and proposals in RPN through the max pooling layer.

3.2   Region Proposal Network

The region proposal network of the proposed algorithm is transplanted directly from Faster R-CNN. As shown in the Fig. 2, the convolution layer with 3×3 kernel size is used on the feature map to simulate the sliding window operation. Taking the center point of the original image area corresponding to each sliding window as the center, multiple anchor boxes with different scales and proportions are generated. In this paper, five scales of $\{32^2, 64^2, 128^2, 256^2, 512^2\}$ and three proportions of $\{1:1, 1:2, 2:1\}$ are generated for each corresponding center point on the original image, with a total of 15 anchor boxes. Every time the sliding window slides to a position, a one-dimensional vector is generated. After two convolution operations with the 1×1 kernel size, the probability of the foreground or background of each anchor box and the regression parameters of the anchor boxes are output respectively. By further filtering out the anchor boxes with out of bounds and low probability, the left proposals representing the foreground regions are output to the detection network.

3.3 Detection Network

In the detection network, firstly, the coordinates of the proposals are projected onto the feature maps output from the feature extraction network. The feature matrices are segmented on the feature maps according to the coordinates, and the ROI pooling layer is used for flattening. Then, the predicted position and label information are obtained through the regression and soft max layer respectively. Finally, the redundant outputs are removed by the post-processing method.

The post-processing algorithm adopts soft non-maximum suppression (Soft-NMS) algorithm [18]. The algorithm is implemented for bird's nest and suspended foreign objects respectively. Soft-NMS algorithm reduces the confidence of possible redundant detection boxes, then eliminates the detection boxes with confidence lower than the established threshold. The pseudo code of Soft-NMS is shown in Table 1 and can be represented as the following process:

1) Rank all confidence levels in set $\mathcal{S}$ from high to low. After sorting, it is assumed that the confidence of the mth detection box is the highest, which is called b_m. At this time, add the mth detection box to set $\mathcal{M}$.
2) Add the detection box in set $\mathcal{M}$ to set $\mathcal{D}$ and remove b_m from set $\mathcal{B}$. Then, check the remaining detection boxes in set $\mathcal{B}$, and multiply the confidence of each remaining detection box by the weight function $f\left(iou\left(\mathcal{M}, b_i\right)\right)$. In this paper, a linear weight function $s_i = s_i\left(1 - iou\left(\mathcal{M}, b_i\right)\right)$ is selected to lower the confidence of detection boxes which have a large overlap with b_m. Where iou is the ratio of intersection and union of two detection boxes.
3) Repeat 1) and 2) until all detection boxes in set $\mathcal{B}$ are placed in set $\mathcal{D}$. Then, the detection boxes with confidence lower than the threshold are regarded as redundant detection boxes and removed.

Table 1. The pseudo code of Soft-NMS. Where $\mathcal{B}$ and $\mathcal{D}$ are the detection boxes before and after the Soft-NMS algorithm, respectively. $\mathcal{S}$ is the score corresponding to all detection boxes.

$$\textit{Input} : \mathcal{B} = \{b_1, b_2, \ldots, b_N\}, \mathcal{S} = \{s_1, s_2, \ldots, s_N\}$$
$$\textit{Output} : \mathcal{D}, \mathcal{S}$$

$\textit{Begin}$

$\mathcal{D} \leftarrow \{\}$

$\textit{while } \mathcal{B} \neq \textit{empty do}$

  $m \leftarrow argmax\mathcal{S}$

  $\mathcal{M} \leftarrow b_m$

  $\mathcal{D} \leftarrow \mathcal{D} \cup \mathcal{M}; \mathcal{B} \leftarrow \mathcal{B} - \mathcal{M}$

  $\textit{for } b_i \textit{ in } \mathcal{B} \textit{ do}$

  $s_i \leftarrow s_i f\left(iou\left(\mathcal{M}, b_i\right)\right)$

  $\textit{end}$

$\textit{end}$

$\textit{return } \mathcal{D}, \mathcal{S}$

$\textit{end}$

3.4   Loss Function

The loss function using in the proposed algorithm is directly transplanted into the loss function of Faster R-CNN. It mainly includes two different parts: the loss function of RPN and detection network.

The loss function of RPN can be described as Eq. 4:

$$L\left(\{p_i\},\{t_i\}\right) = \frac{1}{N_{cls}}\sum_i L_{cls}\left(p_i,p_i^*\right) + \lambda\frac{1}{N_{reg}}\sum_i p_i^* L_{reg}\left(t_i,t_i^*\right) \qquad (4)$$

$$L_{cls} = -\log\left(p_i\right) \qquad (5)$$

where L_{cls} represents the foreground and background classification loss, and is expressed as the cross-entropy loss, as shown in Eq. 5. L_{reg} is a preliminary bounding box regression loss expressed as smooth L_1 loss. p_i represents the probability that the ith anchor box will be predicted as foreground. When the sample used for loss calculation is a positive sample, the value of p_i^* is 1. Otherwise, p_i^* is 0. t_i is the parameter of bounding box regression, which is using for predicting the ith anchor box. t_i^* denotes the ground truth corresponding to the ith anchor box. N_{cls} represent the number of samples in one batch. N_{reg} represents the number of locations where anchor boxes are created in the feature maps.

The loss of detection network can be expressed as Eq. 6:

$$L\left(p, u, t^u, v\right) = L_{cls}(p, u) + \lambda[u \geq 1]L_{loc}\left(t^u, v\right) \qquad (6)$$

where L_{cls} denotes the classification loss and is denoted as the cross-entropy loss. L_{loc} represents the regression loss, and is denoted as smooth L_1 loss. p is the probability of the predicted target output by the soft max layer. u is the ground truth of the corresponding class. t^u is the regression parameter of the class u predicted by the bounding box regressor $\left(t_x^u, t_y^u, t_w^u, t_h^u\right)$. $\mathcal{V}$ corresponds to the regression parameters of the ground truth (v_x, v_y, v_w, v_h).

4   Experiment

4.1   Dataset

The dataset used in the experiment is 2701 transmission line images collected by UAV. The dataset includes 1542 bird's nest images and 1159 suspended foreign object images. All ground truth boxes are marked by the substation staff for several months, and saved as Pascal VOC dataset format in the form of XML file. All transmission line images collected by UAV are in high-quality PNG format. Some sample images and corresponding annotations are shown in the Fig. 4, where the ground truth results are manually marked in the green boxes.

2701 UAV images are made into train set and test set in the ratio of 8 : 2 at random. The train set and test set consist of 2160 and 541 UAV images

Fig. 4. Some sample UAV images and corresponding annotations in the dataset.

respectively. During the experiment, the strict separation of train set and test set is always maintained.

4.2 Training Process

In the experiment, random flipping, random clipping, brightness transformation and affine transformation are used for data enhancement to increase the number of train set to 5 times of the original train set. The two groups of images after data enhancement are shown in Fig. 5. Data enhancement can simulate different brightness, UAV camera damage or signal interference, so as to improve the robustness in actual detection.

In the experiment, the pretrained weight of ResNet50 on ImageNet [19] dataset is used to partially initialize the model, and KaiMing initialization is used for the initialization of added CBAM structure. We uses two GPU NVIDIA geforce GTX 2080 for training our model, and the whole network architecture is trained end-to-end. The learning rate is set to 0.02. The warm up strategy is adopted in the first 500 iterations, and the decay of learning rate is set at the 8th and 16th epoch. Batch size is set to 16, SGD gradient descent strategy is adopted, momentum is set to 0.9 and the weight decay is set to 0.0001. A total of 24 epochs are trained. The process of training is shown in Fig. 6.

Fig. 5. Two groups of images after data enhancement. The first image of each group is the original image.

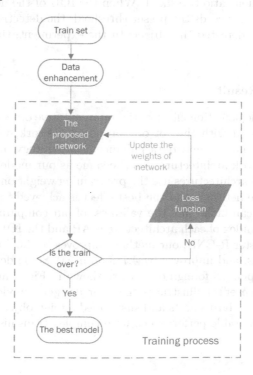

Fig. 6. The representation of training process.

4.3   Evaluation Method

AP is the most reliable and commonly used performance index in the evaluation of object detection algorithm, which can take into account the accuracy and recall of detection at the same time. The calculation process of AP is as follows. Suppose that there are N detection boxes for a class finally output by the network. Firstly, the N detection boxes are sorted according to the confidence, and the accuracy and recall of the top 1, top 2... Top N detection boxes are calculated respectively to generate the P-R curve.

The calculation of precision can be expressed as Eq. 7:

$$p = \frac{TP}{TP + FP} \tag{7}$$

The calculation of recall can be described as Eq. 8:

$$r = \frac{TP}{TP + FN} \tag{8}$$

where TP is the number of partial detection boxes that successfully match a ground truth box. FP is the number of partial detection boxes that fail to match a ground truth box. FN indicates the number of ground truth boxes that cannot be detected. The criterion to judge the successful detection is the intersection and union ratio threshold. When the IOU of the detection box and the ground truth box exceeds the preset threshold, the detection box is deemed to have successfully detected the object. In the experiment, the IOU threshold is set to 0.5.

4.4   Detection Results

In this summary, the detection effect of the representative object detection architectures are compared with that of our proposed network, which is shown in Table 2. The same dataset and data enhancement strategy are adopted in all representative network architectures, as the same as our model. In the training process, all network architectures use the pre training weight on ImageNet to initialize the backbone network, and the best effect is achieved through parameter adjustment, which can make sure the validness of our comparative experiment.

The evaluation index of each architecture is AP and the IOU threshold is 0.5. Compared with Faster R-CNN, our method improves by 4.6% and 4.9% respectively in two classes, and improves by 4.7% on mAP. Some detection results of bird's nest and suspended foreign object are shown in Fig. 7 and Fig. 8, respectively. The detection effect illustrates that our designed model can accurately classify and locate the bird's nests and suspended foreign objects, and the model can also have considerable performance with complex transmission line scene.

Table 2. Comparison of AP with different algorithms.

Algorithm	Backbone	AP%(nest)	AP%(foreign object)	mAP%
Faster R-CNN	ResNet50	82.6	67.3	75.0
Faster R-CNN	ResNet101	84.3	69.2	76.8
Cascade R-CNN	ResNet50	84.9	70.1	77.5
Cascade R-CNN	ResNet101	85.7	71.4	78.6
Proposed algorithm	ResNet-CBAM	87.2	72.2	79.7

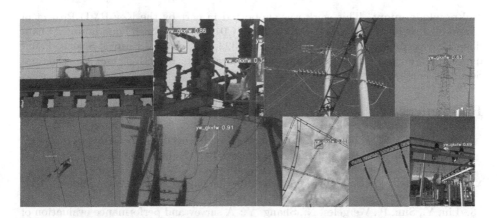

Fig. 7. Detection effect of suspended foreign object.

Fig. 8. Bird's nest detection effect.

5   Conclusion

In this paper, a two-stage object detection algorithm is proposed for the detection of bird's nests and suspended foreign objects in transmission lines. A feature extraction network based on attention mechanism is designed to better consider the context information of small targets in UAV images, which can improve the detection effect of two kinds of small target defects in transmission lines. A post-processing algorithm based on Soft-NMS is proposed to reduce the loss of small targets in detection. Our research has shown that the proposed method achieves 4.7% improvement in AP with 0.5 IOU threshold compared with Faster R-CNN. We believe that the proposed method can be used for transmission line inspection, and will perform better when using more data for training.

Acknowledgement. This work was supported in part by the Key R&D Project of Shandong Province under Grant No. 2022CXGC010503, the Youth Foundation of Shandong Province under Grant No. ZR202102230323, the National Natural Science Foundation for Young Scientists of China under Grant No. 61903155, and the Doctoral Scientific Fund Project under Grant No. xbs1910.

References

1. Ding, Y., et al.: Bird-related fault analysis and prevention measures of ±400 kV Qinghai-Tibet DC transmission line. Energy Rep. **7**, 426–433 (2021)
2. Zaidi, S.S.A., Ansari, M.S., Aslam, A., Kanwal, N., Asghar, M., Lee, B.: A survey of modern deep learning based object detection models. Digital Sign. Process. **162**, 103514 (2022)
3. Liu, Y., Sun, P., Wergeles, N., Shang, Y.: A survey and performance evaluation of deep learning methods for small object detection. Expert Syst. Appl. **172**, 114602 (2021)
4. Lin, T.-Y., et al.: Microsoft COCO: common objects in context. In: Fleet, D., Pajdla, T., Schiele, B., Tuytelaars, T. (eds.) ECCV 2014. LNCS, vol. 8693, pp. 740–755. Springer, Cham (2014). https://doi.org/10.1007/978-3-319-10602-1_48
5. Pang, Y., Wang, T., Answer, R.M., Khan, F.S., Shao, L.: Efficient featurized image pyramid network for single shot detector. In: Proceedings of the IEEE/CVF Conference on Computer Vision and Pattern Recognition, pp. 7336–7344 (2019)
6. Yang, X., Liu, Q., Yan, J., Feng, Z., He, T.: R3Det: refined single-stage detector with feature refinement for rotating object. arXiv preprint arXiv:1908.05612 (2019)
7. Hu, J., Shen, L., Sun, G.: Squeeze-and-excitation networks. In: Proceedings of the IEEE Conference on Computer Vision and Pattern Recognition, pp. 7132–7141 (2018)
8. Woo, S., Park, J., Lee. J.Y., Kweon, I.S.: CBAM: convolutional block attention module. In: Proceedings of the European Conference on Computer Vision (ECCV), pp. 3–19 (2018)
9. Hu, J., Shen, L., Albanie, S., Sun, G., Vedaldi, A.: Gather-excite: exploiting feature context in convolutional neural networks. Adv. Neural Inf. Process. Syst. **31**, 9423–9433 (2018)
10. Liu, B., Huang, J., Lin, S., Yang, Y., Qi, Y.: Improved YOLOX-S abnormal condition detection for power transmission line corridors. In: 2021 IEEE 3rd International Conference on Power Data Science (ICPDS), pp. 13–16 (2021)

11. Zheng, X., Jia, R., Gong, L., Zhang, G., Dang, J.: Component identification and defect detection in transmission lines based on deep learning. J. Intell. Fuzzy Syst. **40**(2), 3147–3158 (2021)
12. Fan, P., et al.: Defect identification detection research for insulator of transmission lines based on deep learning. J. Phys. Conf. Ser. **1828**, 012019. IOP Publishing (2021)
13. He, K., Zhang, X., Ren, S., Sun, J.: Deep residual learning for image recognition. In: Proceedings of the IEEE Conference on Computer Vision and Pattern Recognition, pp. 770–778 (2016)
14. Lin, T.Y., Dollár, P., Girshick, R., He, K., Hariharan, B., Belongie, S.: Feature pyramid networks for object detection. In: Proceedings of the IEEE Conference on Computer Vision and Pattern Recognition, pp. 2117–2125 (2017)
15. Ioffe, S., Szegedy, C.: Batch normalization: accelerating deep network training by reducing internal covariate shift. In: International Conference on Machine Learning, pp. 448–456. PMLR (2015)
16. Nair, V., Hinton, G. E.: Rectified linear units improve restricted Boltzmann machines. In: ICML (2010)
17. Ren, S., He, K., Girshick, R., Sun, J.: Faster R-CNN: towards real-time object detection with region proposal networks. Adv. Neural Inf. Process. Syst. **39**(6), 1137–1149 (2015)
18. Bodla, N., Singh, B., Chellappa, R., Davis, L.S.: Soft-NMS-improving object detection with one line of code. In: Proceedings of the IEEE International Conference on Computer Vision, pp. 5561–5569 (2017)
19. Deng, J., Dong, W., Socher, R., Li, L.J., Li, K., Li, F.F.: ImageNet: a large-scale hierarchical image database. In: 2009 IEEE Conference on Computer Vision and Pattern Recognition, pp. 248–255. IEEE (2009)

An Improved Weighted QMIX Based on Weight Function with Q-Value

Tianyu Li and Fei Han[✉]

School of Computer Science and Communication Engineering, Jiangsu University, Zhenjiang, Jiangsu, China
hanfei@ujs.edu.cn

Abstract. Multi-agent reinforcement learning has developed rapidly in recent years. WQMIX, which bases on the Q-value decomposition is one of the most prominent algorithms recently. However, the weight function in WQMIX is a little bit simple. This function divides the actions into two parts by one condition. And because of the binary weighting of action space, the convergence speed of the mixing network is slower than QMIX's convergence speed. To solve these problems, this paper proposes a new weight function. With this new weight function, the actions of the agents can be divided into four different categories by three conditions. The leveled actions produce different loss and the mixing network can optimize the loss function faster than the regular WQMIX. We have also given the proof that the weighted actions can divide the result of the loss function. We conducted experiments on two maps in the SMAC multi-agent reinforcement learning framework. The experiments show that in the 5m_vs_6m map, MWQMIX got an average reward of 11.6 and the maximum reward of 26. The algorism got the maximum win rate of 56%. In the more complex map, 3s5z_vs_3s5z, MWQMIX got mean reward of 12.37 and the maximum reward of 18. In this map the algorism got the maximum of 9.4%. The results show that our algorism performs better than VDN, QMIX and WQMIX in both two maps.

Keywords: Deep learning · Multi-agent reinforcement learning · Q-value decomposition · WQMIX

1 Introduction

Deep reinforcement learning is produced from the combination of the reinforcement learning and the deep learning. Reinforcement learning describes a method of learning by rewarding and punishing the agent to complete a task without specific instruction [1]. In order to solve the problem of the traditional reinforcement learning such as the high dimension of the state space or the action space and too many vectors in these spaces, deep learning has been combined with it, which has also developed rapidly in the decades. In the last five to ten years, deep reinforcement learning has received a lot of attention among machine learning and artificial intelligence researchers, and many algorithms have emerged. Q-Learning [2] in model-free reinforcement learning

H. Zhang et al. (Eds.): NCAA 2022, CCIS 1638, pp. 442–453, 2022.
https://doi.org/10.1007/978-981-19-6135-9_33

is one of the algorisms that have a good effect, and the algorithm itself is very suitable for improvement with deep neural networks, so the deep Q-value network (DQN) was proposed by Mnih et al. [3].

With the development of research hotspots such as control science and robotics and autonomous driving, single-agent reinforcement learning has gradually become difficult to deal with the complex application scenarios. Therefore, reinforcement learning that regards multiple agents as one system to jointly accomplish a certain goal was presented, called the multi-agent reinforcement learning [4]. Compared with the single-agent reinforcement learning, the difficulties of multi-agent are:

(1) unstable environment
(2) Each agent can only obtain local information of the system
(3) The agent's individual goals and the system's global goals are not necessarily consistent
(4) The number of agents increases and the operation is complicated

The early multi-agent reinforcement learning mechanically applied the DQN to each agent in the multi-agent system, which caused the difficult training problem and a large number of parameters in the network. In order to solve these problems, the concept of Q value decomposition is proposed, that is, how to solve the influence of each agent on the team's Q value in the process of centralized training and decentralized execution. Currently, most of the multi-agent reinforcement learning algorithms are based on how to solve the problem of Q value decomposition. Tan et al. presented the Independent Q-learning (IQL) [5], which means every agent in the system has independent reinforcement learning algorism. Tampuu et al. proposed the method to combine the IQL and DQN [6]. Gupta et al. presented an algorism [7] combined DQN, TRPO [8], DDPG [9], IQL and Recurrent Neural Network (RNN) [10]. At the same year, Sunehag et al. proposed the Value-decomposition networks (VDN) [11]. In 2018, Rashid et al. presented QMIX [12], which is another value decomposition method. One year later, Son et al. proposed QTRAN [13], which is an improved version of VDN and QMIX. In 2020, Rashid et al. proposed another algorism: WQMIX [14]. WQMIX add a piecewise function based on QMIX in order to deal with the complex state space. According to the author, WQMIX performs better in complex environments but needs longer time to train the mixing network.

The disadvantage of WQMIX is that the convergence speed is slow and the weight function is too simple. In order to solve the problem of WQMIX, this paper proposes a more complex weight function. Weighting allows the mixing network to perform more targeted optimization of the loss function.

2 Preliminaries

2.1 VDN

Treating the MARL as the combination of several independent agents RL causes credit assignment problem. For some agents, the team reward may be brought by teammates' actions. If some partners learn the correct policy, they will perform more positively for higher team reward. That causes other agents get spurious reward and become lazy. So, the VDN parts the Q value in summation way (1).

$$Q_{tot} = \sum_{i=1}^{n} Q_i \tag{1}$$

With the Q_i divided by the VDN, agents make their own decisions by their local observation. The VDN gives each agent a Q_i to get new action by its memory and actions (2).

$$a^i = \arg max_{a^{i'}} Q_i(h^i, a^{i'}) \tag{2}$$

After dividing Q value into independent Q_i, agents make decision independently but the whole system learns Q function centralized. Because Q_i cannot be learned independently, VDN still needs to learn Q function from the regular Q-learning rule. Thus, all the agents in the network need to share their messages to the trainer. In some complex cases, large number of agents will bring more parameters than simple cases, which causes huge learning efficiency decrease. For making independent decisions, the VDN completely decomposes the overall Q function in a simple summation way, which makes the fitting ability of the multi-agent Q network very limited.

2.2 QMIX

The VDN is a template of centralized training and decentralized execution (CTDE). But the summation decentralization is simple and limited. Thus, QMIX uses neural network to decompose the state-action value function Q_{tot} (3).

$$Q_{tot} = MLP(Q_1, \ldots, Q_n) \tag{3}$$

However, using regular neural network as MLP causes nonmonotonicity problem. The monotonicity problem means the actions calculated by the distributed training strategy and the actions calculated by the overall Q function need to be consistent in terms of 'optimal performance' (4).

$$\arg max_u Q_{tot}(\tau, u) = \begin{pmatrix} \arg max_{u^1} Q_1(\tau^1, u^1) \\ \cdots \\ \arg max_{u^n} Q_n(\tau^n, u^n) \end{pmatrix} \tag{4}$$

If the algorithm cannot make this equation hold, the distributed training strategy will not get the largest Q_{tot}, which means the policy is not the best one. The VDN holds Eq. (5). Furthermore, it conforms a stronger condition:

$$\frac{\partial Q_{tot}}{\partial Q_i} = 1 \tag{5}$$

because (6)

$$Q_{tot} = \sum_{i=1}^{n} Q_i. \tag{6}$$

The Eq. (6) is a sufficient and unnecessary condition of the above one. Actually, it is still too strong for the monotonicity optimal performance. A relaxed condition can be like (7).

$$\frac{\partial Q_{tot}}{\partial Q_i} > 0. \tag{7}$$

In order to hold the inequation, QMIX uses a special neural network to replace the MLP as the mixing network. The expression of the mixing network is (8).

$$Q_{tot}(\tau, u) = W_2^T Elu\left(W_1^T Q + B_1\right) + B_2, \tag{8}$$

where Q is an n-dimensional vector. Because both W_1 and W_2 are calculated by a set of hypernetworks activated by the absolute value function, the derivative function of this expression can hold the inequation (9).

$$\frac{\partial Q_{tot}}{\partial Q} = \left(\frac{\partial Elu\left(W_1^T Q + B_1\right)}{\partial Q}\right)^T W_2 = \left(\frac{\partial Elu\left(W_1^T Q + B_1\right)}{\partial\left(W_1^T Q + B_1\right)} \cdot W_1\right)^T W_2 \tag{9}$$

In this equation, every element in W_1 and W_2 are larger than 0, and because of

$$0 < \frac{\partial Elu(x)}{\partial x} \leq 1 \tag{10}$$

every element in the derivative function is larger than 0, which means the mixing network can fit to the best policy.

2.3 Weighted QMIX

The mixing network using for recompositing value holds a relaxed condition, which is a sufficient and unnecessary condition. So, in some cases, the QMIX still cannot fit the value functions. In order to simplify the problem, Rashid supposes that all agents observe the global state S, then the function space of the Q_{tot} can be defined as the Q^{mix} (11).

$$Q^{mix} = \{Q_{tot}|Q_{tot}(s, u)$$
$$= f_s(Q_1(s, u_1), \ldots, Q_n(s, u_n)), \frac{\partial f_s}{\partial Q_a} \geq 0, Q_a(s, u) \in R\} \tag{11}$$

With this function space, the objective of the QMIX can be regarded as an optimization issue denoted by T_{Qmix}^* (12),

$$\arg\min_{q \in Q^{mix}} \sum_{u \in U} T^*(Q_{tot}(s, u) - q(s, u))^2, \forall s \in S \tag{12}$$

where T^* is the Bellman Optimal Operator. The QMIX operator is to find the point most closed to the Q function in the Q^{mix} function space. If the real Q function is not in the Q^{mix} function space, the operator will not fit to the real Q but find the suboptimal solution in the function space. To break out the limitation of the Q^{mix} function space, Rashid weights every action value and get a weighted operator (13).

$$\arg \min_{q \in Q^{mix}} \sum_{u \in U} T^* w(s, u)(Q_{tot}(s, u) - q(s, u))^2 \tag{13}$$

It is easy to find that when $w(s, u) \equiv 1$ the weighted operator becomes back to the T^*_{Qmix} operator. The weight function is defined as a piecewise function (14).

$$w(s, u) = \begin{cases} 1 & u = u^* = \arg\max_u Q(s, u) \\ \alpha & otherwise \end{cases} \tag{14}$$

With the weight function, the QMIX fitted Q function can be mapped to a nonmonotonic function space, which can perform better than the regular Q function. However, the weight function is a simple piecewise function. The fitting ability of the network can be limited by it. So, to improve the fitting ability of the weighted QMIX, we present a more complex weight function.

3 The Proposed Method

Basically, the mixing network is to optimize the QMIX operator T^*_{Qmix}. And the weighted QMIX gives the QMIX operator a piecewise function as the weight function in order to divide the value range of the QMIX operator. Eligible actions will produce higher Q values and the other actions will cause lower Q values. So that the WQMIX can select those eligible actions frequently for higher score. However, the weight function divides all the actions into only two parts. That causes there is only one condition to select those actions bringing high Q value. In some cases, the action does not fit the condition but still gives the agent a high reward. The performance of this action can be underestimated because the piecewise function gives it a lower weight. The author gives two conditions but they are used in two different functions. During the experiments we found that in the full training procession the actions some time prefer one of the conditions randomly and some time fit another one. So, we combine the two conditions into one piecewise function (15), called the algorism mixed WQMIX, or MWQMIX.

$$w(s, u) = \begin{cases} 1 & \hat{u} = u^* and Q_{tot}(s, u) < Q(s, u) \\ \alpha & Q_{tot}(s, u) < Q(s, u) \\ \beta & \hat{u} = u^* \\ \theta & otherewise \end{cases} \tag{15}$$

With this new piecewise function, the loss function of the mixing network, which is equal to the weighted QMIX operator, can change into different expressions because of the different conditions the action conformed.

Proposition. For any action u in A and any s in S, the weighted $Q(s, u)$ can change the loss function into four different expressions which have different ranges.

Proof. Because

$$u^* \in \arg\max Q \tag{16}$$

and

$$\hat{u} \in \arg\max \Pi_w Q, \tag{17}$$

when $\hat{u} = u^*$, there has (18).

$$Q_{tot}(s, \hat{u}) = Q(s, \hat{u}) \tag{18}$$

In this case, the loss function of the mixing network changes into the expression (19).

$$loss = w \sum_{u \neq u^*} (Q(s, u) - Q_{tot}(s, u))^2 \tag{19}$$

When $Q_{tot}(s, u) < Q(s, u)$, that means the Q value of the action has been underestimated. Although in this case the u^* may not equals to the $\hat{u}$, the action still can be a good option for the agent under the state s. Those fitting condition actions make the loss function changes into another expression (20).

$$loss = (Q(s, u^*) - Q_{tot}(s, u^*))^2 + w \sum_{u \neq u^*} (Q(s, u) - Q_{tot}(s, u))^2 \tag{20}$$

The author of weighted QMIX put this condition into another independent piecewise function. However, during the experiments, we found that when the action fits to the condition $u = u^*$, it may not fit another condition. Also, when the action's Q value fits to $Q_{tot}(s, u) < Q(s, u)$, it may not equal to u^*. But we still found that many actions fitted both two conditions. We think these actions bring higher Q value for the whole team. No matter how the loss function changes, there is a weight in it. When the action fits the first condition in our piecewise function, according to the new piecewise function, the weight equals to 1 and the loss function is the following expression.

$$loss_1 = \sum_{u \neq u^*} (Q(s, u) - Q_{tot}(s, u))^2 \tag{21}$$

If those actions fit the second condition, because $\hat{u} \neq u^*$, the actions are divided into two parts. For those occasionally equaled to u^* actions, the loss value for them is

$$(Q(s, u) - Q_{tot}(s, u))^2 \tag{22}$$

For those actions not equal to u^*, because in this case the value of the weight function is α, the loss is (23).

$$\alpha \sum_{u \neq u^*} (Q(s, u) - Q_{tot}(s, u))^2 \tag{23}$$

These two parts combines the whole loss function:

$$loss_2 = (Q(s, u^*) - Q_{tot}(s, u^*))^2 + \alpha \sum_{u \neq u^*} (Q(s, u) - Q_{tot}(s, u))^2 \tag{24}$$

When the actions only fit $u^* = \hat{u}$ but do not fit $Q_{tot}(s, u) < Q(s, u)$, the loss function is same as the $loss_1$. But we think because these actions do not fit the second condition, the Q value they earned is lower than those actions that fit both two conditions. To make these two parts of actions different, we give the loss function a weight and the loss function becomes (25).

$$loss_3 = \beta \sum_{u \neq u^*} (Q(s, u) - Q_{tot}(s, u))^2 \tag{25}$$

For the rest actions that cannot fit to any conditions, the loss function is still combined by two parts. For those equal to u^*, the loss function is

$$(Q(s, u) - Q_{tot}(s, u))^2 \tag{26}$$

And for the actions that occasionally equal to u^*, the loss function is the summation of the square (27).

$$\theta \sum_{u \neq u^*} (Q(s, u) - Q_{tot}(s, u))^2 \tag{27}$$

The loss function just like the $loss_2$, but because the weights are different, so the two loss functions cause different consequences.

$$loss_4 = (Q(s, u^*) - Q_{tot}(s, u^*))^2 + \theta \sum_{u \neq u^*} (Q(s, u) - Q_{tot}(s, u))^2 \tag{28}$$

Although the loss functions are described in different expressions, but basically all the expressions have weights to make loss different, we can change these four weights to influence the differences between these loss values.

$$loss = \begin{cases} \sum_{u \neq u^*} (Q(s, u) - Q_{tot}(s, u))^2 \\ (Q(s, u^*) - Q_{tot}(s, u^*))^2 + \alpha \sum_{u \neq u^*} (Q(s, u) - Q_{tot}(s, u))^2 \\ \beta \sum_{u \neq u^*} (Q(s, u) - Q_{tot}(s, u))^2 \\ (Q(s, u^*) - Q_{tot}(s, u^*))^2 + \theta \sum_{u \neq u^*} (Q(s, u) - Q_{tot}(s, u))^2 \end{cases} \tag{29}$$

In summary, when the weight function is (15), the loss function becomes (29).

4 Experiments and Discussion

4.1 SMAC

The StarCraft Multi-Agent Challenges (SMAC) uses the StarCraft game as the multi-agent experiment framework [15]. In each time step, agents observe the surrounding environment in their sight. The agent's field of vision is an area with a radius of 9 squares, so the agent can only observe those lived allies acting in the sight. Each agent has a feature vector to describe it state in a single time step. The vector has six dimensions, distance, relative x, relative y, health, shield and unit_type included. The distance, relative x and relative y are used for describing the relative position of the nearest enemy. The health, shield and unit_type are used for describing the enemy's characteristics. Furthermore,

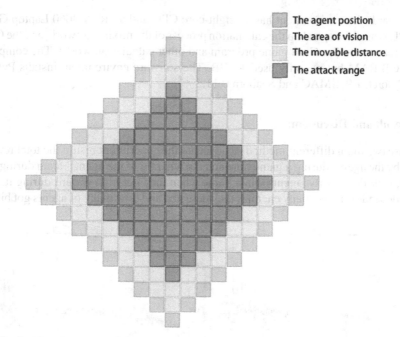

The agent position
The area of vision
The movable distance
The attack range

Fig. 1. The Manhattan distance of the vision, the movable distance and the attack range

the agent-controlled unit's health, shield and unit_type can be observed. The agent can get the previous actions of those allays that in its sight. It can also get the topographic features, height and walkability included, with a radius of 8 blocks (see Fig. 1).

The agent action space is discrete. For those alive agents, the actions space contains six actions: move to four directions, attack and stop. For those agents that health is not positive, there is only one action called no-op is executable, which cannot be executed by the living agent.

4.2 Environments

We used two different environments of the SMAC to test the algorism: 5m_vs_6m and 3s5z_vs_3s6z. The 5m_vs_6m has five friendly units and six enemies. The 3s5z_vs_3s6z has eight allays and nine opponent units. Both sides contain two different types of units and both the enemy sides dominant in number. The unit called 'm' only has health, which means when the health comes to zero the unit goes dying. But in another environment, both 's' and 'z' have shield. When a unit who has shield get under attack, they lost their shield firstly. If the unit survives from an attack and keeps this safe state for a moment, the lost shield can recover again. Furthermore, 'z' is a kind of melee units while 's' can attack targets far away from it. So, the situation of the 3s5z_vs_3s6z is more complex than the 5m_vs_6m.

The hardware environment has an eight-core CPU and an RTX 3060 Laptop GPU. The GPU is used for speed up the calculation process of the mixing network, and the CPU is used for run the StarCraft game program and other calculation works. The computer has 16 GB RAM but we only used 14 GB. The software environment installs Python 3.8.8, Pytorch 1.9, SMAC and Seaborn 0.11.2.

4.3   Result and Discussion

There are two main different methods to evaluate the experiment result: the total reward earned by the agents during a game time and the win rate of the agents' team during one training episode. Mostly, agents win a game when they get high reward during it. The reward depends on how many enemies they defeated and how many of agents got bitten.

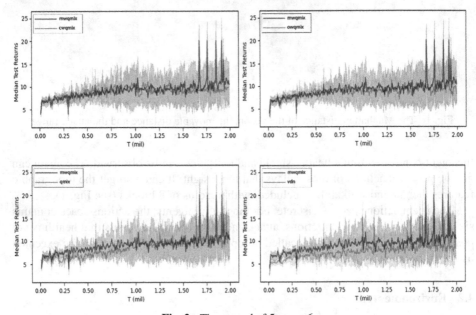

Fig. 2. The reward of 5m_vs_6m

Figure 2 shows that in 5m_vs_6m map, comparing with CWQMIX and OWQMIX, MWQMIX gets a similar mean reward, but reaches higher Maximum reward at the rear 0.5 million train steps. Comparing with QMIX, during the first 1 million training steps, MWQMIX takes the lead but at last QMIX gets a close mean reward. VDN has the lower reward during the whole training process, which is caused by its simple Q value dividing method. As we discussed in the second part, the linear summation function limited it fitting ability, which makes it cannot react as good as other algorism in this complex situation.

Table 1. The win rate of 5m_vs_6m

Algorism	Mean reward	Maximum reward	First win step (mil)	Maximum win rate (%)
MWQMIX	**11.6**	**26**	**0.32**	**56**
CWQMIX	11.27	17	0.19	34
OWQMIX	11.43	15	0.24	37
QMIX	10.97	22	0.21	52
VDN	7.48	18	0.26	56

In Table 1 we can find that MWQMIX has the highest maximum win rate, but it is the last algorism to get the first win. Comparing with the WQMIX, QMIX and VDN have higher maximum win rate. We think that is because the regular WQMIX weights for actions and that causes the mixing network converges slower than QMIX and VDN.

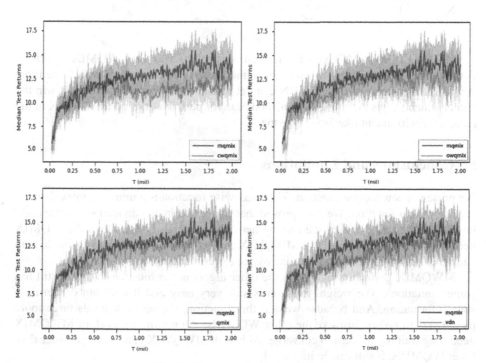

Fig. 3. The reward of 3s5z_vs_3s6z

In the 3s5z_vs_3s6z map, all the algorisms get lower reward than in another map because 3s5z_vs_3s6z is harder for the agents (see Fig. 3). It can be found that MWQMIX gets higher maximum reward than the rest of the algorisms. We can see that CWQMIX and OWQMIX have not converged yet because at the last one million training steps the rewards are not stable. Still, the reason is that these two algorisms weight the actions and converge slower than the other algorism.

Table 2. The win rate of 3s5z_vs_3s6z

Algorism	Mean reward	Maximum reward	First win step (mil)	Maximum win rate (%)
MWQMIX	**12.37**	**18**	**0.3**	**9.4**
CWQMIX	11.46	16	0.21	9.3
OWQMIX	12.28	17	1.14	6.2
QMIX	14.76	15	–	0
VDN	13.82	15	–	0

In Table 2, MWQMIX gets the highest maximum. The three WQMIX algorisms get lower mean reward but can win the game sometime. Comparing with these three WQMIX, although QMIX and VDN can get higher mean reward, but cannot win the game, which shows that these two methods cannot perform as good as WQMIX in the complex environment like 3s5z_vs_3s6z.

5   Conclusions and Limitations

We present a new weight function. This piecewise function is a more complex than the regular weight function. We have proved that this function can divide the loss function into for parts, and it makes the mixing network converging faster than the WQMIX networks. Finally, we proved that MWQMIX can get higher reward and win rate in both simple environment and complex map by the experiments.

MWQMIX performs better than the other algorisms in most cases, but it still has some limitations. The weight function is still very easy and it does not concern the state information. And because we add this function, the network needs three more parameters to be set. Just like the other WQMIXs, the mixing network of MWQMIX need long training process to converge. Although MWQMIX can converge faster than other WQMIXs, it still can be improved.

Acknowledgments. This work was supported by National Natural Science Foundation of China under Grant nos. 61976108 and 61572241.

References

1. Sutton, R.S., Barto, A.G.: Reinforcement Learning: An Introduction. MIT Press, Cambridge (2018)
2. Watkins, C.J.C.H., Dayan, P.: Q-learning. Mach. Learn. **8**(3), 279–292 (1992)
3. Mnih, V., Kavukcuoglu, K., Silver, D., et al.: Human-level control through deep reinforcement learning. Nature, **518**(7540), 529–533 (2015)
4. Yang, Y., Wang, J.: An overview of multi-agent reinforcement learning from game theoretical perspective. arXiv preprint arXiv:2011.00583 (2020)
5. Tan, M.: Multi-agent reinforcement learning: Independent vs. cooperative agents. In: Proceedings of the Tenth International Conference on Machine Learning, pp. 330–337 (1993)
6. Tampuu, A., Matiisen, T., Kodelja, D., et al.: Multiagent cooperation and competition with deep reinforcement learning. PLoS ONE **12**(4), e0172395 (2017)
7. Gupta, J.K., Egorov, M., Kochenderfer, M.: Cooperative multi-agent control using deep reinforcement learning. In: Sukthankar, G., Rodriguez-Aguilar, J. (eds.) AAMAS 2017. LNCS, vol. 10642, pp. 66–83. Springer, Cham (2017). https://doi.org/10.1007/978-3-319-71682-4_5
8. Schulman, J., Levine, S., Abbeel, P., et al.: Trust region policy optimization. In: International Conference on Machine Learning, pp. 1889–1897. PMLR (2015)
9. Lillicrap, T.P., Hunt, J.J., Pritzel, A., et al.: Continuous control with deep reinforcement learning. arXiv preprint arXiv:1509.02971 (2015)
10. Zaremba, W., Sutskever, I., Vinyals, O.: Recurrent neural network regularization. arXiv preprint arXiv:1409.2329 (2014)
11. Sunehag, P., Lever, G., Gruslys, A., et al.: Value-decomposition networks for cooperative multi-agent learning. arXiv preprint arXiv:1706.05296 (2017)
12. Rashid, T., Samvelyan, M., Schroeder, C., et al.: QMIX: monotonic value function factorisation for deep multi-agent reinforcement learning. In: International Conference on Machine Learning, pp. 4295–4304. PMLR (2018)
13. Son, K., Kim, D., Kang, W.J., et al.: QTRAN: learning to factorize with transformation for cooperative multi-agent reinforcement learning. In: International Conference on Machine Learning, pp. 5887–5896. PMLR (2019)
14. Rashid, T., Farquhar, G., Peng, B., et al.: Weighted QMIX: expanding monotonic value function factorisation for deep multi-agent reinforcement learning. Adv. Neural Inf. Process. Syst. **33**, 10199–10210 (2020)
15. Samvelyan, M., Rashid, T., De Witt, C.S., et al.: The starcraft multi-agent challenge. arXiv preprint arXiv:1902.04043 (2019)

An Improved Ensemble Classification Algorithm for Imbalanced Data with Sample Overlap

Yafei Zhang and Fei Han[⊠]

School of Computer Science and Communication Engineering, Jiangsu University, Zhenjiang, Jiangsu, China
hanfei@ujs.edu.cn

Abstract. Classification of imbalanced data remains an important topic of machine learning. Existing and recent literature showed that class overlap had a higher negative impact on the performance of learning algorithms. The learning task becomes more difficult when there is data overlap in imbalanced data sets. Under the background of unbalanced data, this paper aims at the problem of sample overlap, and proposes an ensemble classification method of imbalanced data, namely unbalanced overlapping random forest (IORF). We consider the problem of sample overlap in imbalanced data classification and introduce the coefficient of sample difficulty to measure the importance of each training sample. When generating different data subsets according to the weighted bootstrap method, pay more attention to overlapping samples. Finally, the generated data subset is used to train the diverse decision tree for ensemble. In addition, in order to prevent the repeated selection of minority class samples from causing over-fitting of the classifier to the minority class samples, a data enhancement method based on Gaussian perturbation is proposed to reduce the over-fitting of the classifier to the overlapping minority class samples. The experimental results show that the proposed method can further improve the classification performance.

Keyword: Ensemble learning · Data imbalance · Sample overlap · Data enhancement · Random forest

1 Introduction

In a data set, if the number of samples of some classes significantly exceeds the number of samples of other classes, the data set is called imbalanced. Classes with majority class samples are called majority classes, and classes with a small number of samples are called minority classes [1]. For example, medical diagnosis [2], intrusion detection [3], text classification [4], fault detection and diagnosis [5] are all imbalanced data. In these fields, we hope to correctly classify minority samples, that is, the classification value of minority samples is higher. However, when dealing with imbalanced data sets, traditional classifier algorithms tend to favor the majority class data, resulting in poor classification performance for a few classes.

© The Author(s), under exclusive license to Springer Nature Singapore Pte Ltd. 2022
H. Zhang et al. (Eds.): NCAA 2022, CCIS 1638, pp. 454–468, 2022.
https://doi.org/10.1007/978-981-19-6135-9_34

There are usually two ways to classify imbalanced data. One is to use some method to rebalance data at the data level. The commonly used method is resampling, which is divided into over-sampling, under-sampling and mixed sampling methods. Second, at the algorithm level, it mainly includes single classifier, integrated classifier and cost sensitive. Cost-sensitive methods are used to increase the ability of classifier to classify minority class samples by paying more attention to minority class samples. However, the two methods do not consider the problem of class overlap [6, 7].

At the data level, the main solution is to balance the distribution of sample quantity through resampling to improve the classification performance of the classifier. Random over-sampling and under-sampling are methods for repeated sampling of some samples in minority class samples and random extraction of some samples in majority class samples [8]. However, there are some problems in using these methods. For example, repeated selection of minority class samples will increase the redundancy of samples, while random removal of part of the majority class samples may cause partial information loss, resulting in poor performance of the classifier. In order to solve the problem of random undersampling, typical undersampling methods include EasyEnsemble [9] and Balanceccade [10].

At the algorithm level, cost sensitive methods are mainly used to deal with the imbalance problem. Studies show that these methods are superior to sampling-based methods in some applications [11, 12]. In imbalanced learning, the commonly used cost sensitive boosting method is cost sensitive Boosting. Sun et al. [13] proposed AdaCost based on AdaBoost.

Integrated classifier can achieve better prediction performance by integrating multiple weak classifiers, and it has been proved that it can achieve better effect of specific gravity sampling when processing imbalanced data [14, 15]. According to the different integration methods, it can be divided into boosting integration and Bagging integration. Although integration methods were not originally proposed to address class imbalances, they can be well applied to the classification of imbalanced data by combining them with the data-level methods mentioned earlier.

Although many methods have been proposed to solve the problem of category imbalance, most of the relevant researchers focus on category imbalance. However, class imbalance is not the only reason for the poor learning performance of classifiers. Overlapping data occurs in areas where data from different categories overlap, making it difficult to distinguish between the two categories.

When classifying data, the performance of learning algorithm can be improved effectively by dealing with the problem of class overlap in imbalanced data [16]. Mehmood et al. [17] divide the multi-class dataset into overlapping and non-overlapping region. The overlapping region is further filter into the Critical and less Critical region depending upon their sample contribution in the overlapped region. Chen et al. [18] propose clustering-based adaptive decomposition and editing-based diversified oversamping procedure (CluAD-EdiDO), to solve the problems of class imbalance and overlapping simultaneously. Fernandes et al. [19] proposed a muti-class imbalanced learning (EVINCI). EVINCI uses a multiobjective evolutionary algorithm (MOEA) to evolve a set of samples taken from an imbalanced dataset. It selectively reduces the concentration of less representative instances of the majority classes in the overlapping areas while selecting samples that produce more accurate models. Tao et al. [20] propose a SVDD (Support Vector Data Description) boundary and DPC (Density Peaks Clustering) clustering technique-based oversampling approach (SVDDDPCO) for handling imbalanced and overlapped data. The method identifies those misclassified majority or few minority instances by the boundary as potential overlapped or noisy ones and eliminates them. Yuan et al. [21] proposed a weighted bootstrap method to generate sub-data sets containing different local information and trained various base classifiers for integration.

In order to solve the problem of poor classification effect due to the difficulty of overlapping data classification in imbalanced data classification, a random forest algorithm is proposed based on local sampling for overlapping regions. Neighbor algorithm to find out its first by k is in the overlapping area of the sample, and then use a subset of the set of rules to select different center, according to the sample distance subset of center distance and sample set different weights for different samples of overlapping degree, according to the sample weight random samples to join the training subsets, using subset training classifier. At the same time, in order to eliminate the problem of over-fitting of overlapping minority class samples caused by repeated selection of overlapping minority class samples, a data processing method based on Gaussian perturbation is proposed. Half of the difference of each feature between the disturbed sample and the nearest sample is taken as the variance of gaussian noise, and the noise features generated are processed out of bounds, and finally a relatively safe sample with Gaussian disturbance is obtained. The method not only increases the diversity of overlapping minority class samples, but also avoids the further deepening of sample overlap. Experiments show that the proposed method has better performance than other advanced methods.

2   The Proposed Method

Although many researchers have put forward a lot of opinions on imbalanced data classification and sample overlap, only a few studies have focused on the overlap in imbalanced data classification. So far, many researchers have introduced ensemble learning into imbalanced data classification, and achieved good results, but failed to solve the impact of the overlapping problem. According to the characteristics of data overlap, we try to increase the diversity of ensemble classifier and reduce the impact of sample overlap on classification performance.

2.1   The Steps of the Method

The random forest algorithm is an integration algorithm based on Bagging. It generates multiple training subsets through bootstrap technology, obtains multiple decision trees through training, and finally integrates multiple decision trees through some strategy to obtain the final prediction result. Common decision tree algorithms include C4.5, ID3 and CART, in which the Gini index is used to select the optimal feature in CART tree and determine the optimal binary segmentation point of the feature. In this paper, random forest is selected as the target classifier, and a solution is proposed to solve the problem of imbalanced data and data overlap. The overall process of the algorithm is shown in Algorithm 1.

Algorithm 1

Input: 1. trainning set $D = (x, y) = \{(x_1, y_1), (x_2, y_2), \ldots, (x_n, y_n)\}$, test set x

2. number of subsetst: t

Output: classification result: y_pred

1. The subset of data is obtained through algorithm 2: $C = \{c_1, c_2, \ldots, c_t\}$

2. **for** i=1,…,t **do**

3.   Select data subset ci

4.   Algorithm 3 is used for gaussian perturbation of data subset

5.   When training the nodes of the decision tree model, some features are

6.   randomly selected, and the optimal feature is selected as the feature of the

7.   decision tree node division according to gini index to create a decision tree.

8. **end for**

9. The decision tree is combined with the majority voting method to obtain the ensemble classifier

10. Predict test set X using the ensemble classifier

Compared with the traditional random forest algorithm, the main difference between our algorithm and the traditional random forest algorithm lies in step 1 and Step 4. The traditional random forest algorithm obtains the data subset through bootstrap method, while this method obtains the data subset through algorithm 2. Before using the training subset to train the decision tree, gaussian disturbance processing is performed on the data subset. Other steps are the same as those of random forest algorithm.

2.2 Generation of the Data Sub-set with Weighted Bootstrap Technique

As mentioned above, when data is classified, data overlap is also a big problem in addition to data imbalance which will affect the classification results. Here, k-nearest Neighbors are used to define the degree of overlap of sample X, which is expressed in Eq. (1). Where k represents the number of neighbors, and d_k represents the number of different categories in k neighbor samples of X. Therefore, the more heterogeneous samples around sample X, the higher the overlap degree of sample X.

$$overlap\,degress(x_i) = \frac{d\_k+1}{k+1} \tag{1}$$

In imbalanced data classification, the number of minority class samples is rare but often has important value. In order to measure the importance of minority class samples, we use the imbalance ratio IR to quantify the degree of imbalance between majority class samples and minority class samples. IR is represented by Eq. (2), where $N_{minority}$ represents the number of minority class samples and $N_{majority}$ represents the number of majority class samples.

$$IR = \frac{N_{majority}}{N_{minority}} \tag{2}$$

Since both data imbalance and sample overlap will reduce the classification performance of the classifier, we should consider these two factors at the same time to preliminarily define the degree of difficulty for the sample to be correctly learned by the classifier. Here we use the difficulty coefficient (HTL) to quantify how much attention each sample needs to be given. HTL is represented by Eq. (3), where X represents the training sample. It can be seen from this that the higher the sample overlap is, the more difficult it is to learn and more worthy of attention. Meanwhile, minority class samples need more attention than majority class samples.

$$HLT(x) = \begin{cases} overlap\,degree(x) * IR & x \in minority\,class \\ overlap\,degree(x) & x \in majority\,class \end{cases} \tag{3}$$

As mentioned above, improving ensemble diversity can improve the classification effect of ensemble classifier. In order to obtain base classifier with diversity, different training subsets can be generated for training. To achieve this, we can generate a diverse subset of data by selecting a center of mass with different data distributions and then selecting samples around that center of mass. In early studies, this idea was realized by Armano [22] using the iterative process of centroid probability. Initially, in the first iteration, all samples x are assigned the same probability ($P_1^{(u)} = 1/D$), where $P_t^{(u)}(x)$ represents the probability that sample X is chosen as the t times centroid, and a sample is chosen at random as the first centroid u_1. In the next iteration, the probability of all samples being selected as the center of mass is updated using a distance-based strategy. Equation (4) is used to update the sample probability, and Eq. (5) is used to normalize the sample probability. According to Eq. (4), when the sample is close to the centroid in the t generation, the probability of being selected as the centroid in the t + 1 generation decreases. This strategy can make the centroid distributed in different regions. Finally, in each iteration, the sample with the highest probability is selected as the center of mass and its probability is set to 0. The distribution of different centroids can guide the base classifier to focus on different regions of the sample, thus improving the diversity of ensemble classifier.

$$Q_{t+1}^{(u)}(x) = P_t^{(u)}(x) * \log(1 + ||x - u_t||) \tag{4}$$

$$P_{t+1}^{(u)}(x) = \frac{Q_{t+1}^{(u)}(x)}{\sum_{i=1}^{|D|} Q_{t+1}^{(u)}(x_i)} \tag{5}$$

However, in the face of imbalanced data, the selected partial centroids are redundant because they are in non-overlapping regions and contribute less to classification. Yuan et al. [21] added the hard-to-learn coefficient expressed by Eq. 3 into the process of centroid probability iteration and replaced Eq. (4) with Eq. (6), so that the selected centroid was always located in the sample overlap region, thus improving the quality of centroid selection.

$$Q_{t+1}^{(u)}(x) = P_t^{(u)}(x) * \log(1 + ||x - u_t||) * HTL(x) \tag{6}$$

After obtaining multiple centroids through iteration, we can construct subsets of data around each centroid. Inspired by Yuan et al., we use Eq. (7) to express that the weight of each instance x_i in the selection of the t-th subset is $w_{i,t}$. According to the Equation, the samples with greater overlap degree and closer to the center of mass are given higher weights, which emphasizes the importance of local overlap regions. At the same time, minority class samples were multiplied by IR imbalance ratio to balance the impact of data imbalance.

$$w_{i,t} = \begin{cases} \exp\left\{-\frac{1}{2} * \left[(x_i - u_t) * V * (x_i - u_t)^T\right]\right\} * IR * \left(1 + \frac{overlaps(i)}{k}\right) \\ \qquad x_i \in minority\ class \\ \exp\left\{-\frac{1}{2} * \left[(x_i - u_t) * V * (x_i - u_t)^T\right]\right\} * \left(1 + \frac{overlaps(i)}{k}\right) \\ \qquad x_i \in majority\ class \end{cases} \tag{7}$$

Finally, the sequence of data subsets with diversity can be obtained through the method of weight sampling. The specific algorithm is listed in Algorithm 2.

Algorithm 2: Generating data subset
Input: trainning set D, iterations t
Output: data subset C= $\{c_1, c_2, ..., c_t\}$
Initialize: $C = \emptyset, P_1^{(u)} = 1/D$
  for i=1,...,t **do**
      if i=1 **then**
          The sample with the largest coefficient of difficulty HLT is selected as the center of the first subse
      else
          Alculate $Q_i^{(u)}(x)$ for all samples by Equation (6)
          The probability of obtaining all samples $P_j^{(u)}(x)$ by normalizing $Q_i^{(u)}(x)$ values by Equation (5)
          Select the sample with the highest probability $P_j^{(u)}(x)$ as the subset center
          Set subset center $P_j^{(u)}(x)$=0
          The weights of all samples were calculated by equation (7) and (8), and samples were randomly selected to add subset c_i according to the weights
          Add c_i to C
    End if
  End for

2.3 Gaussian Perturbation of Data Subset

When minority class samples are repeatedly sampled, it is easy to cause over-fitting of the classifier to minority class samples. The traditional oversampling method cannot deal with the overlapping minority class samples selected repeatedly well. A data enhancement method is proposed based on Gaussian perturbation For each minority class sample x_i in the subset, if its sampling times k is greater than 2, then the distance between the sample and each feature of the nearest sample is calculated, denoted as $d_i = \{d_{i,1}, d_{i,2}, \ldots, d_{i,m}\}$, where $d_{i,j}$ represents the difference between the ith sample and the jth feature of the nearest sample. Set the variance of gaussian function $\sigma = d_{ij}/2$, the mean u = 0, generate noise $g_i = \{g_{i,1}, g_{i,2}, \ldots, g_{i,m}\}$, in order to prevent the generation of noise amplitude is too large ($|g_{i,j}| > d_{i,j}$).To cross the generated noise, setting $g_{i,j} = 2 * d_{i,j} - |g_{i,j}|$. . The generated noise sample is obtained by combining the generated noise with the original sample ($z_i = x_i + g_i$). A new subset of data is obtained by replacing the repeatedly selected k-2 minority class samples in the data set with the samples processed by Gaussian perturbation. The specific process is expressed in Algorithm 3.

Algorithm 3: Gaussian perturbed

Input: data subset c= $\{(x_1, y_1), (x_2, y_2), ..., x_n, y_n\}$

Output: new data subset

For minority class samples, if the number of sampling k is greater than 2

    Calculate the distance from each feature of the nearest sample $d_i = \{d_{i,1}, d_{i,2}, ..., d_{i,m}\}$

    Set the variance of the Gaussian function σ= $d_{ij}/2$, mean value μ=0.

    Generate k - 2 gaussian noise $g_i = \{g_{i,1}, g_{i,2}, ..., g_{i,m}\}$

    For each characteristic of the gaussian noise, if have crossed ($|g_{i,j}| > d_{i,j}$), set $g_{i,j} = 2 * d_{i,j} - |g_{i,j}|$

    Get the gaussian perturbation sample $z_i = x_i + g_i$

    Add the gaussian perturbation sample (z_i, y_i) to the new data subset NC.

Add gaussian perturbation sample to new data subset NC for the minority class samples whose sampling times k is less than 2 and majority class samples, directly add the data subset NC

---

3 Experiments and Discussion

3.1 Experimental Settings and Performance Measures

The 20 data sets used in the experiment were all from the keel database, as shown in Table 1. In order to compare the performance of the algorithm on the data sets with different degree of imbalance, we select the data sets with different degree of balance and number the data sets. The comparison algorithms used are adaCost (AC), BalancedRandomForeast (BRF), EasyEnsemble (EE), RUSBoost (RUB), Balanced Bagging (BB) and Overlapping imbalanced Sensitive Random forest (OIS_RF). CART tree was selected for all basic classifiers, the maximum feature number was the square root of the number of data features, and the number of base classifiers was 41. According to Yuan et al. [21], the influence of sample overlap degree depends more on distinguishing the difference of sample overlap degree, rather than determining the definite value of overlap degree. Therefore, the k value in Eq. (1) is 5. In order to avoid the contingency of experimental results, the experiment used 5-fold cross validation, repeated 10 times on each data set, and the results were averaged.

For imbalanced data, the accuracy rate cannot accurately measure the performance of the classifier. If the classifier divides all the samples into most classes, it can also obtain high accuracy, but the classification performance of a few classes is poor. Therefore, in this paper, we adopt AUC and F-Measure (F1) to evaluate the performance of the comparison algorithm. AUC considers the true positive rate and false positive rate and is used to evaluate the overall performance of the algorithm. F-measure (F1) is used to evaluate the classification performance of minority class samples.

Table 1. The experimental datasets

Dataset	Sign	Size	Attr	IR
ecoli0vs1	A1	220	7	1.85
wisconsin	A2	683	9	1.86
vehicle1	A3	846	18	2.90
glass0123vs456	A4	214	9	3.20
newthyroid2	A5	215	6	5.14
segment0	A6	2308	19	6.02
glass6	A7	214	9	6.38
glass015vs2	A8	172	9	9.12
glass2	B1	214	9	11.59
cleveland0vs4	B2	177	13	12.62
ecoli0146vs5	B3	280	6	13.00
ecoli4	B4	336	7	15.80
Zoo3	B5	101	16	19.20
glass5	B6	214	9	22.78
lymphography-normal-fibrosis	B7	148	18	23.67
abalone3vs11	C1	502	8	32.47
kr-vs-k-zero_vs_eight	C2	1460	6	53.07
poker89vs6	C3	1485	10	58.40
kr-vs-k-zero_vs_fifteen	C4	1460	6	53.07
poker8vs6	C5	101	10	85.88

3.2 Influence of k Value

According to Eq. (7), k value is an important parameter to define sample weight. In general, the larger the k value is, the less influence the overlap degree has on the sample weight, and vice versa. Therefore, the selection of different k values will affect the sample distribution in the sample subset, thus affecting the diversity between base classifiers and classification performance. Here we study the effect of k value on the performance of the proposed algorithm. Table 2 lists the f1 value and AUC value of the algorithm when the parameter k value is {1, 3, 5, 7}. Where, the black font represents the results with better performance in the same data set, while those with little difference are not marked. Finally, The Times of better performance of different K values in different data sets are counted.

Table 2. Classification performance as evaluated by F1 (%) and AUC (%) in different k values

	F1				AUC			
	K = 1	K = 3	K = 5	K = 7	K = 1	K = 3	K = 5	K = 7
A1	98.96	98.92	98.96	98.96	98.36	98.33	98.36	98.36
A2	**95.97**	**95.95**	95.88	95.8	**97.26**	**97.23**	97.16	97.11
A3	**64.04**	63.24	62.81	63.33	**77.87**	77.12	76.71	77.1
A4	87.75	**88.09**	87.67	87.06	**94.12**	94.01	93.97	93.56
A5	93.08	93.43	**93.56**	93.36	96.13	96.1	**96.25**	96.19
A6	98.42	**98.45**	98.42	98.39	99.18	99.22	99.2	99.2
A7	87.53	87.41	87.68	**88.84**	93.14	92.73	92.92	**93.36**
A8	**51.65**	44.73	48.85	47.93	**77.28**	71.66	72.45	72.56
B1	38.85	38.25	**40.39**	38.23	**70.56**	68.38	68.99	67.04
B2	**63.10**	60.47	62.6	60.35	**84.29**	80.81	82.44	78.53
B3	72.7	71.7	70.85	**73.83**	89.31	88.1	87.35	**89.46**
B4	**80.03**	78.49	77.76	79.01	**92.58**	92	91.52	91.61
B5	**57.33**	55.33	54.67	54.67	**79.17**	79.06	79.06	79.05
B6	**94.4**	91.15	90.27	90.93	**99.66**	98.46	98.46	98.51
B7	70.67	71.67	**73.33**	69.33	84.65	**87.08**	86.58	85.44
C1	**97.71**	97.14	**97.71**	97.14	99.92	**99.9**	99.92	99.9
C2	**99.27**	98.03	98.91	98.24	**99.99**	99.15	99.33	99.31
C3	**80.63**	75.48	77.75	76.19	**84.8**	81	82.4	81.2
C4	100	100	100	100	100	100	100	100
C5	57.73	54.07	**58.1**	56.9	74	73.17	**74.67**	74
COUNT	10	3	5	2	11	3	2	2

As can be seen from Table 2 that for different k values, their performance on different data sets is different. The reason is that different data sets have different data distributions. When focusing on overlapping samples, some data sets yield better results, while others do not. For most datasets, paying attention to overlapping samples can improve the classification performance of the classifier. In the following experiment, for k value in Eq. (7), we will set k = 1 for comparison experiment.

3.3 Influence of the Numbers of Classifiers

In the process of obtaining data subset in Sect. 2, centroids with different distributions are obtained iteratively, so as to obtain sample subset with local distribution. Theoretically, collecting multiple local information with different distribution can obtain the overall information of the data and produce better performance.

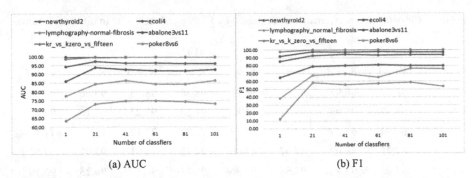

(a) AUC   (b) F1

Fig.1. Classification performance of IORF in difference number of classfiers.

In order to further understand the relationship between classifier number and integration performance, we test our IORF method in different number of classifiers – 1, 21, 41, 61, 81, and 101 - on the six datasets with different imbalance ratio: newthyroid2-5.14ecoli4-15.80, lymphography_normal_fibrosis- 23.67, abalone3vs11–32.47, kr_vs_kzero_vs_fifteen-53.07, poker8vs6- 85.88. The results of Fig. 1 (a) show that when the number of classifiers is less than 41, AUC shows an upward trend and then tends to be stable. The classification performance of some data sets decreased slightly with the increase of the number of classifiers, such as ecoli4, lymphography_normal_fibrosis. The performance of F1 is similar to that of AUC. As the number of classifiers increases, the classification performance does not improve significantly.

3.4 Performance

To compare the effectiveness of the proposed algorithm, we compare the performance of the proposed algorithm with six imbalanced data classification algorithms, including F1 and AUC. The data sets and parameter Settings are described in Sect. 3.1.

In order to verify the effectiveness of the two methods proposed in this paper, we set up two groups of comparison tests. The algorithm without gaussian perturbation of the data subset is expressed as IORF1. Formula (7) Eliminates 1 + overlaps(I)/k and carries out Gaussian perturbation algorithm to the data subset, which is expressed as IORF2. The algorithm adopted by both methods is expressed as IORF.

Table 3, 4 shows the results of all algorithms in the two indicators. Black words indicate the highest classification performance of each corresponding data set. Each w/t/l triplet indicates how many datasets the classifier shown at the top of the column had a Higher/SAME /lower effect than the algorithm proposed in this paper.

Table 3. Classification performance comparisons by F1 (%)

	AC	BRF	EE	RUB	BB	OIS_RF	IORF	IORF1	IORF2
A1	98.01	98.96	98.96	98.96	98.96	98.96	98.96	98.92	98.90
A2	92.13	95.94	95.92	95.59	**96.36**	95.83	95.97	95.40	95.93
A3	50.12	65.46	**65.88**	48.18	64.95	62.21	64.04	62.21	64.40
A4	85.46	88.28	88.50	82.93	**89.14**	87.23	87.75	87.31	87.48
A5	87.39	92.83	90.03	**93.79**	92.63	93.12	93.08	92.38	93.55
A6	97.35	97.60	97.91	**99.10**	97.89	98.40	98.42	98.55	98.40
A7	79.83	82.06	78.14	75.61	81.41	87.45	87.53	**87.65**	87.49
A8	25.99	32.38	44.23	24.05	38.00	34.54	**51.65**	37.58	50.21
B1	21.96	33.98	36.54	27.11	32.11	28.54	38.85	33.52	**40.13**
B2	45.89	51.00	55.99	51.33	60.00	61.51	**63.10**	60.98	61.78
B3	69.56	64.76	59.47	70.81	65.97	68.28	72.70	69.40	**73.56**
B4	67.40	61.70	60.95	81.92	74.42	78.91	**80.03**	75.02	79.44
B5	51.80	14.79	14.71	39.05	17.89	47.33	**57.33**	50.00	**57.33**
B6	51.53	35.26	43.02	66.43	45.52	92.67	94.40	**95.20**	88.40
B7	26.33	41.33	53.33	51.33	55.33	**70.67**	**70.67**	70.00	68.00
C1	**100.00**	87.62	87.62	87.62	87.62	97.14	97.71	97.43	97.71
C2	88.19	46.30	67.48	83.14	61.64	97.96	**99.27**	98.39	98.60
C3	15.74	6.33	9.95	35.26	19.50	38.74	**80.63**	42.87	76.48
C4	100.00	73.08	98.46	100.00	94.29	100.00	100.00	100.00	100.00
C5	12.68	3.23	5.05	17.30	3.97	19.80	57.73	23.40	**59.26**
Rank	9	8	7	6	5	4	1	3	2
W/T/L	(1, 1, 18)	(2, 1, 17)	(2, 1, 17)	(3, 2, 15)	(3, 1, 16)	(1, 3, 16)		(3, 1, 16)	(5, 3, 12)

The data in Table 3, 4 are analyzed by symbolic test method, and the proposed algorithm is significantly superior to other comparison algorithms in the evaluation indicators F1 and AUC. The reason is that the proposed algorithm focuses on the boundary minority class samples, making each subset focus on different sample overlapping regions, which not only promotes the classifier's recognition ability of boundary samples, but also improves the diversity among base classifiers. In addition, gaussian perturbation is applied to minority class samples repeatedly selected in the sample to avoid the over-fitting of the classifier to minority class samples and improve the classification performance of the classifier. In order to further verify the role of the two methods proposed in this paper in the overall approach, two versions of our method are proposed, namely IORF1(emphasis on overlap) and IORF2(Gaussian perturbation sample). By analyzing the experimental results of F1 and AUC, we can see that IORF1 and IORF2 both perform well.

Table 4. Classification performance comparisons by AUC (%)

	AC	BRF	EE	RUB	BB	OIS_RF	IORF	IORF1	IORF2
A1	97.38	98.36	98.36	98.36	98.36	98.36	98.36	98.33	**98.43**
A2	93.82	97.26	97.26	96.48	**97.58**	97.08	97.26	96.69	97.24
A3	66.55	79.74	**79.69**	65.58	78.15	75.54	77.87	75.63	78.00
A4	90.27	94.64	94.62	87.05	94.74	93.05	**94.82**	93.00	93.43
A5	92.23	96.13	94.60	95.44	95.16	94.52	**96.31**	94.35	96.02
A6	98.51	99.09	99.12	**99.72**	99.14	99.09	99.18	99.15	99.18
A7	87.61	90.84	91.42	83.19	92.23	91.83	**93.14**	92.39	92.64
A8	59.41	72.63	**82.80**	57.80	70.46	62.85	77.28	64.70	73.05
B1	58.20	**80.43**	79.22	60.71	70.56	61.75	70.56	63.94	68.39
B2	71.02	86.17	**91.16**	71.81	88.00	75.72	84.29	76.38	79.20
B3	85.40	87.50	88.65	84.04	85.58	82.52	89.31	83.75	**89.85**
B4	82.72	93.38	91.67	**94.05**	93.41	87.93	92.58	84.90	92.11
B5	79.17	69.24	63.92	75.82	64.12	75.27	**79.70**	76.75	79.26
B6	74.91	90.24	92.93	95.37	93.17	97.71	**99.66**	98.33	96.51
B7	64.48	81.49	82.52	82.89	83.41	**85.18**	84.65	85.11	83.79
C1	**100.00**	99.48	99.48	99.48	99.48	99.90	99.92	99.91	99.92
C2	95.20	97.73	99.06	99.55	98.81	99.15	**99.99**	98.99	99.32
C3	57.17	66.98	72.80	69.38	70.78	63.20	**84.80**	65.00	81.60
C4	100.00	99.40	99.98	100.00	99.91	100.00	100.00	100.00	100.00
C5	55.99	57.79	69.71	58.90	57.55	57.17	74.00	59.25	**76.00**
Rank	9	6	4	8	3	7	**1**	5	2
W/T/L	(1, 1, 18)	(4, 2, 14))	(4, 2, 14)	(2, 2, 16)	(4, 2, 14)	(1, 2, 17)		(1, 1, 18)	(4, 1, 15)

4   Conclusions

In this paper, an improved ensemble learning algorithm is proposed for classification of imbalanced data. In order to solve the problem that it is difficult for classifiers to classify overlapping samples in imbalanced data classification, we introduce sample overlap degree and imbalanced ratio to measure the learning difficulty of different samples. Taking overlapping samples with different distributions as the center, diversity sample subset is selected. At the same time, a data enhancement method based on Gaussian perturbation is proposed to avoid over-fitting of overlapping minority class samples by classifier caused by repeatedly selected minority class samples. In this method, gaussian perturbation is applied to minority class samples repeatedly sampled to reduce the over-fitting of the classifier to the boundary samples. The performance of ensemble classifier for imbalanced data classification is improved by combining the above two methods.

Acknowledgments. This work was supported by National Natural Science Foundation of China under Grant nos. 61976108 and 61572241.

References

1. Tao, X., Li, Q., Guo, W., et al.: Self-adaptive cost weights-based support vector machine cost-sensitive ensemble for imbalanced data classification. Inf. Sci. **487**, 31–56 (2019)
2. Hassan, M.M., Huda, S., Yearwood, J., et al.: Multistage fusion approaches based on a generative model and multivariate exponentially weighted moving average for diagnosis of cardiovascular autonomic nerve dysfunction. Inform. Fus. **41**, 105–118 (2018)
3. Tan, X., Su, S., Huang, Z., et al.: Wireless sensor networks intrusion detection based on SMOTE and the random forest algorithm. Sensors **19**(1), 203 (2019)
4. Li, Y., Guo, H., Zhang, Q., et al.: Imbalanced text sentiment classification using universal and domain-specific knowledge. Knowl. Based Syst. **160**, 1–15 (2018)
5. Han, S., Choi, H.J., Choi, S.K., et al.: Fault diagnosis of planetary gear carrier packs: a class imbalance and multiclass classification problem. Int. J. Precis. Eng. Manuf. **20**(2), 167–179 (2019)
6. Lin, W.C., Tsai, C.F., Hu, Y.H., et al.: Clustering-based undersampling in class-imbalanced data. Inf. Sci. **409**, 17–26 (2017)
7. Douzas, G., Bacao, F., Last, F.: Improving imbalanced learning through a heuristic oversampling method based on k-means and SMOTE. Inf. Sci. **465**, 1–20 (2018)
8. Dong, X., Yu, Z., Cao, W., et al.: A survey on ensemble learning. Front. Comp. Sci. **14**(2), 241–258 (2020)
9. Drummond, C., Holte, R.C.: C4. 5, class imbalance, and cost sensitivity: why under-sampling beats over-sampling. In: Workshop on learning from imbalanced datasets II, vol. 11, pp. 1–8 (2003)
10. Liu, X.Y., Wu, J., Zhou, Z.H.: Exploratory undersampling for class-imbalance learning. IEEE Trans. Syst. Man Cybern. Part B (Cybern.) **39**(2), 539–550 (2008)
11. Ma, Y., Zhao, K., Wang, Q., et al.: Incremental cost-sensitive support vector machine with linear-exponential loss. IEEE Access **8**, 149899–149914 (2020)
12. Yang, K., Yu, Z., Wen, X., et al.: Hybrid classifier ensemble for imbalanced data. IEEE Trans. Neural Netw. Learn. Syst. **PP**(99), 1–14 (2019)
13. Sun, Y., Kamel, M.S., Wong, A.K.C., et al.: Cost-sensitive boosting for classification of imbalanced data. Pattern Recognit. **40**(12), 3358–3378 (2007)
14. Sun, T., Jiao, L., Feng, J., et al.: Imbalanced hyperspectral image classification based on maximum margin. IEEE Geosci. Remote Sens. Lett. **12**(3), 522–526 (2014)
15. Mellor, A., Boukir, S., Haywood, A., et al.: Exploring issues of training data imbalance and mislabelling on random forest performance for large area land cover classification using the ensemble margin. ISPRS J. Photogramm. Remote Sens. **105**, 155–168 (2015)
16. Vuttipittayamongkol, P., Elyan, E., Petrovski, A.: On the class overlap problem in imbalanced data classification. Knowl. Based Syst. **212**, 106631 (2021)
17. Mehmood, Z., Asghar, S.: Customizing SVM as a base learner with AdaBoost ensemble to learn from multi-class problems: a hybrid approach AdaBoost-MSVM. Knowl. Based Syst. **217**, 106845 (2021)
18. Chen, X., Zhang, L., Wei, X., et al.: An effective method using clustering-based adaptive decomposition and editing-based diversified oversamping for multi-class imbalanced datasets. Appl. Intell. **51**(4), 1918–1933 (2021)
19. Fernandes, E.R.Q., de Carvalho, A.C.: Evolutionary inversion of class distribution in overlapping areas for multi-class imbalanced learning. Inf. Sci. **494**, 141–154 (2019)

20. Tao, X., Chen, W., Zhang, X., et al.: SVDD boundary and DPC clustering technique-based oversampling approach for handling imbalanced and overlapped data. Knowl. Based Syst. **234**, 107588 (2021)
21. Yuan, B.W., Zhang, Z.L., Luo, X.G., et al.: OIS-RF: a novel overlap and imbalance sensitive random forest. Eng. Appl. Artif. Intell. **104**, 104355 (2021)
22. Armano, G., Tamponi, E.: Building forests of local trees. Pattern Recognit. **76**, 380–390 (2018)

Load Forecasting Method for Park Integrated Energy System Considering Multi-energy Coupling

Xin Huang, Xin Ma, Yanping Li[✉], and Chunxue Han

School of Information and Electrical Engineering, Shandong Jianzhu University, Jinan 250101, China
maxin20@sdjzu.edu.cn, Liyanping0531@126.com

Abstract. The load volatility of the park integrated energy system (PIES) is large, and multiple energy sources are deeply coupled, so accurate multivariate load prediction has become an inevitable choice to improve the operational efficiency and reliability of the PIES. Based on this, this paper proposes a method for predicting the cold, heat and electricity loads of the PIES considering the coupling relationship of each energy source. Firstly, the coupling characteristics between multiple loads in the system and the influence of meteorological factors on the loads are analyzed by using Spearman correlation coefficients; second, the gated recurrent network (GRU) is used as the primary prediction method, with an attention mechanism (AM) added to increase the model's prediction accuracy. Finally, the feasibility of the proposed technique is tested using numerous comparison models. The algorithm's results show that it has an RMSE of 1.981 and an MAE of 1.414, both of which are lower than the comparison model's error and have a higher prediction efficiency.

Keywords: Park integrated energy system · Multivariate load forecasting · Attention mechanism · Gated recurrent unit

1 Introduction

In recent years, energy and environmental issues have attracted widespread attention from all over the world and all walks of life. The integrated energy system, as the physical carrier of the new generation of energy [1], may achieve graded energy usage, effectively increase the comprehensive energy utilization efficiency, and reduce the park's carbon emissions. Figure 1 depicts the system structure diagram. The integrated energy system can be classified into three levels based on the scale of service users: inter-district, regional, and park. Among them, Park Integrated Energy System (PIES) has a small user scale, poor system robustness and large load fluctuation [2, 3]. In the operation and scheduling of combined cooling, heating, and electricity systems, as well as energy management, precise load forecast is critical. Therefore, how to effectively solve the fluctuation and coupling of multi-energy load existence and accurately realize multi-energy load prediction has become a hot research topic nowadays [4].

H. Zhang et al. (Eds.): NCAA 2022, CCIS 1638, pp. 469–481, 2022.
https://doi.org/10.1007/978-981-19-6135-9_35

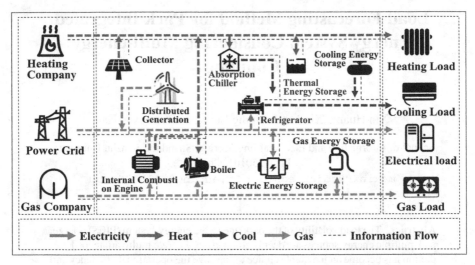

Fig. 1. Integrated energy system framework.

Currently, PIES single type of load forecasting has achieved remarkable results, In terms of electric loads, a model based on a mixed-mode decomposition method is proposed in the literature [5] for decomposing the total electric load signal, and the model has a short-term electric load forecast accuracy of 99%. In terms of cold load prediction, the literature [6] predicts cold load by using an elephant swarm optimization (EHO) algorithm combined with a multilayer perceptron neural network (MLP). In terms of thermal load, the literature [7] proposes a PCLA model that has been shown to outperform previously published thermal load prediction models.

However, various types of energy sources in the PIES are deeply coupled and interact with each other, and the dynamic characteristics of various types of energy sources are very different, so the load prediction method for a single type of energy source cannot be extended to the field of multi-energy prediction. The load prediction approach for a single energy source is unable to effectively characterize the strong coupling relationship between numerous energy sources, severely compromising the prediction results. As a result, domestic and foreign scholars have conducted research on predicting multiple types of loads, and literature [8] used autoregressive moving average with external input models to anticipate cold, heat, and electricity loads, and literature [9] proposed a hybrid model combining multiple regression models to predict cold loads and electricity loads. The aforementioned literature mainly uses a single model or a hybrid model for prediction, and ignores the relationship of numerous loads coupling, so the Pearson correlation coefficient approach is used in the literature [10] to examine the dynamic coupling connection between various loads, and this method demonstrates that incorporating the multi-energy coupling relationship can increase forecast accuracy significantly. Therefore, this paper takes into account the nonlinear relationship between multivariate loads, dissects the multi-energy coupling characteristics using the correlation analysis method, and adopts a model that integrates multiple prediction methods to solve the problem of single method model's low prediction accuracy.

Based on the above analysis, this paper uses Spearman correlation coefficients to analyze the coupling of multivariate loads, and adopts an attention-based gated cyclic unit to fit the hot and cold electric load curves, using the more relevant influencing factors as inputs to improve the model's ability to mine time series information. The arithmetic example validates the model employed in this paper, demonstrating that the strategy can improve the multivariate load forecasting model's performance.

2 Correlation Analysis Method

The variety of loads in the PIES, the tight coupling presented among the loads, as well as the fluctuation of the loads under the influence of meteorological factors, make the prediction more complicated and cause difficulties [11, 12]. For this reason, when carrying out multivariate load forecasting, it is necessary to dig deeper into the variation pattern of the load as a way to define suitable input variables, so as to enhance the multivariate load forecasting accuracy.

Considering the nonlinear relationship between multivariate load and meteorological factors in PIES [13], Spearman correlation coefficient is used in this paper to measure the strength of the correlation between multivariate load and load and meteorological factors.

A nonparametric measure of the dependence of two variables is the Spearman correlation coefficient [14], commonly known as the Spearman rank correlation coefficient. The Spearman correlation coefficient is calculated using the equation below. Table 1 offers an example of how to calculate the data's rank order difference.

$$\rho = 1 - \frac{6 \sum_{i=1}^{n} d_i^2}{n(n^2 - 1)} \tag{1}$$

where, d_i^2 is the level difference between sequences X_i and Y_i; n is the number of data; ρ takes values in the range of $[-1, 1]$.

Table 1. Example of data rank difference.

Position	Original X	Rank	Original Y	Rank	Rank difference
1	10	3	7	4	1
2	345	1	12	3	2
3	45	2	33	2	0
4	2	4	145	1	3

3 Multivariate Load Prediction Model

3.1 Gated Recurrent Neural Network

In Recurrent Neural Networks (RNN), Long-Short-Term Memory (LSTM) was proposed and used to overcome the problem of gradient explosion and disappearance [15, 16, 17] and then Junyoung Chung et al. [18] suggested the Gated Recurrent Unit Network (GRU), which consists of only two gates: update and reset. Figure 2 depicts the GRU structural diagram. Compared with the LSTM with three gates, it has fewer parameters and faster convergence, which can largely improve the training efficiency.

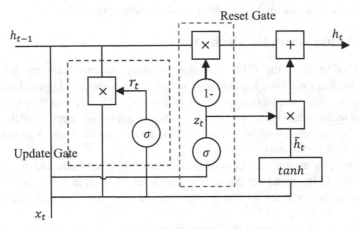

Fig. 2. Architecture of GRU system.

The update gate aids the model in identifying how much data from the past should be delivered to the future. It is calculated using the equation below.

$$z_t = \sigma\left(W^{(z)}x_t + U^{(z)}h_{t-1}\right) \tag{2}$$

The reset gate specifies how much information from the past should be forgotten. The equation below is used to determine the reset gate.

$$r_t = \sigma\left(W^{(r)}x_t + U^{(r)}h_{t-1}\right) \tag{3}$$

The new memory content is controlled by the reset gate, also called the candidate state at the current moment, which is calculated by the following equation:

$$\tilde{h}_t = tanh(r_t \cdot Uh_{t-1} + Wx_t) \tag{4}$$

By multiplying the update gate with the preceding moment and adding the information kept by the current memory to the final memory, the final output content is calculated. The following is the calculating procedure:

$$h_t = (1 - z_t) \cdot \tilde{h}_t + z_t \cdot h_{t-1} \tag{5}$$

where, σ is the sigmoid activation function, which accepts values between 0 and 1; tanh is the activation function, which accepts values between -1 and 1; x_t is the input vector; h_{t-1} is the information at the previous moment; $W^{(z)}$, $U^{(z)}$, $W^{(r)}$, $U^{(r)}$ are the weight matrices; U and W are the parameter matrices.

3.2 Attention Mechanism

Attention Mechanism (AM) allows us to focus our attention on the information that is relevant to the outcomes and avoid wasting time in the background noise [19]. As a result, the AM algorithm can improve the model's performance by learning more information from the input sequence and focusing on the important aspects. Figure 3 depicts its structural diagram.

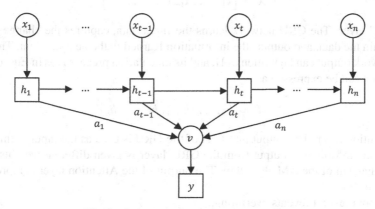

Fig. 3. Structure of attentional mechanism.

Where, x_i in the figure is the input of the GRU neural network, h_i denotes the output of the GRU network model, a_i is the value of the AM's attention probability density function on the output of the GRU hidden layer, and y is the output value of the GRU network after adding the Attention mechanism.

In the AM algorithm, the attention weight matrix α and the feature vector v are calculated as:

$$e_{t,k} = u_s \cdot \tanh(w_s h_t + b_s) \tag{6}$$

$$\alpha_{t,k} = \frac{\exp(e_{t,k})}{\sum_{k=1}^{l} e_{n,k}} \tag{7}$$

$$v = \sum_{t=1}^{n} \alpha_{t,k} h_t \tag{8}$$

where, $e_{t,k}$ is the unnormalized weight matrix; w_s, b_s and u_s are the randomly initialized AM weight matrix, bias and time series matrix, respectively.

3.3 GRU Multivariate Load Prediction Model Based on the Attention Mechanism

By combining GRU, which mines the variation pattern of time series, and Attention, which focuses on the key information, the model is able to mine the key features and give higher weights to the input information of long time series to improve the prediction accuracy. Figure 4 depicts the model structure.

(1) Input layer: The data is preprocessed as input variables after the input factors have been determined. Assuming that there are eventually n variables as inputs, the following equation can be used to express them:

$$X = \left[x_1, \ldots, x_{t-1}, x_t, \ldots, x_n\right]^T$$

(2) GRU layer: The GRU network learns the input data, captures the change pattern within the data, and outputs the information learned to the next moment. The GRU network output can be written as H, and its calculation process is as in Eqs. (2)–(5), then it can be expressed as:

$$H = \left[h_1, \ldots, h_{t-1}, h_t, \ldots, h_n\right]^T$$

(3) Attention layer: The output matrix H of the GRU is used as the input to this layer, and the information output from the GRU layer is given different weights by the mechanism of the AM algorithm. The output of the Attention layer is represented by Eq. (8).
(4) Dropout layer. Prevents overfitting.
(5) Output layer. The output layer uses the Attention layer's output as an input, and after the completely connected layer, the output layer calculates the final output y.

4   Analysis of Algorithms

4.1   Data Pre-processing and Parameter Setting

The data used in this paper is a PIES of an office building built by Trnsys software, simulating the cooling load, heat load and electrical load for one year. The data also contains meteorological data consisting of temperature (in °C), humidity (in %), solar radiation intensity (in W/m^2), and wind speed (in m/s). The granularity of the dataset is 24 data collection per day, i.e. one point every 1h. A total of 8760 data points were collected, with 7008 being regarded as training sets and 1752 being considered as test sets.

Because the data units differ and span a large range, the min-max normalization approach [20] was employed to scale the values to between −1 and 1. This made it easier to train the model network, and the calculation formula was:

$$x_n = \frac{x - x_{mean}}{x_{max} - x_{min}} \tag{9}$$

where, x is the original load data; x_n is the normalized data; $x_{mean}, x_{max}, x_{min}$ represent the data's mean, maximum, and minimum values, respectively.

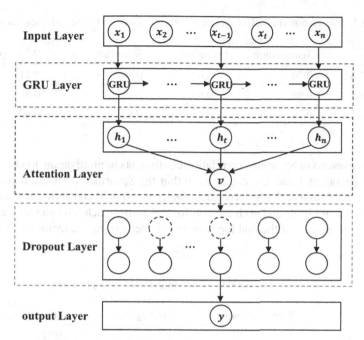

Fig. 4. PIES multivariate load short-term forecasting model.

Root Mean Square Error (RMSE) and Mean Absolute Error (MAE) are utilized as indicators to estimate the soundness of the model in this research toward evaluate and analyze the accuracy of the prediction outcomes. The following is the calculating procedure:

$$RMSE = \sqrt{\frac{1}{n} \sum_{i=1}^{n} (a_i - b_i)^2} \tag{10}$$

$$MAE = \frac{\sum_{i=1}^{n} |a_i - b_i|}{n} \tag{11}$$

where, n is the number of samples contained in the data set, a_i is the true value of each prediction task at moment i, and b_i is the predicted value of each prediction task at moment i.

4.2 Correlation Analysis

Meteorological factors affect each type of load in the PIES, causing randomness and volatility, and changes in one type of load can affect the other types of load.

This paper employs Spearman correlation coefficients to calculate and assess the degree of influence of meteorological conditions on each load, as well as to analyze the coupling between loads. Table 2, Table 3 shows the calculation results of Spearman's correlation coefficient.

Table 2. Correlation coefficients between multiple loads and meteorological factors.

	Temperature	Humidity	Solar radiation	Wind speed
Electrical load	0.70	−0.83	0.94	0.71
Cooling load	0.98	−0.63	0.81	0.66
Heating load	0.73	−0.68	0.65	0.60

From the results of Spearman correlation coefficients of multivariate loads and meteorological factors in Table 2, it can be seen that the Spearman correlation coefficients of electrical loads and each meteorological factor reach 0.70 and above, the correlation coefficients of cold loads and each meteorological factor reach 0.63 and above, and the correlation coefficients of thermal loads and each meteorological factor reach 0.60 and above.

Table 3. Correlation coefficient between multiple loads.

	Electrical load	Cooling load	Heating load
Electrical load	1	0.74	0.72
Cooling load	0.74	1	−0.68
Heating load	0.72	−0.68	1

From the Spearman correlation coefficients between the loads in Table 3, it can be seen that the correlation coefficient between the electric load and the cold load is 0.74 and the correlation coefficient with the heat load is 0.72, showing a strong correlation; while the Spearman's coefficient between the cold load and the heat load is −0.68, also showing a strong correlation, due to the difference between the heating season and the heating season, resulting in a negative correlation between the two.

In summary, it demonstrates that the various forms of loads in PIES are tightly connected, and that meteorological factors have a significant impact on load variation. Therefore, in this paper, each meteorological factor is used as an influencing factor, along with the historical data of multivariate loads as the input data set, as a way to improve the accuracy of the prediction model.

4.3 Analysis of Load Forecasting Results

The PIES integrates cooling, heating and electricity systems, and the loads are closely coupled with each other; therefore, not only the influence of meteorological factors but also the coupling factors among the loads should be considered when making multivariate load prediction. Two scenarios are set up to verify the effect of the model provided in this paper:

Scene 1: GRU-Attention load prediction model considering multivariate load coupling factors;
Scene 2: GRU-Attention load forecasting model without considering the coupling factors of multiple loads.

Taking a 24-h period of a day as an example, Fig. 5, Fig. 6 and Fig. 7 show the predicted versus true values of the electric, thermal and cooling loads under these two scenes, respectively; Fig. 8 depicts the load forecast errors in these two scenes.

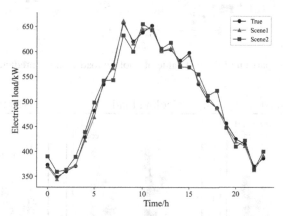

Fig. 5. Comparison of the real value of electrical load with the predicted value.

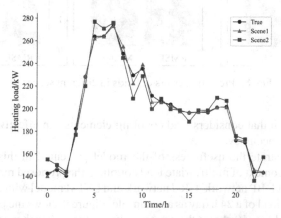

Fig. 6. Comparison of the real value of heating load with the predicted value.

As can be seen from Fig. 8, compared with Scene 2, the RMSE of electric load in Scene 1 is reduced by 4.312 and MAE by 3.992; the RMSE of cold load is reduced by 4.995 and MAE by 3.371; the RMSE of thermal load is reduced by 1.451 and MAE by 1.723. The error in the scene with multiple load coupling is lower than the error in the model without multivariate load coupling for each type of load, indicating that the

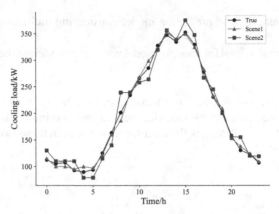

Fig. 7. Comparison of the real value of cooling load with the predicted value.

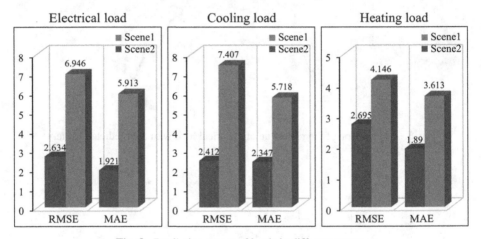

Fig. 8. Prediction errors of loads in different scenes.

prediction approach that considers load coupling elements can effectively improve the prediction model's accuracy.

To further illustrate the usefulness of the model presented in this paper for load forecasting in the context of multivariate load coupling, the proposed model is compared with ARIMA model, BP network, GRU network and LSTM model with Attention added, taking the electric load of a 24 h day as an example. Figure 9 shows the prediction results of each model, and Fig. 10 shows the error results analysis of each model.

From Fig. 10, the RMSE and MAE of the proposed prediction model, GRU-Attention, are lowered by 7.809 and 7.024, respectively, when compared to GRU without the attention mechanism, demonstrating that including the attention mechanism can successfully optimize the model to improve the model's prediction accuracy; In addition, the proposed model is compared to the classical models ARIMA and BP neural network models, which are commonly used for time series prediction, and the proposed model's prediction accuracy has obvious advantages, although the training and prediction times

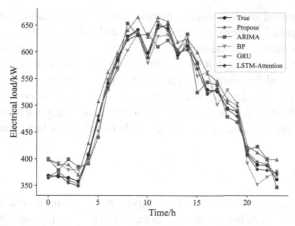

Fig. 9. Electric load forecasting results under different models.

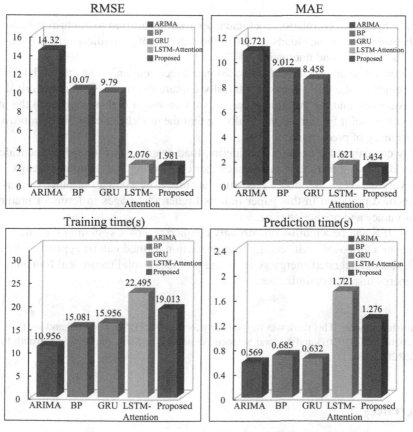

Fig. 10. Comparison results of different prediction models.

increase, the proposed model's overall performance is better than other models; additionally, the proposed model's prediction effect and LSTM-Attention are compared. The difference is minor, but the former takes 3.482 and 0.445 s less to train and predict than the latter, indicating that GRU's underlying structure is more streamlined than LSTM's, and the model is more efficient at the same level of prediction accuracy. As a result, the model described in this study can effectively exploit the coupling relationship between various load time series, improving prediction accuracy for multiple loads with periodicity and coupling, and saving model prediction time for large training data.

5   Conclusion

In this article, a gated cyclic unit network model based on an attention mechanism is employed for multivariate load prediction in the PIES, taking into consideration the tight coupling of cooling, heating, and electrical demands. The consideration of arithmetic examples might lead to the following conclusions.

(1) The Spearman correlation coefficient approach is utilized to confirm the coupling between multivariate loads as well as the presence of a significant effect of climatic conditions on load fluctuation.
(2) Using an attention mechanism-based gated recurrent unit network model for load prediction, the GRU network can fully capture the time-series of historical load sequences, and the Attention mechanism can assign higher weights to the information useful for the output results, so that the model can effectively improve the accuracy of prediction.
(3) By comparing the cases of considering load coupling factors and not considering load coupling factors, it can be seen that considering load coupling factors effectively improves the model prediction accuracy; by comparing other models, the method proposed in this paper reflects higher advantages in terms of prediction accuracy and prediction time.

IES will be a multi-load, strongly coupled complex system in the future. This paper's proposed multi-energy load prediction method can be applied to different levels of integrated energy systems and provides a solid theoretical foundation for energy dispatch optimization.

Acknowledgement. This work was supported by the Key R&D Program of Shandong Provincial (No. 2020CXGC010201) and Natural Science Foundation of Shandong Province Youth Project (No. ZR2021QF011).

References

1. Li, J., Liu, J., Yan, P., Li, X., Zhou, G., Yu, D.: Operation optimization of integrated energy system under a renewable energy dominated future scene considering both independence and benefit: a review. Energies **14**(4), 1103 (2021)

2. Yuan, J., Wang, L., Qiu, Y., Wang, J., Zhang, H., Liao, Y.: Short-term electric load forecasting based on improved extreme learning machine mode. Energy Rep. **7**(S7), 1563–1573 (2021)
3. Zhu, J., Dong, H., Zheng, W., Li, S., Huang, Y., Xi, L.: Review and prospect of data-driven techniques for load forecasting in integrated energy systems. Appl. Energy **321**, 119269 (2022)
4. Zhao, J., Chen, L., Wang, Y., Liu, Q.: A review of system modeling, assessment and operational optimization for integrated energy systems. Sci. China Inf. Sci. **64**(9), 1–23 (2021)
5. Hu, Y., Li, J., Hong, M., Ren, J., Man, Y.: Industrial artificial intelligence based energy management system: integrated framework for electricity load forecasting and fault prediction. Energy **244**(PB), 123195 (2022)
6. Moayedi, H., Mu'azu, M.A., Foong, L.K.: Novel swarm-based approach for predicting the cooling load of residential buildings based on social behavior of elephant herds. Energy Build. **206**(C), 109579 (2020)
7. Chung, W.H., Gu, Y.H. and Yoo, S.J.: District heater load forecasting based on machine learning and parallel CNN-LSTM attention. Energy **246**, 123350 (2022)
8. Tang, Y., Liu, H., Xie, Y., Zhai, J., Wu, X.: Short-term forecasting of electricity and gas demand in multi-energy system based on RBF-NN model. In: Proceedings of the International Conference on energy Internet, p. 136–141 (2017)
9. Niu, D., Yu, M., Sun, L., Gao, T., Wang, K.: Short-term multi-energy load forecasting for integrated energy systems based on CNN-BiGRU optimized by attention mechanism. Appl. Energy **313**, 118801 (2022)
10. Li, A., Xiao, F., Zhang, C. and Fan, C.: Attention-based interpretable neural network for building cooling load prediction. Appl. Energy **299**, 117238 (2021)
11. Zheng, J., et al.: Multiple-load forecasting for integrated energy system based on copula-DBiLSTM. Energies, **14**(8), 2188 (2021)
12. Wang, X., Wang, S., Zhao, Q., Wang, S., Fu, L.: A multi-energy load prediction model based on deep multi-task learning and ensemble approach for regional integrated energy systems. Int. J. Electr. Power Energy Syst. **126**(PA), 106583 (2021)
13. Liu, D., Wang, L., Qin, G., Liu, M.: Power load demand forecasting model and method based on multi-energy coupling. Appl. Sci. **10**(2), 584 (2020)
14. Huang, Y., Li, C.: Accurate heating, ventilation and air conditioning system load prediction for residential buildings using improved ant colony optimization and wavelet neural network. J. Build. Eng. (2020). Prepublish
15. Wang, Z., Hong,T., Piette, A.: Building thermal load prediction through shallow machine learning and deep learning. Appl. Energy **263**, 114683 (2020)
16. Liao, Z., Huang, J., Cheng, Y., Li, C., Liu, P.X.: A novel decomposition-based ensemble model for short-term load forecasting using hybrid artificial neural networks.. Appl. Intell. (2022). Prepublish
17. Hou, T., et al.: A novel short-term residential electric load forecasting method based on adaptive load aggregation and deep learning algorithms. Energies **14**(22), 7820 (2021)
18. Chung, J., Gülçehre, Ç., Cho, K.H., Bengio, Y.: Empirical evaluation of gated recurrent neural networks on sequence modeling. CoRR,2014,abs/1412.3555 (2014)
19. Lin, J., Ma, J., Zhu, J., Cui, Y.: Short-term load forecasting based on LSTM networks considering attention mechanism. Int. J. Electr. Power Energy Syst.**137**, 107818 (2022)
20. Zhang, Z., Hong, W.C., Li, J.: Electric load forecasting by hybrid self-recurrent support vector regression model with variational mode decomposition and improved cuckoo search algorithm IEEE Access **8**, 14642–14658 (2020)

Vehicle Re-identification via Spatio-temporal Multi-instance Learning

Xiang Yang[1], Chunjie Li[2,3,4,5], Qingwei Zeng[1], Xiu Pan[3,4,5], Jing Yang[6],
and Hongke Xu[6(✉)]

[1] Hebei Xiongan Rongwu Expressway Limited Company, Shijiazhuang, Hebei, China
[2] School of Transportation, Southeast University, Nanjing, China
[3] Hebei Provincial Communications Planning, Design and Research Institute Co., Ltd.,
Shijiazhuang, Hebei, China
[4] Research and Development Center of Transport Industry of Self-driving Technology,
Shijiazhuang, Hebei, China
[5] Hebei Engineering Research and Development Center of Vehicle-Road-Cloud Integration,
Shijiazhuang, Hebei, China
[6] Chang'an University, Xi'an, China
{2020132075,xuhongke}@chd.edu.cn

Abstract. Vehicle re-identification is a cross-camera vehicle retrieval method. Compared with the method of manually retrieving surveillance video to realize vehicle re-identification, the method based on deep learning uses computer to achieve cross-camera matching of target vehicles, which saves labor costs, so it has high practical application value in the field of intelligent transportation. Common vehicle re-identification algorithms achieve re-identification by making the characteristics of different pictures of the same id vehicle tend to be consistent. Generally, these methods rely on manually annotated datasets. However, the accuracy of the model may decrease due to the possibility of labeling errors when manually labeling datasets, especially large-scale datasets. To solve the above problems, this paper proposes a vehicle re-identification algorithm based on spatio-temporal multi-instance learning. It uses a multi-instance bag to train a feature extraction model and pays attention to the features of the entire multi-instance bag and ignores the features of a single instance. So it can handle the problem of mislabeling in the dataset. Experimental results show that the model is feasible: on the VeRi dataset, the model can achieve 33.3% mAP and 67.9% Rank-1 accuracy.

Keywords: Deep learning · Multi-instance learning · Intelligent transportation · Vehicle re-identification

1 Introduction

Vehicle re-identification technology plays an important role in the fields of intelligent transportation [17], urban traffic computing [18], intelligent monitoring [8], etc. It has high practical value. Especially in the field of intelligent transportation, vehicle re-identification algorithms have played an important role in suspicious vehicle search, cross-domain vehicle tracking, traffic behavior analysis, automatic toll collection systems, and vehicle counting. As shown in Fig. 1, vehicle re-identification is a task of searching the candidate set for other images of the same vehicle as a given vehicle image from the query set. Different from tasks such as vehicle detection and tracking, the vehicle re-identification problem can be regarded as a retrieval problem of approximately duplicate images [11]. In the vehicle re-identification data set, the images of vehicles are captured by different cameras. Due to the different monitoring angles of different cameras, the captured vehicle pictures are also different. Therefore, the vehicle re-identification task is different from the general image retrieval task.

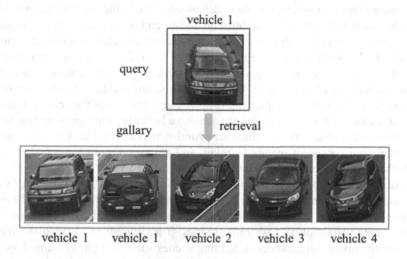

Fig. 1. Schematic diagram of the re-identification task.

Regarding the problem of vehicle re-identification, the current common vehicle re-identification algorithms mainly rely on the feature extraction model to extract the features of the information carried by the image itself, and then according to the labels of the data set, the features of the pictures are belonging to the same category as close as possible, and the different categories are enlarged. The distance between the features of the image is used to train a feature extractor that can better express the apparent characteristics of the vehicle, so as to realize the re-identification of the vehicle. For instance, Yang et al. [16] used convolutional neural networks to extract visual features from images of complete vehicles and parts, performed fine-grained classification and attribute preprocessing, and used a 3D model-based method to match and fit 2D images for use in Vehicle re-identification; Tu et al. [13] proposed a discriminative feature representation for vehicle recognition based on spatio-temporal cues. Through two-stream structure and spatio-temporal metrics, a multi-granular visual representation was constructed, and robust features were constructed in the embedding space. Liu et al. [10] proposed a method based on the fusion of attributes and color features, respectively extracting texture features, color features and semantic features from the image, and performing weighted summation according to different weights, which is conducive to the fusion of feature similarity for vehicle re-identification. Zhu et al. [24] designed a four-way deep learning network for extracting four-way depth features of vehicle images, compressing the basic feature maps into horizontal, vertical, diagonal and anti-diagonal orientation feature maps, which were normalized and connected to new features. Liu et al. [9] proposed a region-aware deep model to extract features from a series of local regions and jointly use vehicle identity, type and color to train the model. Yan et al. [15] proposed a multi-task deep learning framework based on multi-dimensional information to simultaneously complete vehicle classification and similarity ranking. Vehicle identity and vehicle attributes such as model and color are used In vehicle classification to classify and retrieve vehicles. Wei et al. [14] coupled vehicle image representation generation, dependency modeling, and attention modules for subtle visual information to establish an end-to-end hierarchical attention classification model. Zhou et al. [20] proposed a perspective-based attention inference model, which divided the body appearance into five parts: front, front side, side, rear, and rear side, and designed a multi-view generation network to generate feature vectors from various perspectives of the vehicle, learn to generate multi-view features from single-view features to learn more comprehensive appearance features of the vehicle. Hsu et al. [4] propose a stronger baseline method to achieve a better feature representation ability, and combine shift-invariant convolutional neural network with ResNet backbone network to enhance consistent feature learning, and propose a multi-layer feature fusion module that combines mid-to-high level features.

To sum up, most of the common vehicle re-identification methods aim to fully exploit and utilize the information carried by the images, and the accuracy of the model depends heavily on the accuracy of the annotation of the dataset. However, due to the lack of camera resolution in real traffic scenes, it is difficult to distinguish different vehicles with similar appearances in lighting scenes such as night or rainy days. It is difficult to manually label the data set, and the possibility of wrong labels in the data set is relatively high. The general vehicle re-identification model accuracy may be affected. In order to solve the problem that the above models rely heavily on the label accuracy of the dataset, a vehicle re-identification model based on spatio-temporal multi-instance learning is proposed. The multi-instance learning method is used to improve the model's anti-interference ability to a few wrong labels and train a more robust feature extraction model.

The main contributions of this paper are as follows:

Contribution 1: This paper provides an idea of extending the scope of application of multi-instance learning from the original binary classification task to the multi classification task, which may provide some references for subsequent work.

Contribution 2: This paper designs a vehicle re-identification model based on spatio-temporal multi-instance learning, which combines instance-level loss and multi-instance packet-level loss for training, and provides a model that pays more attention to the overall characteristics of the cluster.

2   Multi-instance Learning

2.1   Introduction of Multi-instance Learning

Multi-instance learning [21] refers to a learning problem that uses multi-instance bags as training units. Its purpose is to establish a multi-instance classifier through the learning of multi-instance bags with classification labels, and use the classifier to predict bag or instance without labels.

Multi-instance learning belongs to the field of weakly supervised learning, in multi-instance learning, suppose there is a set of data, this set of data only has two categories [22]: positive and negative, several pieces of this set of data form a multi-instance bag B:

$$B \in \{I_1, I_2, I_3 \ldots\}. \tag{1}$$

Among them, the multi-instance bag B has a label, and a single piece of data I_n itself does not necessarily carry a label.

When a multi-instance bag is marked as a positive bag, there is at least one data classified as positive in the multi-instance bag. When a multi-instance bag is marked as negative, all data in the multi-instance bag are classified as negative data. Therefore, the probability that the multi-instance bag is negative can be obtained by multiplying the probability that all instances are judged to be negative. Suppose the probability that an instance I_n is judged as a positive instance is p_n, then the probability of a multi-instance bag being judged as a negative bag is P_-:

$$P_- = \prod_{j=0}^{j=l} -log_{10}(1 - p_n^j). \tag{2}$$

Among them, l is the number of samples in the bag, $n \in (0, N]$, N is the number of bags.

The loss for a multi-instance bag can be expressed as:

$$loss = -log_{10}(1 - P_-). \tag{3}$$

The purpose of multi-instance learning is to train a classifier that can correctly predict the category of samples with unknown labels by learning the multi-instance bags with positive and negative labels [23].

Currently, multi-instance learning is mainly used in object detection and other fields. In the transportation domain, multi-instance learning is mainly used for vehicle detection problems. For instance, Cao et al. [1] design a weakly supervised multi-instance learning algorithm to learn instance-level vehicle detectors from weak labels, This approach only requires group annotations at the region level without explicitly labeling the bounding boxes of vehicles. Multi-instance learning is rarely used in the field of vehicle re-identification.

Generalization to the Multi Class Case. At present, common multi-instance learning methods can handle the binary classification problem well such as in the field of target detection, it is determined whether a certain area of an image is a certain target. But for multi-classification problems, multi-instance learning cannot handle it. However, the vehicle re-identification problem can be transformed into a multi-classification problem of cross-view images. In order to deal with the vehicle re-identification problem with the multi-instance learning method, we propose a multi-instance learning method that extends the multi-instance learning method suitable for the binary classification task to the multi-instance learning method for the multi-classification task.

In order to extend the application range of multi-instance learning from binary classification problems to multi-classification problems, we redefine the positive and negative of multi-instance bags: Assuming that there is a set of data, this set of data can be divided into k categories, and extract several components from this set of data to make up the multi-instance bag B:

$$B \in \{I_1, I_2, I_3 \ldots\}, \tag{4}$$

where I_n is an instance belonging to the $n^{th} (n < k)$ class. If the number of instances in B is more than the total number of instances in other categories, then B is the nth positive bag. If the number of instance in B is less than the total number of instance in other categories, then B is the negative bag of the nth category. When B is the nth positive bag, it is also the negative multi-instance bag of other categories. Let B be the nth positive multi-instance bag, and the weight of the instance I_j belonging to the ith class is S_j^i, then the features of the multi-instance bag B can be expressed as F_B:

$$F_B = \sum_{j=0}^{j=L} (S_j^i \cdot f_j), \tag{5}$$

where L is the length of the multi-instance bag, and f_j is the feature corresponding to the instance I_j.

The probability that the multi-instance bag B is predicted to the nth positive multi-instance bag by the multi-instance bag level classifier is expressed as P_B^n, The loss function of the multi-instance bag adopts the cross-entropy loss function, which is expressed as:

$$loss = -log_{10}(P_B^n). \tag{6}$$

3   Vehicle Re-identification via Spatio-Temporal Multi-instance Learning

In view of the fact that the training unit used in multi-instance learning is a multi-instance bag rather than the instance itself, which can well avoid the influence of unknown labels, this paper adopts multi-instance learning to build a vehicle re-identification model.

As shown in Fig. 2 about the main frame of the vehicle re-identification model, our model first builds a multi-instance bag based on the temporal information annotated by the dataset. Time stamp and camera number from the camera which carrying

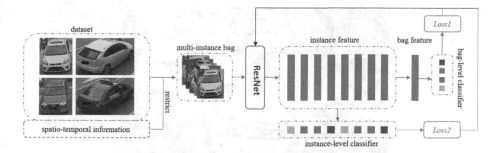

Fig. 2. The main frame of the vehicle re-identification model.

the time and space information of the vehicle. However, many vehicle re-identification methods ignore this part of the information. After completing the construction of the multi-instance bag according to the constraints of space and time information, Each instance in the multi-instance bag will be input into a feature extraction model such as ResNet to extract corresponding deep features. After that, one branch of the model is to compute the eigenvectors of the bag from the instance eigenvectors, and calculate the multi-instance bag-level loss based on the bag-level classifier results. Another branch computes instance-level loss directly from instance-level classifier results. The feature extraction model is trained with the sum of the multi-instance bag-level loss and the instance-level loss. This process is described in detail next.

3.1 Construct of the Multi-instance Bag

We use spatio-temporal information to constrain the sampling process and build a multi-instance bag for training. As shown in Fig. 3, in vehicle re-identification via on spatio-temporal multi-instance learning, the nth positive multi-instance bag refers to the fact that there are more pictures with the identity of the nth type of vehicle in the multi-instance bag than pictures which with the identity of other types of vehicles. On the contrary, the negative multi-instance bag of the nth class refers to the negative multi-instance bag in which the images with the identity of the nth class of vehicles are less than the pictures of the other class vehicles. The benchmark image will be set for each multi-instance bag, to collect positive multiple-instance bags, we compare the vehicle identity information of the benchmark image with the vehicle identity information of other images in the dataset. When we collect negative multiple-instance bags, the spatio-temporal information is introduced to constrain the sampling, the picture we will acquired and the reference picture which is another vehicle identity are located in the same camera node and during the same shooting period. That is, using the spatio-temporal constraints to collect difficult negative samples to construct a negative multi-instance bag, and improve the recognition performance of the model for difficult negative samples.

Positive mult-instance bag [   ]

Negative mult-instance bag [   ]

Fig. 3. Visualization of positive or negative multi-instance bag.

3.2 Training with Multiple-Instance Bags

Vehicle re-identification based on spatio-temporal multi-instance learning constructs two types of classifiers: instance-level classifier and multi-instance bag-level classifier. Among them, the main function of the instance-level classifier is to construct multi-instance bag features. During training, input several multi-instance bags into the feature extraction model (The feature extraction model used in this paper is ResNet50), feature extracted for each instance I in the multi-instance bag, and input the instance-level classifier to predict the score s of each instance relative to the class to which it belongs to the multi-instance bag, for the instance inside each bag, find the score of an instance based on the bag label, compute the instance-level loss, the instance-level loss we called $loss1$ is the cross-entropy loss:

$$loss1 = -ylog_{10}(s), \qquad (7)$$

where y is the label of the instance.

Use softmax function to convert the score of the instance into the weight of each instance feature in the bag feature composition, if the bag label is the jth class, the instance weights in the bag are S_j:

$$S_j = \frac{e^{S_j}}{\sum_{j=0}^{n} e^{S_j}}, \qquad (8)$$

where s_jn is the score of the nth instance predicted by the instance-level classifier to belong to the jth class.

Calculate the feature of the bag according to the feature of the instance and the corresponding weights of each instance F_B:

$$F_B = \sum_{j=0}^{j=L}(S_j \cdot f_j). \tag{9}$$

Among them, L is the number of instances in the multi-instance bag, S_j is the score of the bag label category of the jth instance in the multi-instance bag B, and f_j is the feature of the jth instance.

The multi-instance bag-level loss is related to the positive and negative of the bag. The multi-instance bag-level loss can be expressed as $loss2$:

$$loss2 = \begin{cases} -log_{10}(P_B) & B \in positive \quad bag \\ -log_{10}(1 - P_B) & B \in negative \quad bag. \end{cases} \tag{10}$$

4 Experiment Results and Analysis

4.1 Datasets

Experiments are performed on the **VeRi dataset**. The VeRi dataset contains over 50,000 real-world images of 776 vehicles. The images were captured over 24 h by 20 cameras covering 1.0 square kilometers. The training set of the VeRi dataset has 37,778 vehicle images from 576 vehicle identities, the query set and test set have 200 vehicle identities with 1678 and 11579 vehicle images. This dataset is one of the most commonly used datasets in the field of vehicle re-identification. In order to prove that our model has a certain resistance to unknow labels, we deliberately rewrite the labels of 1000 random pictures in the training set of the VeRi dataset into unknow labels.

4.2 Implementation Details

We adopt the ResNet-50 [3] pretrained on ImageNet [2] as the backbone. We set the batchsize to 16, and the iterations in the entire training process is 10000. We set the bag length from 2 to 6. The feature dimension is 2,048. The adam optimizer with a weight decay of 0.0005 is used to train the network in 60 epochs. The learning rate is initially set as 0.0003 and decreased to one-tenth of every 10 epochs. We resize all input images 255×128 for training. During testing, we apply the inferred backbone to extract features, and then apply the Euclidean distance for retrieval.

To demonstrate the effectiveness of our model, we first set up comparative experiments on the unmodified Veri dataset. The experimental results are shown in Table 1. It can be seen from the table that our model has a rank-1 accuracy of 67.9%, prove that our method is effective. Considering that multi-instance learning actually falls into the category of weakly supervised learning. The accuracy of the vehicle re-identification model is not suitable for comparison with the current common supervised learning methods of vehicle re-identification or unsupervised vehicle re-identification. This comparative experiment is only used to verify the feasibility of our method.

In addition, To demonstrate that our method does work on datasets with missing labels, we set up a comparative experiment where we randomly selected a part of the images from the training set and marked them as blank ids, and performed experiments on this mislabeled dataset. As shown in Table 2, it shows our experimental results. Experimental results show that our model can still retain most of the accuracy when the dataset lacks a small number of labels.

Table 1. Performance comparison of different methods on VeRi dataset (%).

Method	Rank-1	Rank-5	Rank-10
LOMO [5]	40.0	71.1	81.5
MLAPG [6]	41.6	71.9	83.1
VGGTripletloss [7]	43.6	69.6	77.8
CnnEmbedding [19]	45.8	71.1	79.2
PCB [12]	48.9	73.2	80.8
Our method	67.9	82.7	88.1

Table 2. Results on VeRi dataset with missing partial labels (%)

Error	Rank-1	Rank-5	Rank-10	Rank-20	mAP
0	67.9	82.7	88.1	92.3	33.3
500	66.7	81.6	87.5	92.3	28.9
1000	66.4	80.8	86.8	92.0	28.7

4.3 Ablation Studies

In this section, we conduct some ablation studies on VeRi dataset to evaluate the effectiveness of the proposed component in vehicle re-identification via spatiotemporal multi-instance learning.

Effect of Bag Length. In order to study the effect of multi-instance bag length on the performance of the vehicle re-identification model. We set up a comparative experiment to compare the model performance by changing the bag length. The experimental results are shown in Table 2. From the Table 2 we can see that the size of bag length significantly affects the performance of the model. When the bag length is less than 4, the larger bag length is, the better the model performance is. When bag length is greater than 4, the accuracy of the model is inversely proportional to the size of bag length. When bag length takes 4, we get the highest mAP, which is 33.3%. According to the experimental results in Table 2, we can think that the length of the multi-instance bag is strongly related to the accuracy of the vehicle re-identification model. When the length of the multi-instance bag is small, due to the large number of negative samples in the bag, it is easy to mask the features of positive samples and result in low vehicle re-identification model accuracy. When the length of the multi-instance bag is too large, it may be difficult for the vehicle re-identification model to learn valuable information because the difference between positive samples from different perspectives is greater than the difference between positive samples and negative samples from the same perspective. Only when the length of the multi-instance bag is properly selected, the model can achieve higher accuracy.

Table 3. Effect of bag length in VeRi (%)

Length	Rank-1	Rank-5	Rank-10	Rank-20	mAP
2	59.1	76.8	83.6	89.8	23.9
3	59.6	77.1	84.6	90.8	25.4
4	67.9	82.7	88.1	92.3	33.3
5	58.7	76.8	83.4	88.4	20.7
6	50.3	68.7	76.5	83.1	15.4

Effect of Bag-Level Loss. As mentioned above, vehicle re-identification via spatio-temporal multi-instance learning consists of two parts: instance-level classifier and multi-instance bag-level classifier. Therefore, we analyze the effectiveness of the instance-level loss and bag-level loss. In comparative experiments, we investigate the impact of bag-level classifiers on model performance by removing bag-level loss. Table 3 shows the effect of bag-level classifiers. From the Table 3 we can see that the accuracy of the model drops significantly after removing bag-level loss, but the model is still able to retain most of the accuracy. We can think that the main role of bag-level loss is to help the model to be more robust. The multi-instance bag-level classifier is obtained by classifying the features of the multi-instance bag and using the bag-level loss to train and optimize, the features of the multi-instance bag are obtained by the weighted summation of all the instance features in the bag, which reflects the overall features of a multi-instance bag and reduces the probability of misclassification of some difficult positive samples. Training a multi-instance bag-level classifier is to minimize the distance between pairs of difficult positive samples. The multi-instance bag-level loss improves the accuracy of vehicle re-identification models by handling difficult positive instances.

Table 4. Effect of bag-level loss in VeRi (%)

Contrast	Rank-1	Rank-5	Rank-10	Rank-20	mAP
Effect	53.2	72.8	81.3	88.2	21.8
Best	67.9	82.7	88.1	92.3	33.3

Effect of Instance-Level Loss. On the contrary, as a comparative experiment, in order to study the effect of instance-level classifiers on the performance of the model, we also cancel the instance-level loss, and the experimental results are shown in Table 4. From Table 4, we can see that the accuracy of the vehicle re-identification model has dropped by a large part, and the model performance is less than half of the previous one. We can conclude that instance-level loss plays a major role in training the model. In fact, this result is as expected. Instance-level classifiers are optimized by classifying vehicles with different identities and trained with instance-level loss. When we designed the instance-level classifier, the main task we assigned it was to process samples of vehicles with different identities. The instance-level loss increases the distance between features of images of vehicles with different identities by optimizing the instance-level classifier, which enable the model to determine the identity of the vehicle and gain the ability to re-identify the vehicle.

Table 5. Effect of bag-level loss in VeRi (%)

Contrast	Rank-1	Rank-5	Rank-10	Rank-20	mAP
Effect	29.5	50.1	58.8	69.0	11.5
Best	67.9	82.7	88.1	92.3	33.3

5   Conclusions

This paper proposes a vehicle re-identification method via spatio-temporal multi-instance learning, which generalizes multi-instance learning to the multi-classification domain, and trains a robust vehicle feature extraction model through multi-instance bag-level loss and instance-level loss. Compared with other methods, this model pays more attention to the sample distribution of the cluster as a whole rather than the characteristics of the samples themselves. Therefore, the model has high stability in the face of possible mislabeling of manually labeled datasets. We conduct extensive experiments, and the experimental results demonstrate that our method is more resistant to mislabeling than other methods.

Acknowledgment. This work was supported by Science and Technology Project of Hebei Provincial Department of Transportation under Grant No. RW-202008.

References

1. Cao, L., et al.: Weakly supervised vehicle detection in satellite images via multi-instance discriminative learning. Pattern Recogn. **64**, 417–424 (2017)
2. Deng, J.: A large-scale hierarchical image database. In: IEEE Computer Vision and Pattern Recognition (2009)
3. He, K., Zhang, X., Ren, S., Sun, J.: Deep residual learning for image recognition. In: IEEE Conference on Computer Vision and Pattern Recognition, pp. 770–778 (2016)
4. Hsu, C.C., Hung, C.H., Jian, C.Y., Zhuang, Y.X.: Stronger baseline for vehicle re-identification in the wild. In: IEEE Visual Communications and Image Processing, pp. 1–4 (2019)
5. Liao, S., Hu, Y., Zhu, X., Li, S.Z.: Person re-identification by local maximal occurrence representation and metric learning. In: IEEE Conference on Computer Vision and Pattern Recognition, pp. 2197–2206 (2015)

6. Liao, S., Li, S.Z.: Efficient PSD constrained asymmetric metric learning for person re-identification. In: IEEE International Conference on Computer Vision, pp. 3685–3693 (2015)
7. Liu, H., Tian, Y., Yang, Y., Pang, L., Huang, T.: Deep relative distance learning: tell the difference between similar vehicles. In: IEEE Conference on Computer Vision and Pattern Recognition, pp. 2167–2175 (2016)
8. Liu, W., Zhang, Y., Tang, S., Tang, J., Hong, R., Li, J.: Accurate estimation of human body orientation from RGB-D sensors. IEEE Trans. Cybern. **43**(5), 1442–1452 (2013)
9. Liu, X., Zhang, S., Huang, Q., Gao, W.: RAM: a region-aware deep model for vehicle re-identification. In: IEEE International Conference on Multimedia and Expo, pp. 1–6 (2018)
10. Liu, X., Liu, W., Ma, H., Fu, H.: Large-scale vehicle re-identification in urban surveillance videos. In: IEEE International Conference on Multimedia and Expo, pp. 1–6 (2016)
11. Mei, T., Rui, Y., Li, S., Tian, Q.: Multimedia search reranking: a literature survey. ACM Comput. Surv. **46**(3), 1–38 (2014)
12. Sun, Y., Zheng, L., Yang, Y., Tian, Q., Wang, S.: Beyond part models: Person retrieval with refined part pooling (and a strong convolutional baseline). In: European Conference on Computer Vision, pp. 480–496 (2018)
13. Tu, J., Chen, C., Huang, X., He, J., Guan, X.: Discriminative feature representation with spatio-temporal cues for vehicle re-identification. arXiv preprint arXiv:2011.06852 (2020)
14. Wei, X.-S., Zhang, C.-L., Liu, L., Shen, C., Wu, J.: Coarse-to-fine: a RNN-based hierarchical attention model for vehicle re-identification. In: Jawahar, C.V., Li, H., Mori, G., Schindler, K. (eds.) ACCV 2018. LNCS, vol. 11362, pp. 575–591. Springer, Cham (2019). https://doi.org/10.1007/978-3-030-20890-5_37
15. Yan, K., Tian, Y., Wang, Y., Zeng, W., Huang, T.: Exploiting multi-grain ranking constraints for precisely searching visually-similar vehicles. In: IEEE International Conference on Computer Vision, pp. 562–570 (2017)
16. Yang, L., Luo, P., Change Loy, C., Tang, X.: A large-scale car dataset for fine-grained categorization and verification. In: IEEE Conference on Computer Vision and Pattern Recognition, pp. 3973–3981 (2015)
17. Zhang, J., Wang, F., Wang, K., Lin, W., Xu, X., Chen, C.: Data-driven intelligent transportation systems: a survey. IEEE Trans. Intell. Transp. Syst. **12**(4), 1624–1639 (2011)
18. Zheng, Y., Capra, L., Wolfson, O., Yang, H.: Urban computing: concepts, methodologies, and applications. ACM Trans. Intell. Syst. Technol. **5**(3), 1–55 (2014)
19. Zheng, Z., Zheng, L., Yang, Y.: A discriminatively learned CNN embedding for person re-identification. ACM Trans. Multimedia Comput. Commun. Appl. **14**(1), 1–20 (2017)
20. Zhou, Y., Shao, L.: Aware attentive multi-view inference for vehicle re-identification. In: IEEE Conference on Computer Vision and Pattern Recognition, pp. 6489–6498 (2018)
21. Zhou, Z.H.: Multi-instance learning: a survey. Department of Computer Science 1 (2004)
22. Zhou, Z.H., Sun, Y.Y., Li, Y.F.: Multi-instance learning by treating instances as non-IID samples. In: International Conference on Machine Learning, pp. 1249–1256 (2009)
23. Zhou, Z.H., Zhang, M.L.: Solving multi-instance problems with classifier ensemble based on constructive clustering. Knowl. Inf. Syst. **11**(2), 155–170 (2007)
24. Zhu, J., et al.: Vehicle re-identification using quadruple directional deep learning features. IEEE Trans. Intell. Transp. Syst. **21**(1), 410–420 (2019)

Multivariate Time Series Imputation with Bidirectional Temporal Attention-Based Convolutional Network

Yanzhuo Lin and Yu Wang[✉]

School of Mechanical Engineering, Xi'an Jiao Tong University, Xianning Xi Road No. 28,
Xi'an 710049, People's Republic of China
ywang95@xjtu.edu.cn

Abstract. The problem of missing data in time series will make the process of analysis much more tough and challenging. Imputation of missing values in multivariate time series can effectively solve this problem. Recurrent neural networks (RNNs) are widely used in sequential data due to their properties of sequential modeling. However, RNN has some problems such as gradient and long calculation time. In recent years, time series modeling has been fully developed utilizing a feedforward model based on convolutional networks and an attention mechanism, which has the advantage of parallelism over RNNs. This paper proposes a multivariate time series imputation model (BTACN) based on Temporal Convolutional Networks (TCN) and attention mechanism. Multivariate time series features were extracted by bidirectional TCN, and then attention was weighted to capture the long-term and short-term dependence of time series. Minimizing both reconstruction and imputation loss is used to train the model. Experiments on real datasets and simulated datasets reveal the superiority of the proposed method in terms of imputation performance.

Keywords: Missing data · Time series · Multivariate · Neural network

1 Introduction

Multivariate time series data are omnipresent in a wide range of applications, for example, meteorological observations [1, 2], medical logs [3], traffic flow data [4], economic and financial marketing [5], etc. Time series modeling and prediction is a topic that researchers have been interested in for a long time. However, due to sensor failure, communication failure, or privacy concerns, multivariate time series often have partial or many missing values in the real-world environment, which has a direct impact on the accuracy of downstream tasks (classification or regression, etc.)

In data-driven modeling, missing data is a prevalent issue. The statistical analysis will be skewed by the absence of data, resulting in incorrect results. Furthermore, because many data modeling techniques assume complete information for all variables included, missing data invalidates many of them. The simplest way to solve the missing data

© The Author(s), under exclusive license to Springer Nature Singapore Pte Ltd. 2022
H. Zhang et al. (Eds.): NCAA 2022, CCIS 1638, pp. 494–508, 2022.
https://doi.org/10.1007/978-981-19-6135-9_37

problem is to delete it directly. However, it is prone to fallacy in analytical conclusions [6] because missing data usually contains potentially important information.

In recent years, many methods to fix missing data problems for multivariate time series have been studied, which contain statistics and machine learning approaches. The missing data is replaced with probable values generated by substituting values from the available observed variables in a statistically based technique [7]. Common statistical approaches include linear imputation, mean imputation, and last observation carried forward (LOCF). In addition, traditional machine learning algorithms are extensively applied for missing time series imputation. For instance, Rubul Kumar Bania et al. [8] use the K-nearest neighbor (KNN) imputation method for missing value treatment and eventually achieve the classification of medical data. Multiple imputations by chained equations (MICE) [9] and EM [10, 11] are two other common data imputation algorithms. However, these imputation methods consider data as a sequence of points. Studies [12] show that there are potential patterns and correlations between the data collected in each time step of the time series. In general, these imputation approaches are unable to capture adequate time information, resulting in poor imputation performance.

AI (artificial intelligence)-based methods are widely investigated and applied to time series imputation as technology advances. Researchers use the neural network which has great power of function approximation to automatically learn data features and complete time series data imputation. Due to the cyclic autoregressive structure and the unique performance of the recurrent neural network in sequence modeling [13], Che et al. [14] use a gated recurrent unit (GRU) variant, to handle missing value problems for time series classification. J. Yoon et al. [15] proposed a Multi-directional Recurrent Neural Network that interpolates within data streams and imputes across data streams. Another RNN-based imputation model with good performance is BRITS [16]. The imputed values are treated as variables of RNN graph and can be effectively updated during the backpropagation. However, the RNN-based method has two flaws that restrict its adaptability in reality. One is that, because of its sequential structure, the next time step must wait for the previous time step to finish before proceeding, preventing substantial parallelization of the training and evaluating processes [17]. The second issue is that these methods still struggle to capture sufficient sequential information, especially for long sequence modeling [18].

Recently, many different neural networks for sequence modeling have been developed by academics. Bai et al. [18] apply the convolution to the modeling of sequential problems and build a Temporal Convolutional Network (TCN), which achieves results equivalent to or exceeding that of the RNN model. TCN takes advantage of CNN's capacity to calculate information independently of the preceding time and thus has parallelism. TCN exploits causal convolution, dilated convolution and residual block to learn sequence features as well as avoid future information leaks. However, TCN does not learn distance position dependencies within the sequence, nor does it extract internal relevant information from the input. Transformer [17], which is built on self-attention to construct global relationships between inputs and outputs, is another exemplary paradigm for sequence modeling. Transformer is initially used in the field of natural language processing (NLP) and now has a wide range of applications in other scenarios. DeepMVI was introduced by Bansal et al. [19] for missing value imputation in multidimensional time-series data. Their model includes a Transformer with a convolutional window feature and a kernel regression. Ma et al. [20] use cross-dimensional self-attention (CDSA) to impute missing values in geo-tagged data, especially spatiotemporal datasets, using three dimensions (time, location, and measurement). However, Transformer requires a large number of data sets and high computational power requirements, and the plain encoder is very effective at capturing short-term dependencies.

In this paper, we propose a new imputation model (BTACN) for multivariate time series data. We still use the seq2seq model basis. As illustrated in Fig. 2, our model consists of three parts. The first part learns the features of time series through bidirectional TCN, and then they are weighted and fused with the original input through the attention mechanism in the second part. Finally, the output is carried out through a weight combining block. We design an attention-based encoder with temporal convolutional layers that can capture the long-term and short-term dependence of time series and achieve parallel characteristics. By combining reconstruction loss and imputation loss, our model is well trained. To validate our proposed method, we conduct experiments on two real data sets and one simulated data set. Meanwhile, the impact of the missing rate on the imputation performance was analyzed experimentally. Experimental results show that our model can achieve better performance compared with several baseline imputation approaches.

The remainder of this paper is organized as follows. In Sect. 2, the proposed BTACN model is described in depth. Then, the experimental part and discussion are presented in Sect. 3. Section 4 presents the conclusion and future work of this research study.

2 Methodology

2.1 Notions

Since we focus on multivariate time series imputation, we define a multivariate time series $X = \{x_1, x_2, \ldots, x_t, \ldots, x_T\} \in \mathbb{R}^{T \times D}$ as a sequence with D variables of T observations. The t-th observation $x_t = \{x_t^1, x_t^2, \ldots, x_t^d, \ldots x_t^D\} \in \mathbb{R}^D$ contains all D variables and x_t^d denotes the value of d-th dimension variable in time step t. When x_t^d is missing, we note it as NAN, which is the abbreviation of "not a number". In addition, we define a binary mask vector $M \in [0, 1]^{T \times D}$ to represent the missing values and observed values in X, and m_t^d can be given by

$$m_t^d = \begin{cases} 0 & \text{if } x_t^d \text{ is missing} \\ 1 & \text{if } x_t^d \text{ is observed} \end{cases} \tag{1}$$

The form of the multivariate time series and its corresponding masking vectors can be demonstrated in Fig. 1.

In our paper, we take comprehensive consideration of reconstruction loss and imputation loss. Inspired by denoising autoencoder, which aims to gain robust features against missing data or noise, for each training sample and epoch, the input time series X will be corrupt manually by elementwise multiplication $\odot$ to obtain destroyed time series $\hat{X}$

$$\hat{X} = X \odot p = \{\hat{x_1}, \hat{x_2}, \ldots, \hat{x_t}, \cdots \hat{x_T}\} \in \mathbb{R}^{T \times D} \tag{2}$$

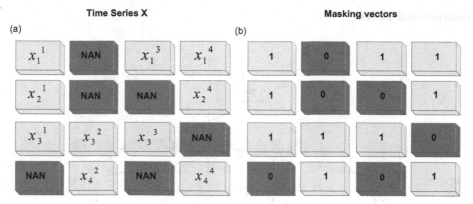

Fig. 1. A simple demonstration of the time series representation and its corresponding mask vector: (a) time-series X and (b) mask vector.

where $p \in [0, 1]^{T \times D}$ represent the binary random mask vector. After that, we can obtain new masking vectors $\hat{M} = \{\hat{m}_1, \hat{m}_2, \ldots \hat{m}_t, \ldots, \hat{m}_T\} \in [0, 1]^{T \times D}$ for $\hat{X}$. In addition, to facilitate later calculation of the imputation error, we denote $\hat{p} = \sim p \in [0, 1]^{T \times D}$ (called evaluating mask), where $\sim$ represents elements of the not. The math definitions of $\hat{M}$ is given by:

$$\hat{m}_t^d = \begin{cases} 0 & \text{if } \hat{x}_t^d \text{ is missing} \\ 1 & \text{if } \hat{x}_t^d \text{ is observed} \end{cases} \tag{3}$$

2.2 Model Structure

The proposed imputation model is shown in Fig. 2. The time series with missing values is input into three sub-blocks (Bidirectional TCNs, Attention Mechanism, and Weight Combining block) in turn for unsupervised imputation tasks.

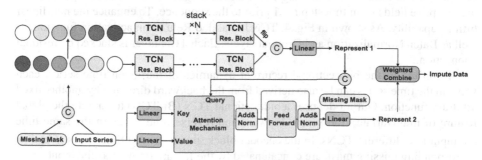

Fig. 2. The BTACN model architecture.

Bidirectional TCNs

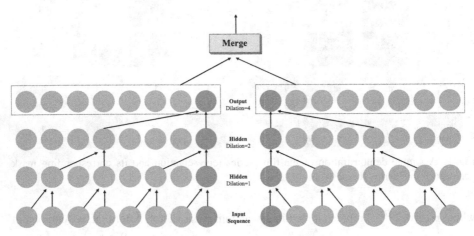

Fig. 3. The representation of Bi-TCN.

In TCN structure, there are three main implementation mechanisms: causal convolution, dilated convolution, and residual connection. Among them, causal convolution can make the input of the current moment t of the sequence only come from the previous $t - 1$ moment, while empty convolution increases the receptive field of the convolution, enabling TCN to effectively model the long sequence with a relatively small number of layers. Consider that our time series is expressed as, let's define a filter $f : \{0, \cdots k - 1\} \rightarrow \mathbb{R}$, the formula for the dilated convolution of the moment t is as follows:

$$F(t) = (X^{(1,\cdots,t)\times D} *_d f)(t) = \sum_{i=0}^{k-1} f(i) \cdot X^{(t-d\cdot i)\times D} \tag{4}$$

where d is the dilation factor, k is the filter size, and $t - d \cdot i$ accounts for the direction of the past. The schematic diagram of dilated causal convolution is shown in the left figure in Fig. 3, where dilation factors d are 1, 2, 4 and filter size $k = 2$. At timestep t it has all the receptive fields from timestep $t - 1$ prior to the sequence. To enhance the non-linear fitting capability. As shown in Fig. 4, TCN has added Relu as an activation function, as well as Batch Normalization and Dropout layers. Each TCN layer is stacked by residual connection.

According to the bidirectional recurrent dynamics of the given time series, each value in the time series can be also derived from the backward direction by another fixed arbitrary function. Our model uses a bidirectional TCN (Bi-TCN) to capture the global feature of multivariate time series. Forward and reverse time series of the same time are input into different TCNs. In the encoder block, the destroyed time series $\hat{X}$ and its corresponding missing mask are concatenated by the feature domain as the input:

$$\kappa = concatenate(\hat{X}, \hat{M}) \in \mathbb{R}^{T \times 2D} \tag{5}$$

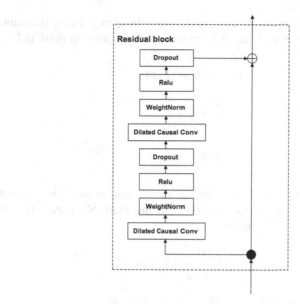

Fig. 4. TCN residual block.

And Bi-TCN Receives a time series from the forward and its corresponding reverse, which can be given as:

$$Z_1 = TCN(\kappa_1, \kappa_2, \cdots, \kappa_t, \cdots \kappa_T) \tag{6}$$

$$Z_2 = TCN(\kappa_T, \kappa_{T-1}, \cdots \kappa_t, \cdots, \kappa_1) \tag{7}$$

$$\mathbb{Z} = concatenate(Z_1, \overleftarrow{Z_2}) \tag{8}$$

$$T_1 = \mathbb{Z} \cdot W_Z + b_Z \tag{9}$$

where Z_1 represents the features from the sample time series, and Z_2 represents the features of the corresponding sequence in reverse order, both of which have dimensions of $T \times 2D$. We fuse Z_1 and Z_2 in the feature domain. Note that before the fusion, to align the feature moments (the same as the initial moments), we reverse order Z_2 (represented as $\overleftarrow{Z_2}$) in Eq. (8).

Attention Mechanism
Through the encoder of bidirectional TCN, time-series features are extracted. The attention mechanism can learn the correlation between time steps of multivariate time series. We use the attention mechanism to weight the obtained features with the original input and further extract the global long-term features of the time series. The attention mechanism multiplies the Query matrix with the transpose of the Key matrix, obtains the weights by normalizing them with the Softmax function, and finally multiplies the

weights with the Value matrix. In the proposed model, the Query matrix is obtained by splicing the output of the bidirectional TCN encoder with the missing mask in Eq. (10).

$$Q = concatenate(T, \hat{M}) \tag{10}$$

$$K = \hat{X} \cdot W_1 + b_1 \tag{11}$$

$$V = \hat{X} \cdot W_2 + b_2 \tag{12}$$

Equation (11) and Eq. (12) apply a linear transformation to destroyed input $\hat{X}$ to obtain the Key matrix and Value matrix with the learnable parameters $W_1 \in \mathbb{R}^{D \times 2D}$, $W_2 \in \mathbb{R}^{D \times 2D}$, $b_1 \in \mathbb{R}^{2D}$, and $b_2 \in \mathbb{R}^{2D}$.

$$H = Softmax\left(\frac{QK^\top}{\sqrt{d_k}}\right)V = \omega \cdot V \tag{13}$$

$$N = ReLU(H \cdot W_H + b_H) \cdot W_B + b_B \tag{14}$$

$$R = N \cdot W_N + b_N \tag{15}$$

$$T_2 = LeakyReLU(R \cdot W_R + b_R) \cdot W_s + b_s \tag{16}$$

Equation (13) shows the weighted output $H \in \mathbb{R}^{T \times 2D}$ after the attention mechanism, and $\omega \in \mathbb{R}^{T \times T}$ represent the weight. Also, a residual connection is made between Q and H. Specific calculation of the feed-forward layer is illustrated by Eq. (14). The linear transformation and activation function gives the same representation $T_2 \in \mathbb{R}^{T \times D}$ as the original dimension of X.

Weight Combining Block

$$U = concatenate(\omega, \hat{M}) \tag{17}$$

$$\xi = Sigmoid(U \cdot W_\xi + b_\xi) \tag{18}$$

$$X_{OUT}\{o_1, o_2, \cdots o_t, \cdots o_T\} = (1 - \xi) \odot T_1 + \xi \odot T_2 \tag{19}$$

Inspired by [21], we use learnable weights applied to and the output of Bidirectional TCNs block and Attention Mechanism block for weight combination. Equation (17) concatenate the weight ω of attention block and the missing mask $\hat{M}$. And then the combining weight ξ will be obtained after linear transformation and activation function successively in Eq. (18), where $W_\xi \in \mathbb{R}^{(T+D) \times D}$ and $b_\xi \in \mathbb{R}^D$ are learnable parameters. As illustrated in Eq. (19), the final output X_{OUT} is given by weight combination.

Loss Function

Considering that our task is the imputation task of time series and the encoder-decoder structure used, the loss function in our proposed method consists of two parts: reconstruction loss and imputation loss. We choose the MAE (Mean Absolute Error) $\mathcal{L}_{MAE}$ to train the model. The object of reconstruction loss $\mathcal{L}_{Reconstruction}$ is calculated by the output of the model X_{OUT}, the input multivariate time series $\hat{X}$ after artificial random missing as well as the missing mask matrix $\hat{M}$. As for the imputation loss $\mathcal{L}_{\text{Imputation}}$, the consideration is the output of the model X_{OUT}, the original time series X and the evaluating mask $\hat{p}$. The MAE and the two loss functions above-mentioned are given by Eq. (20), Eq. (21), and Eq. (22) respectively, where $\odot$ represents elementwise multiplication. And the total loss $\mathcal{L}$ is accumulated by $\mathcal{L}_{Reconstruction}$ and $\mathcal{L}_{\text{Imputation}}$ in Eq. (23).

$$\mathcal{L}_{MAE}(pred, GroundTruth, mask) = \frac{\sum_i (pred_i - GroundTruth_i) \odot mas\,k_i}{\sum_i mas\,k_i} \tag{20}$$

$$\mathcal{L}_{Reconstruction} = \mathcal{L}_{MAE}(X_{OUT}, \hat{X}, \hat{M}) \tag{21}$$

$$\mathcal{L}_{\text{Imputation}} = \mathcal{L}_{MAE}(X_{OUT}, X, \hat{p}) \tag{22}$$

$$\mathcal{L} = \mathcal{L}_{Reconstruction} + \mathcal{L}_{\text{Imputation}} \tag{23}$$

3 Experiments

3.1 Datasets

Beijing Air Quality Dataset

This air quality dataset [22] contains air quality data collected by sensors at various stations in Beijing from March 1, 2013, to February 28, 2017, and it was sampled every hour. Each time point contains 11 air quality characteristics such as PM2.5, SO2, and NO2. A total of 1.6% of the data in the dataset is missing. We selected the air data from the Gucheng station. For this dataset, the division ratio of the training test and test set is 0.8:0.2, and the first 20% of the training set is divided into additional test sets. In the training set, the first 20% are delineated as the validation set to verify the performance of the model. All the training, validation and test data sets were continuously divided. All data are divided into sliding windows with a size of 36 h. Since the true value of the missing value is not available for comment, a 20% artificial mask is carried out on both the validation set and test set, and the artificial mask is used as ground truth for the experimental evaluation of the test set.

Geo-Magnetic Field and WLAN Dataset

This dataset is a geomagnetic field and WLAN dataset from the wristband and smartphone datasets for indoor positioning [23]. The synchronized data is retrieved from the user's smartwatch and smartphone, and contain Wi-Fi and geomagnetic features, as well as inertial sensor data (pitch, gyroscope, etc.) for two activities performed in the same environment. The dataset is a complete multivariate time series (including 9-dimensional feature values) and does not contain missing values. Therefore, the dataset is processed for 20% artificial missing values. The data are divided into training, validation, and test sets at 60%, 20%, and 20%. Same as the air quality dataset, the division of the time series is continuous. All data were divided continuously and the sliding window is 144 in size. When evaluating the imputation performance of the model, the artificial data of the test set are used as ground truth to calculate the evaluation metrics.

NCAA2022 Dataset

NCAA2022 dataset [24] is a simulation time series dataset. According to the frequency domain characteristics, four typical datasets with non-stationary, periodic, impulsive, and chaotic characteristics are transformed into four types of problems, i.e., low-pass, high-pass, band-pass, and band-stop categories. In this paper, the time series with three channels ('XYZ') from each of the above four problems are selected as the experimental data set for imputation. Each type of problem is divided into training sets, validation sets, and test sets. The classification rules are as follows: the first 60% of continuous time-series data is the training set, the last 20% of data is the test set, and the rest is the verification set. In addition, the sliding window is 100 in size. Due to the missing data in the dataset could not being obtained, the artificial random mask was performed on the dataset and it would be the Ground Truth for evaluating the imputation effect on the test set. The experimental data are described in detail in the following table. The algorithm robustness of complex tasks could be described by the comprehensive performance on different problems (Table 1).

Table 1. Data description of the NCAA2022 dataset.

Frequency characteristics	Data length	Dimension	Variable name
Low-pass	7500	3	X, Y, Z
High-pass	7500	3	X, Y, Z
band-pass	7500	3	X, Y, Z
band-stop	7500	3	X, Y, Z

3.2 Baseline Methods

In our paper, we selected the following baseline methods for comparative experiments.

Mean: The average value of each feature in the whole time domain is used for the simple imputation of the missing value.

KNN: K-nearest neighbor algorithm is adopted to search the value near the missing value and average all the neighbors for imputation.

EM (Expectation Maximization): calculate maximum likelihood estimation from incomplete test data and perform probability interpolation for missing data.

MF (Matrix Factorization): missing value imputation is carried out by Matrix decomposition and Matrix completion.

TCN [18]: In the experiment, single-directional TCN is used for interpolation of missing values.

M-RNN [15]: M-RNN divides missing data estimation into two parts: interpolation block and imputation block, and bi-directional RNN is used to model medical time series data sets, and its effect is validated on real data sets.

BRITS [16]: BRITS also use bi-directional RNN. Different from M-RNN, missing values are regarded as RNN graph. At the same time, BRITS consider the correlation between features and carry out reasonable weighted output.

3.3 Experimental Setup

We ran the experiment on the two real datasets and one simulation dataset mentioned above, and to properly evaluate the results, MAE, Mean Absolute Error (MAE), Mean Relative Error (MRE), and Root Mean Square Error (RMSE) were used to comprehensively verify the accuracy of missing value imputation.

$$MAE = \frac{\sum_i |pred_i - GroundTruth_i|}{N} \tag{24}$$

$$MRE = \frac{\sum_i |pred_i - GroundTruth_i|}{\sum_i |GroundTruth_i|} \tag{25}$$

$$RMSE = \sqrt{\frac{1}{N} \sum_i (|pred_i - GroundTruth_i|)^2} \tag{26}$$

For the mean filling, KNN, EM, MF, and other filling methods, we use the recommended parameters based on the Python Package Fancyimpute[1] and Impyute[2] to carry out experiments.

The batch size is set to 64, and we choose an AdamW optimizer with a decay of learning rate (The initial learning rate is $1e-3$) to train our model. All the baseline methods and our model adopt an early stopping strategy which is based on the error of the validation set we divided. When the MAE of the validation set no longer decreases for ten consecutive epochs, the early stop policy is executed. All models are trained with GPU GTX A4000.

Table 2. Performance comparisons amongst approaches on two real data sets and metrics are presented in the order of MAE /MRE /RMSE. The smaller the number, the better. The bolded values are the best.

Method	Datasets	
	Beijing air quality dataset	Geo-magnetic and WALN dataset
Mean	0.4285/0.6316/0.7545	0.2883/0.4250/0.4403
KNN	0.2166/0.3192/0.6139	0.1071/0.1579/0.2287
EM	0.5823/0.8581/0.9901	0.3881/0.5721/0.5828
MF	0.3055/0.4502/0.6153	0.1939/0.2857/0.2974
MICE	0.3687/0.5457/0.5468	0.1148/0.1692/0.2186
TCN	0.2200/0.3249/0.5822	0.1233/0.1817/0.2092
M-RNN	0.1546/0.2181/0.5139	0.0973/0.1435/0.1739
BRITS	0.1486/0.2207/0.5107	0.0828/0.1171/0.1535
Ours	**0.1417/0.2118/0.5052**	**0.0724/0.1067/0.1468**

3.4 Experimental Results

In the two real data sets above, it can be seen from Table 2 that the simple mean filling method cannot accurately fill the missing values of time series, while the machine learning algorithm can greatly improve the imputation accuracy. Among the traditional machine learning baseline methods, KNN has achieved high imputation accuracy. The bidirectional deep learning method performs well in both datasets, and the proposed method achieves better imputation accuracy than all the baseline methods in this paper.

Table 3 shows the comparison of the imputation performance between the proposed method and the baseline method on the NCAA2022 dataset. There are three evaluation indexes (MAE/MRE/RMSE). The smaller these three indexes are, the better the imputation performance of the model will be. The performance of imputation in four different tasks reflects the robustness of the proposed method. For the traditional machine learning algorithm, KNN imputation is better than MF and MICE in the two types of time series, while the EM algorithm has a large error on the whole. Compared with the traditional machine learning method, the imputation performance of ordinary TCN is greatly improved in MAE/MRE/RMSE, but not as good as that of the model based on the bidirectional recurrent neural network (M-RNN and BRITS). Except for the proposed method, the BRITS model performs best among other baseline methods. In terms of low pass, high pass, band pass, and band stop data, the proposed method is all superior to baseline models. Figure 5 shows the RMSE comparisons of various methods over low-pass, high-pass, band-pass, and band-stop time series data.

---

[1] https://github.com/iskandr/fancyimpute.
[2] https://impyute.readthedocs.io/en/master/

Table 3. Performance comparison amongst approaches on NCAA2022 dataset. Metrics are presented in the order of MAE /MRE /RMSE. The smaller the number, the better. The bolded values are the best.

Frequency characteristics	Low-pass	High-pass	Band-pass	Band-stop
Mean	0.757/0.809/0.964	0.629/0.968/0.947	0.635/1.019/1.036	0.775/0.815/0.978
KNN	0.579/0.619/0.904	0.833/1.281/1.195	0.741/1.189/1.217	0.565/0.594/0.905
EM	1.013/1.083/1.304	0.976/1.501/1.277	1.037/1.665/1.409	1.004/1.056/1.267
MF	0.616/0.659/0.819	0.650/0.100/0.960	0.645/1.036/1.038	0.626/0.659/0.832
MICE	0.607/0.649/0.836	0.717/1.104/1.035	0.704/1.130/1.077	0.605/0.636/0.838
TCN	0.087/0.092/0.140	0.241/0.372/0.449	0.255/0.410/0.503	0.114/0.121/0.161
M-RNN	0.081/0.087/0.132	0.202/0.311/0.358	0.178/0.286/0.352	0.093/0.089/0.135
BRITS	0.061/0.063/0.120	0.160/0.245/0.321	0.162/0.259/0.285	0.088/0.092/0.149
Ours	**0.055/0.058/0.087**	**0.131/0.202/0.283**	**0.125/0.201/0.225**	**0.061/0.064/0.091**

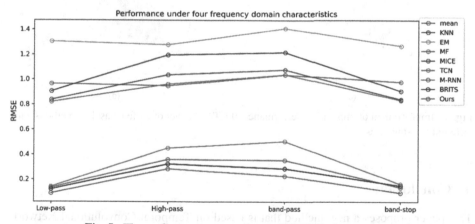

Fig. 5. The performance comparisons of RMSE in imputation.

To further analyze the effect of the missing rate on the imputation performance, the Beijing Air Dataset is given a larger missing rate (40%, 60%, 80%) for the training set, validation set, test set, and experiments on the effect of different missing rates are conducted on the proposed method (BTACN) and the baseline methods, respectively. As shown in Fig. 6, it highlights the improvement of the imputation performance of the proposed method compared with the other five baseline methods for different missing rates. Subplot (a) shows the imputation MAE statistics of six imputation methods including BTACN with different missing rates. In subplots (b), (c), and (d), the value at each branch indicates the extent to which the imputation performance (MRE and RMSE statistics) of BTACN can be improved compared to the baseline methods (Unit 100%). It is obvious that the interpolation performance of the proposed method has a significant improvement over the traditional machine learning methods such as KNN for all missing rates, and it is optimal among all the baseline methods.

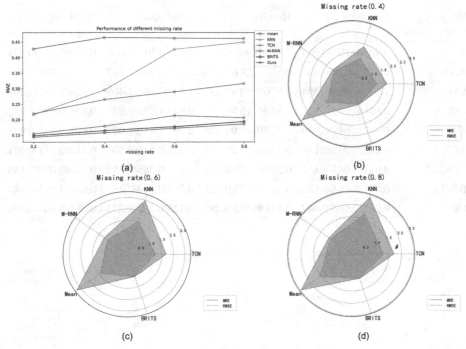

Fig. 6. Improvement of imputation performance of BTACN over other five baseline methods with different missing rates.

4 Conclusion

This paper proposes a new method that is based on Temporal Convolutional Network and attention mechanism for multivariate time series imputation. The method extracts effective features of time series through bidirectional TCN, and further weights them through the attention mechanism to obtain the final output. The proposed model can learn the long-term and short-term dependence of multivariate time series and realize parallel training on sequence data to impute the missing values more effectively. Through experiments on two real datasets and simulated dataset imputation, the performance of the proposed method is improved compared with other baseline methods. The future research direction is to improve the model to realize online imputation for missing values of multivariate time series and to design an end-to-end application framework for practical engineering problems.

Acknowledgments. The method proposed in this paper won second place in the NCAA 2022 Competition for Time Series Prediction.

References

1. Yi, X., Zheng, Y., Zhang, J., Li, T.: ST-MVL: filling missing values in geo-sensory time series data. In: Proceedings of the 25th International Joint Conference on Artificial Intelligence (2016)
2. Deng, R., Chang, B., Brubaker, M.A., Mori, G., Lehrmann, A.: Modeling continuous stochastic processes with dynamic normalizing flows. Adv. Neural Inf. Process. Syst. **33**, 7805–7815 (2020)
3. Esteban, C., Hyland, S.L., Rätsch, G.: Real-valued (medical) time series generation with recurrent conditional GANs (2017)
4. Zhang, J., Zheng, Y., Qi, D.: Deep spatio-temporal residual networks for citywide crowd flows prediction. In: Thirty-first AAAI Conference on Artificial Intelligence (2017)
5. Hsieh, T.-J., Hsiao, H.-F., Yeh, W.-C.: Forecasting stock markets using wavelet transforms and recurrent neural networks: an integrated system based on artificial bee colony algorithm. Appl. Soft Comput. **11**, 2510–2525 (2011)
6. Mohamed, A.K., Nelwamondo, F.V., Marwala, T.: Estimating missing data using neural network techniques, principal component analysis and genetic algorithms. In: Proceedings of the Eighteenth Annual Symposium of the Pattern Recognition Association of South Africa (2007)
7. Nissen, J., Donatello, R., Van Dusen, B.: Missing data and bias in physics education research: a case for using multiple imputation. Phys. Rev. Phys. Educ. Res. **15**, 020106 (2019)
8. Bania, R.K., Halder, A.: R-Ensembler: a greedy rough set based ensemble attribute selection algorithm with kNN imputation for classification of medical data. Comput. Methods Programs Biomed. **184**, 105122 (2020)
9. White, I.R., Royston, P., Wood, A.M.: Multiple imputation using chained equations: issues and guidance for practice. Stat. Med. **30**, 377–399 (2011)
10. Rahman, M., Islam, M.Z: Missing value imputation using a fuzzy clustering-based EM approach. Knowl. Inf. Syst. **46**, 389–422 (2016)
11. Rumaling, M.I., Chee, F.P., Dayou, J., Hian Wui Chang, J., Soon Kai Kong, S., Sentian, J.: Missing value imputation for PM 10 concentration in sabah using nearest neighbour method (NNM) and expectation-maximization (EM) algorithm. Asian J. Atmos. Environ.(AJAE) **14** (2020)
12. Yang, S., Dong, M., Wang, Y., Xu, C.: Adversarial recurrent time series imputation. IEEE Trans. Neural Netw. Learn. Syst. (2020)
13. Goodfellow, I., Bengio, Y.: Deep Learning, vol. 1, no. 2. MIT Press, Cambridge (2016)
14. Che, Z., Purushotham, S., Cho, K., Sontag, D., Liu, Y.: Recurrent neural networks for multivariate time series with missing values. Sci. Rep. **8**, 1–12 (2018)
15. Yoon, J., Zame, W.R., van der Schaar, M.: Estimating missing data in temporal data streams using multi-directional recurrent neural networks. IEEE Trans. Biomed. Eng. **66**, 1477–1490 (2018)
16. Cao, W., Wang, D., Li, J., Zhou, H., Li, L., Li, Y.: BRITS: bidirectional recurrent imputation for time series. Adv. Neural Inf. Process. Syst. **31** (2018)
17. Vaswani, A., et al.: Attention is all you need. Adv. Neural Inf. Process. Syst. **30** (2017)
18. Bai, S., Kolter, J.Z., Koltun, V.: An empirical evaluation of generic convolutional and recurrent networks for sequence modeling (2018)
19. Bansal, P., Deshpande, P., Sarawagi, S.: Missing value imputation on multidimensional time series (2021)
20. Ma, J., Shou, Z., Zareian, A., Mansour, H., Vetro, A., Chang, S.-F.: CDSA: cross-dimensional self-attention for multivariate, geo-tagged time series imputation (2019)
21. Du, W., Côté, D., Liu, Y.: SAITS: self-attention-based imputation for time series (2022)

22. Zhang, S., Guo, B., Dong, A., He, J., Xu, Z., Chen, S.X.: Cautionary tales on air-quality improvement in Beijing. Proc. R. Soc. A Math. Phys. Eng. Sci. **473**, 20170457 (2017)
23. Barsocchi, P., Crivello, A., La Rosa, D., Palumbo, F.: A multisource and multivariate dataset for indoor localization methods based on WLAN and geo-magnetic field fingerprinting. In: 2016 International Conference on Indoor Positioning and Indoor Navigation (IPIN), pp. 1–8. IEEE (2016)
24. https://dl2link.com/ncaa2022/calls/competition/

Author Index

Printed in the United States
by Baker & Taylor Publisher Services

Printed in the United States
by Baker & Taylor Publisher Services